Chaotic Dynamics

An Introduction Based on Classical Mechanics

Since Newton, a basic principle of natural philosophy has been determinism, the possibility of predicting evolution over time into the far future, given the governing equations and starting conditions. Our everyday experience often strongly contradicts this expectation. In the past few decades we have come to understand that even motion in simple systems can have complex and surprising properties.

Chaotic Dynamics provides a clear introduction to chaotic phenomena, based on geometrical interpretations and simple arguments, without in-depth scientific and mathematical knowledge. Examples are taken from classical mechanics whose elementary laws are familiar to the reader. In order to emphasise the general features of chaos, the most important relations are also given in simple mathematical forms, independent of any mechanical interpretation. A broad range of potential applications are presented, ranging from everyday phenomena through engineering and environmental problems to astronomical aspects. It is richly illustrated throughout, and includes striking colour plates of the probability distribution of chaotic attractors.

Chaos occurs in a variety of scientific disciplines, and proves to be the rule, not the exception. The book is primarily intended for undergraduate students in science, engineering and mathematics.

TAMÁS TÉL is Professor of Physics at Eötvös University, Budapest, Hungary. He has served on the editorial boards of the journals *Nonlinearity* and *Fractals* and the advisory board of *Chaos*. His main fields of interest are non-linear dynamics, statistical mechanics, fluid dynamics and environmental flows.

MÁRTON GRUIZ teaches and studies physics at Eötvös University, Budapest, Hungary. His main area of research is in non-linear dynamics and he is involved in teaching chaotic phenomena.

Chaotic Dynamics

An Introduction Based on Classical Mechanics

Tamás Tél and
Márton Gruiz
Eötvös University, Budapest

Translated by Katalin
Kulacsy

Three-dimensional graphics
by Szilárd Hadobás

CAMBRIDGE UNIVERSITY PRESS
Cambridge, New York, Melbourne, Madrid, Cape Town, Singapore, São Paulo

Cambridge University Press
The Edinburgh Building, Cambridge CB2 2RU, UK

Published in the United States of America by Cambridge University Press, New York

www.cambridge.org
Information on this title: www.cambridge.org/9780521839129

Previous edition published in Hungarian as *Kaotikus Dinamika* by Nemzeti Tankönyvkiadó, Budapest, 2002

English translation (by Katalin Kulacsy) first published 2006

Printed in the United Kingdom at the University Press, Cambridge

A catalogue record for this publication is available from the British Library

ISBN-13 978-0-521-83912-9 hardback
ISBN-10 0-521-83912-2 hardback

ISBN-13 978-0-521-54783-3 paperback
ISBN-10 0-521-54783-0 paperback

Contents

Colour plates

I. Chaotic attractor of an irregularly oscillating body (a driven non-linear oscillator; Sections 1.2.1 and 5.6.2 and equation (5.85)) on a stroboscopic map (Fig. 1.4), coloured according to the visiting probabilities. The colour change from red to yellow denotes less than 8% of the maximum of the distribution. Between 8 and 30% is depicted by a colour change from yellow to white. Above 30% is represented by pure white. The picture contains 1000×1000 points.

II. Chaotic attractor of a driven pendulum (Sections 1.2.1 and 5.6.3 and equation (5.89)) on a stroboscopic map (Fig. 1.8), coloured according to the visiting probabilities. Dark green represents up to 4% of the maximum; medium green to yellow represents between 4 and 50%; bright yellow represents 50% and above. The picture contains 1000×1000 points.

III. Basins of attraction of the three equilibrium states (point attractors marked by white dots) of a magnetic pendulum (Sections 1.2.2 and 6.8.3, Eqs. (6.36) and (6.37)) in the plane of the initial positions parallel to that of the magnets, with zero initial velocities. The friction coefficient is 1.5 times greater than in Fig. 1.10; all the other data are unchanged. The fractal part of the basin boundaries appears to be only slightly extended. The distance between neighbouring initial positions is $1/240$ (the resolution is 1280×960 points).

IV. Basins of attraction of a magnetic pendulum swinging twice as fast as the one in Fig. 1.10 (all the other data are unchanged). The fractal character of the basin boundaries is pronounced.

V. Basins of attraction of the magnetic pendulum in Plate IV, but the fixation point of the pendulum is now slightly off the centre of mass of the magnets (it is shifted by 0.2 units diagonally). The strict symmetry of Plate IV has disappeared, but the character has not changed.

VI. Basins of attraction of a magnetic pendulum for the same friction coefficient as in Plate III, but for a pendulum swinging four times as fast and placed closer to the plane of the magnets.

VII. Basins of attraction of two stationary periodic motions (limit cycle attractors marked by white dots) of a driven pendulum (Sections 1.2.2 and 5.6.3 and equation (5.89)) on a stroboscopic map (like the one in

Fig. 1.13). The lighter blue and red hues belong to initial conditions from which a small neighbourhood of the attractors is reached in more than eight periods. The resolution is 1280×960 points.

VIII. Overview of the frictionless dynamics of a body swinging on a pulley (Sections 1.2.3 and 7.4.3 and equation (7.36)) on a Poincaré map. The mass of the swinging body is smaller than that in Fig. 1.17. The dotted region is a chaotic band, whereas closed curves represent regular, non-chaotic motions. Rings of identical colour become mapped onto each other. Their centres form higher-order cycles (yellow, green and blue correspond, for example, to two-, five- and six-cycles, respectively). The chaotic band pertains to a single, the rings to 24 different, initial condition(s).

IX. Mirroring Christmas-tree ornaments (Section 1.2.4). The four spheres touch, and their centres are at the corners of a tetrahedron. The one on the right is red, that on the left is yellow, and the ones behind and on top are silver. The picture shows the reflection pattern of the flash. (Photographed by G. Maros.)

X. Christmas-tree ornaments mirrored in each other (Section 1.2.4). All four spheres are now silver, but the surfaces tangent to the spheres are coloured white, yellow and red, and the fourth one (towards the camera) appears to be black. The insets show the reflection images at the centres of the tetrahedron from slightly different views. (Photographed by P. Hámori.)

XI. Droplet dynamics in a container with two outlets (Sections 1.2.5 and 9.4.1 and equations (9.24)). The top left picture exhibits the initial configuration of the square-shaped droplet and the state after one time unit (outlets are marked by white dots). The top right, bottom left and bottom right panels display the situation after two, three and four time units, respectively. The part of the droplet that has not yet left the system becomes increasingly ramified, and the colours become well mixed. The picture contains 640×480 points.

XII. Natural distribution of the roof attractor (the same as that of Fig. 5.44(b) from a different view). The grid used to represent the distribution is of size $\varepsilon = 1/500$. The colour of each column of height up to 2% of the maximum is red, it changes towards yellow up to 50%, and is bright yellow beyond 50%.

XIII. Natural distribution on the chaotic attractor of the driven non-linear oscillator presented in Plate I. Blue changing towards red is used up to 10% of the maximum, then red changing towards yellow up to 60%, and bright yellow beyond.

XIV. Natural distribution on the chaotic attractor of the driven pendulum presented in Plate II. Dark green is used up to 2% of the maximum, then medium green up to 5%, green changhing towards yellow up to 30%, and bright yellow beyond.

XV. Natural distribution on the chaotic attractor of an oscillator kicked with an exponential amplitude ($\varepsilon = 1/800$). Same as Fig. P.23(a) from a different view (the distribution is mirroring on the plane of the map).

XVI. Natural distribution on the chaotic attractor of an oscillator kicked with a sinusoidal amplitude. Same as Fig. P.23(b) but seen from below ($\varepsilon = 1/200$). Blue changing towards green is used up to 2% of the maximum visible value, then green changing towards yellow up to 20%, and bright yellow beyond.

XVII. Natural distribution on the roof saddle (the one presented in Fig. 6.13(c), from a different view). Colouring is the same as in Plate XII.

XVIII. The distribution of Plate XVII from the top view and with different colouring.

XIX. Natural distribution on the parabola saddle (the one presented in Fig. 6.13(d) but from a different view). Blue is used up to 4% of the maximum visible value, then blue changing to red up to 21%, red changing towards yellow up to 70%, and bright yellow beyond.

XX. Natural distribution of a conservative system. The distributions in three chaotic bands (from 10^6 iterations), coloured in different shades of green, of the kicked rotator (i.e. of the standard map with the parameter of Fig. 7.7(c)) are shown. The distribution is uniform within each band. The numerical convergence towards the smooth distribution is much slower near the edges, which are formed by KAM tori, than in the interior. The yellow barriers represent a few quasi-periodic tori.

XXI. Escape regions and boundaries in the three-disc problem; Section 8.2.3. Scattering orbits with the first bounce on the left disc are considered. The initial conditions on the $(\theta_n, \sin\varphi_n)$-plane are coloured according to the deflection angle, θ, of the outgoing orbit: red for $0 < \theta < 2\pi/3$; yellow for $-2\pi/3 < \theta < 0$; and blue for $|\theta| > 2\pi/3$. The escape boundaries contain a fractal component, which is part of the chaotic saddle's stable manifold shown in Fig. 8.12(a). Note that all three colours accumulate along the fractal filaments of the escape boundaries! (A property that also holds for the basin boundaries shown in Plates IV–VI.) The left (right) inset is a magnification of the rectangle marked in the middle (in the left inset), and illustrates self-similarity.

XXII. Natural distribution on the attractor of Lorenz's global circulation model (permanent winter, see Section 9.3, Fig. 9.19, from a different view). Colouring is similar to that in Plate XII, but in blue tones.

XXIII. Sea ice advected by the ocean around Kamchatka from the NASA archive (http://eol.jsc.nasa.gov/scripts/sseop/photo.pl?mission=STS045&roll=79&frame=N).

XXIV. Plankton distribution around the Shetland Islands, May 12, 2000, from the NASA archive (http://disc.gsfc.nasa.gov/oceancolor/scifocus/oceanColor/vonKarman_vortices.shtml).

XXV. Ozone distribution (in parts per million, ppm) above the South Pole region at about 18 km altitude on September 24, 2002. The chemical reactions leading to ozone depletion are simulated over 23 days in the flow given by meteorological wind analyses during a rare event when the so-called 'polar vortex' splits into two parts.[1]

XXVI. The same as Plate XXIII but for the distribution of HCl (in parts per billion, ppb). Note that both distributions are filamental but that the concentrations of the two substances are different.[1]

XXVII (Front cover illustration) Natural distribution on the chaotic attractor of the non-linear oscillator of Plate XIII with a blueish-white colouring.

XXVIII (Back cover illustration) Natural distribution on the chaotic attractor of an oscillator kicked with a parabola amplitude (on the parabola attractor). Colouring is the same as in Plate XIX.

[1] Groop, J.-U., Konopka, P. and Müller, R. 'Ozone chemistry during the 2002 Antarctic vortex split', *J. Atmos. Sci.* **62**, 860 (2005).

Preface

> We have just seen that the complexities of things can so easily and dramatically escape the simplicity of the equations which describe them. Unaware of the scope of simple equations, man has often concluded that nothing short of God, not mere equations, is required to explain the complexities of the world.
>
> . . . The next great era of awakening of human intellect may well produce a method of understanding the *qualitative* content of equations.
>
> *Richard Feynman in 1963, the year of publication of the Lorenz model*[1]

The world around us is full of phenomena that seem irregular and random in both space and time. Exploring the origin of these phenomena is usually a hopeless task due to the large number of elements involved; therefore one settles for the consideration of the process as noise. A significant scientific discovery made over the past few decades has been that phenomena complicated *in time* can occur in simple systems, and are in fact quite common. In such *chaotic* cases the origin of the random-like behaviour is shown to be the strong and non-linear interaction of the few components. This is particularly surprising since these are systems whose future can be deduced from the knowledge of physical laws and the current state, in principle, with arbitrary accuracy. Our contemplation of nature should be reconsidered in view of the fact that such deterministic systems can exhibit random-like behaviour.

Chaos is the complicated temporal behaviour of simple systems. According to this definition, and contrary to everyday usage, chaos is not spatial and not a static disorder. Chaos is a type of *motion*, or more generally a type of temporal evolution, dynamics. Besides numerous everyday processes (the motion of a pinball or of a snooker ball, the auto-excitation of electric circuits, the mixing of dyes), chaos occurs in technical, chemical and biological phenomena, in the dynamics of illnesses, in

[1] R. P. Feynman, R. B. Leighton and M. Sands, *The Feynman Lectures on Physics, Vol. II.* New York: Addison-Wesley, 1963, Chap. 40, pp. 11, 12.

elementary economical processes, and on much larger scales, for example in the alternation of the Earth's magnetic axis or in the motion of the components of the Solar System.

There is an active scientific and social interest in this phenomenon and its unusual properties. The motion of chaotic systems is complex but understandable: it provides surprises and presents those who investigate it with the delight of discovery.

Although numerous books are available on this topic, most of them follow an interdisciplinary presentation. The aim of our book is to provide an introduction to the realm of chaos related phenomena within the scope of a single discipline: classical mechanics. This field has been chosen because the inevitable need for a probabilistic view is most surprising within the framework of Newtonian mechanics, whose determinism and basic laws are well known.

The material in the book has been compiled so as to be accessible to readers with only an elementary knowledge of physics and mathematics. It has been our priority to choose the simplest examples within each topic; some could even be presented at secondary school level. These examples clearly show that almost all the mechanical processes treated in basic physics become chaotic when slightly generalised, i.e. when freed of some of the original constraints: chaos is not an *exceptional*, rather it is a *typical* behaviour.

The book is primarily intended for undergraduate students of science, engineering, and computational mathematics, and we hope that it might also contribute to clarifying some misconceptions arising from everyday usage of the term 'chaos'.

The book is based on the material that one of us (T. T.) has been teaching for fifteen years to students of physics and meteorology at Eötvös University, Budapest, and that we have been lecturing together in the last few years.

Acknowledgements

First of all, we wish to thank Péter Szépfalusy, who established chaos research in Hungary, without whose determining activity and personal influence this book could not have been brought into existence.

One of us (T. T.) is indebted to Ch. Beck, Gy. Bene, W. Breymann, A. Csordás, G. Domokos, R. Dorfman, G. Eilenberger, M. Feigenbaum, H. Fujisaka, P. Gaspard, R. Graham, P. Grassberger, C. Grebogi, G. Györgyi, B.-L. Hao, E. Hernandez-Garcia, I.M. Jánosi, C. Jung, H. Kantz, Z. Kaufmann, G. Károlyi, Z. Kovács, Y.-L. Lai, L. Mátyás, A. de Moura, Z. Neufeld, E. Ott, Á. Péntek, O. Piro, I. Procaccia, A. Provenzale, P. Richter, O. Rössler, I. Scheuring, G. Schmidt, J. C. Sommerer, K.G. Szabó, D. Szász, Z. Toroczkai, G. Vattay, J. Vollmer, J. Yorke, and G. Zaslavsky, who all provided joint work or were included in useful discussions, which greatly determined his view on the subject.

The book owes much to the reviewers of the Hungarian edition, P. Gnädig and F. Kun, who greatly contributed to making it more understandable by their innumerable constructive suggestions.

We are grateful to our colleagues, B. Érdi, G. Götz, Gy. Muraközy, J. Pap and G. Stépán, and students, I. Benczik, G. Csernák, P. Hantz, J. Kiss, S. Kiss, B. Muraközy, G. Orosz, J. Schneider, A. Sótér and A. Suli, for carefully reading and commenting on parts of the book. M. Hóbor, Gy. Károlyi and K. G. Szabó also helped us with their advice on programming and text editing.

Special thanks are due to K. Kulacsy for the translation, to Sz. Hadobás for the aesthetic presentation of the three-dimensional pictures, to G. Károlyi, G. Maros and P. Hámori for the photographs, and to Knorr Bremse (L. Palkovics) and to the *Journal of Atmospheric Science*, as well as the authors, J.-U. Groß, P. Konopka and R. Müller, for their permission to use their figures. The authors would like to thank NASA for the use of Fig. 9.32 and Plates XXIII and XXIV. This work has been supported by Grant OTKA TSO44839.

We would like to express our thanks to the editorial staff of Cambridge University Press; in particular our editors, Simon Capelin and Vince Higgs, and the copy-editor, Irene Pizzie.

How to read the book

The first part of the book presents the basic phenomena of chaotic dynamics and fractals at an elementary level. Chapter 1 provides, at the same time, a preview of the five main topics to be treated in Part III.

Part II is devoted to the analysis of simple motion. The geometric representation of dynamics in phase space, as well as basic concepts related to instability (hyperbolic points and stable and unstable manifolds), are introduced here. Two-dimensional maps are deduced from the equations of motion for driven systems. Elementary knowledge of ordinary differential equations, of linear algebra, of the Newtonian equation of a single point mass and of related concepts (energy, friction and potential) is assumed.

Part III provides a detailed investigation of chaos. The dynamics occurring on chaotic attractors characteristic of frictional, dissipative systems is presented first (Chapter 5). No preliminary knowledge is required upon accepting that two-dimensional maps can also act as the law of motion. Next, the finite time appearance of chaos, so-called transient chaotic behaviour, is investigated (Chapter 6). Subsequently, chaos in frictionless, conservative systems is considered in Chapter 7, along with its transient variant in the form of chaotic scattering in Chapter 8. Chapter 9 covers different applications of chaos, ranging from engineering to environmental aspects.

Problems constructed from the material of each chapter (many also require computer-based experimentation) motivate the reader to carry out individual work. Some of the solutions are given at the end of the book; the remainder appear (in a password-protected format) on the following website: www.cambridge.org/9780521839129.

Topics only loosely related to the main train of ideas, but of historical or conceptual interest, are presented in Boxes. Some important technical matter (for example numerical algorithms, writing equations in dimensionless forms) are relegated to an Appendix. A bibliography is given at the end of the book, and it is broken down according to topics, chapters and Boxes.

In order to emphasize the general aspects of chaos, the most important relations are also given in a formulation independent of mechanics (see Sections 3.5, 4.7, 5.4, 6.3, 7.5 and 8.4). The description of motion occurs primarily in terms of ordinary differential equations, and we concentrate on chaos from such a mathematical background. Irregular dynamics generated by other mathematical structures, which do not represent real phenomena, are thus beyond the scope of the book. The case of one-dimensional maps is mentioned therefore as a special limit only. This approach might provide a useful introduction to chaos for all disciplines whose dynamical phenomena are described by ordinary differential equations.

The book is richly illustrated with computer-generated pictures (24 of which are in colour), not only to provide a better understanding, but also to exemplify the novel and aesthetically appealing world of the geometry of dynamics.

Part I

The phenomenon: complex motion, unusual geometry

Chapter 1
Chaotic motion

1.1 What is chaos?

Certain long-lasting, sustained motion repeats itself exactly, periodically. Examples from everyday life are the swinging of a pendulum clock or the Earth orbiting the Sun. According to the view suggested by conventional education, sustained motion is always regular, i.e. periodic (or at most superposition of periodic motion with different periods). Important characteristics of a periodic motion are: (1) it repeats itself; (2) its later state is accurately predictable (this is precisely why a pendulum clock is suitable for measuring time); (3) it always returns to a specific position with exactly the same velocity, i.e. a single point characterises the dynamics when the return velocity is plotted against the position.

Regular motion, however, forms only a *small part* of all possible sustained motion. It has become widely recognised that long-lasting motion, even of simple systems, is often *irregular* and does not repeat itself. The motion of a body fastened to the end of a rubber thread is a good example: for large amplitudes it is much more complex than the simple superposition of swinging and oscillation. No regularity of any sort can be recognised in the dynamics.

The irregular motion of simple systems, i.e. systems containing only a few components, *is called chaotic*. As will be seen later, the existence of such motion is due to the fact that even *simple* equations can have very *complicated* solutions. Contrary to the previously generally accepted view, the simplicity of the equations of motion does not determine whether or not the motion will be regular.

Understanding chaotic motion requires a non-traditional approach and specific tools. Traditional methods are unsuitable for the description

Table 1.1. *Comparison of regular and chaotic motion.*

Regular motion	Chaotic motion
self-repeating	irregular
predictable	unpredictable
of simple geometry	of complicated geometry

of such motion, and the discovery of the ubiquity of chaotic dynamics has become possible through *computer-based experimentation*. Detailed observations have led to the result that chaotic motion is characterised by the *opposite* of the three properties mentioned above: (1) it does not repeat itself, (2) it is unpredictable because of its sensitivity to the initial conditions that are never exactly known, (3) the return rule is complicated: a complex but regular structure appears in the position vs. velocity representation. The differences between the two types of dynamics are summarised in Table 1.1.

The properties of chaotic systems are unusual, either taken individually or together; the most efficient way to understand them is by considering particular cases. In the following, we present the chaotic motion of very simple systems on the basis of numerical simulations, which are unavoidable when studying chaos. It should be emphasised that all of our examples are discussed for a unique set of parameters, and that slightly different choices of the parameters could result in substantially different behaviour. These examples also serve to classify different types of chaos and help in developing the new concepts necessary for a detailed understanding of chaotic dynamics.

1.2 Examples of chaotic motion

1.2.1 Irregular oscillations, driven pendulum – the chaotic attractor

Objects mounted on spring suspensions (for example car wheels and spin-dryers) oscillate. Because of the losses that are always present due to friction or air drag, these oscillations, when left alone, are damped and ultimately vanish. Sustained motion can only develop if energy is supplied from an external source. The supplied energy can be a more or less periodic shaking, i.e. the application of a driving force (caused by interactions with pot-holes in the case of the car wheel and by the uneven distribution of clothes in the spin-dryer), as indicated schematically in Fig. 1.1.

Fig. 1.1. Model of driven oscillations: a body of finite mass is fixed to one end of a weightless spring and the other end of the spring is moved sinusoidally with time.

As long as the displacement is small, the spring obeys a *linear* force law to a good approximation: the magnitude of the restoring force is

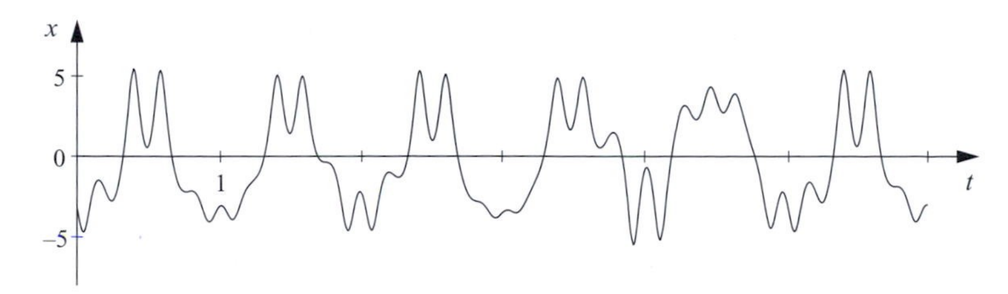

Fig. 1.2. Irregular sustained oscillations of a point mass fixed to the end of a stiffening spring (a driven non-linear oscillator), driven sinusoidally in the presence of friction.

proportional to the elongation. In this case the sustained motion is regular: it adopts the period of the driving force. If the natural period of the spring is close to that of the driving force, then the amplitude may become very large and the well known phenomenon of *resonance* develops. For large amplitudes, however, the force of the spring is usually no longer proportional to the elongation; i.e., the force law is *non-linear*. Resonance is therefore a characteristic example for the appearance of non-linearity.

For non-linear force laws, the restoring force increases more rapidly or more slowly than it would in linear proportion to the elongation: we can speak of stiffening or softening springs, respectively. Whichever type of non-linearity is involved, the sustained state of the driven oscillation may be chaotic. A qualitative explanation is that the spring is not able to adopt exactly the sinusoidal, harmonic motion of the forcing apparatus, since its own periodic behaviour is no longer harmonic. Thus, the sustained dynamics follows the driving force in an averaged sense only, but always differs from it in detail (instead of the uniform hum of the car or the spin-dryer, an irregular sound can be heard in such situations). Neither the amplitude nor the frequency is uniform: the sustained motion does not repeat itself regularly; it is chaotic.

Figure 1.2 shows the motion of a body fixed to the end of a stiffening spring and driven sinusoidally.[1] It can clearly be seen that there is no repetition in the displacement vs. time curve; i.e., the motion is *irregular*.

Slightly different initial conditions result in significant differences in the displacement after only a short time (Fig. 1.3): the dynamics is unpredictable. This figure also shows that the long-term behaviour is of a similar nature in both cases: the two motions are equivalent in a *statistical* sense.

[1] The precise equations of motion of the examples in this section can be found in Sections 5.6.2 and 5.6.3.

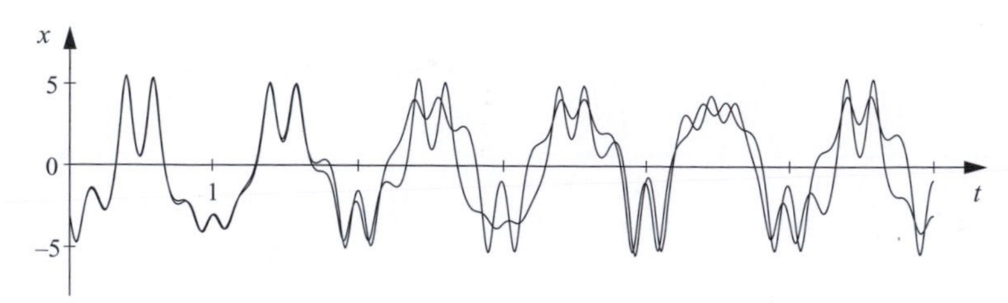

Fig. 1.3. Two sets of motion which started from nearly identical positions. The small initial difference increases rapidly: the motion is sensitive to the initial conditions and therefore it is unpredictable.

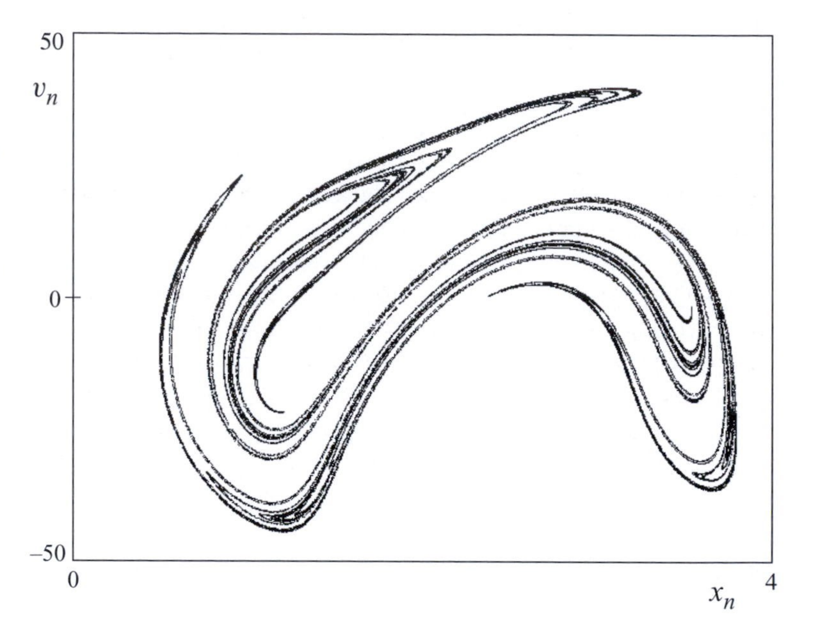

Fig. 1.4. Pattern resulting from a sustained non-linear oscillation in the velocity vs. position representation, using samples taken at time intervals corresponding to the period of the driving force. The position and velocity co-ordinates of the *n*th sample are x_n and v_n, respectively.

An interesting structure reveals itself when we do not follow the motion continuously, but only 'take samples' of it at equal time intervals. Figure 1.4 and Plate I have been generated by plotting the position and velocity co-ordinates (x_n, v_n) of the sustained motion at integer multiples, n, of the period of the driving force, through several thousands of periods.

It is surprising that there are numerous values of x_n to which many (according to detailed examinations, an *infinite* number of) different velocity values belong. Furthermore, the possible velocity values corresponding to a single position co-ordinate x_n do *not* form a continuous interval anywhere. The whole picture has a thready, filamentary pattern, indicating that chaos is associated with a definite structure. This pattern is much more complicated than those of traditional plane-geometrical objects: it is a structure called a *fractal* (a detailed definition of fractals will be given in Chapter 2). Remember that a *single point* would correspond to a periodic motion in this representation. Chaotic motion is therefore infinitely more complicated than periodic motion.

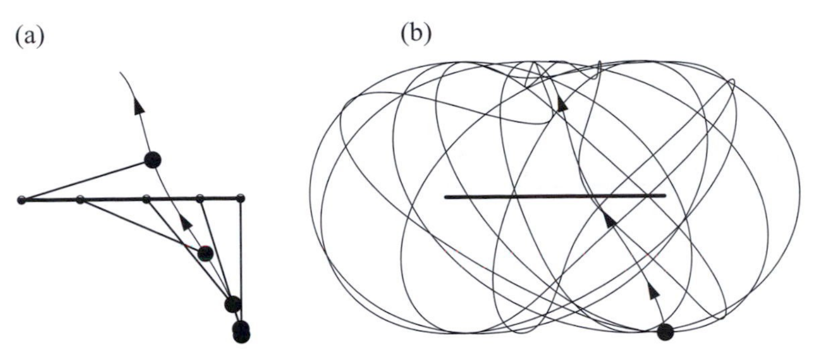

Fig. 1.6. Motion of a driven pendulum. (a) The pendulum a few moments after starting from a hanging state (over the first half period). (b) The path of the end-point of the pendulum for a longer time: the pendulum swings irregularly and often turns over. The horizontal bar indicates the interval over which the suspension point moves.

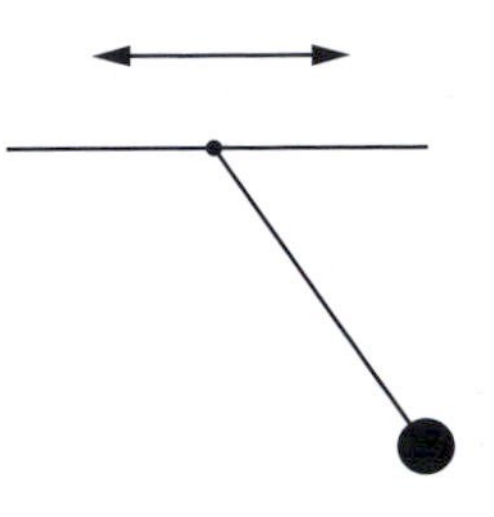

Fig. 1.5. Driven pendulum: the pendulum is driven by the periodic movement of its point of suspension in the horizontal plane.

Another example is the behaviour of a driven pendulum (Fig. 1.5). The large-amplitude swinging of a traditional simple pendulum is non-linear, since the restoring force is not proportional to the deflection angle but to the sine of this angle. Without any driving force, the swinging ceases because of friction or air drag: sustained motion is impossible. The pendulum can be driven in different ways. We examine the case when the point of suspension is moved horizontally, sinusoidally in time. In order to avoid the problem of the folding of the thread, the point mass is considered to be fixed to a very light, thin rod. With a sufficiently strong driving force, the motion may become chaotic. Figure 1.6 shows the path of the pendulum in the vertical plane.

Note that the pendulum turns over several times in the course of its motion. The 'upside down' state is especially unstable, just like that of a pencil standing on its point. Two paths of the pendulum starting from nearby initial positions remain close to each other only until an unstable state, an 'upside down' state, separates them. Then one of them turns over, while the other one falls back to the side it came from (Fig. 1.7). The reason for the unpredictability is that the motion passes through a series of *unstable states*.

The structure underlying the irregular motion can again be demonstrated by following the motion initiated in Fig. 1.6 for a long time and taking samples from it by plotting the position (angular deflection) and velocity (angular velocity) co-ordinates (x_n, v_n) at intervals corresponding to the period of the driving force (Fig. 1.8 and Plate II).

In a frictional (dissipative) system, sustained motion can only develop if some external energy supply (driving) is present. Regardless of the initial state, the dynamics converges to some sustained behaviour that will therefore be called an attracting object, or an *attractor* (for the

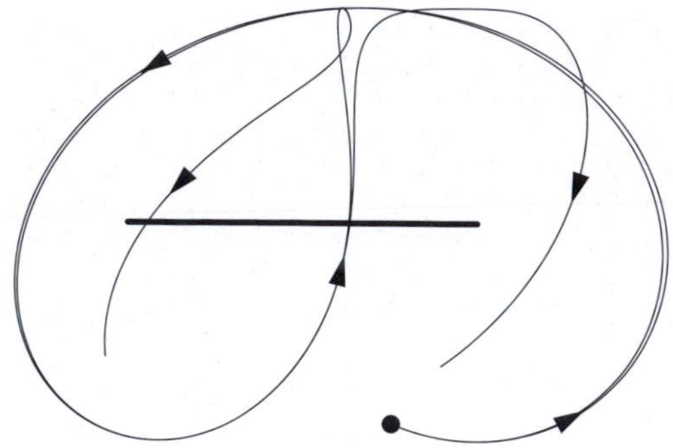

Fig. 1.7. Separation of the paths of two identical driven pendulums starting from nearby points while passing an unstable state. The notation is the same as in Fig. 1.6. The arrows show the direction in which the end-points of the pendulums move.

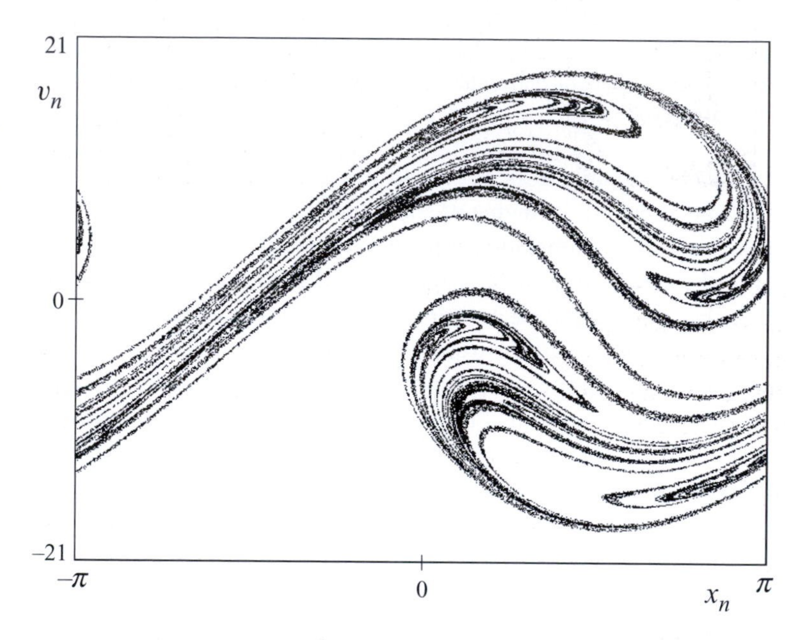

Fig. 1.8. Pattern resulting from a chaotic driven pendulum (chaotic attractor) obtained by plotting the state of the pendulum in the position–velocity co-ordinates at integer multiples of the driving period.

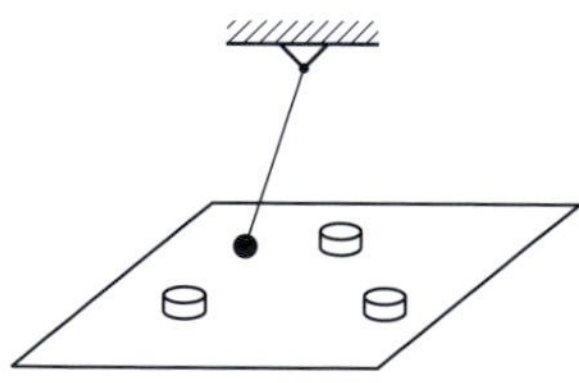

Fig. 1.9. The magnetic pendulum: magnets are fixed to the table and a point mass attracted by the magnets is fixed to the end of the thread. The pendulum ultimately settles in an equilibrium state pointing towards one of the magnets, but only after some irregular, chaotic motion.

exact definition, see Section 3.1.2). *Simple* attractors correspond either to regular or to ceasing motion. A sufficiently large supply of energy inevitably brings about the non-linearity of the system; the sustained dynamics is then usually irregular, i.e. chaotic. This is accompanied by the presence of a *chaotic attractor*, also called a *strange attractor* because of its peculiar structure. Figures 1.4 and 1.8 display examples of chaotic attractors.

1.2.2 Magnetic and driven pendulums, fractal basin boundary – transient chaos

Consider a pendulum, the end-point of which is a small magnetic body, moving above three identical magnets placed at the vertices of a horizontal equilateral triangle (Fig. 1.9). When the force between the

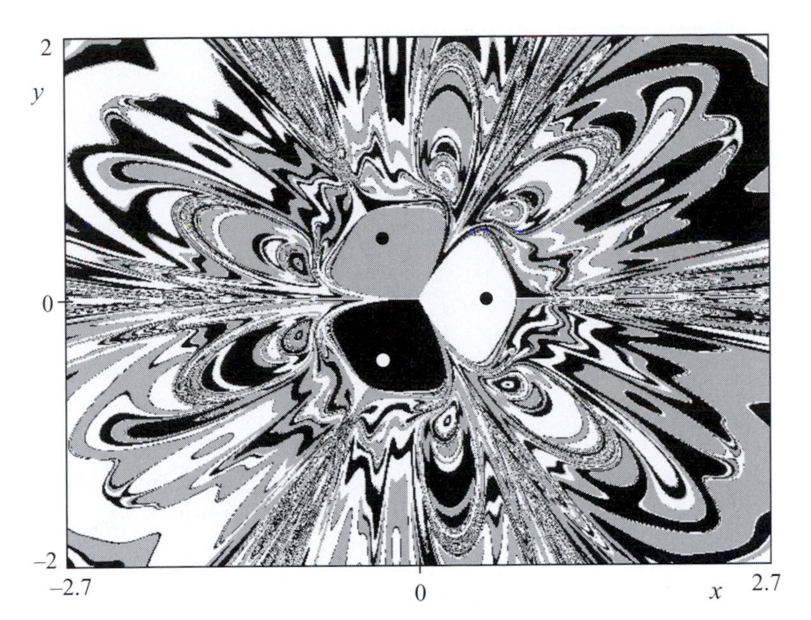

Fig. 1.10. Basin of attraction of the three equilibrium states of the magnetic pendulum (one white and two black dots). Each point on the horizontal plane is shaded according to the magnet in whose neighbourhood the pendulum comes to a rest when starting above that point with zero initial velocity.

end of the pendulum and the magnets is attracting, the pendulum can come to a halt, pointing towards any of the magnets. Thus there are three simple attractors in the system. Starting above any point of the plane, we can use a computer to calculate which magnet the pendulum will be closest to after coming to rest.[2] By assigning three different colours to the three attractors, and to the corresponding initial positions that converge towards them, the whole plane can be coloured. Each identically coloured area is a *basin of attraction*. Surprisingly, the basin boundaries are interwoven and entangled in a complicated manner (see Fig. 1.10 and Plates III–VI); these simple attractors have *fractal basin boundaries*. (Naturally, the close vicinity of each attractor appears in one colour only: the boundaries do not come close to the attractors.)

Motion starting near the fractal boundary remains irregular for a while, exhibiting *transient chaos*, i.e. chaos lasting for a finite period of time (Fig. 1.11), but ultimately it ends up on one of the attractors.

A driven pendulum (Fig. 1.5) may also exhibit transient chaos. When the friction is sufficiently large, the pendulum can exhibit regular sustained motion only. There are two options for the given parameters (see Fig. 1.12, which depicts the paths corresponding to these two simple attractors in the vertical plane). An overall view of the basins of attraction can again be obtained by representing the starting point in the position

[2] The equations of motion of the magnetic pendulum can be found in Section 6.8.3.

Fig. 1.11. Path of the end-point of the magnetic pendulum viewed from above. The motion is irregular before reaching one of the rest positions: it is transiently chaotic. (The fixed magnets are represented by solid black dots.)

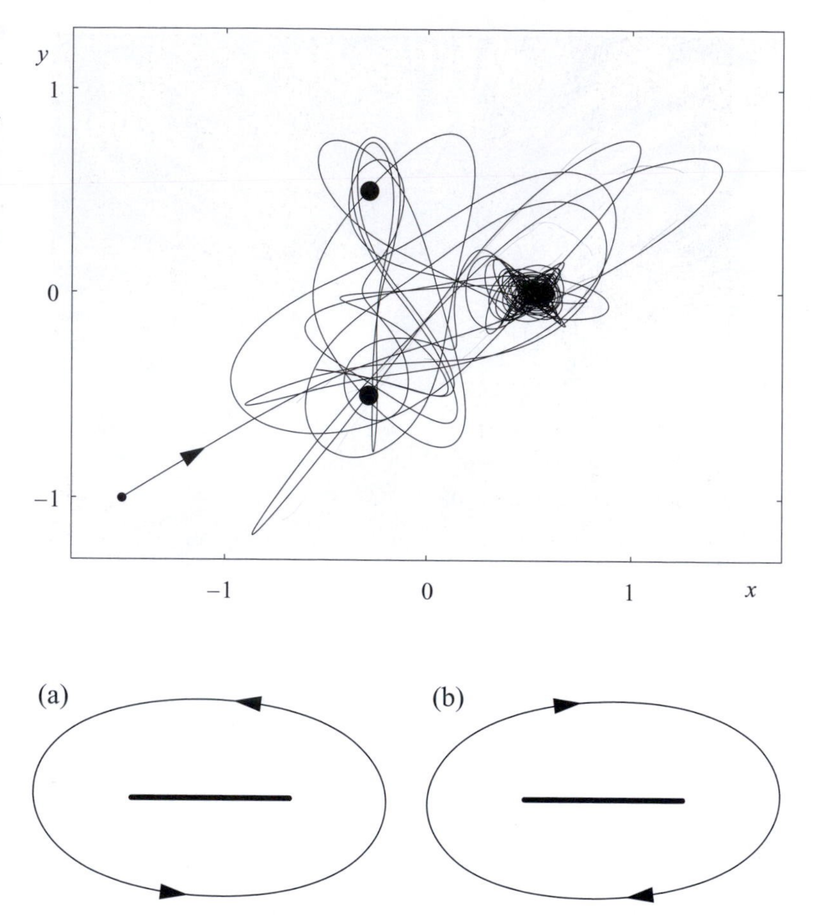

Fig. 1.12. Simple periodic attractors of the driven pendulum: for sufficiently strong friction only these two types of sustained motion exist. All the different initial conditions lead to one of these motions, corresponding to a simple attractor each.

(angular deflection) – velocity (angular velocity) plane in the colour of the attractor which the motion ultimately converges to (Fig. 1.13 and Plate VII).

Motion starting close to the boundary is similar initially to that seen in the case of the chaotic attractor, but it ultimately converges to one of the simple attractors. Irregular dynamics has a finite duration; it is transient. There exist, however, very exceptional initial conditions from which the dynamics never reaches any of the attractors, and is chaotic for any length of time. There exists an infinity of such motion (Fig. 1.14), but the initial conditions that describe these state do not form a compact domain in the plane, but rather a fractal cloud of isolated points called a *chaotic saddle*.

Fig. 1.13. Basins of attraction in the driven pendulum on the plane of initial conditions. The two simple attractors in Fig. 1.12 appear here as points (white and black dots), and the initial states converging towards them are marked in black and white, respectively.

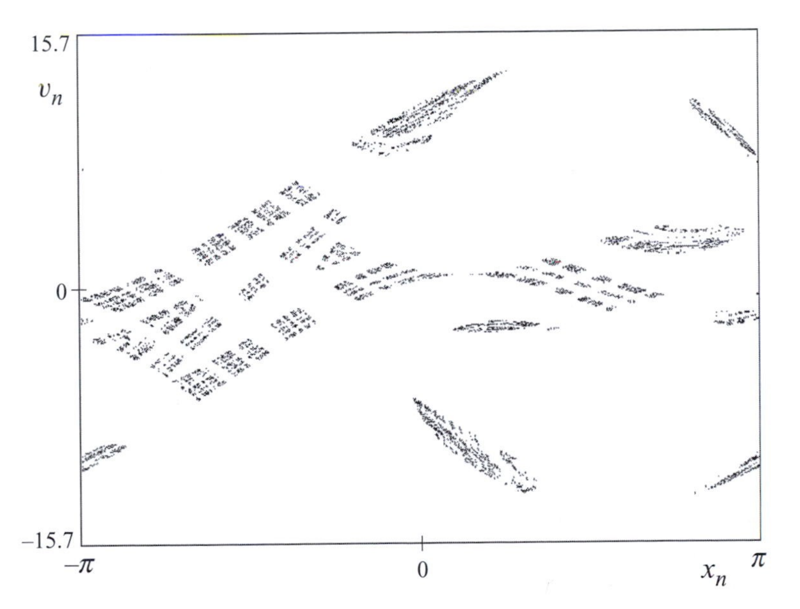

Fig. 1.14. Initial states of the driven pendulum of Fig. 1.13 that never reach either simple attractor: all points shown here are on the basin boundary and, if followed in time, they keep moving between themselves after every period of the driving force. This chaotic saddle is responsible for chaotic dynamics of transient type.

Thus, chaotic dynamics can also occur if the sustained forms of motion are regular, but there are many possible transient routes (chaotic transients) leading to them. In such cases several simple attractors coexist, each with its own basin of attraction defined by the set of initial conditions which converges to the given attractor. The basins of attraction often penetrate each other, and their boundaries can also be filamentary fractal curves. The motion starting from the vicinity of these fractal basin boundaries behaves randomly along the boundary for a long time, as if it is difficult to decide which attractor to choose. During this period of uncertainty the motion is irregular and is bound to fractal structures.

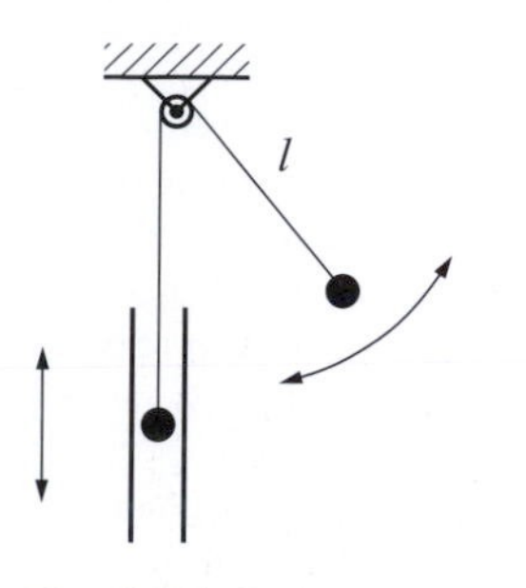

Fig. 1.15. Body swinging on a pulley: two point masses are joined by a thread wound around a pulley of negligible radius, one of them swinging freely in a vertical plane, the other moving vertically only.

1.2.3 Body swinging on a pulley, ball bouncing on slopes – chaotic bands

Let us examine what happens in frictionless (conservative) systems. Consider two point masses joined by a thread wound about a small pulley (see Fig. 1.15). The case when both points can only move vertically is a well known secondary school problem. Here, however, we let one of the point masses swing in a vertical plane (with the thread always stretched for the sake of simplicity). It will be shown that new types of chaotic motion develop under such conditions.[3]

The instantaneous length, l, of the thread of the point mass that can swing is one of the position co-ordinates; the other is the angle of deflection. In the traditional arrangement, where only vertical displacement is allowed, the heavier mass always pulls the other one up, but the situation is much more interesting now. If the swinging body is thrust horizontally with sufficient momentum while the other body moves downwards, then the swinging body turns over several times, the thread shortens, the body spins faster, and thus becomes able to pull the other body upward, even if the latter is the heavier. (It is assumed that the swinging body does not collide with anything and that the thread does not become unattached from the pulley when turning over.) Thus, a long-lasting, complicated, chaotic motion may develop. The path of the swinging body and the length of the thread vs. time are shown in Figs. 1.16(a) and (b), respectively. Again, the paths of the motion starting from nearby initial conditions soon branch off; the motion is unpredictable.

An overview of the motion corresponding to a given total energy can be presented with the help of some sampling technique. The system is not driven in this case, and therefore sampling will not take place at identical time intervals, rather at identical configurations: whenever the swinging body passes through the vertically hanging configuration, the instantaneous length, $l_n \equiv x_n$, of the swinging thread and the rate of change of this length, v_n, will be plotted as one point in the plane. Thus, chaotic motion is represented by a sequence of points jumping around in a disordered manner and dotting a finite region of the plane; This is called the *chaotic band* (Fig. 1.17). Other initial conditions outside of the

[3] The equations of motion of the examples in this section can be found in Sections 7.4.1 and 7.4.3.

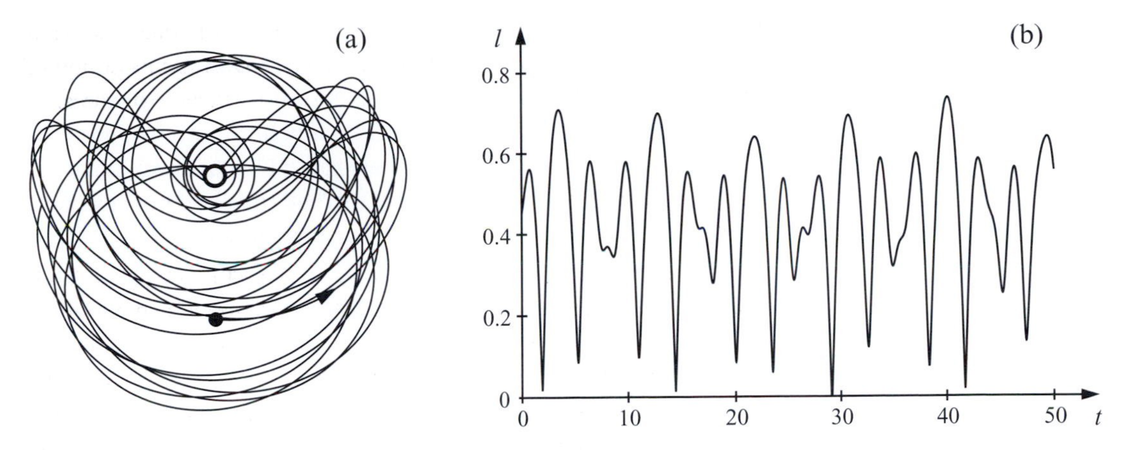

Fig. 1.16. Frictionless motion of a body swinging on a pulley. (a) The spatial path of the swinging body (the initial position is marked by a black dot, the pulley by the centre of an open circle); (b) the dependence of the length of the swinging thread on time within the same time interval.

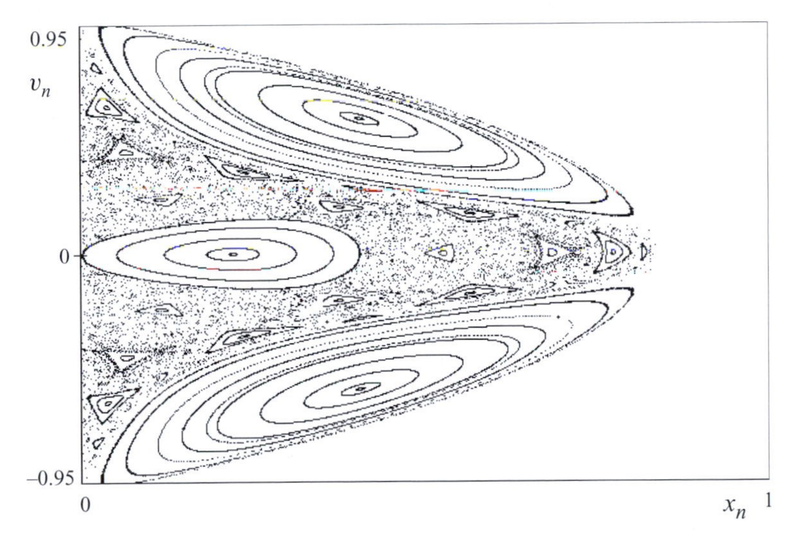

Fig. 1.17. Overview of the motion of a body swinging on a pulley without air drag and at a given total energy, on the basis of samples of length and velocity (x_n, v_n) taken when passing through the vertical position, from the left. The dotted region is a chaotic band, which can be traced out by motion starting from a single initial condition. The sets of closed curves form regular islands.

band may result in a single point, a few points or a continuous line, all of which correspond to regular motion. These objects together usually form closed domains that can be called regular islands. A frictionless chaotic system is characterised by a hierarchically nested pattern of chaotic bands and islands. Together they form a complicated structure of interesting texture, different from the fractals presented so far (see Fig. 1.17 and Plate VIII).

Our second example illustrates the fact that elastic collisions with flat surfaces can also lead to chaotic motion. Maybe the simplest situation

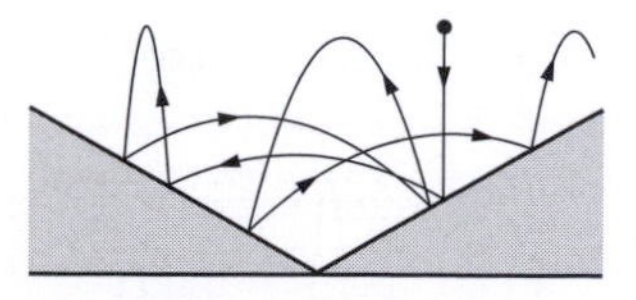

Fig. 1.18. Ball bouncing on two slopes of identical inclination that face each other in a gravitational field.

is the case of an elastic ball bouncing on two slopes that face each other (Fig. 1.18). (A motion very similar to this can be realised in experiments with atoms.) Chaotic behaviour arises because after bouncing back from the opposite slope the ball does not necessarily hit its original position. Non-linearity and inherent instability are caused by the break-point between the slopes. The chaotic motion of two balls dropped from identical heights but slightly different positions soon branches off (Fig. 1.19), just as in the previous examples.

A sampling technique providing a good overview of the dynamics is in this case to plot the two velocity components as points of a plane, at the instant of each bounce (Fig. 1.20).

There is no need to apply driving forces in order to sustain a motion in frictionless systems, since there is no dissipation and energy is conserved. On the other hand, this motion cannot converge to a well defined sustained motion because there are *no* attractors in frictionless, conservative

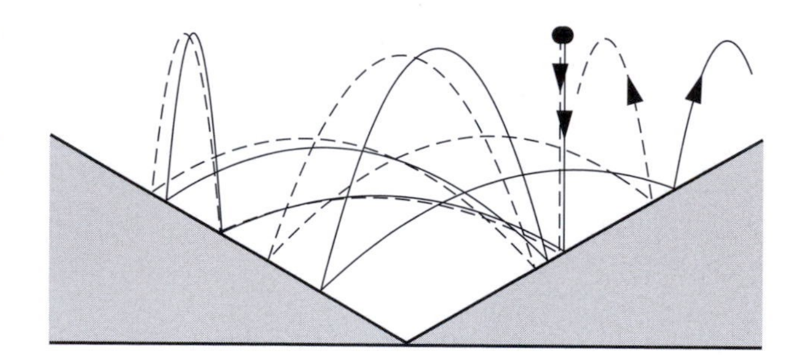

Fig. 1.19. Paths of two balls starting from nearly identical initial positions above the double slope (the continuous line is identical to that drawn in Fig. 1.18). The motion is sensitive to the initial conditions.

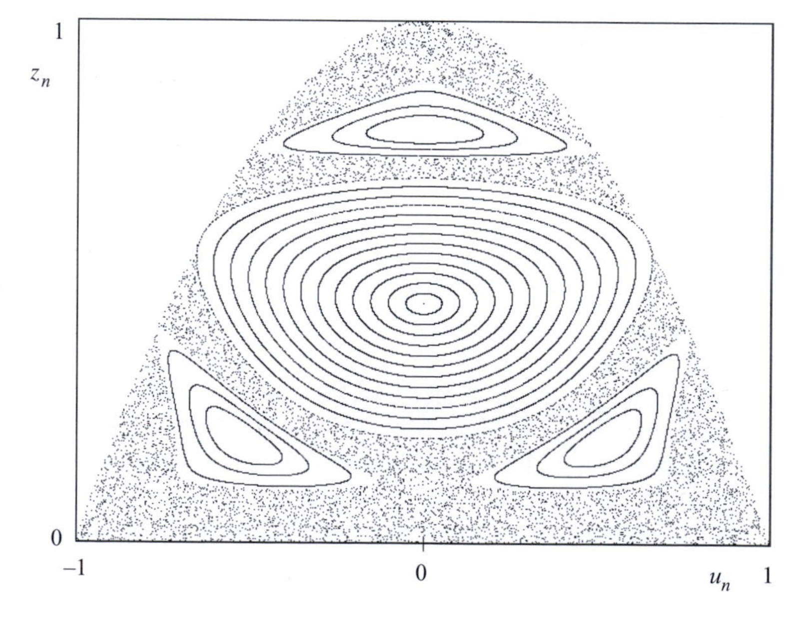

Fig. 1.20. Pattern generated by the possible motions of a ball bouncing on a double slope with given total energy in a representation where the abscissa is the velocity component parallel to the slope (u_n) and the ordinate is the square of the component perpendicular to the slope (z_n) taken at the instance of the nth bounce. The dotted region is a chaotic band. The angle of inclination of the slope is 50°.

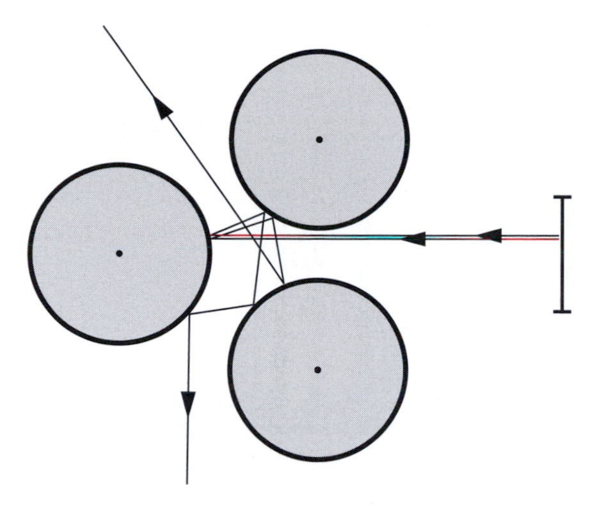

Fig. 1.21. The three-disc problem: particles bouncing perfectly elastically between identical discs fixed at the vertices of a regular triangle. Paths starting from nearby initial points soon diverge.

systems. As a result, the nature of all motion strongly depends on the initial conditions and the total energy. Regular motion corresponds to certain sets of initial conditions, while chaotic motion corresponds to other sets. The initial conditions that lead to chaotic motion form chaotic bands that, contrary to chaotic attractors, are plane-filling objects.

1.2.4 Ball bouncing between discs, mirroring Christmas-tree ornaments – chaotic scattering

Three identical discs are placed at the vertices of a regular triangle in the horizontal plane and a ball is bouncing among them – like in a pinball machine (see Fig. 1.21). The motion is considered to be frictionless; therefore the velocity of the particle is constant during the entire process. Starting from a given point, the motion depends on the initial direction of the velocity vector. Some initial conditions cause the particle to bounce for a *very* long time between the discs; during this time the dynamics of the particle is complicated and aperiodic.[4] Two slightly different initial conditions cause the paths to diverge rapidly (Fig. 1.21); therefore, this motion is also chaotic. The deviation of paths with nearby initial conditions is easy to explain, since the discs act as dispersing mirrors and the angle between the straight sections of the paths increases with each collision. The complicated structure related to the motion manifests itself in several ways. The number of bounces experienced by the particles that start along a segment in a given direction towards the discs strongly depends on the initial position. Some initial conditions lead to many collisions (Fig. 1.22). Moreover, there is an infinity of initial points

[4] A detailed investigation of this problem can be found in Section 8.2.3.

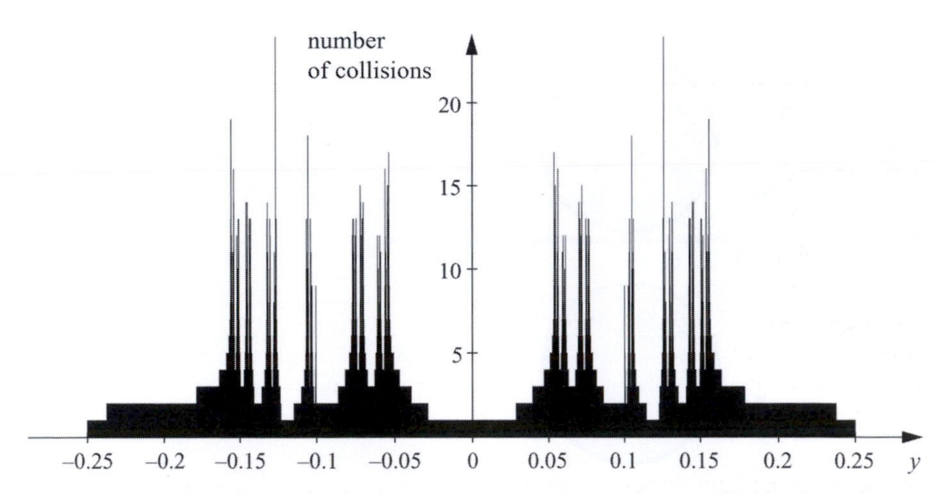

Fig. 1.22. Number of collisions of 20 000 particles starting with unit velocity at right angles to the line segment drawn in Fig. 1.21, as a function of the y co-ordinate. (The centres of the discs are at unit distance from each other.)

from which an arbitrary number of bounces can, in principle, occur (the particles then become trapped among the discs), but these do not form an interval: they form a scattered fractal cloud along the line segment.

Three or four Christmas-tree ornaments in contact with each other reflect light several times before light reaches our eyes. The interesting fractal images resulting from these reflective spheres (Plates IX and X) are examples of everyday consequences of chaotic motion.

The process whereby a significant force is only present in a finite region of a frictionless system is usually called scattering. Such a force can be tested via the motion of particles approaching from large distances. This motion is initially rectilinear, but the force causes the path to curve; then the particle leaves the scattering process and resumes its rectilinear motion, most probably in a new direction. The chaotic nature of the process arises because the motion may become long-lasting and irregular in the region where finite forces are in action. In these cases we speak of *chaotic scattering*. The average lifetime of chaos, similar to the dynamics around fractal basin boundaries, is finite. Even though there are no attractors in this case, the different outgoing states play a role similar to that of simple attractors. Chaotic scattering always involves transient chaos.

1.2.5 Spreading of pollutants – an application of chaos

Chaotic motion occurs in numerous phenomena related to practical applications. One of these is discussed here: the spreading of pollutants

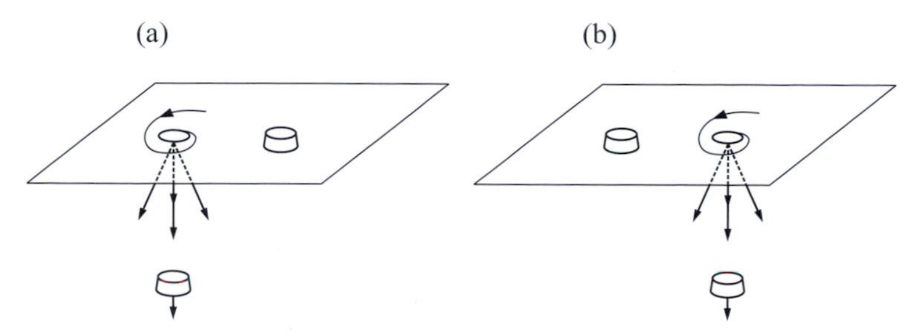

Fig. 1.23. Tank with two outlets. The outlets, when opened alternately, generate chaotic advection in a flat container. (a) and (b) illustrate the flow in the first and second half period, respectively. The flow itself is very simple; the advection of the particles is nevertheless chaotic.

in a flowing medium (air or water). The environmental significance of this matter is obvious.

Consider a large and flat container with two point-like outlets. Water whirls while flowing out. The two outlets are alternately open, each for half a period (Fig. 1.23), yielding a flow periodic in time. We want to know how a dye particle moves in this flow. For the sake of simplicity, it is assumed that the material properties of the dye are identical to that of the liquid; the only difference is the colour. In this case, the motion is determined by the condition that the instantaneous velocity of the particle is identical to that of the liquid. The path of the particle is then easy to follow.[5] Chaos arises because a particle moving towards the open outlet may not reach it within half a period; therefore it starts moving towards the other outlet, but again it may be too late to be drained, and so on. It may thus take a very long time before the particle flows out of the container. Figure 1.24 illustrates two complicated paths starting very close to each other, but leaving the tank via different outlets.

In the context of the spreading of pollution, it is especially important to follow the motion of a dye droplet. This corresponds to the examination of the dynamics of an ensemble of particles, each starting from a certain initial region, the initial shape of the droplet. A surprising discovery is that, despite the chaotic motion of each individual particle, the drop traces out a well defined thready fractal structure (after losing its original compact shape) within a short time (Fig. 1.25 and Plate XI).

The spreading of impurities in the form of filamentary patterns can be observed in numerous phenomena, ranging from oil stains on road surfaces through the mixing of cream in coffee to the propagation of

[5] The equation of motion for this example can be found in Section 9.4.1.

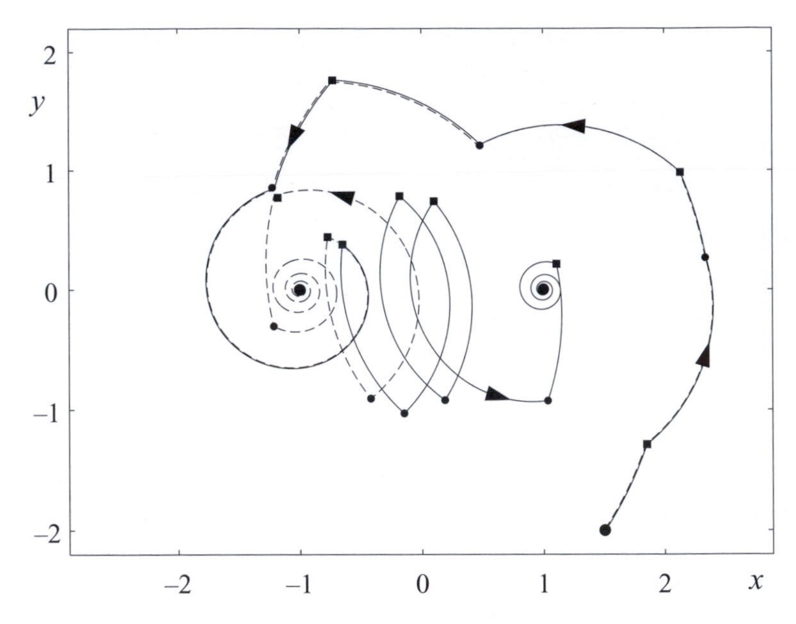

Fig. 1.24. Paths of two dye particles (continuous and broken lines) starting near each other in a tank with two outlets (situated at $(-1, 0)$ and $(1, 0)$). The consecutive black dots (squares) indicate the instants when the left (right) outlet is opened. In the initial instant the left outlet is opened. The time spent in the tank is very different for the two particles.

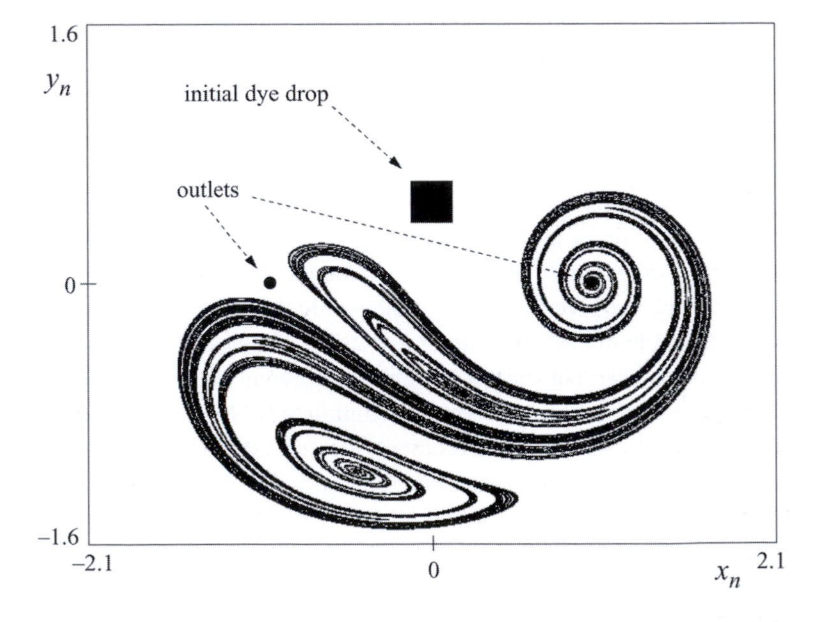

Fig. 1.25. Shape of a dye drop initially and after five periods in the tank with two outlets.

chemical pollution in the atmosphere. This thready structure unmistakably signals the chaotic motion of the individual pollutant particles.

The type of chaos found in the advection problem may depend on the parameters of the system. The problem of the tank with two outlets in the above arrangement is analogous to the problem of the fractal basin boundaries. If the outlets are closed but the alternating whirling motion is sustained by mixers, the so-called blinking vortex model is obtained. In this case there is no outflow that could be the analogue of the simple

Table 1.2. *Comparison of the traditional and phase space representations of dynamics.*

Traditional representation	Phase space representation
instantaneous co-ordinates	point in the phase space
time-dependence ($x(t)$, $v(t)$)	trajectory ($v(x)$)
structure in time	structure in phase space
individual	global

attractors for the advected particle; the chaotic behavior of the dye or impurity particles is therefore the same as that of conservative systems. It may be important to take into account that the density of the known pollutant may not be identical to that of the fluid and/or that the particle is of finite size (for example in the case of aerosols). Consequently, the velocity of the particles usually differs from that of the liquid. It can be shown that advection then corresponds to dissipative systems. The advection dynamics can then have attractors, often even chaotic ones. This implies that pollutant particles accumulate along a fractal pattern on the surface of the fluid. This phenomenon can indeed be observed in lakes, bays and harbours as a direct consequence of chaos!

1.3 Phase space

Our examples have shown that the traditional representation via the displacement or velocity vs. time graphs does not provide a suitable overview of the motion, since, however long the observation time may be, one can always expect some further novel behaviour. The order appearing in chaos does not manifest itself in the position vs. time representation, but rather in the position vs. velocity representation.

The instantaneous *state* of a mechanical system is given by its position and velocity co-ordinates, since the motion can be continued uniquely if one knows these co-ordinates and the dynamical equation. The position and velocity variables define the *phase space* of a system (for more details, see Section 3.5). For motion occurring along a straight line with position x and velocity v, the phase space is the (x, v) plane. The state of the system is represented by a single point in the phase space, and this point wanders, indicating the change of the state, as time passes. The path of the motion in phase space is called the *trajectory* (Fig. 1.26). The trajectory itself does not indicate directly how fast this change is in time. The arrow only shows the direction of the motion. A set of several trajectories, however, provides a global overview of the different possible types of motion of the system (see Table 1.2).

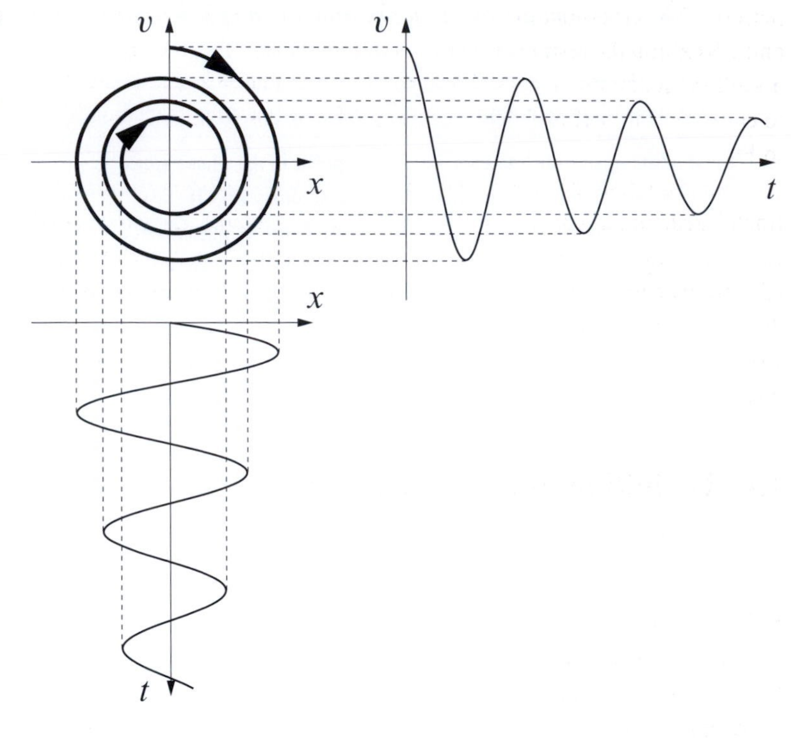

Fig. 1.26. Trajectory in phase space (thick line). The path described by the motion of a particle in phase space can be constructed from the respective projections of the $x(t)$ and $v(t)$ graphs. The direction of time is represented by the arrow on the trajectory.

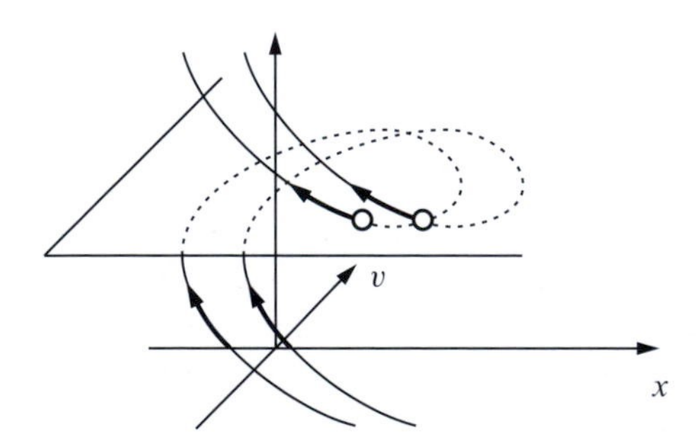

Fig. 1.27. Monitoring trajectories using maps. In higher-dimensional phase spaces, samples are taken on certain sections. The rule relating the co-ordinates of two consecutive intersects of a trajectory with this surface (or equivalent ones) is a map.

Two data points are often insufficient to define the state of a system uniquely; i.e., the phase space is three- or more dimensional (this is always the case with chaos). In such a situation it is useful to take samples from the higher dimensional phase space according to some rule. This is usually done by taking a 'section' of the phase space and recording the points of a trajectory on this section only, as illustrated by the schematic Fig. 1.27. In driven cases it is advisable to 'look at' the system at time instants corresponding to integer multiples of the driving period. This representation is called a *stroboscopic map*. Thus, Figs. 1.4 and 1.8 exhibit the results of stroboscopic mappings. In non-driven cases a section can be defined by the fulfilment of conditions corresponding to certain configurations. This defines a *Poincaré map*, like the one seen in Fig. 1.17.

Our examples have demonstrated that it is in such maps that the fractal structure of chaotic dynamics becomes plausible. Only in special cases (like those of the magnetic pendulum, the mirroring spheres and advection) can fractal structures be observed in real space. Therefore the use of phase space is inevitable as a means of understanding the structure accompanying chaos. (However, phase space is very useful in investigating regular motions also.)

1.4 Definition of chaos; summary

Chaos is a motion, a temporal dynamics of *simple* systems that can be described in terms of a few variables. Such motion is:

- irregular in time (it is not even the superposition of periodic motions, it is really aperiodic);
- unpredictable in the long term and sensitive to initial conditions;
- complex, but ordered, in the phase space: it is associated with a fractal structure.

These properties are so strongly and uniquely bound to chaotic dynamics that they may be used to define 'chaos'. We shall apply this definition throughout the book.

The listed characteristics are present simultaneously: when a simple system is aperiodic over a long time, its evolution must be unpredictable and representable by a fractal structure in suitable co-ordinates. From a traditional view, all three characteristics are novel and surprising. A single common feature underlying them is that the long-term behaviour is random-looking, irregular and therefore it can properly be described by using *probabilistic* concepts only.

On the other hand, not all complicated temporal behaviour can be considered to be chaotic, only those that derive from simple laws. *Noisy* motion is the random behaviour of some component of a system with a great number of constituents (for example the Brownian motion of a particle), which is the consequence of the complicated interaction with the environment (i.e. the other constituents). Chaos is a *bridge* between regular and noisy motion. It differs from regular motion in that it is probabilistic and differs from noise in that its randomness is due to the strong interaction (following from simple laws) of the few constituents, i.e. to the *inherent* dynamics. Noisy motion fills the phase space uniformly, thus fractal structures *cannot* develop.

Table 1.3. *Basic types of chaos and related phenomena and sets.*

	Permanent chaos	Transient chaos
Dissipative	motion on chaotic attractors	chaotic transients towards attractors, fractal basin boundaries (chaotic saddles)
Conservative	motion in chaotic bands	chaotic scattering (chaotic saddles)

The *traditional* investigation of motion concentrates on regular, periodic behaviour, since the applied classical mathematical tools are not suitable for describing chaos. These tools can only indicate chaos in as much they break down and yield meaningless results. The modern approach, supported by numerical investigations, makes it clear that it is regular motion that is exceptional.

Two important classes of chaotic dynamics (so far simply called chaos) are *permanent* and *transient* chaos. In the latter case, only exceptional initial conditions lead to steady chaotic motion; typical initial conditions result in finite time chaotic behaviour (which can last for an arbitrary long time, however). Both classes can occur in frictional (dissipative) systems as well as in frictionless (conservative) systems. The phase space sets underlying different kinds of chaos (chaotic attractors, bands and saddles) are collectively called chaotic sets. The main types of chaotic dynamics are summarised in Table 1.3, and will be studied in detail in Chapters 5–8.

It is also worth discussing the types of chaos from the point of view of the energy input. In non-driven frictional systems, motion ceases and chaos can only be present as a transient (often accompanied by a fractal basin boundary). Driven frictional motions may be related to chaotic attractors. In frictionless cases chaos (both in a chaotic band and in the form of chaotic scattering) might occur without forcing.

1.5 How should chaotic motion be examined?

Before turning to a detailed analysis of motion, we list some instructions worth keeping in mind in what follows, based on the lessons drawn from the examples of this chapter.

- You should understand unstable behaviour (considered to be uninteresting in traditional approaches), even in non-chaotic systems.
- Become acquainted with the phase space representation of and the geometric approach to dynamics and the use of the stroboscopic and Poincaré maps.

- It is pointless to hope that the long-term dynamics can be given analytically in terms of known functions (the infinite series constructed to describe the dynamics do not even converge).
- Solve the equations of motion numerically.
- Proper understanding requires the introduction of new concepts and the search for new theoretical relations.
- Do not forget about the measurement errors that inevitably accompany observation and simulation, and follow their temporal evolution.
- Accept the necessity of using particle ensembles and of describing them by means of probabilistic concepts (distribution, typical behaviour, average).
- Become acquainted with the geometry of fractals.

Box 1.1 Brief history of chaos

The possibility of chaotic motion was first formulated by the French mathematician Henri Poincaré in the 1890s (obviously in a terminology largely different from that used nowadays) in his paper on the stability of the Solar System. Some time later, the Russian mathematician, Sonia Kovalevskaia, proved that the motion of a heavy, asymmetric spinning top is usually chaotic (it is only regular at special values of the moment of inertia). These results were mostly forgotten and only lived on in the first half of the twentieth century due to the work of the American scientist George Birkhoff and his German colleague, Eberhard Hopf, on statistical mechanics and ergodic theory. Independently of these developments, chaotic behaviour was found in certain non-linear electrical circuits during World War II, but the results could not be properly interpreted. As a continuation of the Birkhoff–Hopf line, in the mid 1960s the Russians Andrey Kolmogorov and Vladimir Arnold and the German Jürgen Moser worked out the statement that has since been named after their initials, the KAM theorem, formulating the condition of weak chaotic motion in conservative systems. The investigation of strong chaos became possible due to the appearance of computers. The behaviour related to chaotic attractors occurring in dissipative systems was first described by the American meteorologist Edward Lorenz in 1963. He recognised the unpredictability of chaotic behaviour in connection with the numerical solution of a model named after him. The term ‘chaos’ itself was introduced by the American mathematician James Yorke for the random-looking dynamics of simple deterministic systems in a paper in 1975. The work of the American physicist Mitchell Feigenbaum helped the term become widespread. In 1978 he proved the system-independence, i.e. the so-called universality, of one of the possible routes towards chaos. In the investigation of the statistical properties of chaos, a major role was played by, among others, B. Chirikov, M. Berry, L. Bunimovich, J. P. Eckmann, H. Fujisaka, P. Grassberger, C. Grebogi, M. Hénon, P. Holmes, L. Kadanoff, E. Ott, O. Rössler, D. Ruelle, Y. Sinai, and S. Smale. The possibility of the occurrence of chaos has established a new way of thinking in widely different disciplines (see Box 9.3); this has been pioneered by H. Aref, P. Cvitanović, J. Gollub, A. Libchaber, R. May, C. Nicolis, H. Swinney, Y. Ueda, J. Wisdom, and others.

Chapter 2
Fractal objects

2.1 What is a fractal?

2.1.1 Objects with large surfaces

It is taken for granted that the surface or volume of a traditional geometrical object, for example a sphere or a cube, is well defined. Indeed, filling the object with smaller and smaller cubes leads to better and better approximations, and the total volume of the cubes converges to that of the object in question. It is well known that the surface, S, is proportional to the second, while the volume V is proportional to the third, power of the linear size, L, of the object. Consequently, the surface-to-volume ratio, S/V, is proportional to $V^{-1/3}$. (For plane figures, the ratio of the perimeter, P, to the area, A, is proportional to $A^{-1/2}$.) The surface-to-volume ratio is therefore finite and becomes *smaller* as the size becomes larger. This is why surface phenomena are of little importance compared with volume phenomena for macroscopic systems of traditional geometry.[1]

On the other hand, it is known that there exist macroscopic objects with large surface area. These are always porous, with ramified or pitted surfaces. Effective chemical catalysts, for example, must have a large surface. The need for rapid gas exchange accounts for the large surface-to-volume ratio of the respiratory organs. The surface area of the human lungs (measured at microscopic resolution), for example, is the same as

[1] This is used for example in thermodynamics, when the internal energy of a system is considered to be the sum of the internal energies of the finite volume elements, thus neglecting the interactions of the volume elements across the surfaces.

that of a tennis court (approximately $100\,m^2$), while the volume is only a few litres ($10^{-3}\,m^3$). How small the amount of matter is in candyfloss or in beer froth can be checked with a single mouthful; the numerical value of their surface area is much greater than that of their volume. Such systems obviously do not follow the rule $S/V \sim V^{-1/3}$ which is valid for traditional objects: indeed, our very concept of measuring surface area has to be revised in view of these types of surfaces.

A unique value cannot be assigned to the surface of such ramified systems because the surface area essentially depends on the *resolution*, in other words, on the accuracy of the measurement (see Table 2.1). The numerical value of the surface area increases with the resolution of the observation, through several orders of magnitude. Consider, for example, an island with an embayed coastline: one would try to determine the perimeter of the island using a high-resolution map and counting how many times a certain compass setting fits on the coastline. As smaller and smaller settings are chosen, more bays and peninsulae appear; therefore the number of settings required increases more rapidly than for an object with a smooth perimeter. The resulting perimeter length keeps changing, increasing as the compass setting decreases. The same tendency would be observed when trying to measure the surface area of a mountain by fitting smaller and smaller squares on it.

It is useful to introduce the term *observed* surface or perimeter. Consider squares of size ε or line segments of length ε. The observed surface, $S(\varepsilon)$, or the observed perimeter length, $P(\varepsilon)$, shows how the respective values depend on the resolution, ε. For the sake of simplicity, the resolution is given in units of the linear size, L, of the system. Note that ε is a dimensionless number smaller than unity, since the observed surface or perimeter length should be determined on scales smaller than the total extension. Refining the resolution implies decreasing ε. Experience shows that for fractals the observed surface or perimeter increases as a negative power of the resolution:

$$S(\varepsilon),\ P(\varepsilon) \sim \varepsilon^{-\gamma}, \quad \text{for } \varepsilon \ll 1, \tag{2.1}$$

where γ is a non-trivial (usually non-integer) positive power or exponent.[2] Since no precise value of the surface or the perimeter can be given, they are better characterised by the exponent γ. This latter will prove to be simply related to a quantity called the fractal dimension (see equation (2.5)).

Fractals occurring in reality only approximate the property that relation (2.1) is valid in the limit $\varepsilon \to 0$, partly because the resolution can never be infinitely fine, and partly because fractal properties are usually

[2] The notation $\sim$ expresses proportionality without the prefactor written out.

Table 2.1. *Comparison of traditional and fractal objects. $P(\varepsilon)$ and $S(\varepsilon)$ denote the perimeter length and surface area observed with accuracy ε, respectively.*

Traditional object	Fractal object
perimeter P and surface S exist	P or S is undefined
$P(\varepsilon)$ and $S(\varepsilon)$ convergent	$P(\varepsilon)$ or $S(\varepsilon)$ increases with resolution
smooth on small scales	ramified on small scales

lost or become meaningless below a certain size. (The lungs' surface becomes smooth, i.e. two-dimensional, on the micrometer scale of cells, and measuring the perimeter of an island becomes meaningless at a scale of a few metres where the coastline becomes undefined because of the waves.)

The fact that the perimeter or the surface of a fractal is not defined implies that the object cannot be well approximated with squares or cubes: its structure differs fundamentally from that of traditional objects. On the other hand, the fact that relation (2.1) is valid throughout several orders of magnitude of ε indicates that the object exhibits the same structure when observed at any resolution within this range: fractals are said to be *self-similar*. The surface of the Moon is a good example: the typical crater-dominated structure is characteristic from the millimetre to the 1000 km scale, the latter approaching the Moon's radius. Thus, the surface of the Moon is self-similar throughout nine orders of magnitude. This is why it is necessary to indicate the names or sizes of the main craters in pictures showing the surface of the Moon. As this example illustrates, self-similarity usually does not mean that a magnified view is identical to the whole object, but rather that the character of the patterns is the same on all scales.

As a mathematical model of a coastline, consider a Koch curve. It is constructed as follows. First, a segment (shorter than $1/2$) is removed symmetrically from the centre of a unit interval, then two segments of the same length as the remaining pieces are joined to the new end-points in the shape of a roof (Fig. 2.1(a)). Denoting the length of the remaining pieces by r $(1/4 < r < 1/2)$, the result is a broken line of length $4r$. This process is then repeated with the segments of length r: the resulting new segments are of length r^2 (Fig. 2.1(b)). The essence of the algorithm is an iterative repetition of the same rule applied always to the most recently obtained segments. Meanwhile, the curve becomes more and more broken and its length increases. The curve obtained as the limit of this construction is called the Koch curve. At the nth step of the

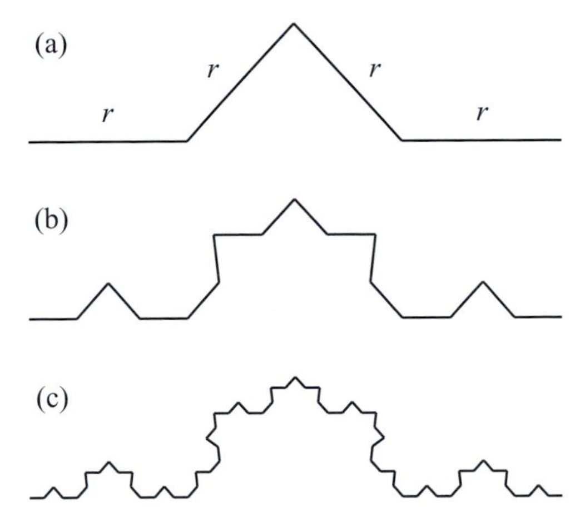

Fig. 2.1. Koch curve. First three steps of the construction with parameter $r = 0.3$.

construction, the length and the number of the segments are r^n and 4^n, respectively; the length of the curve is therefore $P(n) = (4r)^n$. At the same time, this is the observed length of the exact curve when using the resolution $\varepsilon = r^n$. Being interested in the resolution dependence of the length, we express n via the logarithm of ε as $n = \ln\varepsilon / \ln r$.[3] Thus, $P(\varepsilon) = 4^{\ln\varepsilon/\ln r}\varepsilon$, and as a consequence of the identity $4^{\ln\varepsilon} = \varepsilon^{\ln 4}$, we obtain

$$P(\varepsilon) = \varepsilon^{1+\ln 4/\ln r}. \qquad (2.2)$$

The exponent γ defined by (2.1) is thus $\gamma = \ln 4/\ln(1/r) - 1 > 0$. The length of a Koch curve is undefined, since the observed length is a negative power of the resolution. The part of the exact curve which sits on an original segment of length r is a downscaled version of the full curve by a factor of r. The same holds for any segment of length r^m ($m > 1$), with a reduction factor of r^m. Koch curves are prototypical fractals. They are, in addition, self-similar in an exact geometrical sense. Note that the larger the parameter r, and therefore the exponent γ, the more ramified the coastline it models. In the limit $r \to 1/4$, on the other hand, the curve turns into a straight line segment, since the intervals are too short to form triangles. The length of a straight line segment does not therefore depend on the resolution, and exponent γ vanishes as expected.

Problem 2.1 Examine how the perimeter of a traditional object, a circle of unit radius, varies with resolution when it is approximated by the perimeter of an inscribed n-sided regular polygon, the side length ε, being the resolution ($n \gg 1$, i.e. $\varepsilon \ll 1$).

[3] In this chapter the notation ln may be considered to denote the logarithm of arbitrary base.

Problem 2.2 Determine the area of a Koch island formed by three triadic ($r = 1/3$) Koch curves, the first four construction steps of which are as follows:

Show that the area of the Koch island remains finite.

2.1.2 Fractal dimension

Objects with a large surface or perimeter are strongly ramified; the value of their surface area or perimeter length increases with resolution, while the entire object is restricted to a finite region of space. The concepts of the traditionally two-dimensional surface and the traditionally one-dimensional perimeter become meaningless, since these objects penetrate significantly into a space of dimension one more than their own. It is therefore a straightforward generalisation of the concept of dimension to assign *fractional* or *irrational* numbers greater than two (one) but less than three (two) to objects with a large surface area (perimeter length) so that the more ramified the object, the larger the dimension.

To this end, it is worth *covering* the object with cubes of linear size ε. For a traditional body the number of cubes needed to cover it is obviously proportional to ε^{-3}, while for plane figures and lines, the respective numbers of squares of size ε and segments of length ε are proportional to ε^{-2} and ε^{-1}, respectively. The exponent with opposite sign is therefore the dimension itself, which is an integer for traditional objects. A Koch curve, however, can be covered with 4^n segments each of length $\varepsilon = r^n$, and this number can be expressed in terms of ε as $\varepsilon^{-\ln 4/\ln(1/r)}$. Thus, the negative of the exponent is not an integer for objects with large surface area.

To deduce the general definition of fractal dimension, consider a set of points in a Euclidean space of $d = 1$, 2 or 3 dimensions (i.e. along a line, on a plane or in space). Let $N(\varepsilon)$ be the *minimum* number of d-dimensional cubes of linear size ε necessary to cover the object (Fig. 2.2). This number obviously increases with resolution, namely as a negative power, but the exponent, D_0, is not necessarily identical to the dimension, d, of the space. The relation

$$\boxed{N(\varepsilon) \sim \varepsilon^{-D_0}, \quad \text{for } \varepsilon \ll 1,} \tag{2.3}$$

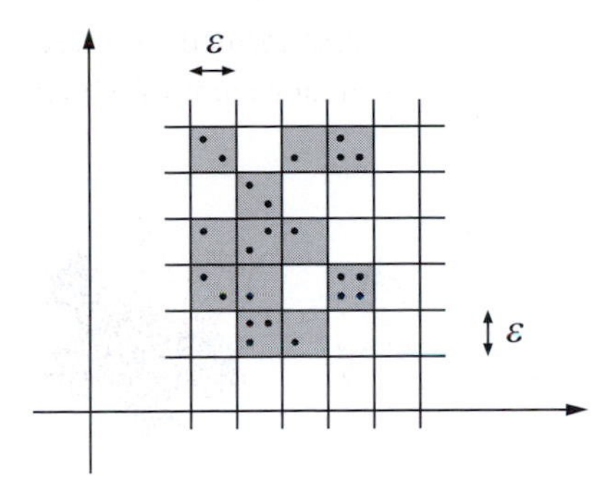

Fig. 2.2. Measuring fractal dimension. A set of points (or any object) is covered with identical d-dimensional cubes of size ε, and the number of cubes containing points (grey boxes) is $N(\varepsilon)$. As the resolution ε decreases, $N(\varepsilon)$ increases according to the relation $N(\varepsilon) \sim \varepsilon^{-D_0}$, where D_0 is the fractal dimension.

defines the fractal dimension, D_0, of the object in question. Rearranging the equation yields

$$D_0 = \frac{\ln N(\varepsilon)}{\ln 1/\varepsilon}, \qquad \text{for } \varepsilon \ll 1. \tag{2.4}$$

The fractal dimension can therefore be extracted from the dependence of the number of covering boxes on resolution. This number equals d for traditional objects. A set is described as fractal if its D_0 is *smaller* than the dimension of the space.[4] The complement set of a fractal is not a fractal. As for (2.1), relations (2.3) and (2.4) have to be fulfilled throughout several orders of magnitude of ε (see Fig. 2.3).

Problem 2.3 Determine the number of boxes required to cover a unit interval and a right-angled isosceles triangle, for which the length of the equal sides is unity. Demonstrate that their respective dimensions defined by (2.3) are indeed 1 and 2.

The relation to the observed perimeter or surface area is now obvious, since this is the number of boxes multiplied by the length or area of a box: $P(\varepsilon) = \varepsilon N(\varepsilon)$ and $S(\varepsilon) = \varepsilon^2 N(\varepsilon)$. The exponent γ in (2.1) is therefore

$$\gamma = D_0 - 1 \quad \text{and} \quad \gamma = D_0 - 2 \tag{2.5}$$

for the perimeter and surface area, respectively. The rate of increase of the observed surface area or perimeter is thus a unique expression of the fractal dimension. Such systems are therefore better characterised by their fractal dimension (or their exponent γ) than by the numerical value of their perimeter or surface area at a certain resolution.

[4] Another class of fractals, fat fractals, will be discussed in Section 2.2.3.

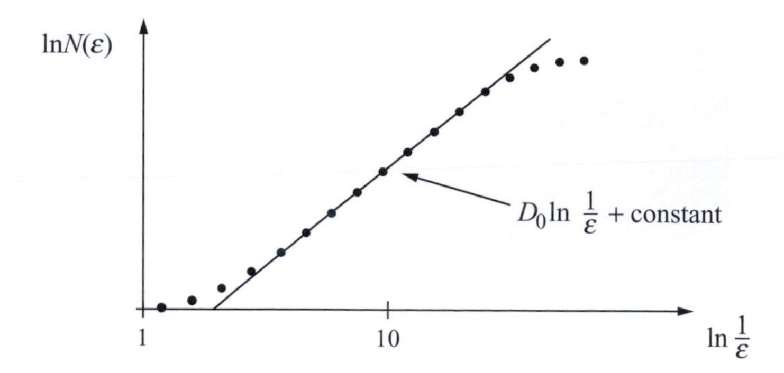

Fig. 2.3. The fractal dimension is the slope of the straight line appearing in the log–log plot of the number of boxes against reciprocal resolution. The curve $\ln N$ vs. $\ln 1/\varepsilon$ deviates from a straight line of slope D_0 both for resolution approaching unity and for very small resolution. The respective reasons are that for coarse resolution no power-law behaviour can be expected, and that on very fine scales new effects set in and the system behaves differently.

The fractal dimension of a Koch curve is, according to (2.4),

$$D_0 = \frac{\ln 4}{\ln (1/r)}, \tag{2.6}$$

a number between 1 and 2. For the triadic case ($r = 1/3$), $D_0 = \ln 4/\ln 3 = 1.262$.[5] As shown, the observed perimeter increases as ε^{1-D_0}, while the observed surface area decreases as ε^{2-D_0} with refining resolution. This implies that on covering a Koch curve with squares of size ε, the area of the object converges to zero and the curve does not fill any part of the plane. On the other hand, the curve is more complicated than any smooth curve, which is reflected by its increasing length and by its dimension being greater than unity.

The dimension as a measure of ramification increases monotonically with the parameter r. The choice $r = 1/4$ corresponds to a smooth line segment, a one-dimensional object. Koch curves with parameter r close to $1/4$ are only slightly ramified (such as, for example, the edge of a slice of cauliflower), while values $D_0 = 1.2 - 1.3$ around $r = 1/3$ correspond to the average dimension of coastlines or to the dimension of a section through the Moon's surface (Fig. 2.4). Values near $r = 1/2$ belong to strongly jagged curves, with dimensions approaching 2 (see Fig. 2.5). Almost plane-filling curves or space-filling surfaces are ubiquitous in Nature. Examples are river networks, with their smaller rivers, brooks and water courses spreading over the tributary basins, or the vascular system of living organisms having a lymphatic system, and the dense foliage of trees.

[5] Irrational numbers are henceforth given up to three decimals.

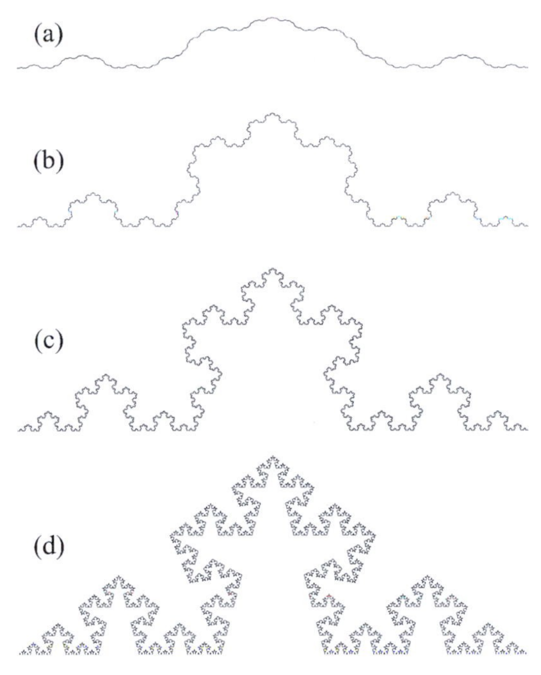

Fig. 2.4. Koch curves with parameters (a) $r = 0.26$, (b) $r = 0.3$, (c) $r = 0.35$, (d) $r = 0.4$. Larger parameters, r, correspond to more jagged curves and higher fractal dimensions. The respective dimensions are $D_0 = 1.029,\ 1.151,\ 1.321$ and 1.513.

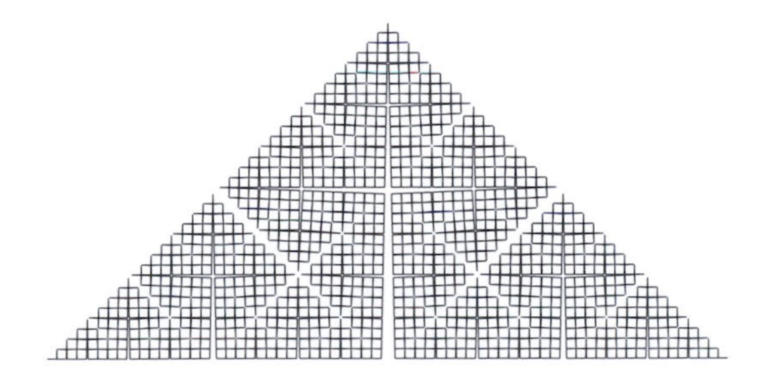

Fig. 2.5. Koch curve with parameter $r = 0.49$ after the first six steps of construction. The exact curve is almost plane-filling; its dimension is $D_0 = 1.943$.

Problem 2.4 Calculate the dimension of this Koch-type fractal:

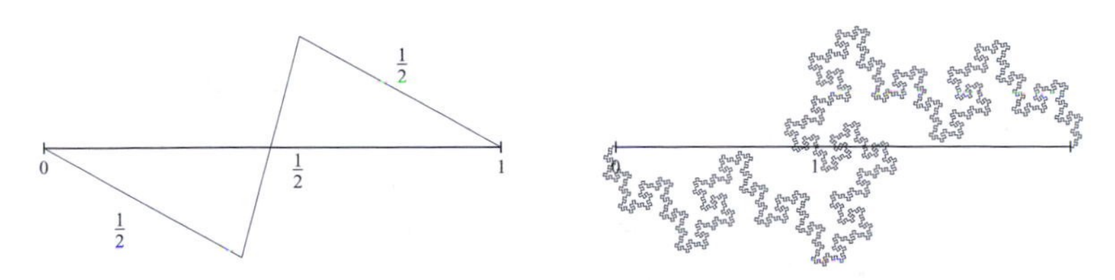

(Measure three segments of length $1/2$ onto the unit segment as shown, and repeat this in a proportionally reduced fashion for the newly obtained segments.)

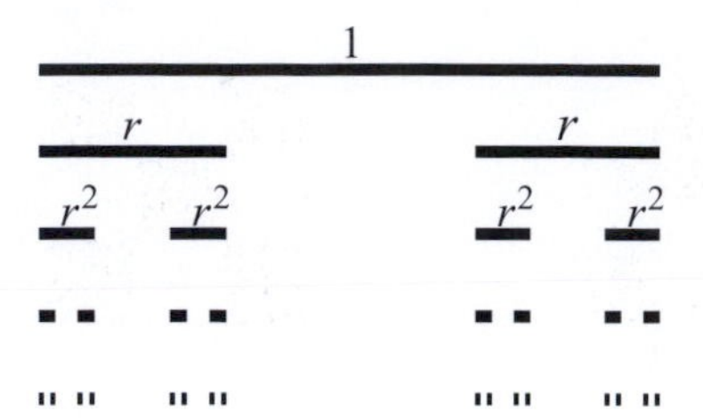

Fig. 2.6. Cantor set: the first four steps of the construction with parameter $r = 0.3$. The dimension of the set is $D_0 = 0.576$.

2.2 Types of fractals

2.2.1 Exactly self-similar fractals

An important group of fractals whose dimension can be expressed by simple formulae is that of exactly self-similar fractals, in which small regions of the fractal are similar to the entire fractal.

One-scale fractals consist of N identical parts, each a copy of the entire fractal reduced by exactly the same factor, $r < 1$. Koch curves are obviously of this type. Another typical example is a Cantor set. This is constructed by preserving the two outer segments of length $r < 1/2$ of a unit interval, then removing the proportional middle segments of the remaining pieces of length r, then of length r^2, etc. (Fig. 2.6). Since the number of segments of length $\varepsilon = r^n$ needed to cover the set is 2^n, the number of covering intervals is $N(\varepsilon) = \varepsilon^{\ln 2/\ln r}$, and

$$D_0 = \frac{\ln 2}{\ln(1/r)}. \tag{2.7}$$

A Cantor set does not form a continuous curve, rather it is a dispersed set of an (uncountable) infinity of points. Accordingly, its dimension is less than unity.

Cantor sets and Koch curves consist of $N = 2$ and $N = 4$ identical parts, respectively, that are exact copies of the entire fractal reduced by a factor r each. Consequently, the fractal dimension of such self-similar objects consisting of N units is expected to be

$$D_0 = \frac{\ln N}{\ln(1/r)}. \tag{2.8}$$

This is easy to see: the minimum number, $N(\varepsilon)$, of boxes corresponding to resolution ε is obviously $NN_1(\varepsilon)$, where $N_1(\varepsilon)$ is the number of boxes needed to cover one part. On the other hand, due to the similarity, this is exactly the number of boxes covering the entire fractal if the size of the boxes is multiplied by $1/r$: $N_1(\varepsilon) = N(\varepsilon/r)$. Altogether, therefore, $N(\varepsilon) = N \cdot N(\varepsilon/r)$. Substituting definition (2.3) yields $Nr^{D_0} = 1$, which is equivalent to (2.8).

Problem 2.5 Determine the dimension of the fractals obtained by repeating a given construction step: (a) a Cantor cloud, (b) a Sierpinski gasket and (c) a snowflake fractal.

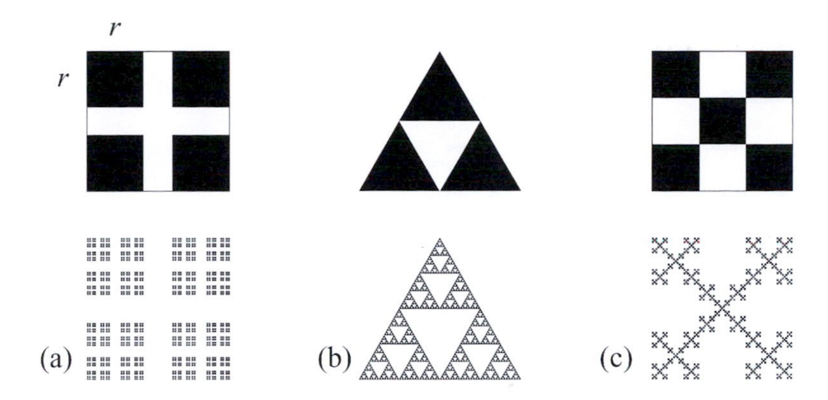

The white parts are removed from the initial black objects (top); the resulting pattern is repeated several times in a self-similar way, which results in the fractals shown at the bottom.

Problem 2.6 Determine the ratio of the observed perimeter to the observed surface area of a Cantor cloud. At what values of the parameter does the perimeter diverge?

Problem 2.7 A simple model of porous materials with a large surface can be obtained by constructing (a) a Menger sponge or (b) a Sierpinski tower. (a) A regular three-dimensional cross consisting of seven identical cubes of size $1/3$ is removed from a unit cube. This is repeated for the remaining smaller and smaller cubes. (b) At the vortices of a regular tetrahedron, four copies of the tetrahedron, reduced by a factor of one-half, are kept. The rest is removed, and this process is repeated again and again. What is the fractal dimension of each of these objects?

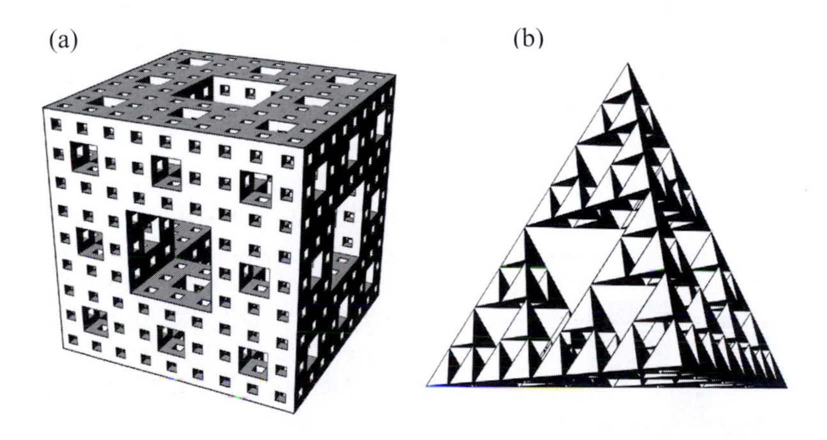

A *multi-scale fractal* is a fractal consisting of N parts, each of which is a copy of the entire fractal reduced by factors $r_j < 1,\ j = 1, 2, \ldots, N$. Then the total number of boxes necessary to cover the fractal is

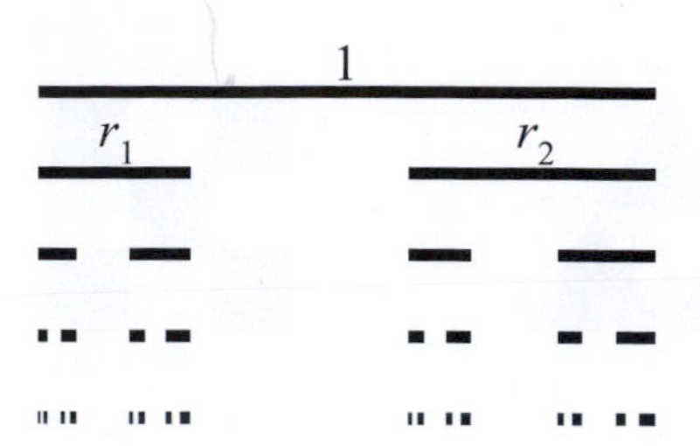

Fig. 2.7. Two-scale Cantor set. The first four steps of the construction with parameters $r_1 = 0.25$ and $r_2 = 0.4$ are shown. Segments are removed repeatedly from the middle of the intervals in such a way that the length of the preserved segments on the left and on the right is r_1 and r_2 times the original length, respectively.

$N(\varepsilon) = \sum_{j=1}^{N} N_j(\varepsilon)$. However, because of the similarity, the number of boxes covering part j is the same as the number of boxes covering the entire fractal if the size of the boxes is multiplied by $1/r_j$: $N_j(\varepsilon) = N(\varepsilon/r_j)$. Thus, $N(\varepsilon) = \sum_{j=1}^{N} N(\varepsilon/r_j)$, and substituting definition in (2.3) yields the following relation:

$$\sum_{j=1}^{N} r_j^{D_0} = 1. \tag{2.9}$$

The fractal dimension is now determined by an implicit equation that can only be solved numerically; there is no explicit formula for D_0.

A simple example of a multi-scale fractal is the two-scale or asymmetric Cantor set. In the first step of its construction, a segment of length r_1 is kept on the left and a segment of length r_2 is kept on the right end of the unit interval. This is repeated for the smaller and smaller remaining segments (see Fig. 2.7). For $r_1 = 0.25$ and $r_2 = 0.4$, for example, the numerical solution of the equation $(0.25)^{D_0} + (0.4)^{D_0} = 1$ yields the fractal dimension $D_0 = 0.605$.

Problem 2.8 Determine the dimension of the two-scale Cantor set in the special case when $r_2 = r_1^2 = 1/4$.

Problem 2.9 Determine the dimension of the two-scale snowflake fractal constructed from a unit square by preserving four squares of size $2/5$ in the corners and a fifth of size $1/5$ in the centre,

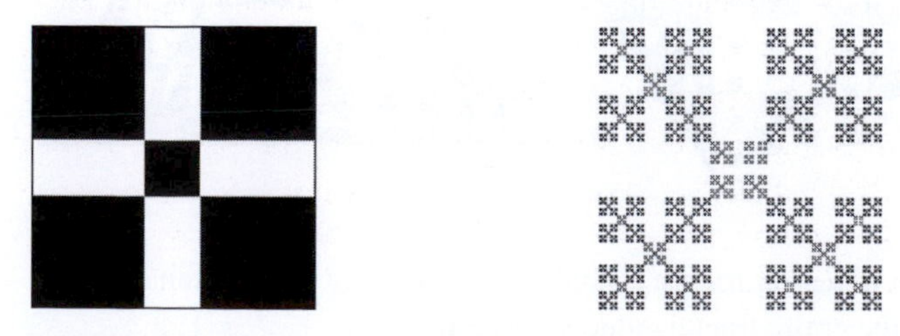

Repeat this for each remaining square.

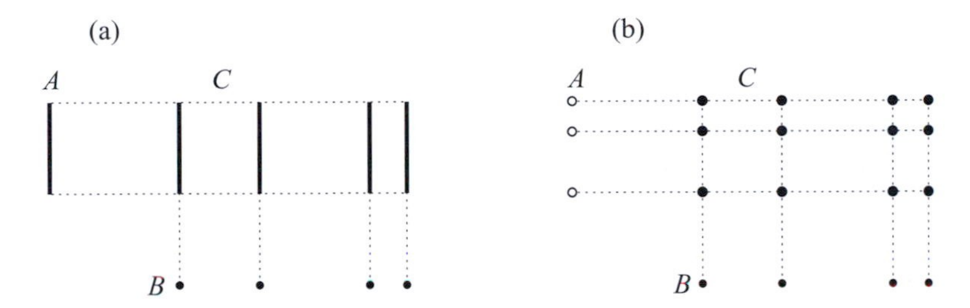

Fig. 2.8. Diagram explaining what is meant by projecting two sets, A and B, together to obtain set C. The composite set C is also called the direct product of components A and B. (a) Component A is a line segment. (b) Component A is the union of three points. Component B is, in both cases, the union of four points.

Fig. 2.9. Cantor filaments: the first four steps of the construction with parameter $r = 0.4$. This construction is similar to that of a Cantor set, but the initial object is not a line segment but a square, and rectangles are removed instead of intervals. The resulting dimension is $D_0 = 1.756$.

2.2.2 Projecting together fractals

There exists an important class of fractals – whether exactly self-similar or not – which can be decomposed into component fractals. This is the case when a fractal is created by projecting two simpler fractals together (see Fig. 2.8).

As a first example, consider the case of Cantor filaments. These are constructed by symmetrically removing a rectangle from the centre of a unit square in such a way that the two remaining rectangles are of width r and of unit height. In the following steps the remaining, narrower and narrower, rectangles, always of unit height, are thinned out in the same manner (Fig. 2.9). The result is an infinite set of parallel unit intervals. When intersected with a horizontal line, a Cantor set of parameter r is obtained. Cantor filaments appear when projecting together a unit interval and a Cantor set of parameter r. In other words they are the direct products of these component sets. When determining the dimension of Cantor filaments, notice that covering the object with squares of size $\varepsilon = r^n$ yields 2^n columns, containing $1/\varepsilon$ boxes each. Thus, the number of covering boxes is $N(\varepsilon) = \varepsilon^{(\ln 2/\ln r - 1)}$. The dimension of Cantor filaments is therefore

$$D_0 = 1 + \frac{\ln 2}{\ln(1/r)}, \tag{2.10}$$

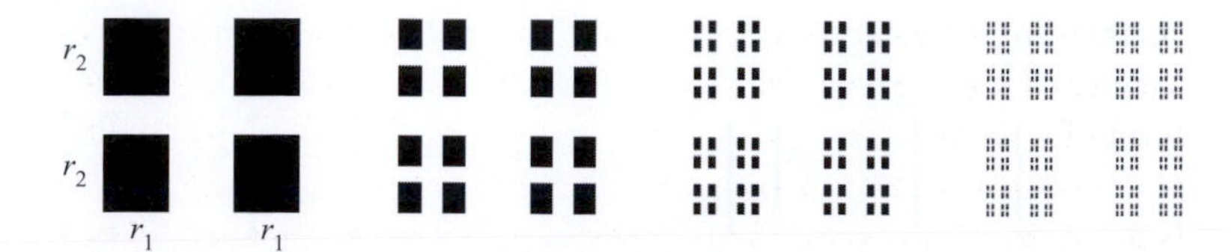

Fig. 2.10. Asymmetric Cantor cloud: the first four steps of the construction with parameters $r_1 = 0.3$ and $r_2 = 0.4$. Rectangles are removed from the unit square in a symmetrical cross shape in such a way that the sides of the preserved rectangles are r_1 and r_2 times the original sides, and this is repeated iteratively.

greater than that of a Cantor set by one. An interesting feature of Cantor filaments is that their observed perimeter increases while their area decreases with refining resolution. The observed perimeter, $P(\varepsilon)$, can be expressed as a *negative* power of the observed area, $A(\varepsilon)$: $P \sim A^{-\beta}$, where β is positive. Thus, the larger the perimeter, the smaller the area!

Problem 2.10 Express the exponent β in terms of the fractal dimension.

An asymmetric Cantor cloud is obtained by removing a cross symmetrically from the centre of a unit square in such a way that four identical rectangles of width r_1 and height r_2 remain (Fig. 2.10). This procedure is then repeated for each remaining rectangle, keeping the proportionality factors r_1 and r_2. The resulting set of points is concentrated in smaller and smaller rectangles. Covering them with squares of size $\varepsilon = r_1^n$ requires $2^n = \varepsilon^{\ln 2/\ln r_1}$ columns. Each of these columns now contains less than $1/\varepsilon$ boxes, since a Cantor cloud is a fractal vertically also: a Cantor set of parameter r_2. Thus, the boxes of size $\varepsilon = r_1^n$ are less densely placed than in the case of a vertically continuous object. Their number in a column can be estimated as $\varepsilon^{-\ln 2/\ln(1/r_2)}$, where $\ln 2/\ln(1/r_2)$ is the fractal dimension of the Cantor set formed vertically. The number of covering boxes is therefore $N(\varepsilon) = 2^n \varepsilon^{-\ln 2/\ln(1/r_2)} = \varepsilon^{\{\ln 2/\ln r_1 - \ln 2/\ln(1/r_2)\}}$. Thus, the fractal dimension is given by

$$D_0 = \frac{\ln 2}{\ln(1/r_1)} + \frac{\ln 2}{\ln(1/r_2)}. \tag{2.11}$$

For $r_1 = r_2 \equiv r$, $D_0 = \ln 4/\ln(1/r)$, formally the same as the dimension of a Koch curve for $r > 1/4$. The two fractals are, however, essentially different, since one of them is a broken line, while the other is a set of points dispersed in a plane. This example shows that fractal dimension is only one measure of an object: identical dimensions do not imply identical objects.

The determination of the dimension of a composite fractal obtained by projecting two arbitrary fractals, embedded in perpendicular straight-line segments, together is based on the fact that the number of squares of size ε needed to cover the composite fractal appears in the form of a product: $N(\varepsilon) = N^{(1)}(\varepsilon)N^{(2)}(\varepsilon)$. Here $N^{(1)}(\varepsilon)$ is the number of columns of width ε perpendicular to the horizontal axis and $N^{(2)}(\varepsilon)$ is the number of bands of height ε parallel to the horizontal axis. Each $N^{(i)}$ increases according to the dimension $D_0^{(i)}$ of the corresponding fractal component: $N^{(i)}(\varepsilon) \sim \varepsilon^{-D_0^{(i)}}$. For the co-projected set this yields $N(\varepsilon) \sim \varepsilon^{-D_0^{(1)}-D_0^{(2)}}$; therefore the total fractal dimension is

$$\boxed{D_0 = D_0^{(1)} + D_0^{(2)}.} \tag{2.12}$$

Thus, the fractal dimension of composite fractals is the *sum* of the dimensions of the components. These dimensions, $D_0^{(i)}$, are often called the *partial fractal dimensions*.

This sum rule is valid not only for fractals embedded in one dimension, but also for the direct product of arbitrary fractals, and the number of components is also arbitrary. The same relation holds for traditional objects, as, for example, a plane is the direct product of two straight lines, and its dimension is in fact $1 + 1$. It is important to emphasise that the sum rule is valid not only for fractals projected along perpendicular lines, but also along arbitrary smooth curves. This is because such a projection is a smooth transformation that can modify the prefactors not written out in (2.3) only, not the exponent of the power law, the dimension.

Problem 2.11 What is the algorithm given in terms of the removal of pieces from a unit cube which leads to the direct product of three Cantor sets of parameter $r = 1/3$ placed along the edges of a unit cube? What is the relation of the observed volume of this set to that of the Menger sponge (Problem 2.7(a)) at the same resolution, $\varepsilon = 3^{-n}$?

2.2.3 Thin and fat fractals

Covering a fractal embedded in d dimensions with d-dimensional cubes of size ε, the *observed volume* $V(\varepsilon)$ of a fractal is given by

$$V(\varepsilon) = \varepsilon^d N(\varepsilon). \tag{2.13}$$

This is a generalisation of the traditional concept of volume.[6] Because of the resolution dependence (2.3) of the number of covering boxes, the

[6] The observed surface area in this generalised concept is $S(\varepsilon) \sim \varepsilon^{d-1}N(\varepsilon)$, which is for $d = 2$, for example, the observed perimeter length.

observed volume varies according to the following rule:

$$V(\varepsilon) \sim \varepsilon^{d-D_0}, \quad \text{for} \quad \varepsilon \ll 1. \tag{2.14}$$

The difference between the dimension of the space and that of the fractal is often called the *co-dimension* of the fractal. The fractal dimension of the objects examined so far is smaller than the dimension of the space; consequently, the observed volume *decreases* with resolution. These fractals may be called *thin fractals*, since their observed volume would vanish in the limit $\varepsilon \to 0$. One can also say that these fractals are not space-filling; in a mathematical sense they are sets of measure zero. They do not possess a volume, but they do have an extension; moreover, their observed surface may tend to infinity. This has been shown for the Koch curve embedded in a plane, where the two-dimensional volume – the surface area – decreases, while the length increases with resolution.

The fact that the fractal dimension of a set equals the dimension of the space, i.e. $D_0 = d$, does not necessarily imply that the set is a traditional object. Even if the observed volume converges to some finite value, V, for sufficiently small resolutions, it may occur that the structure of the object is essentially ramified. Such objects are called *fat fractals*: parts of them appear to be porous, while others look smooth. ('Bulky' plane figures bounded by fractal curves, such as the Koch island in Problem 2.2, are not fat fractals.)

A measurable property of fat fractals is that their observed volume depends on the resolution in such a way that the *deviation* from the exact volume, V, decreases slowly and proportionally to a power of the resolution:

$$\boxed{V(\varepsilon) - V \sim \varepsilon^{\alpha}.} \tag{2.15}$$

Here $\alpha < 1$ the fat fractal exponent, is a positive number. (For traditional objects, α is typically unity; see Problem 2.3.) Exponent α is not a dimension, rather it is the co-dimension of the difference of the observed volume and the volume itself. As (2.15) indicates, the complement of a fat fractal is itself a fat fractal of the same exponent α.

Problem 2.12 Consider a Cantor construction in which the proportion of the interval length removed at step n is λ_n. Given the sequence $(\lambda_1, \lambda_2, \ldots)$, determine the fractal dimension D_0. What is the condition for $D_0 = 1$?

Simple examples of fat fractals can be obtained by modifying the Cantor construction. In the first step the centre third of the unit interval is removed, but afterward it is $1/9$ (and not $1/3$) of the remaining $1/3$ intervals, then $1/27$ of the new segments, etc. At the nth step, $1/3^n$ of the

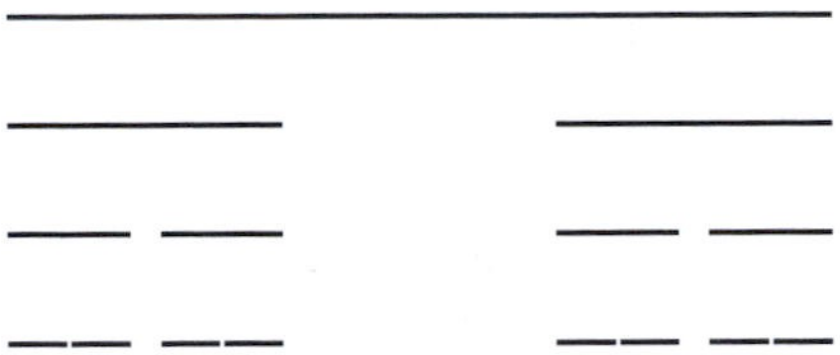

Fig. 2.11. Fat Cantor set. In contrast to a Cantor set, a decreasing proportion of the segments is removed. These proportions are $1/3,\ 1/9,\ 1/27, \ldots, 1/3^n$. The total length of the fat Cantor set constructed in this manner converges to a finite number greater than zero. The dimension is $D_0 = 1$ and the fat fractal exponent, α, is $0 < \alpha < 1$.

Fig. 2.12. Fat Cantor filaments: the first three steps of the construction.

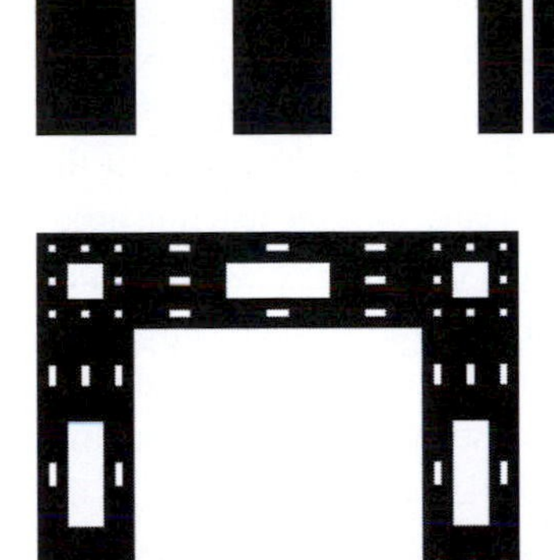

Fig. 2.13. A prototypical fat fractal of the plane. The figure shows the result of the construction described in the text after three steps with $\lambda = 0.6$.

remaining segments is removed (Fig. 2.11). The length of the removed intervals decreases much more rapidly than the triadic Cantor set.

Problem 2.13 Determine the volume V (the length) and exponent α of the fat Cantor set of Fig. 2.11.

Two-dimensional fat fractals can be obtained, for example, by forming the direct product of a fat Cantor set (Fig. 2.11) and an interval as shown in Fig. 2.12.

A more general type of fat fractal can be constructed by cutting out areas of decreasing proportions from a compact object. In the example of Fig. 2.13, a square of size $\lambda < 1$ is cut out in the first step from the middle of a unit square. The continuations of the edges of this square define rectangles of the remaning area. The construction goes on therefore by cutting out a downscaled copy of each rectangle from its middle, and so on. In order to obtain a fat fractal, the reduction factor is chosen to be λ^n in step n. The result is a planar object in which the black and the white regions are intricately interwoven into each other, and both have finite area.

In the following, the attributive of thin fractals will be omitted and these objects will simply be called fractals, while fat fractals will always be accurately named.

Problem 2.14 Determine the fat fractal exponent of the object shown in Fig. 2.13.

2.3 Fractal distributions

In natural phenomena fractals not only mean a mere geometrical structure but also provide stages on which 'something is going on'. Processes occurring on fractals generate *distributions* of certain measurable quantities. These generally become time-independent rather rapidly. The character of such *stationary distributions* depends on the nature of the process going on on the fractal. In the case of catalysts, the reaction product is not necessarily of uniform distribution everywhere; for example, the reaction rate might be higher in certain parts of the fractal and lower in others. If the fractal is a connected set and its material is a conductor, then an electric voltage generated between two points of this structure will produce highly different current intensities at different sites of the fractal. On fractals produced in the course of slow precipitation processes (such as electrolytic precipitation), each point of the surface can be characterised by the probability that a new particle will be bound to that point. If particles move at random along the surface of the fractal, then some points of the object may be visited more often than others. In chaotic systems, it is a consequence of the dynamics that a time-independent distribution develops on some fractal subset of phase space.

Since all these distributions are non-negative, they can be normalised to unity on the entire fractal. Such distributions can therefore be considered as *probability distributions*, and their normalisation expresses the fact that an event certainly occurs somewhere on the entire fractal. Experience shows that distributions developing on fractals are rather inhomogeneous and can themselves be considered as fractals in some sense. It is not only the support of the distribution (the set where the function is non-zero), but also the internal structure of the entire function that exhibits a fractal character. The dimension of this function is, however, different from the dimension D_0 of the support.

In the following, such highly inhomogeneous probability distributions will be investigated for (for the sake of simplicity) fractals

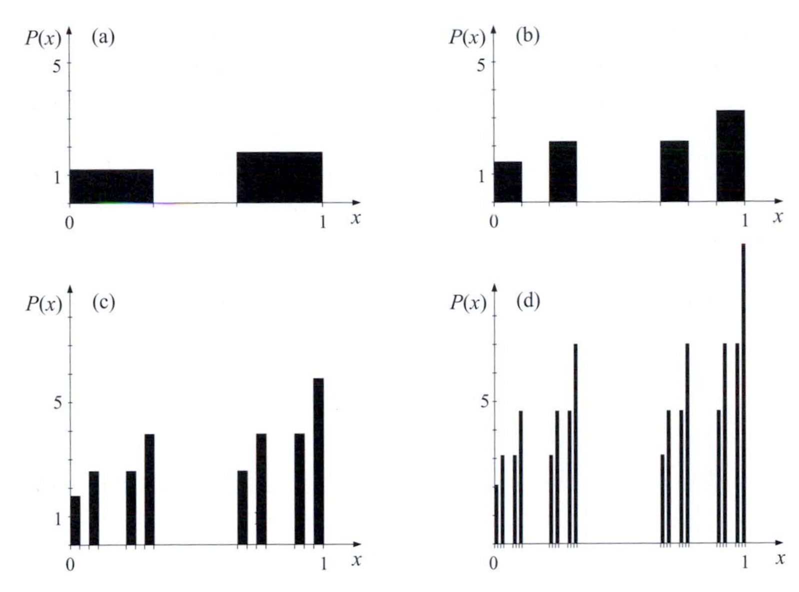

Fig. 2.14. Fractal distribution: first four steps of the construction of a probability distribution on a Cantor set with parameter $r = 1/3$. The parameter of the distribution is $p_1 = 0.4$ ($p_2 = 0.6$). (a) The respective areas of the columns on the left and on the right are p_1 and p_2. (b) The respective areas of the columns from left to right are p_1^2, $p_1 p_2$, $p_2 p_1$ and p_2^2. Proceeding similarly in (c) and (d), the distribution becomes more and more inhomogeneous.

embedded in one dimension. Let the chosen fractal be a triadic ($r = 1/3$) Cantor set, on which different distributions can exist. Consider, for example, one that is constructed in parallel with the Cantor set. In the first step, let the distribution be constant on the interval of length $1/3$ kept on the left, and let the probability of the entire interval be some number $p_1 < 1$. This will be the only parameter of the entire distribution. Let the distribution be constant on the right interval as well; then, the probability of this right interval is obviously $p_2 = 1 - p_1$ (Fig. 2.14(a)). In the next step, the probabilities are divided in the same proportion. This means that the probability of the outermost left interval is p_1^2, that of the two middle intervals is $p_1 p_2$ and that of the outermost right interval is p_2^2 (Fig. 2.14(b)). Note that the total probability of the four intervals is still unity. In the third step, the probabilities of the intervals from left to right are $p_1^3, p_1^2 p_2, p_1^2 p_2, p_1 p_2^2, p_1^2 p_2, p_1 p_2^2, p_1 p_2^2, p_2^3$, respectively (Fig. 2.14(c)). At the nth step, each probability takes on one of the possible values $p_m \equiv p_1^m p_2^{n-m}$, where m is less than or equal to n: $m = 0, 1, \ldots, n$ (Fig. 2.14(d)). The number of intervals for a given parameter, m, equals the number of possible ways of choosing m elements from n elements. Consequently, altogether there are $N_m = \binom{n}{m}$ intervals with probability $p_m = p_1^m p_2^{n-m}$. What are the *typical* intervals in this distribution? For a

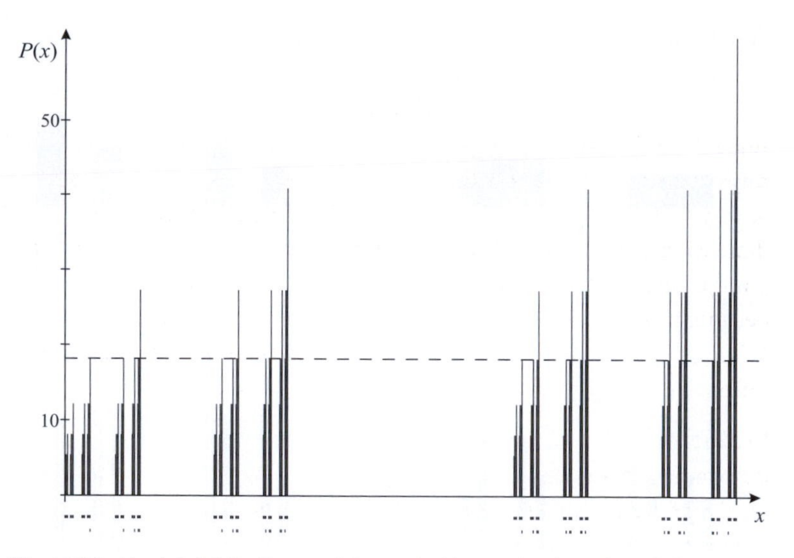

Fig. 2.15. Fractal distribution, and the typical intervals whose fractal dimension is the information dimension D_1. Below the graph both its support and the typical intervals (bottom line) are displayed after $n = 7$ steps of the construction. The height of the columns belonging to the typical intervals is marked by a dashed line. For increasing n, the actual typical intervals provide a more and more dominant proportion of the total probability.

given level n, intervals with the same p_m carrying the maximum possible total probability are considered to be typical.

Without violating generality, probability p_1 can be assumed to be less than p_2: $p_1 < 1/2$. The least probable box is then obviously the leftmost one, while the most probable box is the rightmost one. This latter is, however, not typical, since there is only one such box. Typical boxes are those that are not too improbable, and, in addition, *their number is sufficiently large* to provide the main contribution to the total probability at any fixed $n \gg 1$, i.e. these are the intervals for which $N_m p_m$ is maximum. It turns out that the total probability of these typical boxes is unity (up to a tiny deviation, vanishing for $n \to \infty$)! Therefore they *faithfully* represent the entire distribution (see Problem 2.15). This means that practically no probability is neglected when using the typical intervals. The dimension of the fractal made up of the typical boxes is called the *information dimension* of the distribution and is denoted by D_1 (see Fig. 2.15). The number, N^*, of the typical boxes therefore changes with resolution ($\varepsilon = 3^{-n}$ in our example) as follows:

$$N^*(\varepsilon) \sim \varepsilon^{-D_1}, \quad \text{for} \quad \varepsilon \ll 1. \tag{2.16}$$

The information dimension pertains to a subset of the support, to the typical intervals, therefore it cannot be greater than the fractal dimension:

$$\boxed{D_1 \leq D_0.} \tag{2.17}$$

Equality only holds when the distribution is homogeneous: only uniform distributions can have the same information and fractal dimension. A distribution is *fractal* if relation (2.17) holds as an inequality. The dimension (D_1) of the set that behaves *typically* from a probabilistic point of view is then smaller than the dimension of the entire support[7] because it does not include the very probable or very improbable intervals (Fig. 2.15). The more inhomogeneous a distribution, the greater the difference between the fractal and the information dimensions.

Problem 2.15 Using Stirling's formula ($\ln n! = n \ln n$ for $n \gg 1$), determine the value, m^*, belonging to the most probable intervals at level n of the example shown in Fig. 2.14 and determine the information dimension D_1.

An alternative definition of the information dimension that also accounts for the terminology is particularly well suited for numerical evaluations. It is based on determining the distribution with a finite resolution ε. Assume that the probability, $P_i(\varepsilon)$, of each covering box is known. The index, i, of the boxes ranges from 1 to the total number, $N(\varepsilon)$, of non-empty boxes. Normalisation implies $\sum_{i=1}^{N(\varepsilon)} P_i(\varepsilon) = 1$. The inhomogeneity of a distribution is characterised by its *information content*, the negative average of the logarithm of the probabilites. It assumes its minimum value in uniform cases, which reflects that homogeneous distributions carry the least information.

The information content, $I(\varepsilon) = -\sum_{i=1}^{N(\varepsilon)} P_i(\varepsilon) \ln P_i(\varepsilon)$, of a distribution, $P_i(\varepsilon)$, determined with finite resolution obviously depends itself on the resolution. The finer the resolution, the more ramified, or uneven, the distribution, and the greater its information content, $I(\varepsilon)$. Experience shows that the increase of the information content is proportional to $\ln(1/\varepsilon)$, the prefactor being the information dimension:[8]

$$\boxed{I(\varepsilon) = -\sum_{i=1}^{N(\varepsilon)} P_i(\varepsilon) \ln P_i(\varepsilon) = D_1 \ln(1/\varepsilon), \quad \text{for} \quad \varepsilon \ll 1.} \tag{2.18}$$

[7] The fractal dimension of the set that provides the leading contribution to the (non-normalised) distribution consisting of the qth power (q = any real number) of the interval probabilities can be determined in a similar way. This is the generalised dimension, D_q. Since the weighing power, q, is arbitrary, an *infinity* of dimensions belong to a fractal distribution. This is why fractal distributions are also often called multi-fractals.

[8] An equivalent form is $\exp I(\varepsilon) \sim \varepsilon^{-D_1}$, which resembles (2.3).

The information dimension thus depends not only on the number of covering boxes, but also on *how occupied* they are. The typical boxes, whose number is $N^*(\varepsilon)$, faithfully represent the entire distribution. Their total probability is practically unity, so the probability of each box is $1/N^*$. Using this, the above sum can be written as $\sum_{i=1}^{N^*(\varepsilon)} 1/N^*(\varepsilon) \ln N^*(\varepsilon) \equiv \ln N^*(\varepsilon)$ to a good approximation. This again yields $N^*(\varepsilon) = \varepsilon^{-D_1}$, i.e. D_1 is indeed the fractal dimension of typical boxes. Definitions (2.16) and (2.18) are therefore equivalent.

The information dimension is not only a proper measure of the inhomogeneity of a fractal distribution, but is also a quantity which is easy to handle numerically. The evaluation of the information dimension via (2.18) provides a good approximation of D_1 for much coarser resolutions ε than those at which (2.3) yields reliable values for the fractal dimension. The explanation is that in the first case very improbable boxes, which are difficult to find numerically, do not contribute.

Problem 2.16 Even on continuous supports inhomogeneous distributions can exist with information dimension less than unity. Let the unit interval be divided into three equal parts to which probabilities p_1, $p_2 = 1 - 2p_1$ and $p_3 = p_1$ are assigned, and each part is divided again into three segments, etc. The first four steps of the construction are then $(p_1 = 0.2,\ p_2 = 0.6,\ p_3 = 0.2)$:

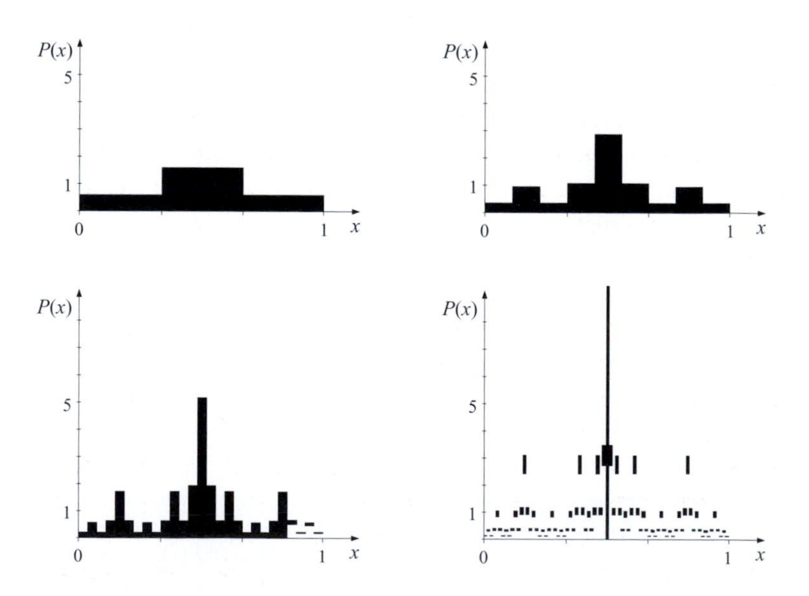

Determine the distribution at the seventh step, and calculate the information dimension.

Table 2.2. *Basic types of chaos and the related fractals.*

	Permanent chaos	Transient chaos
Dissipative	attractors: Cantor filaments	saddles: Cantor clouds fractal boundaries: Cantor filaments
Conservative	bands: fat fractals	saddles: Cantor clouds

2.4 Fractals and chaos

Wide classes of fractals develop from complicated processes often including random elements. Consider, for example, coastlines, the surface of the Moon, catalysts, the lung or the foliage of trees. Our knowledge of most of these processes is still insufficient for an appropriate modelling of the formation of these fractal patterns.

Fractals associated with chaos are exceptional in the sense that they are related to motion, more generally to temporal evolution, and do not include random elements. Their origin is much better understood than that of other fractals. Such fractals are unique consequences of the chaotic nature of the dynamics. Contrary to the fractals examined so far, their construction rule is not a simple recursive algorithm, but *results from the equation of motion* itself. The rule is usually too complicated to make the fractal dimension expressible by means of simple formulae. It is also true that whenever a motion proves to be related to fractal structures, then it is neccesserily chaotic: in chaotic systems we can therefore observe the *unity* of dynamics and geometry. The more chaotic the dynamics, often the more complicated the fractal structure related to it. Regular motion is not related to fractals, but rather to simple structures belonging to the realm of traditional geometry. The fractal patterns of chaos are, however, generally not observable in real space; they become visible in phase space only.

Different types of chaos are associated with different types of fractals (see Table 2.2). Chaotic attractors are objects with a characteristic filamentary structure. If smaller and smaller squares of a chaotic attractor are magnified (see Fig. 2.16) it becomes obvious that the small details are similar to each other. The almost parallel lines resemble Fig. 2.9 presenting Cantor filaments. The structure of *chaotic attractors* therefore corresponds to asymmetric (multi-scale) *Cantor filaments*, i.e. the attractor is locally the direct product of a Cantor set and a smooth curve. Smooth curves correspond to the direction along which nearby points deviate from each other, while the Cantor structure is practically perpendicular to these. The latter is a consequence of dissipation, of the convergence towards attractors.

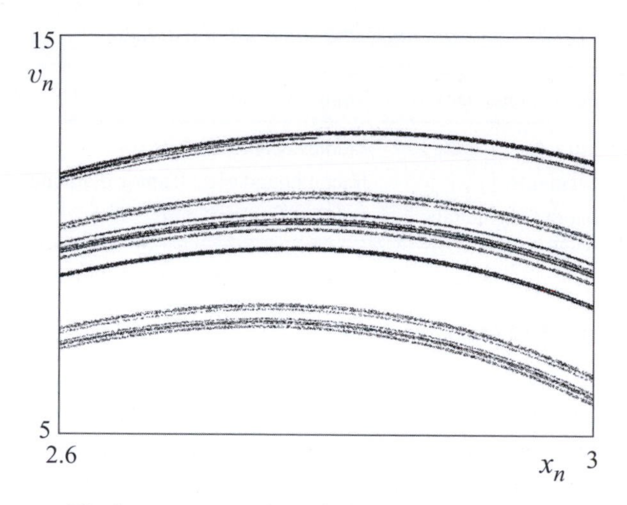

Fig. 2.16. Detail of Fig. 1.4 (chaotic attractor of an anharmonic oscillator) magnified ×10. The filamentary fractal structure also remains present in successive magnifications, providing more and more straight lines.

The long-term motion of a body wandering on a fractal basin boundary between attractors or being subjected to chaotic scattering is even more subtle. The boundary itself is a Cantor filament (see Figs. 1.10 and 1.13) but the origin of chaos is a set more rarefied than this (see Fig. 1.14). The chaotic saddle responsible for *transient chaos* does not include continuous curves but exhibits the structure of an *asymmetric Cantor cloud*, the direct product of two Cantor sets (as illustrated by Fig. 2.10).

Fractal characteristics of the chaos of closed conservative systems are again different. The boundaries of the islands belonging to regular motions are rather ramified; they somewhat resemble Koch curves. A much more significant property is, however, that chaotic bands extend over two-dimensional domains of the plane. On smaller and smaller scales, they are nevertheless interrupted by regular islands, representing regular motion (see Fig. 1.17 and compare with Fig. 2.13). *Chaotic bands* are therefore two-dimensional *fat fractals* (see Table 2.2).

Chaotic motion itself defines a probability distribution. This *natural distribution* gives the probability of visiting different points of the chaotic set in the course of a long-lasting chaotic motion. The natural distributions of both chaotic attractors and saddles is usually highly inhomogeneous, a fractal distribution. Therefore its information dimension, D_1, is less than the fractal dimension, D_0, of the support (the attractor or the saddle). Several examples for such natural distributions will be given in Sections 5.4.4 and 6.3.2 (see also Plates XII–XVI and XVII–XIX, respectively). It is a specific feature of conservative chaos that the natural distribution on chaotic bands is uniform (see Plate XX) and hence not a fractal distribution, although its support is a fat fractal.

The above-mentioned unity of dynamics and geometry manifests itself in a *unique* relation between the information dimension, D_1, and the parameter characterising the deviation of nearby chaotic trajectories, as will be seen in Sections 5.4.6 and 6.3.3.

Box 2.1 Brief history of fractals

The first examples of fractals appeared as mathematical curiosities in the second half of the nineteenth century. The Cantor set was invented by Georg Cantor, the founder of modern set theory, in 1883. Koch curves originated with the Swedish mathematician, Helge van Koch, in 1904. These and other similar examples were considered to have no connection whatsoever with any natural phenomena; nevertheless, they had to be given a place in the edifice of mathematics. A Koch curve is, for example, continuous, but nowhere smooth. Karl Weierstrass found a continuous function that is nowhere differentiable. These discoveries were at first received with displeasure by many, which is well illustrated by the following lines written by Ch. Hermite to T. Stieltjes at the end of the nineteenth century: 'I recoil in fear and loathing from that deplorable evil: continuous functions with no derivatives!'[9] These developments shook classical mathematics to its foundations, and the admission of the new concepts was followed by an advance in abstract directions. Even though the connection between random walk and fractals began to take shape in the first quarter of the twentieth century, half a century was still to pass before the extensive scientific significance of fractals was recognised. This was the achievement of the Polish–French–American mathematician Benoît Mandelbrot (in the period 1977–1982), who also coined the name 'fractal' and worked out the concept of fractal dimension (based on earlier works of F. Hausdorff and A. N. Kolmogorov). The unusual structure present in the phase space of chaotic systems was pointed out by Edward Lorenz in 1963. The exact connection with fractals and the importance of information dimension in the description of chaotic attractors were demonstrated by E. Ott, J. Yorke, D. Farmer, C. Grebogi, P. Grassberger and I. Procaccia during 1981 to 1985. (It is worth noting that the concept of information dimension first appeared in the works of the Hungarian Alfréd Rényi on probability theory in the 1950s.) Subsequently, the fractal concept has spread in science and mathematics due to M. Barnsley, K. Falconer, H.-O. Peitgen, H. E. Stanley, T. Vicsek and numerous others.

[9] This letter can be found at http://www-gap.dcs.st-and.ac.uk/~history/Quotations/ Hermite.html

Part II
Introductory concepts

Chapter 3
Regular motion

The simplest motion occurs in one-dimensional systems subjected to time-independent forces. The most important characteristics of regular (non-chaotic) behaviour will be demonstrated by means of such motion, but this also provides an opportunity for us to formulate some general features. The overview starts with the investigation of the dynamics around unstable and stable equilibrium states, where the essentials already appear in a *linear* approximation. Outside of a small neighbourhood of the equilibrium state, however, *non-linear* behaviour is usually present, which manifests itself, for example, in the co-existence of several stable and unstable states, or in the emergence of such states as the parameters change. We monitor the motion in phase space and become acquainted with the geometrical structures characteristic of regular motion. The unstable states, and the curves emanated from such hyperbolic points, the *stable and unstable manifolds*, play the most important role since they form, so to say, the *skeleton* of all possible motion. In the presence of friction, trajectories converge to the attractors of the phase space. For regular motion, attractors are simple: equilibrium states and periodic oscillations, implying *fixed point attractors* and *limit cycle attractors*, respectively.

3.1 Instability and stability

3.1.1 Motion around an unstable state: the hyperbolic point

Let us start the analysis – contrary to the traditional approach – with the behaviour at and in the vicinity of an unstable equilibrium state.[1]

[1] A general mathematical definition of unstable (and stable) equilibrium states is given in Section 3.5.4.

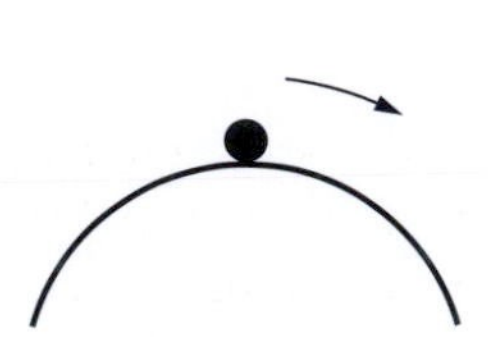

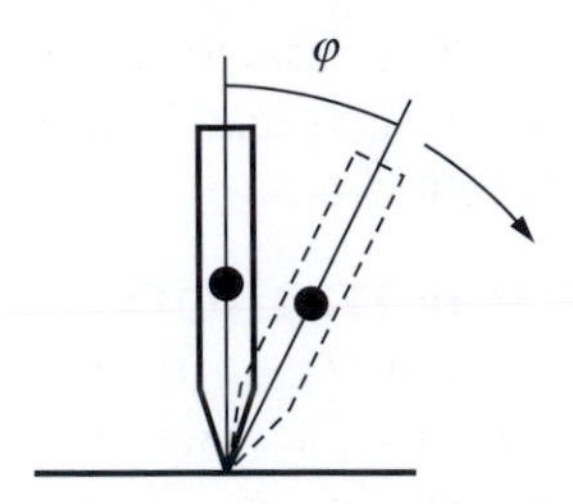

Fig. 3.1. Unstable states: a ball placed on top of a convex surface and a pencil standing on its point. Equilibrium is restricted to one point only, and a displacement from this point, however small, causes the body to deviate at an increasing rate.

An equilibrium state of a body at some position x^* is *unstable* if, when released from a slightly displaced position, the body starts moving further away from x^*. Simple examples are a ball placed onto a convex surface or a pencil standing on its point (Fig. 3.1). In the unstable state itself there is no motion; the velocity vanishes. In the phase space (x, v), an unstable state is thus a point $(x = x^*,\ v = v^* \equiv 0)$, called the *fixed point* for this reason. For the time being, we examine one such point only and place the origin of the co-ordinate system into this point by choosing $x^* = 0$. The force in the neighbourhood of an unstable state (unstable equilibrium point) is always repelling and *increases* with the distance. To understand the essence, it is sufficient to consider a linear force[2] law, i.e.

$$F(x) = s_0^2 x, \tag{3.1}$$

where s_0 is the *repulsion parameter* characterising the strength of the instability. In this section this force law is assumed to be valid for any displacement. We write the coefficient in the form s_0^2 to make its positiveness explicit.

Frictionless (conservative) case

The Newtonian equation of motion states that the acceleration equals the force (per unit mass). Since the acceleration is the second time-derivative,[3] $\ddot{x}$, of the displacement and the only force to act is $F(x)$, the equation of motion is $\ddot{x} = F(x)$. From (3.1) we obtain

$$\ddot{x} = s_0^2 x. \tag{3.2}$$

Problem 3.1 Show that the motion of a pencil standing on its point is described by equation (3.2). Derive the equation of motion valid for

[2] Henceforth by force we understand force per unit mass. Thus, in single-body problems, the mass, m, will not appear.

[3] Time derivation is denoted by the two dots.

small angular deviations, φ, from the vertical position. For the sake of simplicity, assume that the unit mass of the pencil is concentrated at distance l from its point.

As for all linear, homogeneous differential equations with constant coefficients, the solution is exponential. The assumption $x = \exp(\lambda t)$ leads to $\lambda^2 = s_0^2$, i.e. exponent λ can only be the repulsion parameter, s_0, or its opposite, $-s_0$. The general solution is a combination of these basic solutions:

$$x(t) = c_+ e^{s_0 t} + c_- e^{-s_0 t}, \tag{3.3}$$

which yields the velocity

$$v(t) = c_+ s_0 e^{s_0 t} - c_- s_0 e^{-s_0 t}. \tag{3.4}$$

The solution corresponding to a general initial condition, $x(0) = x_0$ $v(0) = v_0$, fulfils $x_0 = c_+ + c_-$, $v_0 = (c_+ - c_-)s_0$, and therefore

$$c_+ = \frac{s_0 x_0 + v_0}{2s_0}, \quad c_- = \frac{s_0 x_0 - v_0}{2s_0}. \tag{3.5}$$

Only a single pair of coefficients, c_+, c_-, is found, which illustrates the uniqueness of the solution.

The phase space trajectories are obtained by eliminating time. Consider the combinations $v - s_0 x$ and $v + s_0 x$, which are, according to (3.3) and (3.4), proportional to $\exp(\mp s_0 t)$. Their product is therefore time-independent,

$$v^2 - s_0^2 x^2 = \text{constant} = v_0^2 - s_0^2 x_0^2, \tag{3.6}$$

for any values x, v in the course of the motion.

The trajectories in the phase space are therefore hyperbolae around the fixed point (Fig. 3.2). This is why such a fixed point is called a *hyperbolic point*.[4] The asymptotes are the straight lines $v = \pm s_0 x$ emanating from the origin. The hyperbolae are thus uniquely determined by the single parameter, s_0, of the dynamics.

Nearly all initial conditions result in a motion along a hyperbola, causing the phase space point to deviate from the origin after a possible initial approach. Note that despite the general moving away, there exist special initial conditions from which the fixed point can be reached. If, for a given positive initial co-ordinate, such a negative initial velocity is chosen which falls on the line

$$v = -s_0 x, \tag{3.7}$$

[4] The term saddle point is also used.

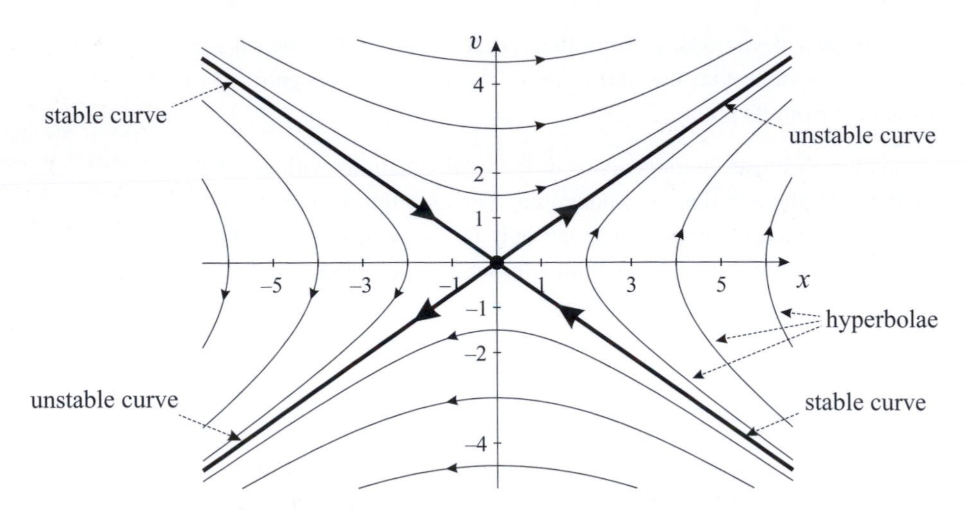

Fig. 3.2. Hyperbolic point and its neighbourhood (frictionless case): shown are a few hyperbolic trajectories (thin lines) and the asymptotes (thick lines). The motion along the trajectories follows the directions marked by the arrows. The hyperbolic point can only be reached along one of the asymptotes, the stable curve. The repulsion parameter is $s_0 = 0.7$.

i.e. if the body is thrust towards the unstable state with a well defined velocity, then in view of (3.3) and (3.5) it exactly reaches the origin according to the law

$$x(t) = x_0 e^{-s_0 t} \tag{3.8}$$

(the pencil stops on its very point). In principle, the approach towards the fixed point requires an infinitely long time, but in practice the body reaches the fixed point (to a very good approximation) after a few multiples of $1/s_0$. The same is valid for the left branch of the asymptote, $v = -s_0 x$, where the initial velocity is positive: in this case, the motion goes in the opposite direction along the asymptote.

Along the other asymptote,

$$v = s_0 x, \tag{3.9}$$

a moving away occurs according to the relation

$$x(t) = x_0 \, e^{s_0 t}; \tag{3.10}$$

it is purely exponential from the very beginning (see (3.3) and (3.5)). The closer the starting point to the fixed point (the smaller x_0), the longer the body remains in the neighbourhood of the hyperbolic point; the closer the initial position of the pencil to the vertical, the longer it takes the pencil to tip over.

The asymptote $v = -s_0 x$ describes the exceptional motion of asymptotically reaching the fixed point. This direction is therefore called

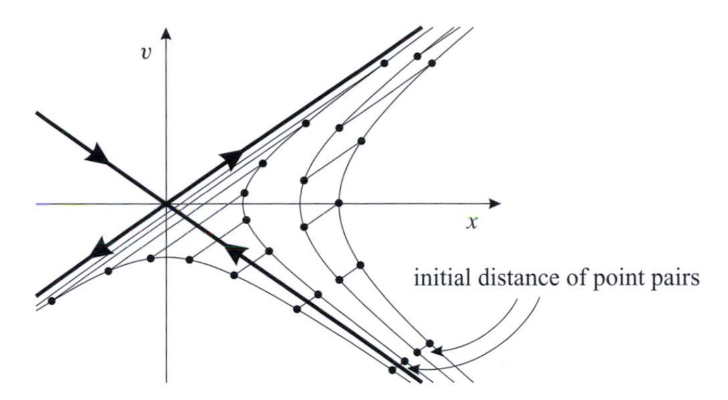

Fig. 3.3. Deviation of point pairs from each other and from the hyperbolic point (the origin). Rapid deviation always occurs along the unstable curve, even when the initial points are on different sides of the stable curve. The repulsion parameter is $s_0 = 0.7$. The distance between the point pairs is plotted at equal times.

the stable direction of the hyperbolic point, as opposed to the unstable direction defined by the other asymptote. Straight-line segments crossing the fixed point in these directions are parts of the *stable and unstable curves* emanating from the fixed points. Figure 3.2 illustrates that the stable and unstable curves divide the phase plane into four quadrants. Note that the stable curve plays the role of a *boundary*. Trajectories emanated 'above' the stable curve imply deviation (the pencil falling) to the right, while those departing 'under' it go in the opposite direction.

Instability in phase space is always related to the appearance of hyperbolic points. The naïve (and false) expectation to see a moving away from an unstable point in all directions of phase space arises because in everyday usage an equilibrium state is called unstable if a body tipped out of that state, *without initial velocity*, moves away from that state. The presence of the stable direction indicates that by taking into account the initial velocity, the situation is less trivial. The clearest picture of an unstable state and its neighbourhood unfolds itself in a phase space representation.

For general initial conditions, after a sufficiently long time ($t \gg 1/s_0$), the exponentially increasing term dominates expression (3.3):

$$x(t) \approx \frac{1}{2}\left(x_0 + \frac{v_0}{s_0}\right) e^{s_0 t}. \tag{3.11}$$

Particles leave the unstable state according to an exponential law. As a consequence, particles starting from nearby initial points also deviate from each other exponentially. If the initial conditions differ by small values δx_0, δv_0 then, according to (3.11), the difference of the position co-ordinates increases as

$$\delta x(t) \approx \frac{1}{2}\left(\delta x_0 + \frac{\delta v_0}{s_0}\right) e^{s_0 t} \quad \text{for} \quad t \gg \frac{1}{s_0}. \tag{3.12}$$

The velocity difference behaves in a similar way, $\delta v(t) \approx s_0 \delta x(t)$. Thus the full distance in phase space increases exponentially as well. Points always depart along the unstable curve (Fig. 3.3). The system is therefore always sensitive to the initial conditions in the vicinity of a hyperbolic point, since nearby trajectories rapidly deviate from each other. Only the points of the stable curve are exceptions to this, since they approach the origin and therefore each other.

Effect of friction

The motion of macroscopic bodies is usually affected by *dissipative*, i.e., energy-consuming, processes. The simplest example of these is *friction* (air drag), a consequence of the interaction with the numerous particles constituting the environment, described by means of simple empirical laws. The friction force is typically a function of the velocity, and is independent of position. For the sake of simplicity, we shall henceforth assume that friction is *directly* proportional to the velocity.[5] Thus, in addition to the position-dependent external force, $F(x)$, a friction force $-\alpha v$ is also present. Here $\alpha > 0$ is the friction coefficient, considered to be constant. The negative sign expresses that friction decelerates. Note that friction of this type does not influence the equilibrium position since there is no motion at that point.

The equation of motion with a finite friction coefficient is of the form

$$\ddot{x} = +s_0^2 x - \alpha \dot{x}. \tag{3.13}$$

Seeking the solution of this homogeneous, non-zero differential equation in the form of $\exp(\lambda t)$, a quadratic equation, $\lambda^2 + \alpha\lambda - s_0^2 = 0$, is obtained, which yields two possible values for λ:

$$\lambda_\pm = -\frac{\alpha}{2} \pm \sqrt{\frac{\alpha^2}{4} + s_0^2}. \tag{3.14}$$

These are real solutions: λ_+ is positive and λ_- is negative. The solution corresponding to the initial conditions, $x(0) = x_0$ and $v(0) = v_0$, is the linear combination of the two exponential terms:

$$x(t) = c_+ e^{\lambda_+ t} + c_- e^{\lambda_- t}, \tag{3.15}$$

and

$$c_+ = \frac{-\lambda_- x_0 + v_0}{\lambda_+ - \lambda_-}, \qquad c_- = \frac{\lambda_+ x_0 - v_0}{\lambda_+ - \lambda_-}. \tag{3.16}$$

The equation of the trajectories is

$$\frac{(v - \lambda_- x)^{\lambda_-}}{(v - \lambda_+ x)^{\lambda_+}} = \text{constant} = \frac{(v_0 - \lambda_- x_0)^{\lambda_-}}{(v_0 - \lambda_+ x_0)^{\lambda_+}}. \tag{3.17}$$

These are also hyperbola-like curves, with two asymptotes (see Fig. 3.4).

[5] 'Static friction', which is independent of velocity, is a simplifying term introduced to describe how extended elastic bodies start sliding from their equilibrium positions. This concept is not required in describing the motion of point-like objects.

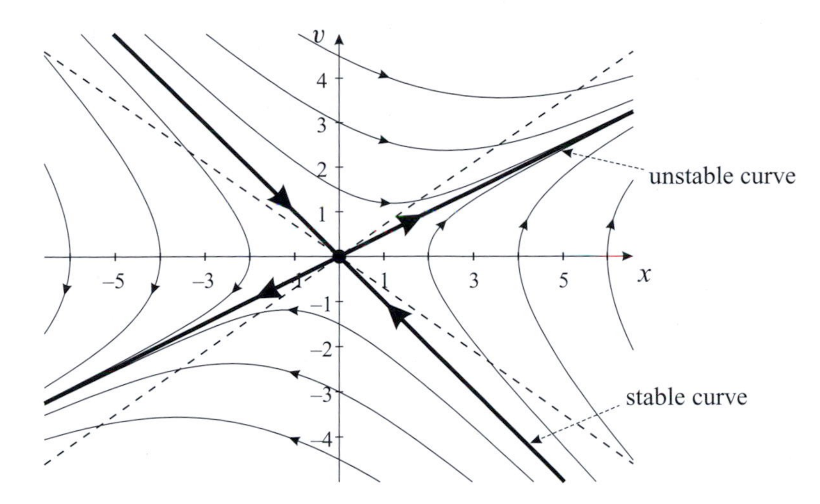

Fig. 3.4. Hyperbolic point in the presence of friction. The character of the structure surrounding the fixed point has not changed, only the asymptotes have turned. (The asymptotes of the frictionless case are marked with dashed lines.) The parameters are $s_0 = 0.7$, $\alpha = 0.5$.

Problem 3.2 Derive the trajectory equation (3.17) from (3.15).

The origin is therefore still called a hyperbolic point of the phase plane. Again there are stable and unstable directions defined by the following asymptotes and displacement rules:

$$v = \lambda_- x, \quad x(t) = x_0 e^{\lambda_- t}, \tag{3.18}$$

and

$$v = \lambda_+ x, \quad x(t) = x_0 e^{\lambda_+ t}, \tag{3.19}$$

as generalisations of relations (3.8) and (3.10). The parameter λ_+ characterising the deviation along the unstable direction is called the *instability exponent*. Note that λ_+ is no longer identical to the repulsion parameter s_0, but depends also on the friction coefficient, thus reflecting the entire dynamics (3.13). With friction, deviation is slower than without; the angle between the unstable direction and the x-axis is therefore smaller than in the frictionless case.

Problem 3.3 Show that the cross formed by the stable and unstable directions turns, for very weak friction ($\alpha/2 \ll s_0$), as a rigid body.

It is of importance that friction has *not* destroyed the hyperbolic character of the fixed point. This property is called the *structural stability* of the hyperbolic behaviour against small changes of the parameters, in the present case against friction.

Again, an exponential rule describes how nearby trajectories deviate from each other:

$$\boxed{\delta x(t), \delta v(t) \sim e^{\lambda_+ t}} \tag{3.20}$$

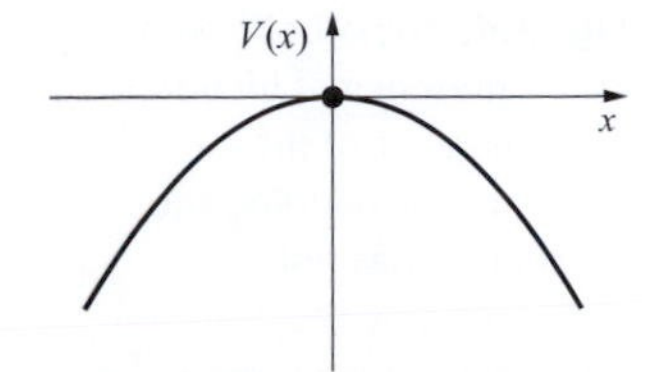

Fig. 3.5. Potential around an unstable state. The unstable behaviour corresponds to the motion 'on top of a hill'.

for $t \gg 1/|\lambda_-|$. It must be kept in mind, however, that exceptional trajectory pairs falling on the stable direction approach each other and the hyperbolic point at an exponential rate, according to the rule $e^{\lambda_- t}$ ($\lambda_- < 0$).

Finally, we note that the concept of *potential* helps to establish a qualitative understanding of the force law, $F(x)$. The potential, $V(x)$, is a function yielding the potential energy per unit mass of a particle at position x. If the force is of a restoring character, the potential increases with the distance x, and vice versa. The force is proportional to the slope of the potential: the general relation between the force, $F(x)$, and the corresponding potential, $V(x)$, is

$$F(x) = -\frac{dV(x)}{dx} \equiv -V'(x). \tag{3.21}$$

If the potential changes only slightly during the motion, it can even be said that a particle moves as if it were travelling on a $V(x)$-shaped relief in a gravitational field. Usually, the vertical velocity component is also important in the motion on a relief, but this is negligible when the potential change is small. Henceforth, the potential–relief analogy should always be applied with this restriction. The potential[6] corresponding to force (3.1) around an unstable state is $V(x) = -s_0^2 x^2/2$ (Fig. 3.5). This

Box 3.1 Instability, randomness and chaos

Very small differences in the neighbourhood of an unstable state (or of its stable curve in phase space) lead to drastically different outcomes. If a pencil is placed onto its point as accurately as possible, it cannot be predicted to which side it will fall when released. This is due to incalculable effects, such as the trembling of our hand or a faint motion of the air. It is therefore a random event for the macroscopic observer whether the pencil tips over to the right or to the left. The motion of a tossed coin is a similar phenomenon (the coin can even fall onto its edge!); its outcome has always been considered as a random process. Instability therefore involves unpredictability and random behaviour, but does *not* imply chaos in itself. The motion of a pencil standing on its point is no longer unstable after it starts falling over, thus it cannot be considered to be chaotic either. Instability is only a *necessary* condition for chaos. The *sufficient* condition is that instability persists for any length of time during the motion; it emerges again and again. *Chaos is sustained instability.* Chaotic motion is only possible if the motion passes through a sequence of unstable states. The presence of an infinity of unstable, hyperbolic states is necessary to render chaos possible. Chaos is an infinite repetition of the behaviour around a simple hyperbolic point. Knowledge of the motion around an unstable point is therefore only an elementary step towards understanding chaos. It has to be completed with the investigation of non-linear effects and of the emergence of instability.

[6] Since the potential is only defined up to a constant, its value at the fixed point can always be chosen to be zero.

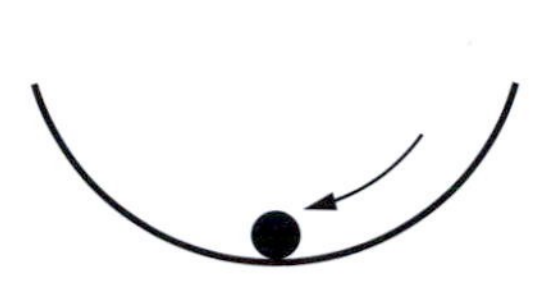

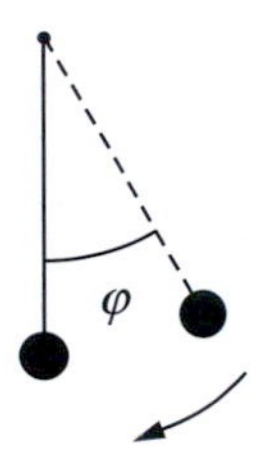

Fig. 3.6. Stable states: a ball placed at the bottom of a concave surface and a pendulum. Rest is only possible at a single point, but, if the system is displaced, it starts moving towards the equilibrium point.

potential therefore actually corresponds to a hill, the top of which (the position $x^* = 0$) is the unstable state, in accordance with the qualitative picture of Fig. 3.1.[7]

3.1.2 Motion around a stable state

An equilibrium state of a body at some point x^* is called *stable* if, when released from a slightly displaced position, the body does not move away: a restoring force acts towards the equilibrium position, x^*. In the presence of friction, in addition, the particle gradually slows down and comes to rest in the equilibrium state. A ball placed in a concave vessel or the swinging of a pendulum around its vertical state ($\varphi = 0$) are simple examples of this (Fig. 3.6). In the phase space (x, v), the stable state is also characterised by a fixed point $(x^*, v^* \equiv 0)$. The position co-ordinate is again chosen to be the origin: $x^* = 0$. The force in the neighbourhood of the stable state (of a stable equlibrium point) is of a restoring character; it increases in the opposite sense to the displacement. In the simplest approach, let the force be of the form

$$F(x) = -\omega_0^2 x, \tag{3.22}$$

where the parameter ω_0 characterises the strength of the attraction. Equation (3.22) describes the linear restoring effect of springs, and the square of the parameter, ω_0, is the force constant per unit mass. The parameter, ω_0, is called the natural frequency, because it determines the period of the oscillation in the frictionless case. It is assumed in this section that this model of attraction is valid for any displacement. The coefficient has been written as $-\omega_0^2$ to make its negative sign explicit.

[7] In the frictionless case, the trajectories are the contours of constant energy per unit mass, $v^2/2 - s_0^2 x^2/2 \equiv \mathcal{E}$; the stable and unstable curves belong to the hill-top value, $\mathcal{E} = 0$.

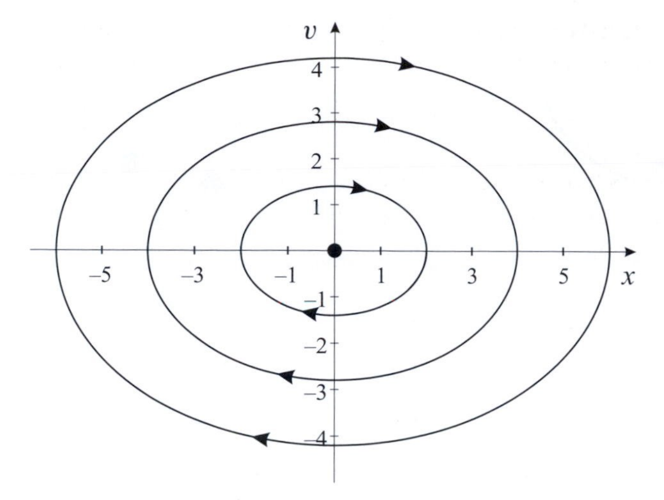

Fig. 3.7. Elliptic point and its neighbourhood. The direction of motion is represented by arrows. The circulation always occurs clockwise (since positive velocities result in positive displacements). The natural frequency is $\omega_0 = 0.7$.

Frictionless (conservative) case: the elliptic point

The equation of motion of the frictionless case is $\ddot{x} = F(x)$, i.e.

$$\ddot{x} = -\omega_0^2 x. \tag{3.23}$$

This is simply the equation of a harmonic oscillation with natural frequency ω_0.

Problem 3.4 Show that the motion of a pendulum around the vertical position is described by equation (3.23) for small displacements.

The solution corresponding to the initial conditions $x(0) = x_0$ and $v(0) = v_0$ is

$$x(t) = x_0 \cos(\omega_0 t) + \frac{v_0}{\omega_0} \sin(\omega_0 t), \tag{3.24}$$

as can be verified by substitution. This can also be written in the form

$$x(t) = A \sin(\omega_0 t + \delta), \tag{3.25}$$

where the amplitude, A, and the phase, δ, are determined by the equations $A^2 = x_0^2 + v_0^2/\omega_0^2$ and $\mathrm{tg}\delta = v_0/(x_0\omega_0)$.[8]

In the phase plane (x, v), the trajectories are ellipses centred at the origin (Fig. 3.7), since taking the square of (3.25) and the corresponding velocity yields

$$v^2 + \omega_0^2 x^2 = v_0^2 + \omega_0^2 x_0^2 = \omega_0^2 A^2 = \text{constant} \tag{3.26}$$

at any instant, t. Such fixed points are therefore called *elliptic* points.

Different initial conditions fall onto different ellipses only if the amplitudes are different. Contrary to a hyperbolic point, the trajectories do

[8] This solution can also be given in the form of (3.15) and (3.16), with $\lambda_\pm = \pm i\omega_0$. Equation (3.24) is recovered by using the relations between exponential functions with imaginary arguments and trigonometric functions.

not leave the neighbourhood of an elliptic point; moreover, the distance between neighbouring trajectories does not increase continually, since

$$\delta x(t) = \delta x_0 \cos(\omega_0 t) + \frac{\delta v_0}{\omega_0} \sin(\omega_0 t). \tag{3.27}$$

The difference therefore sometimes increases and sometimes decreases, but it always remains bounded. Exponential deviation is characteristic of hyperbolic points only.

Effect of friction (dissipative case): point attractors

In the presence of friction, the equation of motion is given by

$$\ddot{x} = -\omega_0^2 x - \alpha \dot{x}. \tag{3.28}$$

Seeking the solution in the form of $\exp(\lambda t)$, a quadratic equation $\lambda^2 + \alpha\lambda + \omega_0^2 = 0$ is obtained, yielding two possible values:

$$\lambda_\pm = -\frac{\alpha}{2} \pm \sqrt{\frac{\alpha^2}{4} - \omega_0^2}. \tag{3.29}$$

The general solution is again of the form

$$x(t) = c_+ e^{\lambda_+ t} + c_- e^{\lambda_- t}, \tag{3.30}$$

$$c_+ = \frac{-\lambda_- x_0 + v_0}{\lambda_+ - \lambda_-}, \quad c_- = \frac{\lambda_+ x_0 - v_0}{\lambda_+ - \lambda_-}. \tag{3.31}$$

Since the real parts of the exponents, $\lambda_\pm$, are always negative, the solution describes a convergence towards the fixed point. What is seen here in a concrete example is a general property of dissipative systems: such systems *forget* their initial conditions. This means that the phase space must have a subset that all trajectories reach. This attracting subset is called an *attractor*. In our case, all the trajectories converge to the origin; consequently, the attractor is a single point.

The elliptic point therefore loses its fundamental properties in the presence of the slightest friction, i.e. it is *structurally unstable*. Consequently, an interesting observation is that, while the nature of the stable dynamics essentially changes when friction is switched on, the behaviour characterising unstable dynamics is only slightly 'deformed' (Section 3.1.1).

The way in which the origin is reached, i.e. the type of the point attractor, depends on the strength of the friction.

Weak damping: spiral attractor

If the friction parameter, $\alpha/2$, is smaller than the natural frequency resulting from the attracting force,

$$\frac{\alpha}{2} < \omega_0, \tag{3.32}$$

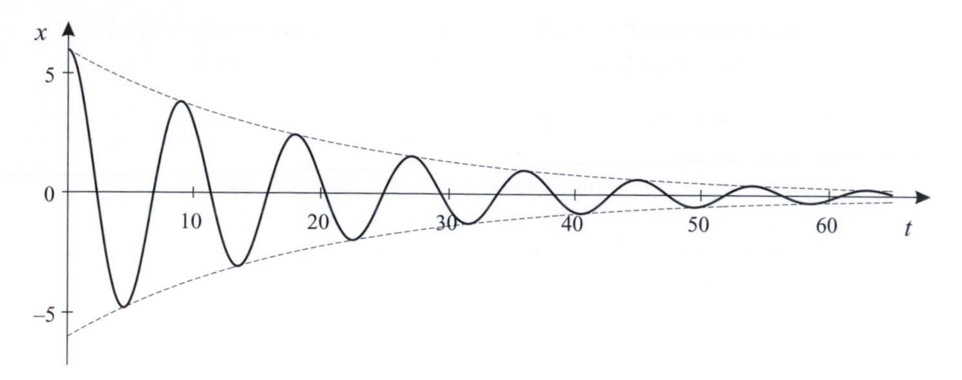

Fig. 3.8. Damped harmonic oscillation with exponentially decreasing amplitude around a stable state in the presence of weak damping. The equation of the dashed lines is $\pm Ae^{-(\alpha/2)t}$. The initial conditions are $x_0 = 6$, $v_0 = 0$, and $\omega_0 = 0.7$, $\alpha = 0.1$.

the expression under the square root of (3.29) is negative, i.e. coefficient $\lambda_\pm$ has an imaginary part. This corresponds to an oscillating decay with frequency

$$\omega_\alpha = \sqrt{\omega_0^2 - \frac{\alpha^2}{4}}. \tag{3.33}$$

Using (3.30), (3.31) and the relations between exponential and trigonometric functions, the solution fulfilling the initial conditions (x_0, v_0) becomes

$$x(t) = x_0 e^{-(\alpha/2)t} \cos(\omega_\alpha t) + \frac{v_0 + (\alpha/2)x_0}{\omega_\alpha} e^{-(\alpha/2)t} \sin(\omega_\alpha t), \tag{3.34}$$

which can also be written as

$$x(t) = Ae^{-(\alpha/2)t} \sin(\omega_\alpha t + \delta). \tag{3.35}$$

This shows that the motion is a harmonic oscillation with an exponentially decaying amplitude (Fig. 3.8). The frequency, ω_α, *decreases*, the oscillations slow, as the friction becomes stronger.

The trajectories in phase space approach the origin along a *spiral*, according to the sign changes of the displacement and the velocity (Fig. 3.9). The origin is therefore called an attracting spiral fixed point or a *spiral attractor*. It is clear from (3.35) that the trajectory converges to the origin *exponentially*. In a mathematical sense, it only reaches it after an infinitely long time, but, due to the rapid decay of the exponential function, the body is practically at rest after a few multiples of the decay time $1/\alpha$.

Problem 3.5 Show that the trajectories around a spiral fixed point form a logarithmic spiral.

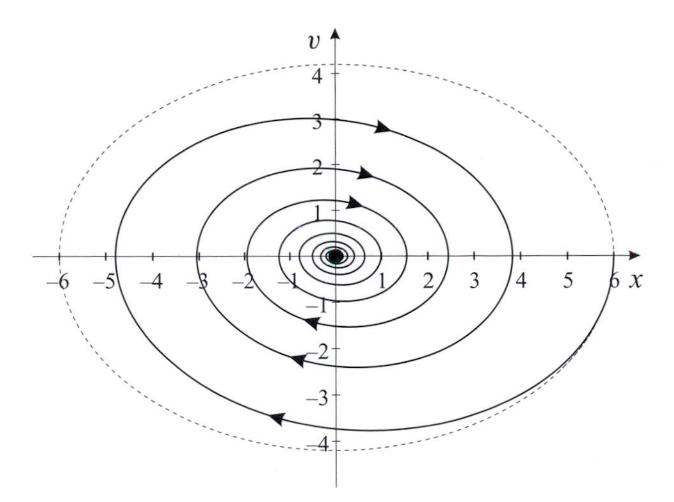

Fig. 3.9. Spiral attractor and its neighbourhood. The dashed line shows the ellipse of the frictionless trajectory. The parameters and the initial conditions are the same as in Fig. 3.8.

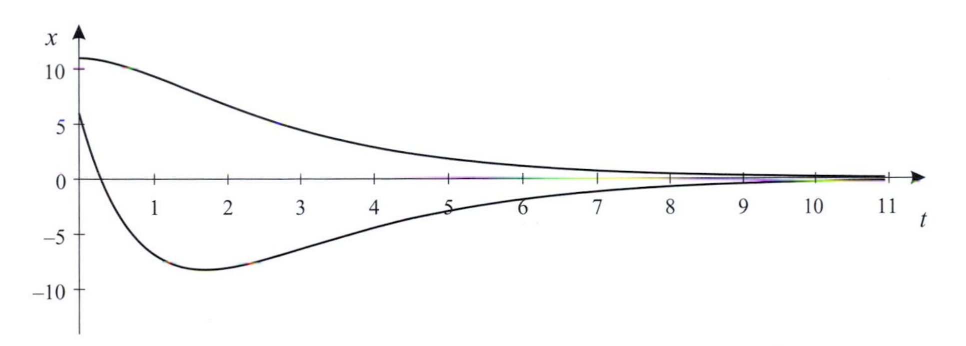

Fig. 3.10. Overdamped oscillations: strong damping around a stable state. In one of the cases ($x_0 = 11,\ v_0 = 0$), the body comes to rest without its displacement changing sign, whereas in the other case ($x_0 = 6,\ v_0 = -25$), it crosses to the opposite side once ($\omega_0 = 0.7,\ \alpha = 1.5$).

Strong damping: node attractor

The period, $2\pi/\omega_\alpha$, of the damped oscillation goes to infinity as $\alpha/2 \to \omega_0$, which indicates that the motion assumes a different character as it leaves the range of weak damping (another structural instability). In the overdamped case, when the coefficient, $\alpha/2$, is greater than the natural frequency ω_0,

$$\frac{\alpha}{2} > \omega_0, \tag{3.36}$$

the exponents $\lambda_\pm$ are real. This results in a decay *without* oscillations (Fig. 3.10). The solution fulfilling the initial conditions (x_0, v_0) is of the form of (3.30) and (3.31), but both exponents are now negative. Strong damping qualitatively means that the motion takes place in a medium

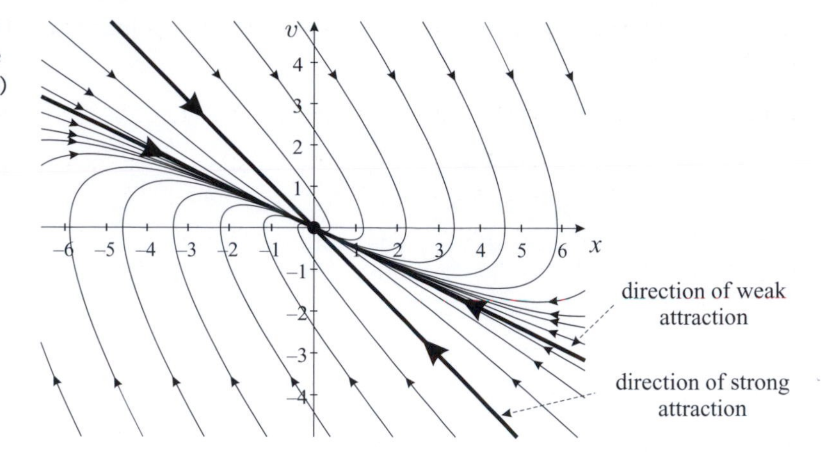

Fig. 3.11. Node attractor and its neighbourhood: some general trajectories (thin lines) and the asymptotes $v = \lambda_{\pm} x$ (thick lines). The parameters are $\omega_0 = 0.7$, $\alpha = 1.5$; $\lambda_- = -1.02$, $\lambda_+ = -0.48$.

which is so dense that not even damped oscillations can develop, and the body stops as soon as possible.

Two special lines appear in the phase plane: the straight lines $v = \lambda_{\pm} x$, along which the decay is described by means of a single exponential (instead of the linear combination of two different exponentials). Since the absolute value of λ_- is greater than that of λ_+, the second term in (3.30) decays more rapidly, and after a long time the first term dominates. The trajectories converge asymptotically to the straight line $v = \lambda_+ x$, along which their time dependence is characterised by a decay with exponent λ_+. Point attractors of this type are called *node attractors* (Fig. 3.11). Points along the straight line $v = \lambda_- x$ are exceptional in the sense that they converge to the origin from the very beginning with a fast exponential decay, according to exponent λ_-. This straight line defines the direction of strong attraction, while the curve $v = \lambda_+ x$ represents that of weak attraction.

Note that all the results concerning motion around a stable point can be obtained from the relations that were valid for the unstable case by using the substitution $s_0 \to i\omega_0$, where ω_0 is real. The directions of strong and weak attraction therefore result from the stable and unstable directions by means of this transformation. Qualitatively speaking, due to the change in the character of the force, the behaviour around a node can be obtained from that around a hyperbolic point by turning the unstable direction from the first and third quadrants of the phase plane into the second and fourth. Simultaneously, of course, the orientation also changes from repulsing into (weak) attracting.

A node attractor, too, is reached at an exponential rate. Thus, the distance between neighbouring point pairs also *decreases* exponentially: $\delta x(t), \delta v(t) \sim \exp(\lambda_+ t)$. An exponential convergence of point pairs can also be seen in the neighbourhood of spiral fixed points. The motion in the neighbourhood of point attractors is thus *not sensitive* to the initial conditions.

Problem 3.6 Determine the trajectories around a node attractor.

The potential corresponding to (3.22), characteristic of a stable state, is $V(x) = \omega_0^2 x^2/2$ (the value at the stable fixed point is chosen to be zero). The potential in question corresponds to a valley called a 'potential well' (Fig. 3.12), and the bottom of the valley, $x^* = 0$, is the stable state, in accordance with Fig. 3.6.[9]

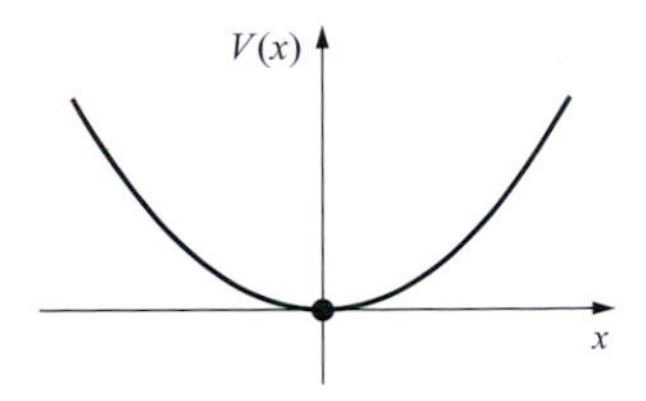

Fig. 3.12. Potential around a stable state. Stable motion occurs 'at the bottom of a valley'.

3.2 Stability analysis

The force $F(x)$ is never an exactly linear function of the position, i.e. the equation of motion is never exactly linear. Before examining the equations of motion,

$$\ddot{x} = F(x) - \alpha\dot{x}, \tag{3.37}$$

for an *arbitrary* force, $F(x)$, in an extended region, it is worth locating the possible equilibrium positions. These can only be fixed points, x^*, where the force vanishes:

$$F(x^*) = 0. \tag{3.38}$$

These are also the extrema of the potential defined by (3.21), where $V'(x^*) = 0$.

The location of the fixed points alone does not provide any information about stability. Even though, in principle, a body placed at point x^* always remains there, in practice numerous small external effects influence it. These move the point slightly off its equilibrium position. The consequences of such small external disturbances can be investigated by following the motion starting from positions slightly displaced from x^*. The question is whether the particle moves further away from the fixed point, i.e. whether the force repels it away from point x^*, or, on the contrary, pulls it back there. If the former holds, the equilibrium position is unstable. Realistic systems cannot permanently remain in such states.

The stability of a fixed point depends on the form of the force law *in a small neighbourhood of the fixed point.* Any smoothly changing force can be approximated around an arbitrary fixed point, x^*, in view of (3.38) by

$$F(x) \approx F'(x^*)(x - x^*) \equiv -V''(x^*)(x - x^*). \tag{3.39}$$

[9] The elliptical trajectories of the frictionless case are contours of constant energy.

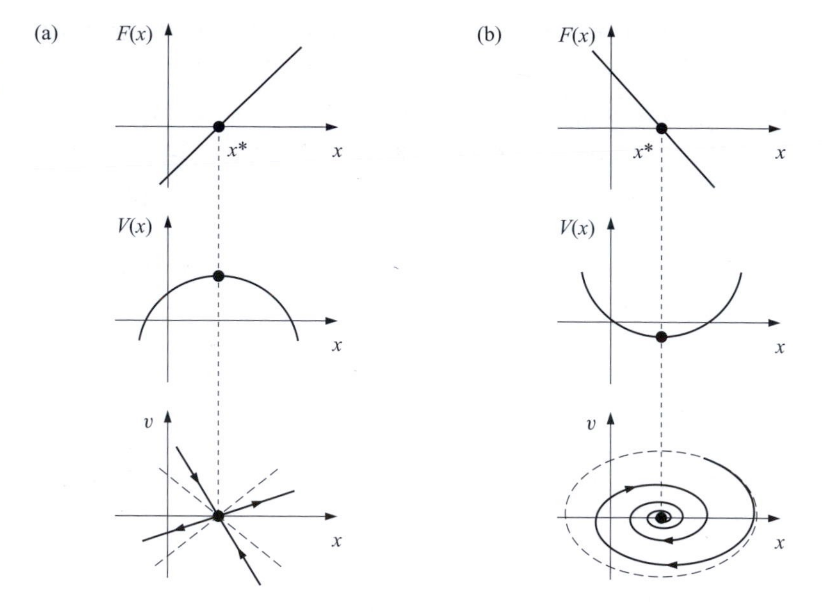

Fig. 3.13. Dependence of the stability of fixed points on the local form of the force law and the potential, and the corresponding phase space structures (trajectories of the frictionless case are marked with dashed lines): (a) unstable, (b) stable state.

This implies that, in general, the force changes linearly for small displacements from the equilibrium position. Expression (3.39) is, in fact, a Taylor-series expansion up to first order. Since $x - x^*$ is small, higher-order terms of the expansion are not included. The stability of the fixed point already follows from this expression: when $F'(x^*)$ is negative, i.e. the potential has a *minimum* (force locally attracting), the equilibrium state is *stable*, and when $F'(x^*)$ is positive, i.e. the potential has a *maximum* (force locally repulsing), the equilibrium state is *unstable*. The application of the potential is useful because it readily provides information about the stability of the fixed point, in accordance with our previous picture of a motion on a relief. While the qualitative nature of the fixed point is determined by the sign of $F'(x^*)$, the measure of the stability or instability is determined by the numerical value of the derivative. The more rapidly the restoring force increases, i.e. the sharper the minimum of the potential, the more stable a state becomes. The parameters s_0 or ω_0 used in the previous sections can always be determined around fixed points, even in the case of non-linear forces. They are given by the derivative in the fixed points:

$$F'(x^*) = -V''(x^*) = s_0^2 \quad \text{or} \quad -\omega_0^2, \tag{3.40}$$

respectively, as shown in Fig. 3.13.

Problem 3.7 Analyse the stability of point $x^* = c$ for the force law $F(x) = ax(x - c)$ in terms of the parameters.

Problem 3.8 Determine the range of validity of the linear approximation (3.39) in terms of both displacement and time. (Take into account higher-order terms in the Taylor expansion.)

Problem 3.9 Estimate for how many seconds the linear approximation (equation (3.1)) is valid for a falling pencil, for different, smaller and smaller, initial angular deviations.

Around unstable states, the validity of the linear equation of motion is always limited to *finite* times only, even for initial conditions very close to the fixed point. The moving point deviates from the fixed point, and thus, sooner or later, leaves the range where the force law, (3.39), is valid. This is why traditional texts are not concerned with motion around unstable states. To us, however, this is a hint to follow the motion into the non-linear range. This raises the fundamental question of how the locally straight stable and unstable curves continue outside the linear regime.

3.3 Emergence of instability

The non-linearity of a force law might be of such a nature that several equilibrium positions become possible, and unstable and stable states co-exist. In this section we investigate how a stable state can become unstable, i.e. how instability emerges under changes of parameter. We shall see that this is usually accompanied by the appearance of one or more *new* stable states.

Due to non-linearity, the results of Section 3.1 are only valid locally, in the vicinity of the stable and unstable states. It is also important to understand the dynamics *globally*, i.e. far away from these points. Knowledge of the phase space structure around the fixed points will nevertheless be of great help: taking into account the fact that trajectories cannot intersect, local properties can be joined together to provide a global view. Generally speaking, we will give a geometric description of non-linear motion in the phase plane. Concepts introduced here will also play an important role in understanding more complicated motions. We show that it is the hyperbolic (unstable) points that organise the global behaviour.

3.3.1 Bistable systems

As a consequence of external changes, an originally stable system may start showing signs of uncertainty; it may become *bistable*. In this case, the single, originally stable, state ceases to exist and two different stable equilibrium states appear. In the neighbourhood of the state that just became unstable, the system has to 'decide' which new stable state to

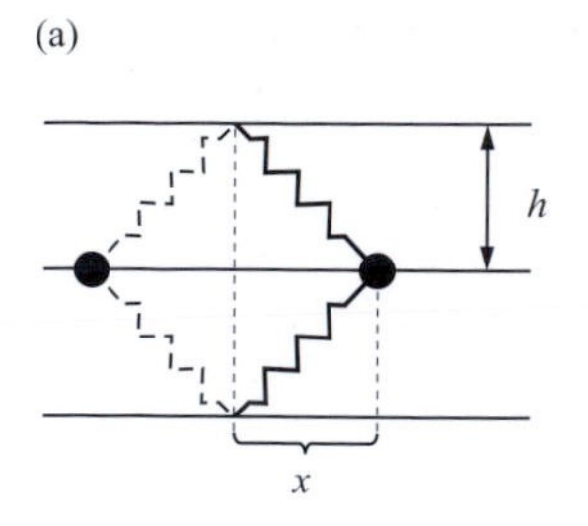

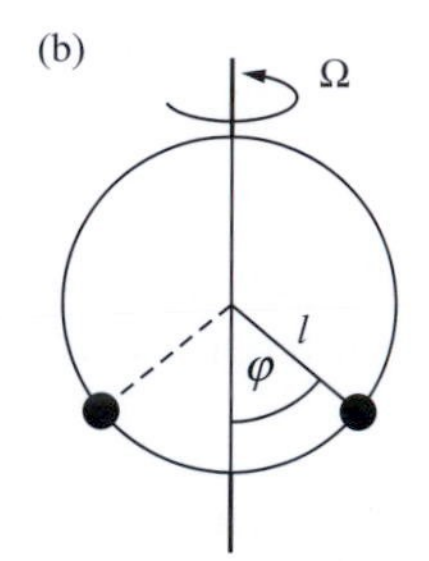

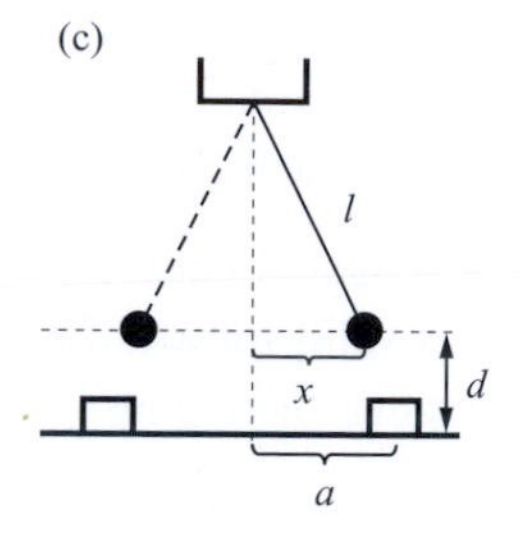

Fig. 3.14. Bistable systems. (a) Point mass fixed to two compressed springs and moving horizontally along a rail. (b) Point mass moving along a ring, rotated at a sufficiently large angular velocity, Ω. (c) Magnetic pendulum near two attracting magnets. In all three cases, the system comes to a halt in one of the two stable positions, which are mirror-images of each other.

approach. As an example, consider a point mass fixed to two springs and moving along a horizontal rail (Fig. 3.14(a)). As long as the points to which the springs are fixed are a long way from each other, the state with zero displacement is the stable equilibrium. When the points get close enough to each other, the mass moves to the right or to the left, and two new stable states develop at positions x^* and $-x^*$. This system is a simple model of the buckling of longitudinally loaded rods. An analogous phenomenon (Fig. 3.14(b)) is the motion of a point mass along a rotated ring, having only one stable state at the bottom while the rotation is slow, but settling in a position characterised by a finite angular deviation when the rotation is fast enough. A third example could be a pendulum with a steel ball as anchor (Fig. 3.14(c)), whose stable equilibrium state is in the vertical position. However, when two attracting magnets are placed near the pendulum, then a displaced state either to the right or to the left will be the stable equilibrium position.

There are numerous similar examples with symmetrically arranged equilibrium states. These can collectively be modelled by the force law:

$$F(x) = -ax(x - x^*)(x + x^*). \qquad (3.41)$$

Here $a > 0$ is a positive parameter, and $x^* > 0$ and $-x^* < 0$ are the position co-ordinates of the stable states on the right and on the left, respectively. The corresponding potential is given by

$$V(x) = -bx^2 + dx^4, \qquad (3.42)$$

where $b = ax^{*2}/2$ and $d = a/4$ are fixed parameters. Since the coefficient of the quadratic term is negative, the potential has a local maximum at the origin that separates two potential wells, and potential (3.42) is therefore called double-welled. Consequently, the behaviour of bistable systems corresponds to motion around two wells separated by a hill in

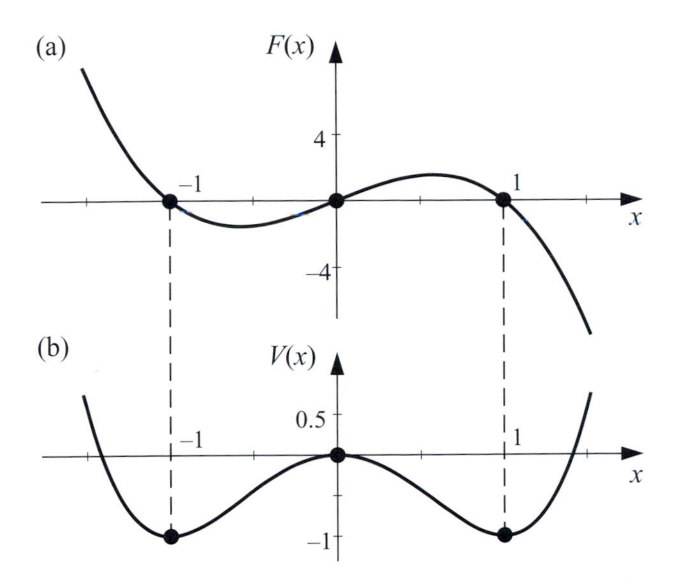

Fig. 3.15. Bistable systems: a general model of the force (a) and of the potential (b). These graphs correspond to functions (3.41) and (3.42), respectively, at parameters $a = 4$, $x^* = 1$ $(b = 2, d = 1)$.

between (Fig. 3.15). At the hill-top, $F'(x_0^* = 0) = ax^{*2}$, the repulsion parameter is therefore given by

$$s_0 = \sqrt{a}x^* = \sqrt{2b}. \tag{3.43}$$

The force is attractive at the bottom of the wells and its derivative is negative: $F'(\pm x^*) = -2ax^{*2}$; consequently, the natural frequency is given by

$$\omega_0 = \sqrt{2a}x^* = 2\sqrt{b}. \tag{3.44}$$

The global behaviour of the system can be constructed from the unstable state at the origin and the stable states at $\pm x^*$.

Problem 3.10 Determine the phase space trajectories of the frictionless bistable system characterised by (3.41).

We discuss the global phase space behaviour for the dissipative case in some detail. The character of the stable state is known to depend on the friction strength; we choose the case of weak friction ($\alpha/2 < \omega_0$).

The fixed points $(x^*, 0)$ and $(-x^*, 0)$ are spiral attractors, and the origin is a hyperbolic point. The shape of the trajectories in their vicinity can be obtained as discussed in the previous section (Fig. 3.13). There are stable and unstable curves in the neighbourhood of the origin (the hyperbolic point). The formulae $v = \lambda_\pm x$ for the asymptotes (the parameters $\lambda_\pm$ follow from (3.14) and (3.43)) are only valid in the immediate vicinity of the origin. Now we investigate how these curves are *continued* when moving away from the origin.

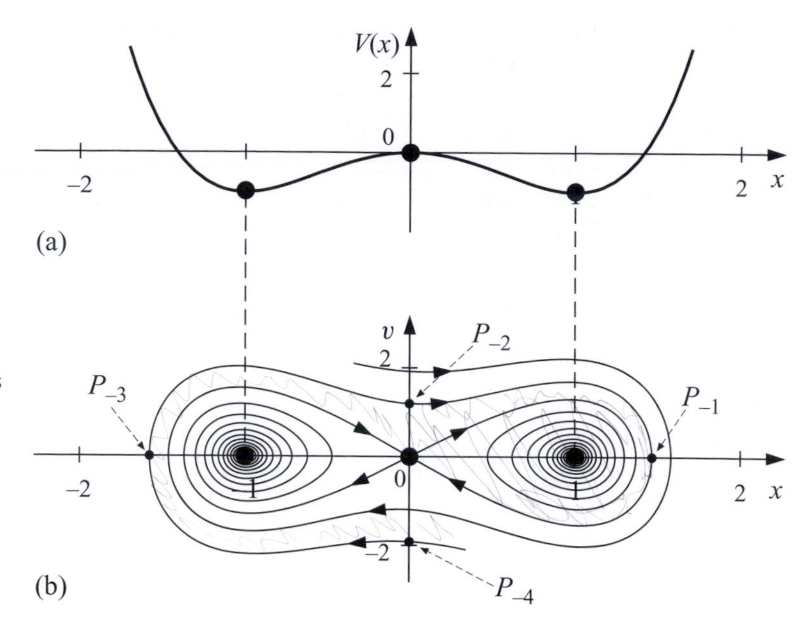

Fig. 3.16. (a) Potential of a bistable system and (b) its global phase space structure. The bottoms of the two wells each represent a spiral attractor, while the hill-top is a hyperbolic point. The trajectory of a body just reaching the hill-top is the stable manifold of this point. The trajectory of a body sliding down from the hill-top with nearly zero initial speed is the unstable manifold which leads to one of the two point attractors ($a = 4$, $x^* = 1$, $\alpha = 0.2$).

In the neighbourhood of the origin, it is the points of the stable curve $v = \lambda_- x$ that move towards the origin. There obviously exist points with this property further away as well: the locally straight segment continues as a bent curve. The set of all phase space points from which particles can reach the hyperbolic point is called the *stable manifold* of the hyperbolic point.

The manifold has *two* branches depending on the side (right or left) from which the origin is reached. The branch emanating to the right slowly bends upwards: the force becomes restoring after passing the bottoms of the wells, and therefore negative initial velocities of smaller and smaller absolute values are necessary for the body to reach the hill-top (Fig. 3.16). The curve then intersects the x-axis. The first point of intersection (P_{-1}) is the position from which a body starting with zero initial velocity slowly climbs up the hill. The stable manifold then bends backwards: a finite positive initial velocity is necessary for a particle thrust to the right to go high enough on the mountain side and to halt exactly on the hill-top after turning back from the mountain. The manifold continues further: the first intersection with the v-axis (P_{-2}) corresponds to the velocity with which a body has to be thrust to the right so that it later stops exactly at its initial position. The manifold then skirts the left well and intersects the negative x-axis (P_{-3}).

Problem 3.11 What is the meaning of the stable manifold's first intersection P_{-4} with the negative v-axis?

The branch emanating to the left is the centrally symmetric image of that emanating to the right. Both branches skirt the wells more and more times, while moving further and further away from the origin. The stable manifold is therefore a curve of *infinite length*.

The *unstable manifold* of the hyperbolic point can be defined in a similar way. This is the trajectory of a body left alone on the hill-top with a negligible initial velocity. It can also be considered as the set of points (x, v) from which the body reaches the hyperbolic point in the course of the time-reversed dynamics (a motion with initial condition (x, v) running in the negative time direction). The unstable manifold in the neighbourhood of the fixed point is a straight line segment whose orientation coincides with the unstable direction of the fixed point. The unstable manifold is thus a curve whose two branches leave the fixed point in opposite directions. Qualitatively speaking, the unstable manifold is the curve along which points from a tiny neighbourhood of the fixed point move away from the fixed point. It is evident that the two branches of the unstable manifold of the origin go into the two attractors, $(x^*, 0)$ and $(-x^*, 0)$. The manifold in the immediate vicinity of the fixed point attractors is a spiral corresponding to the phase space pattern of damped harmonic oscillations.

Two simple attractors co-exist in bistable systems. Trajectories from each point converge towards one of the attractors. The phase space can therefore be split into two parts, depending on the well in which the motion starting from a given point ends up. These are the *basins of attraction* of each of the two attractors. Using an analogy from geography, the basins of attraction are like the tributary basins of rivers. The boundary between the basins of attraction is called the *basin boundary*.

The two branches of the stable manifold of the origin split the phase space into two rolled-up bands: the basins of the two fixed point attractors. The stable manifold of the unstable hyperbolic point forms the basin boundary (Fig. 3.17). A phase space curve that separates trajectories corresponding to motions of different nature is called a *separatrix*. The whole stable manifold is obviously a separatrix. Continuing with our geographical analogy, the separatrix plays the role of a water-shed divide.

By drawing the stable and unstable manifolds of the hyperbolic points and a few other characteristic trajectories, an overview of the entire phase space and of the global dynamics can be obtained. This diagram is often called a *phase portrait*.

It is worth emphasising that not even in the case of this simple problem can the displacement vs. time function (or the trajectory) be described by simple formulae. We recommend that the reader writes a computer program to become acquainted with the dynamical properties.

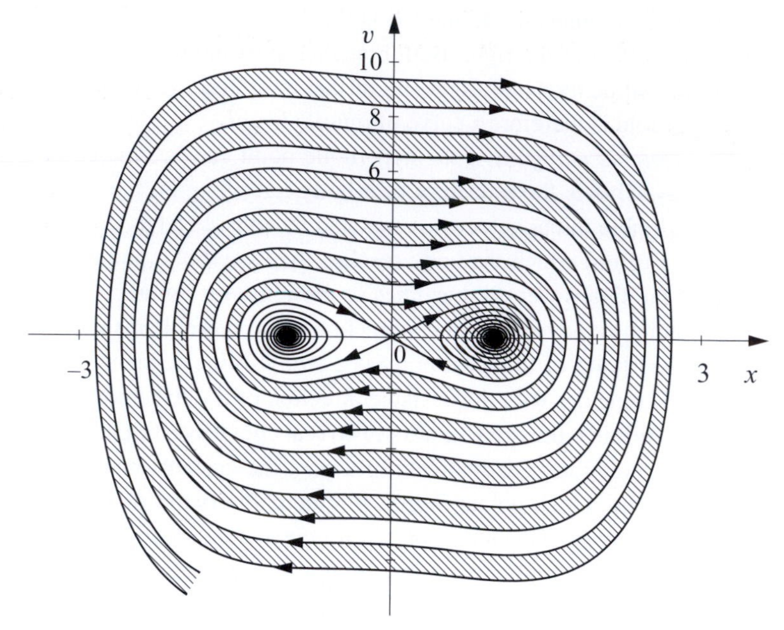

Fig. 3.17. Basins of attraction in a bistable system. The stable manifold of the origin splits the phase space into the basins of the right and the left attractors (shaded and white regions, respectively). The parameters are the same as in Fig. 3.16.

For those who are not familiar with the numerical solution of ordinary differential equations, it is advisable to study the Appendix, specifically Sections A.2, A.3, and A.4.2.

Problem 3.12 Determine the phase portrait of the motion under the force $F(x) = ax(x - c)\,(a > 0,\ c < 0)$, i.e. in the potential well $V(x) = acx^2/2 - ax^3/3$, in both the conservative and the dissipative cases. This force is used as a simple model of ship capsizing. In a strong, steady wind a tall ship tilts to the leeside and might capsize. At the same time, a certain position of the ship tilted to the winward side may be stable. The minimum of the potential well corresponds to this latter state, while its monotonous decrease for $x > 0$ corresponds to capsizing.

3.3.2 Bifurcation

It is interesting to follow the procedure outlining how an unstable state emerges in a system susceptible to bistability. The position co-ordinate is x, and some general parameter of the system, denoted by μ, is changed. It is assumed that the system is symmetrical around some co-ordinate x_0^* and that the force is everywhere attractive for relatively small values of μ. There exists therefore only one stable equilibrium state, at $x^* = x_0^*$. As μ is increased, the fixed point becomes less and less stable and loses its stability at some critical value μ_c. At the same time, two new stable fixed points appear outside of the centre, symmetrically to x_0^*.

Box 3.2 How to determine manifolds numerically

We briefly present a numerical algorithm to construct stable and unstable manifolds. After determining the hyperbolic point, form a small square or disc made up of tens of thousands of points around the fixed point. Since this object contracts in the stable and stretches in the unstable direction as time passes, this ensemble of points appears to be a thin line after a few multiples of the time $1/\lambda_-$, and traces out longer and longer segments of the unstable manifold (Fig. 3.18(a)). The stable manifold can be obtained from the same initial object in a similar way, but now the time step has to be set to a negative value in the program, corresponding to time-reversed dynamics (Fig. 3.18(b)).

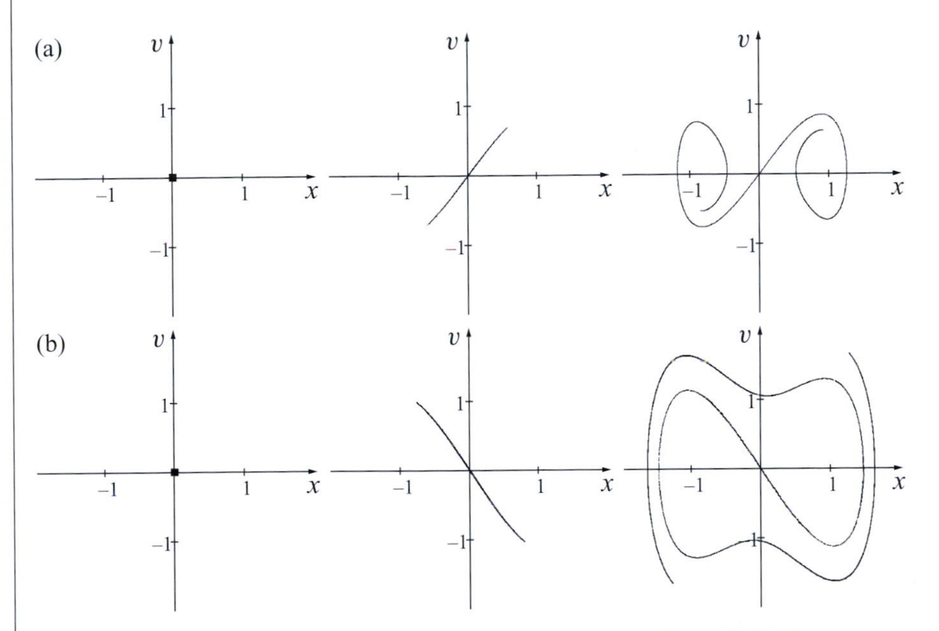

Fig. 3.18. Numerical construction of the unstable (a) and stable (b) manifolds of the origin in the bistable problem ($\alpha = 0.2$, $a = 2$, $x^* = 1$). The initial square of size 0.1 contains 40 000 uniformly distributed points. The figures represent the initial state and the patterns seen after two and six time units (chosen to be $1/\sqrt{b}$; see (3.43)).

Plotting the values of the possible equilibrium states, i.e. the fixed points, vs. μ, a typical structure appears (Fig. 3.19). Three curves are present for $\mu > \mu_c$, which deviate first rapidly and later more slowly from each other as μ increases. Together with the single straight line segment $x^* = x_0^*$ of the range $\mu < \mu_c$, they resemble a fork. This is why the whole process is called *bifurcation*. In the course of bifurcation, a state loses its stability and, simultaneously, new stable states are born. The critical point, $\mu = \mu_c$, is called a *bifurcation point*, and this symmetrical bifurcation process is called a *pitchfork bifurcation*.

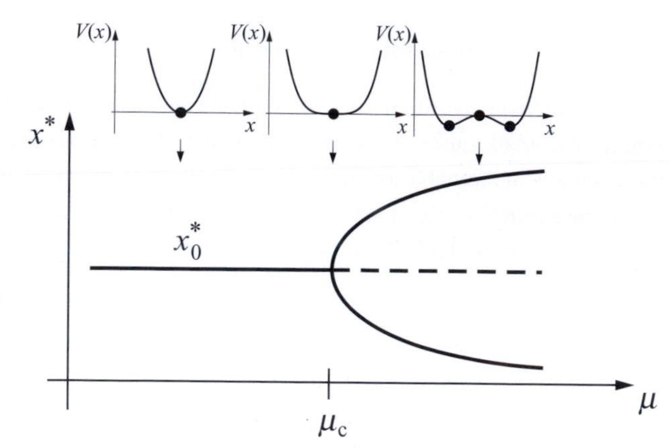

Fig. 3.19. Bifurcation diagram (schematic) of a pitchfork bifurcation. Co-ordinates x^* of the equilibrium states are given in terms of a parameter μ. The dashed line represents the unstable state emerging from a stable state. The shape of the potential is also plotted for selected values of the parameter.

The system characterised by the potential in (3.42) is a good model for pitchfork bifurcations if the parameter $\mu \equiv b$ can be changed in both the negative and positive domains. For $b < 0$, only the origin can be a fixed, stable, point. The bifurcation point is the value $b_c = 0$; beyond this the origin is unstable, but two stable fixed points are present at positions $x^* = \pm\sqrt{b/(2d)}$.

An unambiguous indicator that the instability is approaching is that the frequency of small oscillations around the stable state(s) *decreases*, i.e. the period *increases* when approaching the bifurcation point, regardless of the direction, because the potential becomes flatter and flatter. This phenomenon is called critical slowing down.[10]

Problem 3.13 A body of unit mass is fixed to the ends of two springs of length l_0 and force constant k acting from opposite directions (Fig. 3.14(a)). The body can only move along a horizontal line. The other ends of the springs are fixed outside this straight line, at the same distance, h. Describe the bifurcation process in terms of $\mu \equiv h$. Show that in the vicinity of the bifurcation point the force law is equivalent to (3.41).

Problem 3.14 Determine the frequency, ω_0, of small oscillations around the stable states on both sides of the bifurcation point in Problem 3.13.

Problem 3.15 Determine the bifurcation diagram of a body moving along a rotated ring (Fig. 3.14(b)) in terms of the angular velocity, $\mu \equiv \Omega$, of the rotation.

[10] A pitchfork bifurcation is the analogue of a second-order phase transition of thermodynamics. The best known example of this is magnetic ordering, which appears spontaneously below a critical temperature (no permanent magnetisation is present above this temperature).

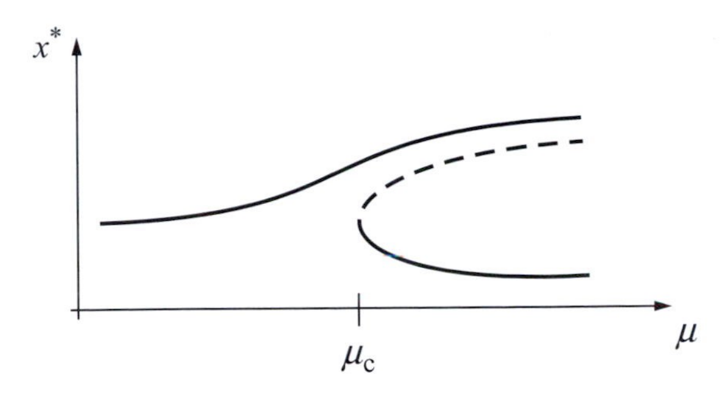

Fig. 3.20. Bifurcation diagram of a distorted pitchfork bifurcation (schematic). In asymmetrical systems, at a value μ_c of the parameter, a second stable equilibrium state arises abruptly, together with an unstable one (dashed line).

Problem 3.16 Work out the equation that determines the equilibrium states of a magnetic pendulum (Fig. 3.14(c)). For simplicity, assume that the force is of Coulomb type and that the pendulum is so long that the motion occurs almost in the horizontal plane, the magnets being at distance d below this plane. What is the condition for an instability?

In systems without symmetry, bifurcations occur in a different way. In such cases the stable state does not cease as the parameter μ increases, but rather two new equilibrium solutions appear above a critical value, μ_c: one stable and one unstable (Fig. 3.20). When decreasing the parameter μ from large values, one of the stable equilibrium states suddenly disappears. A state belonging previously to this branch changes abruptly.[11] Therefore, this phenomenon is sometimes also called a *catastrophe*. The whole bifurcation process is a *distorted* pitchfork bifurcation.

In the course of bifurcations, not only the equilibrium positions are modified, but also the complete phase portrait. Accordingly, the character of the entire dynamics *changes fundamentally*.

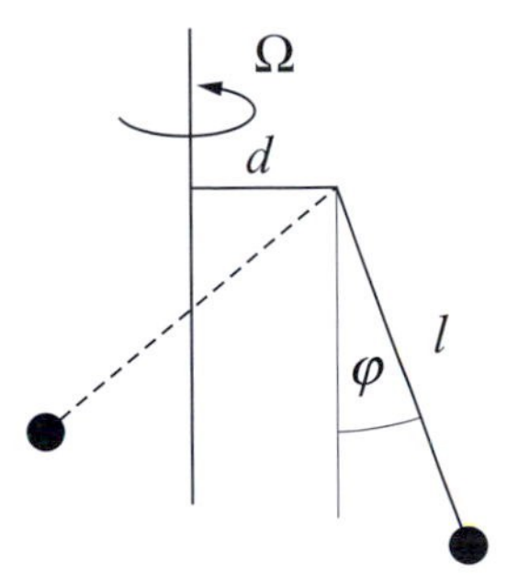

Problem 3.17 Consider a pendulum hanging on a horizontal rod of length d, rotating at angular velocity Ω around a vertical axis (imagine a model of a merry-go-round). Determine the equation yielding the stationary angles of deviation, φ. Show that for a small length, d, of the rod, the system undergoes a distorted pitchfork bifurcation in terms of the angular velocity $\mu \equiv \Omega$.

Problem 3.18 Consider the system shown in Fig. 3.14(a) tilted from the horizontal by a small angle of inclination, α. Show that a distorted pitchfork bifurcation occurs in the system when changing parameter h, which should be identified with μ in this case.

[11] This is a phenomenon analogous to a first-order phase transition of thermodynamics.

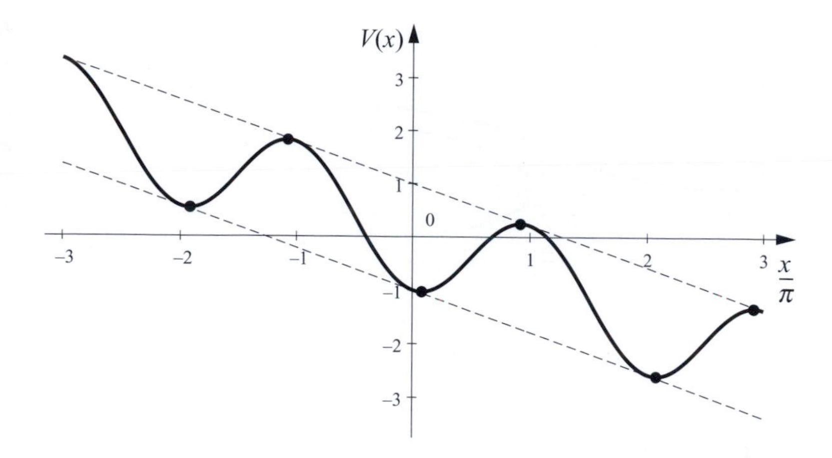

Fig. 3.21. Potential of the motion on a bumpy slope, (3.47), with parameters $A = 1$, $F_0 = 0.25$.

3.4 Sustained periodic motion: the limit cycle (skiing on a slope)

Problem 3.19 Determine the phase portrait of a weakly damped motion under the periodic force $F(x) = -A \sin x$, $A > 0$, whose potential is $V(x) = -A \cos x$. According to the qualitative interpretation of potentials (Section 3.1.1), the problem corresponds to a motion on a wavy surface of period 2π, or on a bumpy road.[12]

Problem 3.20 Work out the phase portrait of Problem 3.19 in the frictionless limit.

We examine a model which describes sliding on a bumpy slope. The force is given by

$$F(x) = -A \sin x + F_0, \tag{3.45}$$

which corresponds to the potential

$$V(x) = -A \cos x - F_0 x. \tag{3.46}$$

Note that $F_0 > 0$ is a parameter of the slope, the force due to the tilt (Fig. 3.21). The dissipative case described by the equation of motion,

$$\ddot{x} = -\alpha \dot{x} - A \sin x + F_0, \tag{3.47}$$

can also be considered as a model of skiing downhill on a bumpy slope, assuming that the drag is proportional to the velocity.[13]

[12] By means of the replacement $x \to \varphi$, the pendulum problem is recovered with φ as the angular deviation. Amplitude A has then to be identified with the ratio of gravitational acceleration, g, to the length, l, of the pendulum.

[13] By means of the replacement $x \to \varphi$, we obtain the problem of a pendulum subjected to a constant torque. The equation of motion also describes the dynamics of the

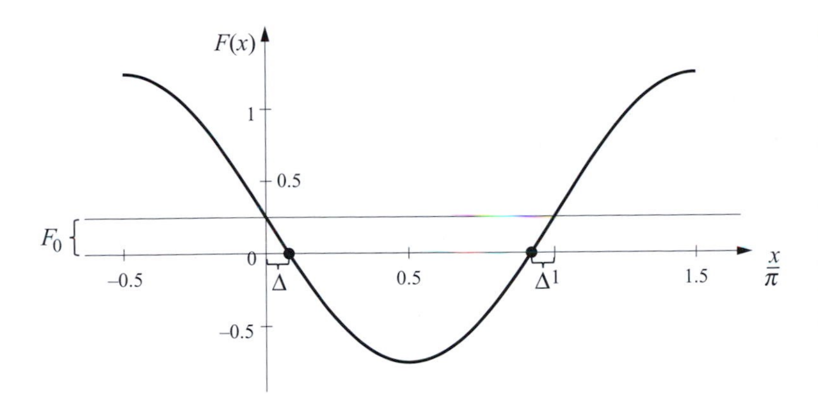

Fig. 3.22. Equilibrium positions along a bumpy slope. The spatially periodic force and the interpretation of the shift Δ ($A = 1$, $F_0 = 0.25$). Remember that fixed points where $F(x)$ decreases and increases are stable and unstable, respectively (cf. Fig. 3.13).

The equilibrium positions, x^*, follow from $F(x^*) = 0$. They are determined by the condition

$$\sin\Delta = F_0/A \tag{3.48}$$

(Fig. 3.22), where $\Delta > 0$ yields the shift of the fixed point from $n\pi$. Since the absolute value of the sine function can only be less than unity, the above equation has a solution only if

$$F_0 \leq A. \tag{3.49}$$

There exists then a set of solutions $x^*_+ = \Delta + 2\pi n$; i.e., the equilibrium states at the bottoms of the dips on the slope. Shifting the points $\pi + 2\pi n$ to the left by this same Δ yields the values $x^*_- = \pi - \Delta + 2\pi n$, the location of the unstable equilibrium states corresponding to local maxima (Fig. 3.22). (If condition (3.49) is not fulfilled, the slope is so steep that no equilibrium positions exist and the skier slides down from everywhere.)

Problem 3.21 Determine the phase portrait of the motion on a bumpy slope in the frictionless limit.

For weak friction, the points x^*_+ are spiral attractors, whereas the points x^*_- are hyperbolic. The left branches of the unstable manifolds of the latter points lead into the nearest spiral attractors (Fig. 3.23).

As long as the friction is not too strong, and for sufficiently steep slopes, a body starting with a negligible initial velocity from the top of a bump to the right not only reaches the next top, but also passes over it with a finite velocity. From then on it passes over all the other bumps. However, its velocity cannot increase arbitrarily, since friction

potential difference occurring in so-called 'Josephson junctions'. Finite velocity corresponds to finite potential drop.

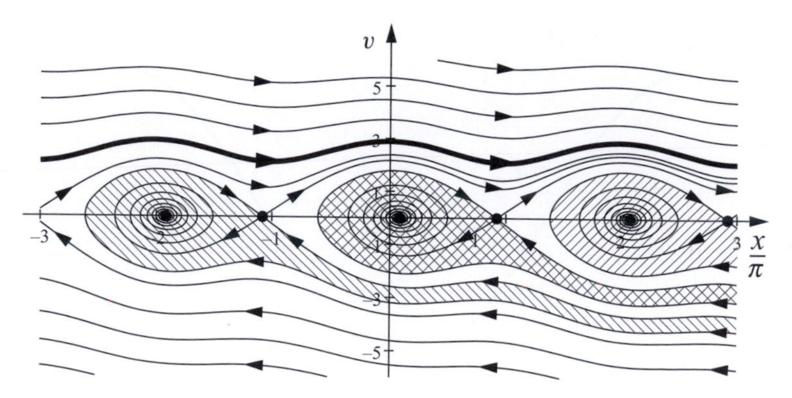

Fig. 3.23. Phase portrait of motion on a bumpy slope with weak friction ($A = 1$, $F_0 = 0.25$, $\alpha = 0.1$). Due to friction, one of the branches of each unstable manifold leads into the attractor in the left hole; the other, however, converges to an undulating curve representing downhill sliding with a bounded speed, i.e. the limit cycle (thick line). The basin of attraction of the limit cycle extends not only into the positive, but also into the negative half-plane. The stable manifolds trace out the basin boundaries of the spiral attractors distinguished by different shading.

increases with increasing velocity. Thus, after a sufficiently long time, a *sustained periodic motion* develops. This cannot be of constant velocity, since the motion has to be faster in the holes than over the bumps. In the analogy with skiing this corresponds to a stationary downhill sliding with pulsating velocity.[14] Such types of motion are represented by a new type of attractor in phase space whose trajectory is a periodic curve. Such a periodic attractor is called a *limit cycle*. Contrary to the point attractors discussed so far, the limit cycle represents a motion that does not cease but remains sustained. This periodic motion develops due to a constant input of energy, ensured in our case by gravitation. On the limit cycle, the average amount of energy consumed by friction is exactly balanced by the decrease of the gravitational potential of the skier. Both point and limit cycle attractors belong to the class of simple attractors.

The limit cycle can be reached both from above (from the direction of large velocities) and from below. The former corresponds to the fact that fast skiers lose speed until they reach the motion prescribed by the limit cycle. On the other hand, the velocity of a skier starting slowly from the top of a bump gradually increases. The right branches of the unstable manifolds thus tend towards the limit cycle.

[14] For a pendulum subjected to a constant torque this implies sustained rotation, while for the Josephson junction this corresponds to the appearance of a sustained potential difference.

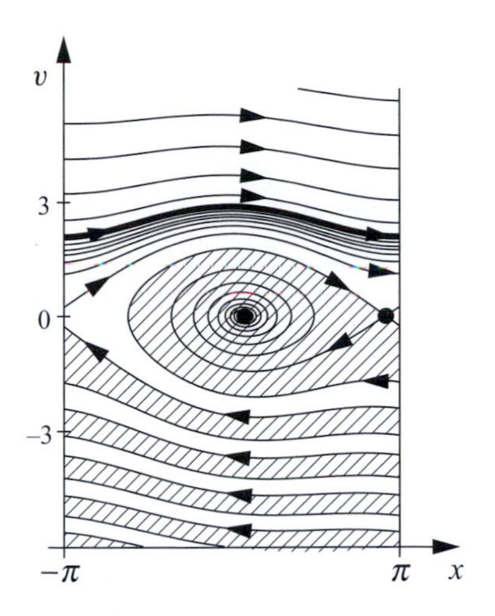

Fig. 3.24. Phase portrait of motion on a bumpy slope in a periodic representation. The lines exiting at position π enter at position $-\pi$ with the same velocity and vice versa. The basins of attraction of the limit cycle (thick line) and of the spiral attractor (black dot) are white and shaded, respectively.

The left branch of the unstable manifold of a hyperbolic point leads into the left neighbouring spiral attractor, which describes coming to a halt in the hole left of the bump. There is therefore an infinity of point (spiral) attractors in co-existence with a limit cycle. The stable manifolds of the hyperbolic points form the basin boundary of the point attractor to their left. Points departing outside these domains all converge to the limit cycle (see Fig. 3.23).

It should be mentioned that, due to the periodicity, identical phase portraits belong to each x interval of length 2π. If we do not wish to distinguish the point attractors from each other, then the whole problem can be analysed on the interval $(-\pi, \pi)$, i.e. it can be restricted to a spatial period. This corresponds to 'rolling up the phase space on a cylinder',[15] as shown in Fig. 3.24. In this representation the spiral attractor reflects that the body comes to a halt somewhere. The trajectories become dense around the limit cycle, illustrating its attracting nature.

Problem 3.22 Find out how the phase portrait of motion on a bumpy slope changes as the friction coefficient increases (as the snow becomes more and more wet).

3.5 General phase space

3.5.1 General definition of phase space

Processes taking place in continuous time are described by means of differential equations. Even if in their original forms these are of higher

[15] This is especially plausible in the case of a pendulum subjected to a constant torque, whose position is defined by the angular deviation φ, which is 2π-periodic.

order (including higher than first-order time derivatives), they can be transformed into a set of *first-order* differential equations by introducing new variables (for example by considering the first derivative as a new unknown). Autonomous differential equations are those that do not explicitly contain time. Consider the autonomous system given by

$$\boxed{\dot{\mathbf{x}} = \mathbf{f}(\mathbf{x}),} \tag{3.50}$$

where $\mathbf{x} \equiv (x_1, x_2, \ldots, x_n)$ is an n-dimensional vector of the unknowns, and the functions $\mathbf{f} \equiv (f_1, f_2, \ldots, f_n)$ are time-independent.[16] Motion of the system implies the time evolution of n independent co-ordinates which can be uniquely represented on the axes of an n-dimensional co-ordinate system. The latter spans the phase space of the problem.

An instantaneous state of the system is represented by a phase space point, $\mathbf{x}$. As the state changes, the point moves and traces out a trajectory.

In the *overwhelming majority* of systems investigated in physical sciences, or more generally in natural sciences (and in economics), the functions f_i change smoothly, i.e. they are differentiable and the derivatives are finite. In the following, differential equations will be understood to be equations with this property. For such equations, given initial conditions always lead to *unique* solutions. This implies that phase space trajectories *cannot intersect* (otherwise two different solutions would belong to the initial condition corresponding to an intersection). The behaviour in phase space is similar to the motion of a fluid, where particle paths cannot cross each other. This is why the vector $\mathbf{f}$ is often said to define a *flow* in phase space. Trajectories can 'collide' at a hyperbolic point, but even there they do not intersect: points belonging to the two branches of the stable manifold tend towards each other, but touch only after an infinitely long time. A system is called *simple* if its dynamics is given in terms of a few variables, i.e. if its phase space is *low-dimensional* (in practice this means a maximum of four to ten variables).

In the phase space representation, time is not present explicitly, and phase space, in principle, contains all possible motions. A single trajectory in itself corresponds to an infinity of individual motions, i.e. all those whose initial condition is a point of the trajectory (see Table 1.2).

In a two-dimensional flow, the dynamics is of the form

$$\dot{x_1} = f_1(x_1, x_2), \quad \dot{x_2} = f_2(x_1, x_2), \tag{3.51}$$

where f_1 and f_2 are arbitrary functions.

[16] Nonautonomous, driven cases are discussed in Chapter 4.

For motion occurring along a straight line of the type investigated so far, the equation of motion (3.37) can be written, by introducing the notation $x \equiv x_1$, $v \equiv \dot{x} \equiv x_2$, in the form of (3.51) with

$$f_1(x_1, x_2) = x_2, \quad f_2(x_1, x_2) = F(x_1) - \alpha x_2. \tag{3.52}$$

Even though mathematically this is not the most general form, the most important types of dynamical behaviour can all be observed within this Newtonian framework.

3.5.2 Dynamics of phase space volumes

The use of phase space raises a question that does not arise in the traditional one-particle approach. How does some subset of the phase space move, and how does its volume change in the course of the motion? By motion of the phase space volume we mean that an arbitrary connected volume is filled continuously with points (initial conditions) and the position of this set of points is recorded in the phase space after time t. Consequently, tracing the motion of a phase space volume corresponds to observing an *ensemble* of particles. The advantage of this method is that, even if certain initial conditions are not typical, the ensemble will show some kind of average behaviour. Investigating the time evolution of phase space volumes is therefore a step towards a statistical description of the dynamics.

Naturally, the shape of an initial object changes in time, and, usually, so does the volume. The rate of change of the phase space volume is easy to determine. Consider a planar problem and a *small* rectangle around point (x_1, x_2). The dimensions of the rectangle along the x_1- and x_2-axes are $\Delta x_1 \equiv x_1' - x_1$ (Fig. 3.25) and $\Delta x_2 \equiv x_2' - x_2$, respectively.

According to (3.51) and Fig. 3.25, the temporal change of the sides is $\Delta \dot{x}_1 = (\partial f_1/\partial x_1)\Delta x_1$ and $\Delta \dot{x}_2 = (\partial f_2/\partial x_2)\Delta x_2$. The time derivative of the phase space volume, $\Gamma = \Delta x_1 \Delta x_2$, is thus given by

$$\dot{\Gamma} = \Delta \dot{x}_1 \Delta x_2 + \Delta \dot{x}_2 \Delta x_1 = \left(\frac{\partial f_1}{\partial x_1} + \frac{\partial f_2}{\partial x_2}\right)\Gamma. \tag{3.53}$$

The phase space *contraction rate*, σ, defined by

$$\dot{\Gamma} = -\sigma \Gamma, \tag{3.54}$$

is thus given by

$$\boxed{\sigma = -\left(\frac{\partial \dot{x}_1}{\partial x_1} + \frac{\partial \dot{x}_2}{\partial x_2}\right) \equiv -\left(\frac{\partial f_1}{\partial x_1} + \frac{\partial f_2}{\partial x_2}\right) \equiv -\mathrm{div}\,\mathbf{f}.} \tag{3.55}$$

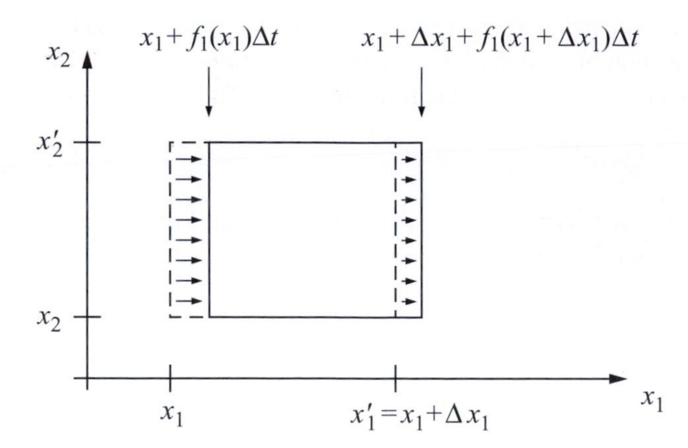

Fig. 3.25. Change of the phase space volume for a small rectangular object in motion occurring along the x_1-axis only. The left and right edges move to positions $x_1 + f_1(x_1)\Delta t$ and $x_1' + f_1(x_1')\Delta t$, respectively, in time Δt. The change in the base length is therefore $(f_1(x_1') - f_1(x_1))\Delta t \approx (\partial f_1/\partial x_1)\Delta x_1 \Delta t$, yielding a rate of change of $\Delta \dot{x}_1 = (\partial f_1/\partial x_1)\Delta x_1$. A similar relation can be derived for the rate of change of the height in motion along the x_2-axis.

The contraction rate is therefore determined by the divergence of the vector $\mathbf{f}$ defining the flow. This result is a generalised form of *Liouville's theorem*. It can be seen from the derivation that the statement is valid for phase spaces of arbitrary dimensions. In a general flow, σ depends on the position in phase space, and its sign is not necessarily constant. Negative σ corresponds to the expansion of the phase space volume.

In the Newtonian system, (3.52), the phase space contraction rate is

$$\sigma = \alpha. \tag{3.56}$$

This implies that only friction, i.e. dissipation, can cause the phase space volume to change. Since α is constant, in the dynamics studied so far, ((3.37) and (3.52)), the phase space contraction rate is independent of the position. According to (3.54), the phase space volume, Γ, then changes exponentially as $\Gamma(t) = \Gamma_0 e^{-\alpha t}$, which implies a fast decrease. Within a few multiples of the decay time, $1/\alpha$, the phase space volume contracts almost to zero (Fig. 3.26). Attractors are therefore always objects of zero phase space volume.

The contraction of the phase space volume is related to irreversibility, and the basic law underlying irreversibility is the *second law* of thermodynamics, which determines the direction of the time evolution of macroscopic systems. An increase of the phase space volume would

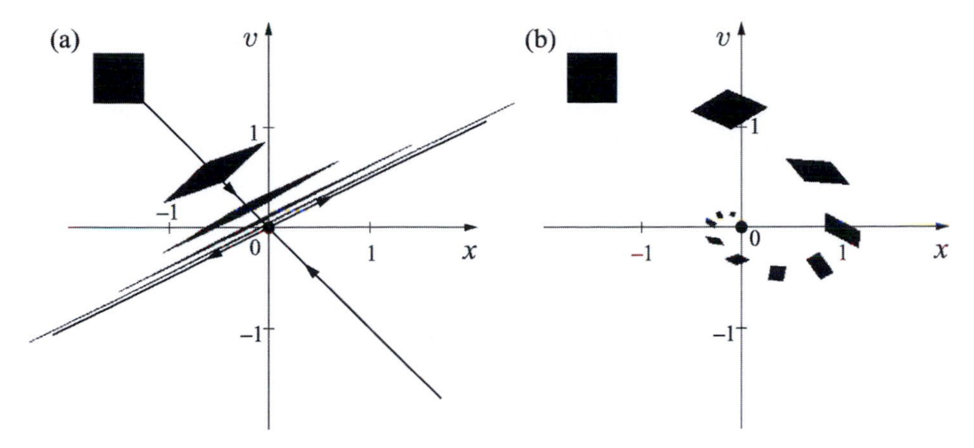

Fig. 3.26. Decrease of phase space volume around (a) a hyperbolic point and (b) a spiral attractor ($\omega_0 = s_0 = 0.7$, $\alpha = 0.5$). In both cases the initial square of size 0.5 is centred at (1.5, −1.5), and its position is presented in every time unit.

only be possible with negative friction. In isolated systems this is impossible due to the second law.[17]

An important special class of motion is that of *conservative* systems, where the phase space volume remains *unchanged*, i.e. Γ = constant. According to (3.56), friction then has to vanish in (3.37), $\alpha = 0$, and $\sigma \equiv 0$ (Liouville's theorem). This, however, does not exclude a significant change in the shape of the phase space object, while leaving its volume unchanged. Lack of friction also means that the total mechanical energy, $\mathcal{E} = v^2/2 + V(x)$, is conserved in time.

The fact that an unstable state in phase space is related to the appearance of hyperbolic points is a consequence of the contraction (or preservation) of the phase space volume. Stretching along the unstable direction is only possible if contraction takes place along the stable one.

Problem 3.23 Determine the phase space contraction rate of the flow (the van der Pol oscillator) defined by the following equations:

$$\dot{x}_1 = x_2, \quad \dot{x}_2 = -x_1 - \alpha(x_1^2 - 1)x_2.$$

[17] Expanding phase space volumes can only occur in exceptional cases when an energy flux originating from its environment is flowing through the system. Then, the state of the system deviates significantly from thermal equilibrium, so much that (at least in certain locations) σ can become negative. Such cases can, of course, only be described by equations different from (3.37).

3.5.3 Time reversal, invertibility

In differential equations describing time evolution, the time arrow can be reversed. Processes occurring in the negative time direction are also *uniquely* determined by the initial conditions. The dynamics is therefore called *invertible*. The inverted form of the general equation (3.50) is given by

$$\dot{\mathbf{x}} = -\mathbf{f}(\mathbf{x}). \tag{3.57}$$

In numerical solutions, this simply corresponds to choosing negative time steps. The phase space contraction rate of the inverted dynamics is the opposite of the original, i.e. $-\sigma$. Time reversal implies that the functions $x_i(t)$ are read backwards between given starting and end-points. Consequently, the shape of the corresponding phase space trajectory does *not* change; only the arrow-head has to be reversed. Phase space therefore contains trajectories of *both* the original and the time-reversed dynamics. We emphasise that invertibility does *not* imply reversibility in the thermodynamical sense. All dissipative processes are irreversible since the system always converges towards an attractor as its ultimate state. Nevertheless, the inverted dynamics (3.57) exists, and describes escape from any small neighbourhood of the attractor of the direct dynamics.

The inverted version of the frictional motion (3.37) is given by

$$\ddot{x} = F(x) + \alpha\dot{x}. \tag{3.58}$$

Here, friction does not slow down the body, rather it speeds it up. This illustrates that although invertibility ensures that the time-reversed motion is uniquely determined, it does not necessarily imply that the time-reversed motion is physically realistic.

In the Newtonian scheme (3.37), it is usual to define time reversal along with a change of sign of the velocity. The corresponding transformation is given by

$$t \to -t, \quad x \to x, \quad v \to -v. \tag{3.59}$$

This implies that besides changing the arrow-head, trajectories are mirrored on the x-axis.

Systems where time-reversed motions are all real motions are called *time-reversal invariant* systems. The motion of such systems, if recorded on a videotape and played backwards, appears to be as realistic as the original; i.e., they are reversible motions in the thermodynamical sense as well. Time-reversal invariance implies a vanishing phase space contraction rate, $\sigma = 0$. The equation of motion, (3.37), with $\alpha = 0$ is invariant under the transformation (3.59). This brings about a new symmetry in phase space: trajectories are *symmetrical* to the x-axis (see, for example, Problems 3.10 and 3.20).

3.5.4 Fixed points and their stability in general two-dimensional flows

The fixed points are the points $\mathbf{x}^*$, where $\dot{\mathbf{x}}^* = \mathbf{0} = \mathbf{f}(\mathbf{x}^*)$, i.e. the points at which the flow vanishes. Suppose we have found some fixed point (x_1^*, x_2^*) of equation (3.51). (In practice, this is not always simple since a non-linear set of equations, $f_1(x_1^*, x_2^*) = 0$, $f_2(x_1^*, x_2^*) = 0$, is to be solved.) The fixed point implies a time-independent state, but typically none of the co-ordinates vanish, since they do not carry the meaning of velocity.

The stability of a fixed point can be determined from the equations valid in the immediate neighbourhood of the point. Assume that the difference $\mathbf{x} - \mathbf{x}^*$ is small, such that it is sufficient to keep only the linear, leading terms. The linearised form of equation (3.51) can be given by

$$(x_1 \dot{-} x_1^*) = a_{11}(x_1 - x_1^*) + a_{12}(x_2 - x_2^*),$$
$$(x_2 \dot{-} x_2^*) = a_{21}(x_1 - x_1^*) + a_{22}(x_2 - x_2^*). \tag{3.60}$$

Using the vector notation $\Delta\mathbf{x} = (x_1 - x_1^*,\ x_2 - x_2^*)$, this linear set of equations becomes

$$\dot{\Delta\mathbf{x}} = A\Delta\mathbf{x}, \tag{3.61}$$

where A is a matrix:

$$A = \begin{pmatrix} a_{11} & a_{12} \\ a_{21} & a_{22} \end{pmatrix} = \begin{pmatrix} \partial f_1/\partial x_1 & \partial f_1/\partial x_2 \\ \partial f_2/\partial x_1 & \partial f_2/\partial x_2 \end{pmatrix}. \tag{3.62}$$

The elements of this *stability matrix* are the partial derivatives, $a_{ij} = \partial f_i/\partial x_j$, evaluated at the fixed point.

With a trial solution of (3.60) in the form $\Delta x_i = u_i e^{\lambda t}$ $(i = 1,\ 2)$, we find that λ has to fulfil the following set of equations:

$$\lambda u_1 = a_{11}u_1 + a_{12}u_2, \quad \lambda u_2 = a_{21}u_1 + a_{22}u_2; \tag{3.63}$$

i.e., λ is an *eigenvalue* of matrix A, since $A\mathbf{u} = \lambda\mathbf{u}$. The condition for the existence of a non-trivial solution of this set of homogeneous linear equations is as follows:

$$\lambda^2 - \lambda \mathrm{Tr}\, A + \det A = 0. \tag{3.64}$$

Here, $\mathrm{Tr}A \equiv a_{11} + a_{22}$ denotes the *trace* of the matrix and $\det A \equiv a_{11}a_{22} - a_{12}a_{21}$ is its determinant. This equation has two different complex solutions, $\lambda_\pm$, whose sum is the trace. On the other hand, the sum

Table 3.1. *Classification of the fixed points of two-dimensional flows according to the real and imaginary parts of $\lambda_\pm$.*

	$\text{Re}\lambda_\pm = 0$	$\text{Re}\lambda_\pm < 0$	$\text{Re}\lambda_- < 0 < \text{Re}\lambda_+$	$\text{Re}\lambda_\pm > 0$
$\text{Im}\lambda_\pm = 0$	(marginal)	node attractor	hyperbolic point	node repellor
$\text{Im}\lambda_\pm \neq 0$	elliptic point	spiral attractor	–	spiral repellor

is the divergence of the linearised equation (3.60); thus,

$$\lambda_+ + \lambda_- = -\sigma \tag{3.65}$$

holds in general. In other words, the negative sum of the eigenvalues is the phase space contraction rate in the vicinity of the fixed point. For conservative systems, $\lambda_+ = -\lambda_-$. An exceptional special case is $\lambda_+ = \lambda_- = 0$. The fixed point is then called *marginal* since it is neither stable nor unstable (the dynamics around it can only be determined from a non-linear analysis).

The nature of a non-marginal fixed point follows from the properties of the eigenvalues (in the following list, the 'pictograms' of the fixed points are also given):

Hyperbolic point: $\lambda_\pm$ is real, $\lambda_+ > 0 > \lambda_-$. The instability exponent defined in Section 3.1.1 is the positive eigenvalue, λ_+, of the general linear problem.

Elliptic point: $\lambda_\pm$ is purely imaginary (this is only possible in conservative cases, with $\sigma = 0$). The natural frequency of Section 3.1.2, ω_0, is the absolute value of the eigenvalues of the general linear problem $\lambda_\pm = \pm i\omega_0$.

Node attractor: $\lambda_\pm$ is real and negative ($\sigma > 0$).

Spiral attractor: $\lambda_\pm$ is complex with real part $\text{Re}\lambda_\pm < 0\,(\sigma > 0)$.

Node repellor:[18] $\lambda_\pm$ is real and positive (only possible in expanding phase spaces with $\sigma < 0$).

Spiral repellor: $\lambda_\pm$ is complex and $\text{Re}\lambda_\pm > 0$ (only possible in expanding phase spaces with $\sigma < 0$).

Repelling fixed points are also listed since the fixed point attractors of a problem always appear as fixed point repellors in the inverted dynamics. For the sake of perspicuity, the different cases are summarised in Table 3.1.

[18] A repellor is a repelling object in phase space.

Nodes and spiral fixed points can only occur in the presence of dissipation, and in such cases they are necessarily attractors. Only hyperbolic points can appear both in conservative and in dissipative systems (even in expanding phase spaces), which illustrates their robustness.

The general solution of (3.60) is always a linear combination of two exponentials with the eigenvalues as the exponents:

$$\mathbf{x}(t) = c_+\mathbf{u}_+e^{\lambda_+ t} + c_-\mathbf{u}_-e^{\lambda_- t}. \tag{3.66}$$

The vectors $\mathbf{u}_\pm$ are the (right) *eigenvectors* of matrix A: $A\mathbf{u}_i = \lambda_i\mathbf{u}_i$ ($i = \pm$). Eigenvectors are the vectors that are only stretched, not rotated, by matrix A, with the stretching rate as the eigenvalue. The coefficients $c_\pm$ are determined by the initial condition, $\mathbf{x}_0 = c_+\mathbf{u}_+ + c_-\mathbf{u}_-$. Even though the eigenvalues and eigenvectors can be complex numbers, (3.66) is always real. A motion described by a linear equation can therefore be decomposed into independent components determined by the eigenvalues and eigenvectors. If the eigenvalues (and therefore the eigenvectors) of the stability matrix, A, are real, the eigenvectors determine directions in phase space. A *single* exponential describes time evolution along these directions. The trajectories *asymptote* to that eigenvector direction which belongs to the greater eigenvalue. In the particular case of a hyperbolic point, the *stable* and *unstable* directions coincide with the eigenvectors belonging to the *negative* and *positive* eigenvalues, respectively. Similarly, the directions of weak and strong attraction of a node coincide with those belonging to the greater and smaller (now both negative) eigenvalues, respectively.

Problem 3.24 Determine the stability matrix, A, of the mechanical problem (3.52) linearised around a fixed point, and show that it yields the same types of fixed points in the above general classification as those obtained in Sections 3.1.1 and 3.1.2.

Problem 3.25 Analyse the stability of the linear problem defined by the matrix

$$A = \begin{pmatrix} 1/2 & 1 \\ a & -2 \end{pmatrix} \tag{3.67}$$

in terms of parameter a.

Problem 3.26 Determine the nature of the fixed point in different regions of the parameter plane defined by det A and σ for non-negative phase space contraction rates: $\sigma \geq 0$.

3.5.5 Phase portraits of general two-dimensional flows

The first step in unfolding the dynamics, (3.51), is to determine its fixed points and their stability. We already know the trajectories characterising

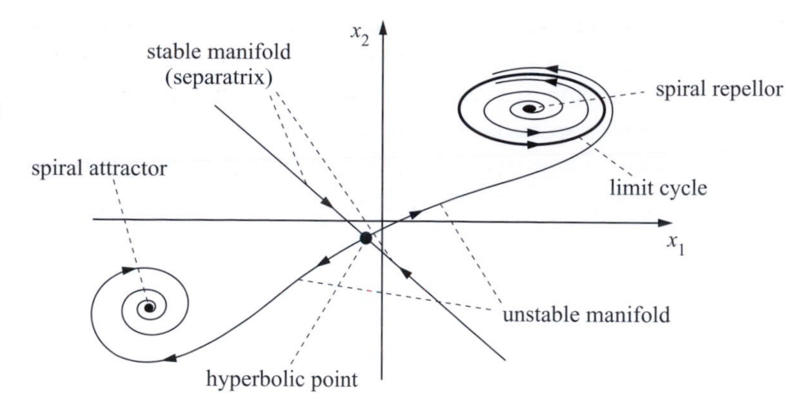

Fig. 3.27. Schematic phase portrait of a general two-dimensional dissipative system. The most complicated possible attractor in a plane, a limit cycle, is represented by a closed curve.

the immediate vicinity of the fixed points (see Sections 3.1.1, 3.1.2 and 3.5.4). Using these as building blocks, a complete phase portrait can be constructed. The stable and unstable manifolds of the hyperbolic points play an especially important role. Both manifolds can be infinitely long curves extending over large domains of phase space. The properties observed in our previous examples are, in general, valid in dissipative systems, the stable manifold always separates different attractors, and it is the basin boundary; the unstable manifold, on the other hand, traces out the way towards the attractors.

The *Poincaré–Bendixson theorem* states that in two-dimensional phase spaces of dissipative systems, the only option for an attractor (besides fixed points) is to be a limit cycle (Fig. 3.27). This property expresses the fact that the phase plane is not spacious enough to allow the appearance of complex trajectories. Such trajectories would have to intersect, which is never possible in phase space. The most complicated attractors of two-dimensional phase spaces are *limit cycles*.

The Poincaré–Bendixson theorem is quite plausible. It is, however, surprising (and cannot be proved mathematically, only demonstrated by means of numerical simulations) that as soon as a *single* dimension is added, the phase space becomes spacious enough to support intricate chaotic dynamics.

In two-dimensional phase spaces there is no need for an extra search for limit cycle attractors, since they can be obtained by simply tracking the unstable manifolds.[19] A qualitative knowledge of these manifolds yields an overall geometrical view of the dynamics, or even of its parameter-dependence, without having to solve the problem in detail. This approach will also be useful in studying chaotic motion.

[19] Or the trajectories emanated from repellors, if there are any.

Problem 3.27 Determine the phase portrait of the system $\dot{x}_1 = x_2$, $\dot{x}_2 = -\alpha x_2 + x_1 - x_1^2 + x_1 x_2$ $(\alpha > 0)$ by means of a numerical simulation of the manifolds of the hyperbolic point. Investigate the dependence on parameter α.

Problem 3.28 By numerically tracking the trajectories around the fixed point, determine the phase portrait of the van der Pol oscillator $\dot{x}_1 = x_2$, $\dot{x}_2 = -x_1 - \alpha(x_1^2 - 1)x_2$ $(\alpha > 0)$.

Chapter 4
Driven motion

The environment of simple systems is often not constant in time and it can influence the system in a periodically changing manner. Think of a body driven by a motor, or of the daily or annual cycle of our natural environment. The change in the environment leads to a *driving* of the system. We study the effect of an external, periodic driving force on a single point mass moving along a straight line. As a consequence of driving, the dimension of the phase space increases. In order to preserve an easy, two-dimensional, visualisation of the motion, it is worth introducing the concept of *maps*. Our aim is to show how different types of motion can be monitored by means of maps. Limit cycles appear as fixed points, limit cycles, corresponding to hyperbolic points in maps, and we formulate their stability conditions. Hyperbolic limit cycles, corresponding to hyperbolic points in maps, and curves emanated from them, the stable and unstable manifolds, constitute the skeleton of the possible motion in maps. We show that continuous time equations of motion can only lead to maps with certain well defined properties. Finally, we formulate what types of systems are candidates for exhibiting chaotic behaviour.

4.1 General properties

4.1.1 Equation of motion

The driving force, besides having an explicit time dependence, might also depend on the instantaneous position.[1] The general form of the

[1] It may also depend on the velocity, but we do not consider these cases.

equation of motion is given by

$$\ddot{x} = F(x) - \alpha\dot{x} + F_d(x, t), \tag{4.1}$$

where the driving force, F_d, is periodic with some period T:

$$F_d(x, t + T) = F_d(x, t). \tag{4.2}$$

Since we consider quantities per unit mass, F_d is the acceleration due to the driving force. Time now appears explicitly on the right-hand side of the equation of motion; therefore it is *non-autonomous*. This seemingly minor modification leads to significant consequences, because, as we shall see, the dimension of the phase space increases by one. As a consequence, a behaviour much more complicated than periodic, a chaotic time evolution, may also appear.

The autonomous systems presented in Chapter 3 follow the same dynamics irrespective of when they are initiated, and this is why we have the freedom to choose the initial time arbitrarily. In contrast, for driven motion the initial value of the driving force does matter. Two or more types of motion may start from the same point (x, v) if this latter belongs to different values of the driving force. Consequently, trajectories in the (x, v)-plane generally intersect, even in the special case when trajectories started at the same time are considered. This is so because the (x, v)-plane is *not* identical to the entire phase space: co-ordinates (x, v) are not sufficient to determine a state uniquely.

4.1.2 Phase space

It follows from the above that in order to characterise the motion uniquely we need also to specify the 'phase' in which the driving force of periodicity, T, acts. To this end we introduce the phase of the driving,

$$\varphi = 2\pi\frac{t}{T} + \varphi_0, \tag{4.3}$$

which is, by definition, an angle, a periodic quantity with period 2π; $\Omega = 2\pi/T$ is called the *driving frequency*.

The driving force can then be expressed in terms of the phase as $F_d(x, \varphi)$. To determine the initial state uniquely, the initial phase, φ_0, is needed too. Introducing the velocity, v, and phase, φ, as new variables, the non-autonomous equation (4.1) can be rewritten as an autonomous system of three first-order differential equations:

$$\dot{x} = v, \quad \dot{v} = F(x) - \alpha v + F_d(x, \varphi), \quad \dot{\varphi} = \frac{2\pi}{T} \equiv \Omega. \tag{4.4}$$

The phase space (see Section 3.5.1) is therefore *three-dimensional*; a state is uniquely determined by three data: x, v and φ. A periodically

Fig. 4.1. Trajectory of a driven motion in its three-dimensional phase space (x, v, φ).

driven motion can thus be faithfully represented by a three-dimensional flow (see Fig. 4.1), where the velocity along the phase axis is constant in time.

An important consequence of driving is that the mechanical energy is not conserved (even if there is no friction): the system sometimes gains and sometimes loses energy (when the external force accelerates or decelerates it, respectively). It is impossible to reach an equilibrium state because of the driving force, the velocity, v, is never zero for a long period of time, and therefore the energy continually changes. There may, however, exist states in which the motion, and consequently the energy, are periodic in time. These sustained motions are *limit cycles*. The simplest ones just take on the period, T, of the driving force, but there may also exist limit cycles whose period is $2T$, $3T$, ..., or, in general, an integer multiple of the period T: nT, where $n > 1$. These are called *higher-order cycles* or periodic orbits (Fig. 4.2).

4.1.3 Stroboscopic maps

Instead of following the full three-dimensional trajectory, it is often useful to study states with a given phase only. We look at the system, or make snapshots of it, at instants which differ by integer multiples of the period T of the driving force. At each instant the position and velocity co-ordinates are registered. These differ from each other in the subsequent pictures by finite values, since the time interval between the snapshots is finite. This procedure can also be considered as taking

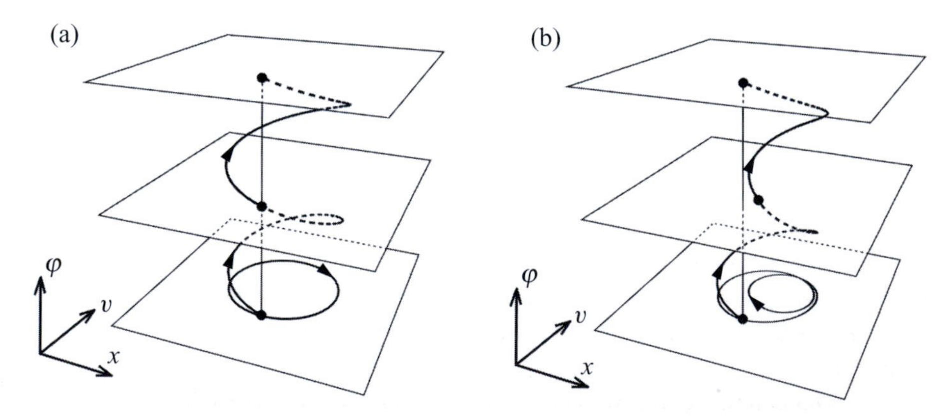

Fig. 4.2. Limit cycles (periodic orbits) in the phase space of a periodically driven system; the planes correspond to states differing by phase 2π (or time T). (a) One-cycle of period T. The intersections of the trajectory and the planes fall above each other. (b) Two-cycle of period $2T$. Only every second point of intersection falls onto the same vertical straight line. The projections on the (x, v)-plane are also indicated.

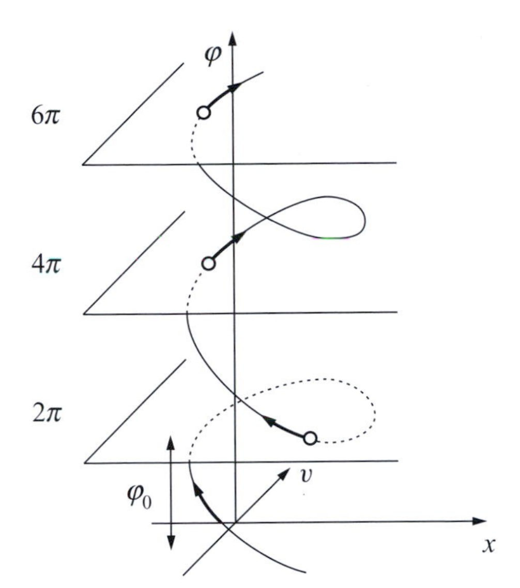

Fig. 4.3. Schematic view of the generation of a stroboscopic map. The trajectory in phase space is intersected by planes perpendicular to the time (phase) axis at time intervals T (at a phase change of 2π).

the intersections of the three-dimensional trajectory with planes of phase $\varphi - \varphi_0 = 2\pi, 4\pi, \ldots, 2\pi n, \ldots$ (see Fig. 4.3).

Let us denote the position and velocity co-ordinates at the nth intersection by x_n and v_n, respectively. The co-ordinates in the nth plane are *uniquely* related to those in the $(n+1)$st plane: the solution of (4.1) with initial conditions x_0, v_0, φ_0 is unique, and the aforementioned co-ordinates are two points along the trajectory. The rule

$$(x_{n+1}, v_{n+1}) = M(x_n, v_n) \tag{4.5}$$

relating the discrete co-ordinates on subsequent snapshots is called a *map*.[2] For the different co-ordinates, the map is given by

$$x_{n+1} = M_1(x_n, v_n), \qquad v_{n+1} = M_2(x_n, v_n), \tag{4.6}$$

where M_1 and M_2 are the functions that define the two components of the map. Point (x_{n+1}, v_{n+1}) is the *image* of point (x_n, v_n), and the application of the map is called *iteration*. This map is actually the *discrete time* form of the differential equation (4.1). It is a difference equation, which always exists, although to determine its specific form is not necessarily easy. For observations taken at integer multiples of T, map (4.5) is the *equation of motion*.

A *stroboscope* is a device providing a periodic flash light; maps of the above kind are therefore called *stroboscopic maps*. Since at the instants of the snapshots the driving is always in the same phase, the form of the

[2] If it is not to be emphasised which iteration step is considered, the notation $(x', v') = M(x, v)$ is used.

Table 4.1. *Comparison of continuous and discrete dynamics. The definitions of quantities $\Lambda_\pm$ and J will be given in Sections 4.5 and 4.6.*

Differential equation	Map
trajectory: a line	trajectory: a sequence of points
limit cycle	fixed point
fractal dimension: D_0	fractal dimension: $D_0 - 1$
stability eigenvalues: $\lambda_\pm$	stability eigenvalues: $\Lambda_\pm$
phase space contraction rate: σ	area contraction rate: J

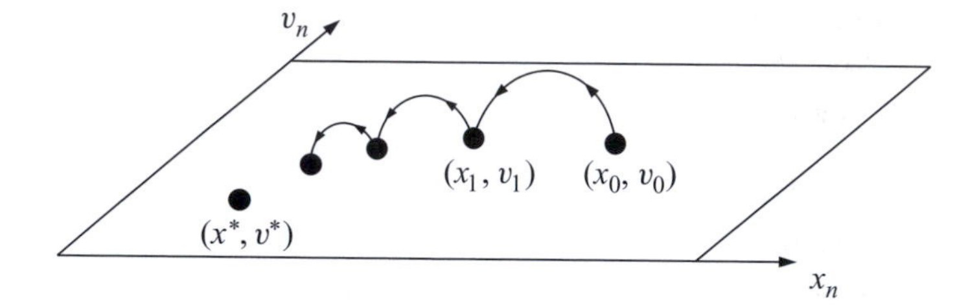

Fig. 4.4. The dynamics on a stroboscopic map consist of jumps: the trajectory is a sequence of discrete points (x_i, v_i), $i = 0, 1, 2, \ldots$ The trajectory shown here converges towards a fixed point (x^*, v^*).

map *does not depend* on the plane on which it is applied: the stroboscopic map is *autonomous*, the rule M itself is *independent* of n, the discrete time.

The advantage of a stroboscopic map is that it is based on the same co-ordinates as those used in non-driven cases. The motion in the phase plane is, however, no longer continuous (Fig. 4.4). Instead of the continuous lines of flows, trajectories on stroboscopic maps are *sequences of points*. It is true in general that the dimension of a given object on a map is *less by one* than the dimension in the flow. So, for example, a limit cycle of period T appears in the stroboscopic map as a single point. The representation of a two-cycle consists of two points jumping back and forth (see Fig. 4.2). The relation between the dynamics represented by flows and by maps is summarised in Table 4.1.

Naturally, a stroboscopic map contains less information than the original flow since the behaviour between two snapshots is not investigated. In spite of this, we obtain a faithful picture of the general character of the motion by following a map. Moreover, the information lost can be regained if, instead of studying a map with a fixed phase, a whole family of maps is considered as a function of the initial phase, φ_0.

The plane of the map is considered as the discrete-time phase space of the system and the motion on it as a discrete trajectory. The use of maps is favourable in many respects, and it is worth turning from the three-dimensional approach to understanding this planar but discrete-time dynamics as illustrated in the following sections.

Two large families of driving will be studied.

(i) *Harmonic*, in other words *sinusoidal, driving*: the driving force of period T is a cosine or sine function of time,

$$F_d(x, t) = f_0(x) \cos(\Omega t + \varphi_0), \tag{4.7}$$

where $\Omega = 2\pi/T$ is the driving frequency. The amplitude, $f_0(x)$, can be an arbitrary function of the instantaneous position. In the simplest case, the amplitude is chosen to be constant.

(ii) *Periodic kicking* corresponds to a succession of instantaneous momentum transfers applied with some period T. No driving force acts between kicks. The momentum transfer, Δv, is assumed to depend on the instantaneous position co-ordinate x only, through a given function $uI(x)$. Consequently, the momentum transfer in such kicked systems is given by

$$\Delta v = uI(x(t)) \quad \text{at} \quad t = \frac{\varphi_0}{\Omega} + nT \tag{4.8}$$

and zero otherwise. Here, u is a typical value of the velocity jump, and $I(x)$ is a dimensionless quantity, which is of order unity for finite position co-ordinates. For the time being, the form of $I(x)$ is not specified.

Problem 4.1 Deduce the expression of the driving force for a periodically kicked system in terms of the delta function, $\delta(t)$ (which is zero for $t \neq 0$, but takes on such a large value at $t = 0$ that its integral over any interval around the origin is unity).

4.2 Harmonically driven motion around a stable state

Around a stable state (i.e. at the bottom of a well) the motion of a body is expected to remain stable in the presence of driving. The system obviously possesses an attractor, due to friction, though the state belonging to it is not an equilibrium, but is rather a time-periodic behaviour. With a driving force of constant amplitude, the equation of motion for small displacements, x, around the original equilibrium state ($x = 0$) is given by

$$\ddot{x} = -\omega_0^2 x - \alpha \dot{x} + f_0 \cos(\Omega t + \varphi_0), \tag{4.9}$$

where f_0 is constant. The natural frequency, ω_0, usually differs from the driving frequency, Ω. For the sake of simplicity, very weak damping is assumed: $\alpha/2 \ll \omega_0$.

With given initial conditions x_0, v_0 at time $t = 0$, the equation of motion possesses a unique solution, $x(t)$, $v(t)$, which is a function of the initial phase φ_0. For long times, $t \gg 1/\alpha$, this solution converges to a *limit cycle attractor*, $x^*(t)$, $v^*(t)$, corresponding to a harmonic oscillation which has taken over the frequency, Ω, of the driving and is of some amplitude A. The geometric representation of this limit cycle is a regular helix, whose projection on the (x, v)-plane is an ellipse with semi-major and semi-minor axes, A and ΩA, respectively (see Fig. 4.2(a)). The particular forms of these quantities is given in Appendix A.1.1.

The stroboscopic map can readily be obtained from the exact solution, $x(t)$, $v(t)$. By choosing φ_0 to be zero, the stroboscopic snapshots are taken at the instants $t = nT$. The map co-ordinates are thus $x_n = x(nT)$ and $v_n = v(nT)$. From the particular forms of $x(t)$ and $v(t)$ (see Appendix A.1.1) we find the stroboscopic map in the following form:

$$x_{n+1} \equiv M_1(x_n, v_n) = x^* + EC(x_n - x^*) + \frac{ES}{\omega_0}(v_n - v^*), \tag{4.10}$$

$$v_{n+1} \equiv M_2(x_n, v_n) = v^* - ES\omega_0(x_n - x^*) + EC(v_n - v^*). \tag{4.11}$$

Here, (x^*, v^*) represents the fixed point of the map corresponding to the limit cycle attractor, and

$$E \equiv e^{-(\alpha/2)T}, \quad S \equiv \sin(\omega_0 T), \quad C \equiv \cos(\omega_0 T). \tag{4.12}$$

Parameter E is the damping factor of the oscillation amplitudes over time, T, due to friction, while S and C characterise the phase of the oscillation at time T.

The deviations $\Delta\mathbf{x}_n = (x_n - x^*, v_n - v^*)$ and $\Delta\mathbf{x}_{n+1} = (x_{n+1} - x^*, v_{n+1} - v^*)$ from the fixed point co-ordinates corresponding to the limit cycle thus fulfil a *linear* map,

$$\Delta\mathbf{x}_{n+1} = L\Delta\mathbf{x}_n, \tag{4.13}$$

governed by the matrix

$$L = \begin{pmatrix} EC & ES/\omega_0 \\ -ES\omega_0 & EC \end{pmatrix}. \tag{4.14}$$

Thus the map inherits the linearity of the equation of motion.

Let us examine how the iterates of a point on the stroboscopic map converge to the fixed point. We assume that the deviations are proportional to the nth power of some number Λ, i.e.

$$\Delta x_n \equiv x_n - x^* = u_x\Lambda^n, \quad \Delta v_n \equiv v_n - v^* = u_v\Lambda^n, \tag{4.15}$$

which is analogous to the exponential behaviour seen in the continuous-time dynamics. Substituting this into (4.10) and (4.11) leads to

$$\Lambda\mathbf{u} = L\mathbf{u}; \tag{4.16}$$

thus, Λ is the *eigenvalue* of matrix L and $\mathbf{u}$ is its (right) eigenvector. The characteristic equation is $(EC - \Lambda)^2 + (ES)^2 = 0$, with two solutions for Λ:

$$\Lambda_\pm = E(C \pm iS) = e^{(-\alpha/2 \pm i\omega_0)T}. \tag{4.17}$$

The absolute value of both eigenvalues is less than unity, and therefore the fixed point is, in fact, attracting.

For a general initial condition, the motion on a stroboscopic map is the linear combination of the powers of the two different eigenvalues:

$$\Delta x_n = c_+\Lambda_+^n + c_-\Lambda_-^n, \tag{4.18}$$

$$\Delta v_n = i\omega_0 c_+\Lambda_+^n - i\omega_0 c_-\Lambda_-^n. \tag{4.19}$$

The initial conditions $(x_0,\ v_0)$ determine the coefficients $c_\pm$ as $c_\pm = \Delta x_0/2\ \pm \Delta v_0/(2i\omega_0)$. Substituting this and the eigenvalues (4.17), the complete solution appears in the real form:

$$\begin{aligned} \Delta x_n &= \Delta x_0 e^{-n(\alpha/2)T} \cos(\omega_0 nT) + \frac{\Delta v_0}{\omega_0} e^{-n(\alpha/2)T} \sin(\omega_0 nT), \\ \Delta v_n &= -\omega_0 \Delta x_0 e^{-n(\alpha/2)T} \sin(\omega_0 nT) + \Delta v_0 e^{-n(\alpha/2)T} \cos(\omega_0 nT). \end{aligned} \tag{4.20}$$

Thus, the distance from the fixed point decreases proportionally to $e^{-(n\alpha/2)T}$ in n steps. At the same time, the points in phase space turn around the fixed point in each step. Even though the points approach the fixed point by jumps, they fall on a spiral-shaped curve (Fig. 4.5).

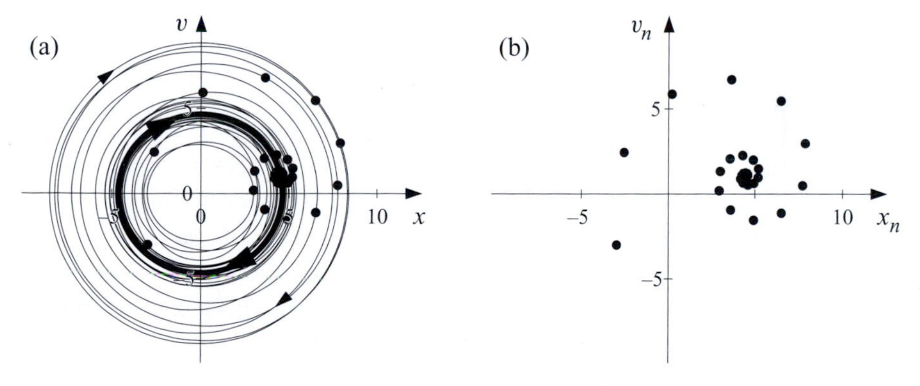

Fig. 4.5. Spiral attractor of a driven harmonic oscillation. (a) Projection of a flow trajectory starting from point $x_0 = -3$, $v_0 = -3$ onto the (x, v)-plane. (b) Points of its stroboscopic map (black dots). The limit cycle is a closed curve on the projection (thick line in (a)) and appears as a fixed point (large black dot) on the stroboscopic map. The parameters are $\alpha = 0.05$, $\Omega = 1$, $\omega_0 = 1.1$, $f_0 = 1$ ($E = 0.855$, $S = 0.120$, $C = 0.993$).

Problem 4.2 Determine the equation of the curve along which points move when approaching the spiral attractor of the stroboscopic map (4.20).

Consequently, this kind of attracting point is called a *spiral attractor* on the map. The motion of the driven system around its limit cycle on the stroboscopic map is thus similar to the motion of a non-driven system around its stable equilibrium state in the flow.

Note that the stroboscopic map is *invertible*. This means that there exists a unique map which describes the time-reversed dynamics. Naturally, this is governed by the inverse matrix, L^{-1}, whose eigenvalues are the reciprocals of the eigenvalues of L. The fixed point is therefore a spiral repellor in the reversed dynamics.

Problem 4.3 Determine the stroboscopic map of a sinusoidally driven harmonic oscillator for arbitrary damping (α comparable with or larger than ω_0). What are the eigenvalues?

4.3 Harmonically driven motion around an unstable state

The motion around an unstable state is also expected to remain unstable in the presence of driving. It may happen, however, that the driving always acts in such phases that it 'pushes back' the body that is ready to move away. As a consequence, there exists a single periodic motion that *never* leaves the vicinity of the original unstable equilibrium state, in contrast with all other motion starting close to the original unstable state. The limit cycle is therefore *unstable* in this case. Its instability is of a *hyperbolic* nature, similar to that of the equilibrium state of a pencil standing on its point.

On the stroboscopic snapshots taken at instants $t = nT$ ($\varphi_0 = 0$) the map is again linear (see Appendix A.1.2). The matrix governing the behaviour around the fixed point (x^*, v^*) is found for very weak damping, $\alpha \ll s_0$, to be

$$L = \begin{pmatrix} EC' & ES'/s_0 \\ ES's_0 & EC' \end{pmatrix}, \tag{4.21}$$

where

$$C' \equiv \text{ch}(s_0 T), \quad S' \equiv \text{sh}(s_0 T). \tag{4.22}$$

The eigenvalues

$$\Lambda_\pm = E(C' \pm S') = e^{(-\alpha/2 \pm s_0)T} \tag{4.23}$$

are now both real; $\Lambda_+ > 1$ and $\Lambda_- < 1$. The eigenvectors are $\mathbf{u}_\pm = (1, \pm s_0)$. The solution belonging to the initial condition (x_0, v_0) is given by

$$\Delta x_n = c_+\Lambda_+^n + c_-\Lambda_-^n, \tag{4.24}$$

$$\Delta v_n = s_0 c_+\Lambda_+^n - s_0 c_-\Lambda_-^n, \tag{4.25}$$

where $c_\pm = \Delta x_0/2 \pm \Delta v_0/(2s_0)$. After a sufficiently long time ($n \gg 1$), the deviation from the fixed point is typically proportional to Λ_+^n:

$$\Delta x_n, \Delta v_n \sim \Lambda_+^n. \tag{4.26}$$

From initial conditions on the line $v_n - v^* = s_0(x_n - x^*)$ ($c_- = 0$), the trajectory always remains on this line, and the distance from the points to the origin increases in each iteration by a factor of Λ_+. In contrast, along the line $v_n - v^* = -s_0(x_n - x^*)$ *convergence* occurs: at each step the point moves closer by a factor of Λ_- ($\Lambda_- < 1$) to the state (x^*, v^*) representing the unstable limit cycle. This type of fixed point map is also called *hyperbolic*, and the above-mentioned two straight lines are the unstable and stable manifolds of the fixed point, respectively (see Fig. 4.6). Note that these directions are exactly the directions of the eigenvectors of matrix L. It is worth emphasising that the lines $v_n - v^* = \pm s_0(x_n - x^*)$ of the manifolds are both *invariant curves*, i.e. curves which are mapped onto themselves in the course of iterations. Not a single point can ever leave these curves. The manifolds again form a cross of straight lines around the hyperbolic point.

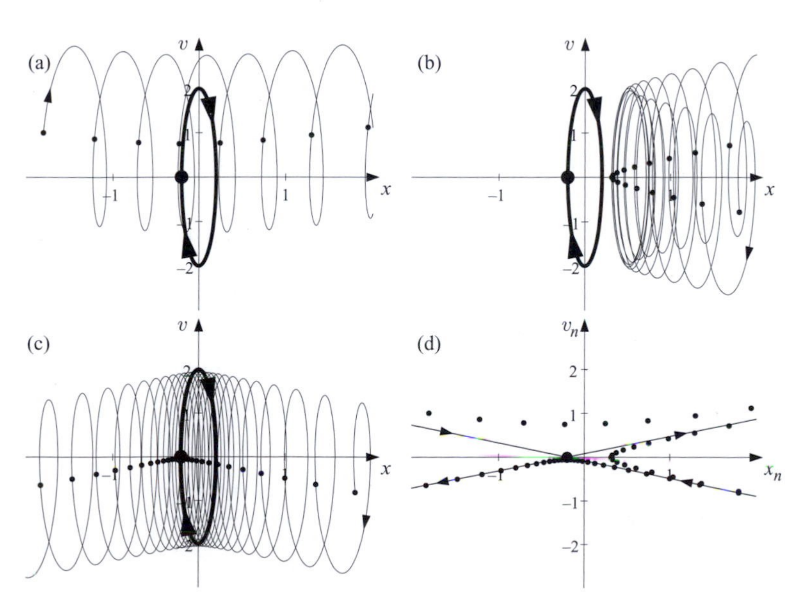

Fig. 4.6. Hyperbolic point of a driven unstable state. In panels (a), (b) and (c) the projections of the flow trajectories started from three different initial conditions ($(x_0, v_0) = (-1.8, 1), (1.8, -0.78)$ and $(1.8, -0.81)$, respectively) are plotted; the thick curves mark the unstable limit cycle. The succession of the stroboscopic points (black dots) is also shown. (d) Representation of the stroboscopic map. The stroboscopic points indicated in (a)–(c) are all shown, along with the stable and unstable manifolds (straight lines) of the hyperbolic point (large black dot). The parameters are $\alpha = 0.01$, $\Omega = 10$, $s_0 = 0.4$, $f_0 = 20$ ($E = 0.997$, $S' = 0.254$, $C' = 1.032$).

Problem 4.4 Determine the equation of the curve along which points move in the vicinity of the hyperbolic point of the map described by (4.24) and (4.25).

The stroboscopic map is invertible in this case as well, and the inverse map is governed by the matrix L^{-1}. In the inverted dynamics, the stable and unstable directions are interchanged, and the points move along them according to the reciprocals of the original eigenvalues. Thus, a hyperbolic point remains hyperbolic in the inverted map.

An interesting general relationship can be uncovered by comparing the eigenvalues of the maps with those of the non-driven damped oscillator or – more generally – of the linearised continuous-time dynamics. In both unstable and stable cases,

$$\Lambda_{\pm} = e^{\lambda_{\pm} T} \tag{4.27}$$

holds, where $\lambda_{\pm}$ are the flow eigenvalues ((3.14) and (3.29)). Relation (4.27) also holds for arbitrary damping (see Problems 4.3 and 4.5). This implies that the behaviour around limit cycles on the stroboscopic maps of driven systems is the *same* as the behaviour around the equilibrium states of non-driven systems. Approaching an attracting limit cycle or deviating from an unstable one happens in the same way as on the (x, v) phase plane of non-driven systems, when observing the latter at integer multiples of some time interval T.[3]

Problem 4.5 Write down the eigenvalues for the stroboscopic map of a sinusoidally driven unstable state for arbitrary damping.

4.4 Kicked harmonic oscillator

Let us now investigate the effect of driving via kicks (see (4.8)) on a weakly damped dynamics around a stable equilibrium state. This corresponds to the motion of a point mass fixed to a spring with natural frequency ω_0, damped with a friction coefficient, $\alpha/2 \ll \omega_0$, and subjected to abrupt momentum transfers at time intervals T (different, in general, from the oscillation period). An abrupt momentum transfer does not change the position. Think of a ball that is kicked. At the instant of the kick, the velocity of the ball changes abruptly, but the ball does not move yet. The displacement, $x(t)$, is a continuous function during the motion, while the velocity, $v(t)$, exhibits discontinuities at intervals

[3] This is due to the property of linear differential equations that the general solution of an inhomogeneous equation contains the general solution of the homogeneous equation (the non-driven case).

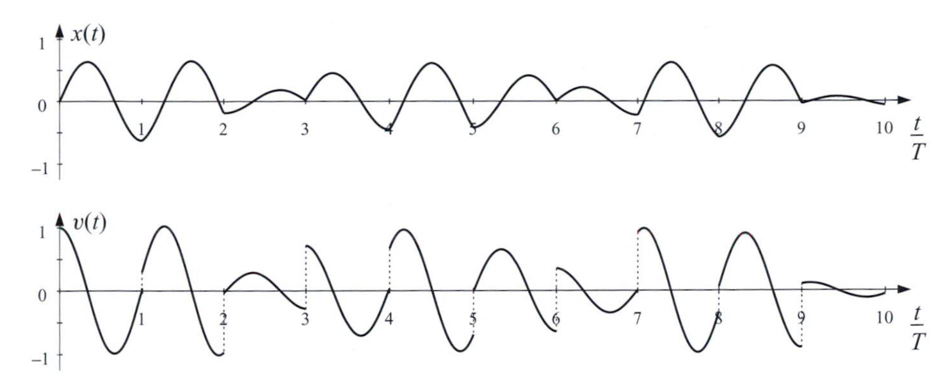

Fig. 4.7. Motion of a kicked oscillator with kicking period T. The amplitude function, called the parabola amplitude, is $I(x) = 1 - ax^2$. The parameters are: $a = 1.8,\ u = 1,\ T = 3\pi/(2\omega_0)$ (the period of the kick is three-quarters of the period of the oscillation), $\alpha T = 0.01$.

T. The magnitude of these jumps is $uI(x)$ (as given by (4.8)). Since the velocity is the derivative of the position with respect to time, the function $x(t)$ *breaks* in the same points (Fig. 4.7), and the difference between the slopes of the tangents from the right and from the left is just $uI(x)$. If function $I(x)$ is neither constant nor linear, but is rather a given non-linear (for example quadratic) function of the position, the sequence of the momentum transfers and the displacement vs. time function become inter-related in a non-trivial, complicated manner. Non-periodic, chaotic motion may arise. This will be analysed in detail in Part III; our aim here is to show that for this simple system the stroboscopic map can be derived *exactly* regardless of the specific form of the kicking amplitude, $I(x)$.

As the kick is instantaneous, the velocity of the body is not unique at that instant (although an arbitrarily short time later it is). This is why we have to decide to consider either the state just before or just after the kick. Both are identically correct; we must, however, apply our choice consistently. We are now going to compare the states *after* the kicks.

For the sake of simplicity, the initial phase is set to zero, $\varphi_0 = 0$, which implies that the first kick comes at time $t = 0$, the second at $t = T$, and so forth. The stroboscopic map is derived in Appendix A.1.3, and, by means of the shorthand notation given in (4.12), it can be written in the form:

$$x_{n+1} = M_1(x_n, v_n) \equiv ECx_n + \frac{ES}{\omega_0}v_n,$$
$$v_{n+1} = M_2(x_n, v_n) = -ES\omega_0 x_n + ECv_n + uI(x_{n+1}). \qquad (4.28)$$

Although this form is valid for very weak damping ($\alpha \ll \omega_0$) only, we emphasise that the map can be derived exactly without any restriction on the damping (see Problem 4.7).

Solely for the sake of compact notation, we choose a special value for the kicking period. Let T be one-quarter of the period, $2\pi/\omega_0$, of the harmonic oscillator, so that $\omega_0 T = \pi/2$. According to (4.12), in this special case $S = 1$ and $C = 0$, and the kicked oscillator map simplifies to

$$x_{n+1} = M_1(v_n) \equiv \frac{E}{\omega_0} v_n,$$
$$v_{n+1} = M_2(x_n, v_n) \equiv -E\omega_0 x_n + uI(x_{n+1}). \tag{4.29}$$

Problem 4.6 Derive the stroboscopic map of the kicked harmonic oscillator taken immediately before the kicks for $T = \pi/(2\omega_0)$.

It is worth writing the map in a dimensionless form. By measuring velocity and length in the units of the typical velocity jump u and of u/ω_0, respectively, this corresponds to the substitution

$$x_n \to \frac{u_0}{\omega_0} x_n, \qquad v_n \to u v_n, \tag{4.30}$$

where the first factors after the arrows are the dimensional units, while the second ones represent the new unitless magnitudes x_n and v_n. Substituting this into (4.29) leads to the dimensionless map:

$$x_{n+1} = E v_n, \qquad v_{n+1} = -E x_n + I(x_{n+1}). \tag{4.31}$$

Here, the natural frequency, ω_0, and the typical kick strength, u, no longer appear. Beside the dimensionless function, I, the damping factor, $E = e^{-(\alpha/2)T} < 1$, is thus the only important parameter of the map.

Note that the inverse of the map exists again. Knowing the point (x_{n+1}, v_{n+1}), its *pre-image*, (x_n, v_n), can be determined. From (4.31) we obtain, after the substitution $n \to n+1$, $n+1 \to n$, the following:

$$x_{n+1} = -\frac{1}{E} v_n + \frac{1}{E} I(x_n), \qquad v_{n+1} = \frac{1}{E} x_n. \tag{4.32}$$

This is the inverted kicked oscillator map, which describes the time-reversed dynamics. The inverse map is unique, irrespective of the specific form of the function $I(x)$ (which itself might be non-invertible). The invertibility is a consequence of the invertibility of the underlying equation of motion.

Problem 4.7 Derive the stroboscopic map of the kicked harmonic oscillator taken immediately after the kicks for arbitrary driving periods and damping coefficients. Also determine the inverse map.

4.5 Fixed points and their stability in two-dimensional maps

Consider a general two-dimensional map,

$$\boxed{x_1' = M_1(x_1, x_2), \quad x_2' = M_2(x_1, x_2),} \tag{4.33}$$

defining the discrete-time evolution of the quantities x_1 and x_2. To avoid multiple indices, the discrete-time variables n and $n+1$ are not written out explicitly; instead, the images are denoted by primes. The fixed points of the map are the pairs (x_1^*, x_2^*), satisfying equations $x_1^* = M_1(x_1^*, x_2^*)$ and $x_2^* = M_2(x_1^*, x_2^*)$. Their stability can be determined from the dynamics of small deviations around the fixed point. It is then sufficient to keep the leading, linear terms in the variables $\Delta x_1 = x_1 - x_1^*$ and $\Delta x_2 = x_2 - x_2^*$. The resulting linearised map is of the form

$$\begin{aligned} \Delta x_1' &= m_{11}\Delta x_1 + m_{12}\Delta x_2, \\ \Delta x_2' &= m_{21}\Delta x_1 + m_{22}\Delta x_2. \end{aligned} \tag{4.34}$$

By introducing the vector notation $\Delta\mathbf{x} = (\Delta x_1, \Delta x_2)$, the linearised map becomes

$$\Delta\mathbf{x}' = L\Delta\mathbf{x}, \tag{4.35}$$

where

$$L = \begin{pmatrix} m_{11} & m_{12} \\ m_{21} & m_{22} \end{pmatrix} = \begin{pmatrix} \partial M_1/\partial x_1 & \partial M_1/\partial x_2 \\ \partial M_2/\partial x_1 & \partial M_2/\partial x_2 \end{pmatrix} \tag{4.36}$$

is called the *stability matrix*, and the derivatives in it are taken at the fixed point $\mathbf{x}^*$.

With a trial solution of (4.34) in the form of $\Delta x_{i,n} = u_i \Lambda^n$ $(i = 1, 2)$, we find that Λ must satisfy the following equations:

$$\begin{aligned} \Lambda u_1 &= m_{11}u_1 + m_{12}u_2, \\ \Lambda u_2 &= m_{21}u_1 + m_{22}u_2. \end{aligned} \tag{4.37}$$

Consequently, Λ is the eigenvalue of matrix L, and is the root of the quadratic equation

$$\Lambda^2 - \Lambda \mathrm{Tr}\, L + \det L = 0. \tag{4.38}$$

The trace of the stability matrix is $\mathrm{Tr}L = m_{11} + m_{22}$.

Table 4.2. *Classification of the fixed points of two-dimensional maps based on the absolute values and the imaginary parts of the eigenvalues.*

	$\|\Lambda_\pm\| = 1$	$\|\Lambda_\pm\| < 1$	$\|\Lambda_-\| < 1 < \|\Lambda_+\|$	$\|\Lambda_\pm\| > 1$
$\mathrm{Im}\Lambda_\pm = 0$	(marginal)	node attractor	hyperbolic point	node repellor
$\mathrm{Im}\Lambda_\pm \neq 0$	elliptic point	spiral attractor	–	spiral repellor

This table is analogous to Table 3.1.

The character of a non-marginal fixed point can be determined from the properties of the eigenvalues (see Table 4.2; the definition of J is given in Section 4.6):

Hyperbolic point: $\Lambda_\pm$ are real, and the absolute value of one of them is greater than unity; $|\Lambda_+| > 1$, $|\Lambda_-| < 1$.

Elliptic point: $\Lambda_\pm$ are complex with absolute values equal to unity; $|\Lambda_\pm| = 1$ (only possible in conservative cases with $J = 1$).

Node attractor: $\Lambda_\pm$ are real with absolute values less than unity; $|\Lambda_\pm| < 1$ ($J < 1$).

Spiral attractor: $\Lambda_\pm$ are complex with absolute values less than unity; $|\Lambda_\pm| < 1$ ($J < 1$).

Node repellor: $\Lambda_\pm$ are real with absolute values greater than unity; $|\Lambda_\pm| > 1$ (only possible in expanding phase spaces with $J > 1$).

Spiral repellor: $\Lambda_\pm$ are complex with absolute values greater than unity; $|\Lambda_\pm| > 1$ (only possible in expanding phase spaces with $J > 1$).

The classification is similar to that of the fixed points in the continuous-time phase plane (Section 3.5.4), if the logarithms of $|\Lambda_\pm|$ are identified with the eigenvalues, $\lambda_\pm$, of the continuous case. The fixed points of two-dimensional maps are therefore of the same type as those of two-dimensional flows.

Elliptic fixed points can occur only in frictionless cases. Nodes and spiral attractors appear only in dissipative systems. Solely hyperbolic fixed points can be present in both conservative and dissipative systems.

The general solution is always a linear combination of two powers:

$$\mathbf{x}_n = c_+\mathbf{u}_+\Lambda_+^n + c_-\mathbf{u}_-\Lambda_-^n. \tag{4.39}$$

Here, vectors $\mathbf{u}_\pm$ are the eigenvectors of matrix L, obeying equation $L\mathbf{u}_\pm = \Lambda_\pm\mathbf{u}_\pm$. The coefficients $c_\pm$ are determined by the initial condition, $\mathbf{x}_0 = c_+\mathbf{u}_+ + c_-\mathbf{u}_-$. If the eigenvalues and eigenvectors are real, the eigenvectors mark directions in the plane of the map along which a single power law describes time evolution. In the case of a hyperbolic point, the stable and unstable manifolds coincide with the lines of the eigenvectors belonging to the eigenvalues less than unity and greater than unity in modulus, respectively.

Problem 4.8 Determine the fixed point – and its stability – corresponding to the limit cycle of a harmonic oscillator kicked with a constant amplitude: $I(x) \equiv I_0$ in (4.31).

4.6 The area contraction rate

An important characteristic of dynamics generated by maps is how a phase space area determined by given initial points evolves in time. Since we are dealing with a discrete-time dynamics, the phase space area may change significantly within a single iteration; in dissipative systems, for example it shrinks. The factor by which a small phase space area decreases in one step is called the *area contraction rate*.

The area contraction rate of a general two-dimensional map, (4.33), at a given point (x_1, x_2) is given by the Jacobian, J, of the derivative matrix, (4.36), of the map linearised at that point (which is now not necessarily a fixed point):

$$\boxed{J(x_1, x_2) \equiv \frac{\partial M_1}{\partial x_1}\frac{\partial M_2}{\partial x_2} - \frac{\partial M_1}{\partial x_2}\frac{\partial M_2}{\partial x_1}}. \tag{4.40}$$

Problem 4.9 Show that the image of an initially rectangular small area is a parallelogram whose area is J times the area of the rectangle, where J is the Jacobian evaluated at the initial position.

The Jacobian plays the same role in maps as does the phase space contraction rate, $\sigma = -\text{div}\,\mathbf{f}$, in continuous-time flows. A Jacobian with absolute value less than unity describes local area contraction, while an absolute value greater than unity would imply expansion. Thus, area preservation only applies for $|J| = 1$. We emphasise that no value of the Jacobian guarantees that the shape of the area remains similar to the initial one. As it will be shown, one of the most conspicuous differences between chaotic and non-chaotic systems is that, in the former case, the phase space objects quickly lose their shape, become convoluted and trace out complicated (fractal) sets, whereas in the non-chaotic case the objects suffer weak deformations only. Both scenarios can happen at the same value of J (provided some other parameters are different).

The behaviour around a fixed point is described by a map, (4.35), linearised at that point. Since the product of the eigenvalues is the determinant, it is generally true that the product

$$\Lambda_+\Lambda_- = J \tag{4.41}$$

is the area contraction rate at the fixed point. The Jacobian of the inverse map is the reciprocal of the Jacobian of the original map.

Evaluating the Jacobian of maps (4.10), (4.11), (4.21), (4.28) and (4.31) presented so far, we find that, in all these cases,

$$J = E^2 = e^{-\alpha T} = e^{-\sigma T}, \tag{4.42}$$

where $\sigma = \alpha$ is the phase space contraction rate of the flow. The Jacobian is thus independent of the location in phase space, an obvious consequence of the constancy of the friction coefficient. The fact that the Jacobian of the kicked oscillator map is constant is due to the non-dissipative nature of the kick, which does therefore not contribute to J. Parameter E appearing in (4.31) thus characterises both the damping and the area contraction rate. It is worth noting that the damping observed in the map can be strong even if friction is weak, provided the period is long enough, since E depends on the product αT only.

Problem 4.10 Classify the fixed point characterised by the stability matrix

$$L = \begin{pmatrix} 1/2 & 1 \\ a & -2 \end{pmatrix}, \tag{4.43}$$

depending on parameter a, in the region where the Jacobian is between zero and unity.

Problem 4.11 Determine the nature of the fixed point of a general two-dimensional map in different regions of the parameter plane defined by $J \equiv \det L$ $(0 < J < 1)$ and Tr L; cf. Problem 3.26.

4.7 General properties of maps related to differential equations

Two-dimensional maps written down arbitrarily are generally not derivable from three first-order ordinary differential equations. If, however, they are, then this implies significant restrictions for the possible forms, which will be presented on the basis of the examples studied so far.

The map must be *invertible*, since the continuous-time differential equation underlying it is invertible (see Section 3.5.3). A necessary condition for invertibility is that the Jacobian be non-zero in the entire plane.

Our examples demonstrate that the Jacobians are positive. Simple examples indicate that in maps with negative Jacobians the orientation of plane figures reverses in each iteration. This is, however, impossible in real systems. All trajectories of the three-dimensional flows investigated move along the phase axis with a constant speed. The corners of a small figure, for example a rectangle, therefore always lie in a plane (the plane $\varphi =$ constant) as the trajectories evolve in time. The vertices move relative to each other within this plane, and the figure deforms into a parallelogram, but the orientation does not change (that could only happen if the area became zero at some time, i.e. if trajectories intersected each other).

The necessary conditions for a map to be derivable from a differential equation are as follows:

- the map should be invertible, i.e. it must possess a unique inverse map;
- its Jacobian, J, should be positive everywhere; and
- J should be less than unity in dissipative systems (it is exactly unity in conservative cases; see Chapter 7).

These conditions also apply to Poincaré maps, to be discussed in detail later.

The global unfolding of the dynamics of a map, (4.33), with the above properties follows similar lines to that of a two-dimensional flow. The first step is to identify the fixed points and their stability. The trajectories around fixed points are already known (see Sections 4.2–4.5), and, using these as building blocks, the whole phase portrait can be constructed. In this process the stable and unstable manifolds of the hyperbolic points again play a particularly important role. The set of points that are mapped asymptotically into a hyperbolic point form the *stable manifold* of that hyperbolic point. The *unstable manifold* of a hyperbolic point is the stable manifold of the same point in the inverted map.

Both manifolds might be infinitely long curves, which expand into significantly large regions of the phase plane. At the same time, these manifolds are *invariant* curves, i.e. curves mapped onto *themselves*. The role of the manifolds is similar in maps to that seen in two-dimensional flows. In dissipative maps the stable manifold always separates different attractors; it is the boundary of the basins of attractions. The unstable manifold, in turn, goes into one of the attractors.

In stroboscopic maps, higher-order cycles may also be present (cf. Fig. 4.2). An n-cycle appears as a set of n distinct points mapped onto each other, and each one returns to its initial position after n steps. All points of an n-cycle are thus fixed points of the n-fold iterated map. Of course, n-cycles can also be unstable. In such cases, the corresponding fixed points of the n-fold iterated map are hyperbolic. This implies that in the n-fold iterated map every distinct point of the cycle possesses a stable and an unstable manifold.

A statement similar to the Poincaré–Bendixson theorem does *not* exist for discrete-time two-dimensional maps. The attractors can therefore be much more complicated than limit cycles; they can even be chaotic. The phase portrait of a general two-dimensional map has a character similar to that of the phase plane (Fig. 3.27), although it is much richer in details (see Fig. 5.47).

Box 4.1 The world of non-invertible maps

When we play any motion recorded on a videotape backwards in time, we witness a unique, although eventually strange, behaviour. This corresponds to the fact that the dynamics (the equation of motion or the map derived from it) is invertible. If the dynamics has no inverse, this is analogous to a situation when the film cannot be rewound. Non-invertibility is often the result of the inverse map being multi-valued. In this case the original phase space cannot be the phase space of the inverted dynamics, because in the latter several trajectories emanate from a single point. In the videotape analogy this corresponds to a strange situation in which, before displaying a picture in the course of the reverse playing, we always have to select one out of several options. The real world, however, is not like this.

Non-invertible maps appear in several mathematical models. These maps do not correspond to differential equations. They cannot model real, physical processes in a faithful manner; they can at most be highly idealised approximants. One of the most widely known examples is the logistic map[4]

$$x_{n+1} = 1 - ax_n^2 \tag{4.44}$$

($a > 0$). Every point, x_{n+1}, can be reached from two pre-images: $x_n = \pm\sqrt{(1 - x_{n+1})/a}$. Thus, the inverse map is given by

$$x_{n+1} = \pm\sqrt{(1 - x_n)/a}. \tag{4.45}$$

This is not unique, since it is important to know which sign is to be chosen. The analytic continuation of the logistic map to the complex plane has similar properties. This quadratic map is most often written in the form $z_{n+1} = C - z_n^2$, where z is a complex number and C is a complex parameter.[5]

Stability properties of non-invertible dynamics essentially differ from those of invertible dynamics. Hyperbolic behaviour cannot exist in non-invertible systems, since this requires the possibility of a unique time reversal; the unstable manifold is always the stable manifold of the inverted dynamics. In non-invertible systems, the unstable states are represented by repellors.

As stroboscopic maps cannot usually be derived exactly, it is often useful to follow the discrete-time behaviour in a model map of a simple form. One should, however, make sure that the map is invertible, otherwise a real process would be modelled in an unrealistic, non-invertible world.

[4] Another often used form of the logistic map is $x_{n+1} = rx_n(1 - x_n)$, $r = \sqrt{1 + 4a} + 1$, which follows from (4.44) with the substitution $x \to x(r - 2)/4 + 1/2$.

[5] The parameters, C, for which the iteration starting from zero does not tend to infinity constitute what is called the *Mandelbrot set*.

4.8 In what systems can we expect chaotic behaviour?

As a consequence of the Poincaré–Bendixson theorem, no two-dimensional flow can be chaotic.

If the dynamics is *linear*, there cannot be chaos, no matter how high the dimension of the system is. A linear motion can be decomposed exactly into components with exponential (or sinusoidal) time dependence, called *normal modes* (see (3.66) and (4.39)). In other words, the time evolution is a linear combination of a finite number of simple, independent modes. This is not so for chaotic systems; chaotic motion *cannot* be decomposed into elementary components.

Necessary conditions for chaotic dynamics to appear are that the system is

- non-linear and that
- its dynamics may be described by at least three independent, first-order autonomous differential equations (a three-dimensional flow).

Chaos may arise in every *non-linear* system whose phase space is *at least three-dimensional*.

According to Section 4.7, two-dimensional invertible maps with positive Jacobians may be related to differential equations. Dynamics described by such maps can therefore also be chaotic. Chaos may thus also be present in any *two-dimensional, non-linear invertible* map.

(Experience shows that chaotic behaviour appears also in non-invertible maps. The simplest mathematical models in which chaotic dynamics is already possible are one-dimensional, non-invertible maps; see, however, Box 4.1.)

A system is called chaotic if parameter values exist at which chaotic motion is possible. Non-linear systems described by at least three-dimensional flows are generally found to be chaotic. Chaoticity is thus not an exception but the rule in such systems. Whether chaotic motion arises or not at a certain parameter *cannot*, however, be decided by inspecting the form of the equations. It can only be revealed via numerical solutions.

Part III

Investigation of chaotic motion

Chapter 5
Chaos in dissipative systems

We begin the detailed study of chaotic behaviour with dissipative systems. We consider permanently chaotic dynamics (cf. Section 1.2.1), and we start our investigations within the framework of a simple 'model' map, the *baker map*. The most important quantities characteristic of chaos will be introduced via this example. The simplicity of the map makes the exact treatment of numerous chaos properties possible, an exceptional feature in the world of chaotic processes. Next we turn to the investigation of a physical system, the kicked oscillator, with different kicking amplitudes. These functions will be chosen in such a way that, in the first case, the attractor is similar to that of the baker map. In the second, the attractor has a different structure and exhibits a general property of chaotic attractors: it appears to be a single continuous curve. The special form of the amplitude function continues to make its exact construction possible. This is no longer so, however, with the third choice, representing a typical chaotic system. The parameter dependence of chaotic systems will also be discussed within the class of kicked oscillators. Based on all these examples, we summarise the most important properties of chaos, first of all at the level of maps. As measures of irregularity, unpredictability and complex phase space structures, we introduce the concepts of topological entropy, Lyapunov exponents and the fractal dimension of chaotic attractors, respectively. Special emphasis will be given to the presentation and characterisation of the natural distribution of chaotic attractors. This is the quantity that expresses the random nature of chaos in a mathematical sense, being, at the same time, the only correct means of characterising such motion over the long term. Then we briefly review how all this appears in flows: we present continuous-time driven

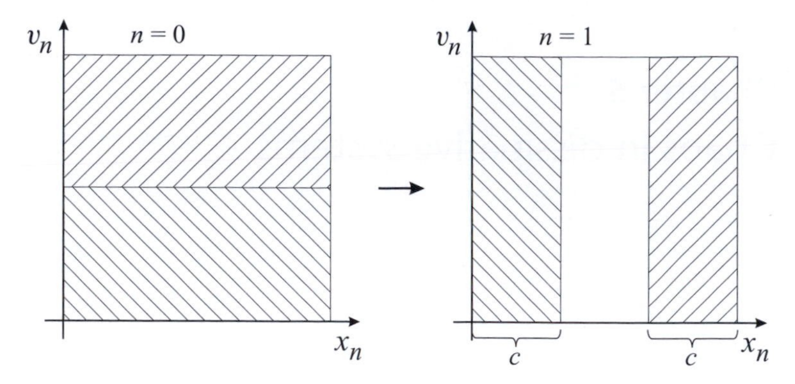

Fig. 5.1. Action of the baker map on the unit square ($c = 1/3$).

systems, for which even the stroboscopic map can only be given in a numeric, rather than an analytic, form. Finally, by means of the example of a water-wheel exhibiting random-like motion due to rain, we investigate the chaotic dynamics of completely different physical origin, described by equations similar to that of the celebrated Lorenz model, which plays an important role in the science of chaos.

5.1 Baker map

5.1.1 Presentation of the map

A baker map (for the origin of the name, see Box 7.1) possesses all the general properties required in Section 4.7 of maps related to differential equations: it is invertible and its Jacobian is positive and not greater than unity. A baker map is a prototype of chaotic dynamical systems. In this field, its role is similar to that of the harmonic oscillator in the realm of regular, non-chaotic motions.

The map can best be defined in terms of its action on its phase space, the unit square. Within one iteration, the square is cut in two equal horizontal bands; these are compressed in one direction by a factor of c and stretched in the other by a factor of 2, and placed at the two opposite sides of the square as shown in Fig. 5.1. In the dissipative case treated in this chapter, contraction is stronger than stretching and the phase space volume shrinks.

In mathematical terms, the map assigns to any point (x_n, v_n) of the plane an image point

$$(x_{n+1}, v_{n+1}) = B(x_n, v_n), \tag{5.1}$$

where the form of B depends on whether the point is situated below or above the critical line $v_n = 1/2$:

$$B(x_n, v_n) = \begin{cases} B_-(x_n, v_n) \equiv (cx_n, 2v_n), & \text{for } v_n \leq 1/2, \\ B_+(x_n, v_n) \equiv (1 + c(x_n - 1), 1 + 2(v_n - 1)), & \text{for } v_n > 1/2, \end{cases} \tag{5.2}$$

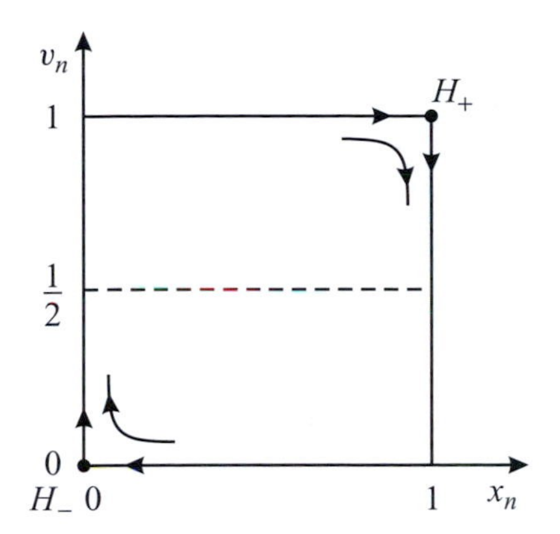

Fig. 5.2. Phase space of the baker map. The form of the map is different below and above the line $v_n = 1/2$, but it is linear in both half squares (see (5.2)). The hyperbolic points, $H_- = (0, 0)$ and $H_+ = (1, 1)$, are also marked, along with their stable and unstable directions. The arrowed curves indicate motion around the fixed points, i.e. jumping along hyperbolae.

and $0 < c < 1/2$. In terms of the components,

$$x_{n+1} = B_1(x_n) \equiv \begin{cases} cx_n, \\ 1 + c(x_n - 1), \end{cases} \qquad v_{n+1} = B_2(v_n) \equiv \begin{cases} 2v_n, \\ 1 + 2(v_n - 1), \end{cases} \tag{5.3}$$

where the top and the bottom rows are valid with the respective conditions $v_n \leq 1/2$ and $v_n > 1/2$.

Mapping (5.2) describes that in one step each point moves twice as far away vertically from and comes c times closer horizontally to the origin or the point $(1, 1)$, depending on whether the point is below or above the line $v_n = 1/2$, respectively (Fig. 5.2). This perpetually repeated *stretching* and *contraction* is an important property of chaotic systems.

The baker map is *piecewise linear*: functions B_- and B_+ (or B_1, B_2) are linear, and non-linearity is introduced into the system by the jump between the two forms of (5.2).[1] It is surprising and remarkable that such a slight non-linearity is in itself *sufficient* for chaos to appear.

Hyperbolic points play a fundamental role in the organisation of motion. To find them, we recall that a fixed point, (x^*, v^*), is left invariant by the mapping $B(x^*, v^*) = (x^*, v^*)$. In the bottom half-square, B_- is valid and, from (5.2), $x^* = cx^*$ and $v^* = 2v^*$, which can only be fulfilled with $x^* = 0$, $v^* = 0$. In the top half-square, the only point that does not move under B_+ is $(1, 1)$. There are therefore only two fixed points: $H_- \equiv (0, 0)$ and $H_+ \equiv (1, 1)$. The baker map (5.2) can also be written

[1] If the transition between the forms B_- and B_+ were not sudden but continuous in a region around $v_n = 1/2$, the map would be strongly non-linear there.

as two maps linearised around the origin and point (1, 1):

$$\begin{pmatrix} x_{n+1} \\ v_{n+1} \end{pmatrix} = \begin{pmatrix} c & 0 \\ 0 & 2 \end{pmatrix} \begin{pmatrix} x_n \\ v_n \end{pmatrix}, \quad \text{for} \quad v_n \leq 1/2, \tag{5.4}$$

$$\begin{pmatrix} x_{n+1} - 1 \\ v_{n+1} - 1 \end{pmatrix} = \begin{pmatrix} c & 0 \\ 0 & 2 \end{pmatrix} \begin{pmatrix} x_n - 1 \\ v_n - 1 \end{pmatrix}, \quad \text{for} \quad v_n > 1/2. \tag{5.5}$$

The dynamics is thus characterised around both fixed points (and in both half-planes) by the matrix

$$L = \begin{pmatrix} c & 0 \\ 0 & 2 \end{pmatrix}. \tag{5.6}$$

The eigenvalues of stability matrix L are $\Lambda_+ = 2$, $\Lambda_- = c < 1$; the fixed points, $H_\pm$, are therefore *hyperbolic* (see Section 4.5). The eigenvectors are $\mathbf{u}_+ = (0, 1)$ and $\mathbf{u}_- = (1, 0)$, and, according to (4.40), the Jacobian (cf. Section 4.6) of the map is $J = \Lambda_+\Lambda_- = 2c \leq 1$.

In Fig. 5.2 the character of the motion around the hyperbolic points is indicated by arrowed curves. These two 'interacting' fixed points organise chaos, acting as a kind of 'mixer' within the enclosed area. Qualitatively speaking, we can also imagine the unit square as a 'box ring', where a point that approaches one of the fixed points gets 'knocked' along the arrowed curve, and upon where, reaching the range of the other fixed point, it receives further hits that push it towards the initial fixed point, and so on.

The *inverse* of the baker map is easy to determine. To an arbitrary point of the unit square, (x_n, v_n), it assigns the point

$$(x_{n+1}, v_{n+1}) = B^{-1}(x_n, v_n), \tag{5.7}$$

where

$$B^{-1}(x_n, v_n) = \begin{cases} (x_n/c,\, v_n/2), & \text{for} \quad x_n \leq 1/2, \\ (1 + (x_n - 1)/c,\, 1 + (v_n - 1)/2), & \text{for} \quad x_n > 1/2. \end{cases} \tag{5.8}$$

The effect of the inverse map depends on whether the observed point is situated to the left or to the right of the critical line $x_n = 1/2$.

5.1.2 Chaos in the baker map

The three characteristic properties of chaotic motion described in Section 1.4 are irregularity, unpredictability and fractal structure. Now we show that motion described by the baker map exhibits all three of these properties.

Consider first an initial condition chosen at random and the trajectory starting from it. Let $x_0 = 1/2$ and $v_0 = 2/\pi^2$; parameter c is chosen to be 1/3. Since $v_0 < 1/2$, B_- has to be taken in the first application of (5.2): $x_1 = c/2 = 1/6 = 0.167$, $v_1 = 2v_0 = 4/\pi^2 = 0.405$. The new

Table 5.1. *Twenty iterates of the baker map from the initial condition $x_0 = 1/2$, $v_0 = 2/\pi^2$, with parameter $c = 1/3$.*

Iteration (n)	x_n	v_n	Iteration (n)	x_n	v_n
1	0.167	0.405	11	0.996	0.012
2	0.056	0.811	12	0.332	0.023
3	0.685	0.621	13	0.111	0.046
4	0.895	0.242	14	0.037	0.093
5	0.298	0.485	15	0.012	0.185
6	0.099	0.969	16	0.004	0.370
7	0.700	0.938	17	0.001	0.740
8	0.900	0.876	18	0.667	0.481
9	0.967	0.753	19	0.222	0.961
10	0.989	0.506	20	0.741	0.923

point still lies in the bottom half-square, implying the use of B_- again: $x_2 = c/6 = 1/18 = 0.056$, $v_2 = 2v_1 = 8/\pi^2 = 0.811$. Now $v_2 > 1/2$; therefore B_+ is valid, and $x_3 = 1 + c(x_2 - 1) = 37/54 = 0.685$, $v_3 = 1 + 2(v_2 - 1) = (16/\pi^2 - 1) = 0.621$, and so on. The map can, in principle, be applied as many times as we like, but in practice it is better to write a short computer program to carry out the iterations.[2] Table 5.1 shows the numerical results.

Noticeably, none of the sequences x_n and v_n contain repetitions, but these sequences are naturally too short for a final judgement to be formed. By means of graphical representations, much longer sequences can be overviewed. Figure 5.3 shows the values of x_n and v_n over the first 300 iterates. The diagrams indicate an *irregular* behaviour. The same would be seen for any other initial condition or parameter.

The other important property of chaotic motion is *unpredictability*. In order to illustrate this, let us consider two nearby points and see how the distance between them increases in the course of consecutive iterations. The initial condition for one of the points is again $x_0^{(1)} = 1/2$, $v_0^{(1)} = 2/\pi^2$, while the other is $x_0^{(2)} = x_0^{(1)} + 10^{-9}$, $v_0^{(2)} = v_0^{(1)} + 10^{-10}$. Figure 5.4 shows the logarithm of the distance vs. the iteration number, n. This quantity increases linearly from the third to the 32nd iterate; the distance therefore increases *exponentially*. The slope of the straight line is 0.693. The motion is not predictable in the long run because of the rapid divergence of trajectories.

The rate of deviation of nearby points is measured by the *Lyapunov exponent*. The vertical component of the distance between two

[2] Readers are encouraged to write their own programs to simulate baker (and other) maps. Appendix A.4.1 provides help with this.

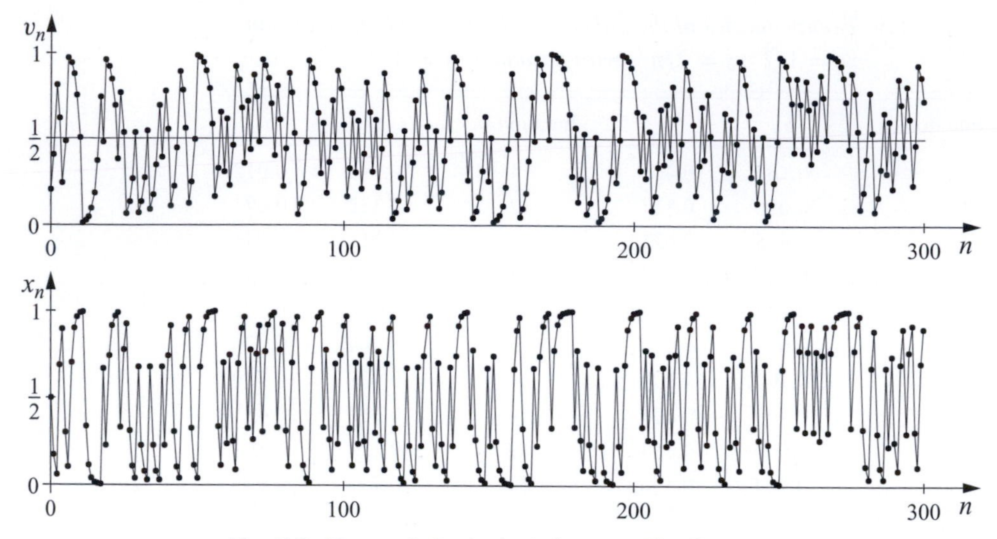

Fig. 5.3. Time evolution in the baker map. The diagram shows no regularity whatsoever: the temporal behaviour of the system is chaotic. The initial condition is $x_0 = 1/2$, $v_0 = 2/\pi^2$ ($c = 1/3$). We have joined up the values in the v_n and x_n diagrams to guide the eye.

Fig. 5.4. Change of the phase space distance, $\Delta r_n = \sqrt{\Delta x_n{}^2 + \Delta v_n{}^2}$ of two points in the course of iterations with an initial distance $\Delta r_0 = 1.005 \times 10^{-9}$ between them. In the logarithmic plot the distance grows linearly up to the 32nd iterate. The difference increases to 10^9 times the original value in 30 steps!

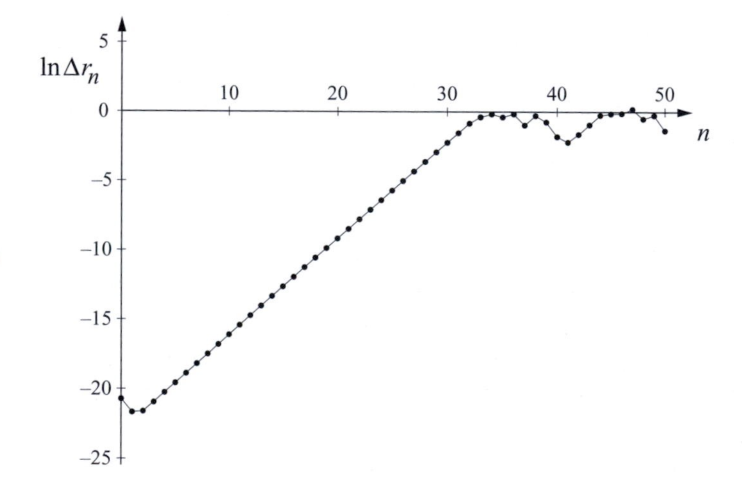

points doubles in one iteration; in n steps, therefore, $\Delta v_n = 2^n \Delta v_0 = e^{(\ln 2)n} \Delta v_0$. The Lyapunov exponent, λ, is read off from the relation $\Delta v_n = \Delta v_0 e^{\lambda n}$ as

$$\lambda = \ln 2 = 0.693. \tag{5.9}$$

In the x-direction, the distance between two nearby points decreases as $\Delta x_n = c^n \Delta x_0$. The negative Lyapunov exponent defined via $\Delta x_n = \Delta x_0 e^{\lambda' n}$ is thus the logarithm of parameter c:

$$\lambda' = \ln c < 0. \tag{5.10}$$

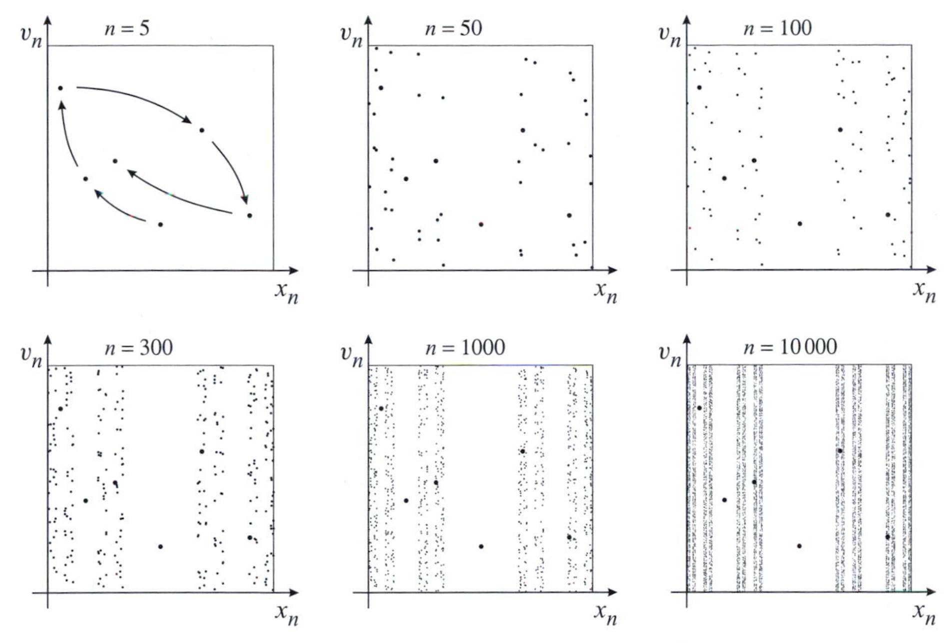

Fig. 5.5. Time evolution of a single trajectory in phase space. Starting from $x_0 = 1/2$, $v_0 = 2/\pi^2$ ($c = 1/3$), the points obtained after 5, 50, 100, 300, 1000 and 10 000 iterations are consecutively plotted in the (x_n, v_n)-plane. Due to the increasing density, the size of the points has been gradually decreased. Arrows show the order of succession of the first five points. These points are marked as larger dots in all the panels.

In the expression of the total phase space distance, $\Delta r_n = \sqrt{\Delta x_n{}^2 + \Delta v_n{}^2}$, the velocity difference increases, while Δx_n decreases, in time. After a few iterations, therefore, Δv_n starts to dominate, and the total distance increases according to the positive Lyapunov exponent, i.e. $\Delta r_n \sim e^{\lambda n}$. The positive Lyapunov exponent therefore yields the growth rate of the full phase space distance between pairs of nearby trajectories (for more details, see Section 5.4.2).

It is worth emphasising that, due to the finite extension of the phase space, this rule of increase is only valid for a finite time (the smaller the initial distance Δr_0, the longer this finite time). In Fig. 5.4 the two points become so much removed from each other after the 32nd iterate that they are in opposite positions within the unit square. Their distance from one another can no longer increase substantially; the logarithmic plot therefore goes into saturation, apart from small fluctuations.

Let us now represent a single trajectory in the phase space of the baker map. As time passes, the points that are apparently jumping randomly trace out more and more clearly a *regular geometric structure* (see Fig. 5.5).

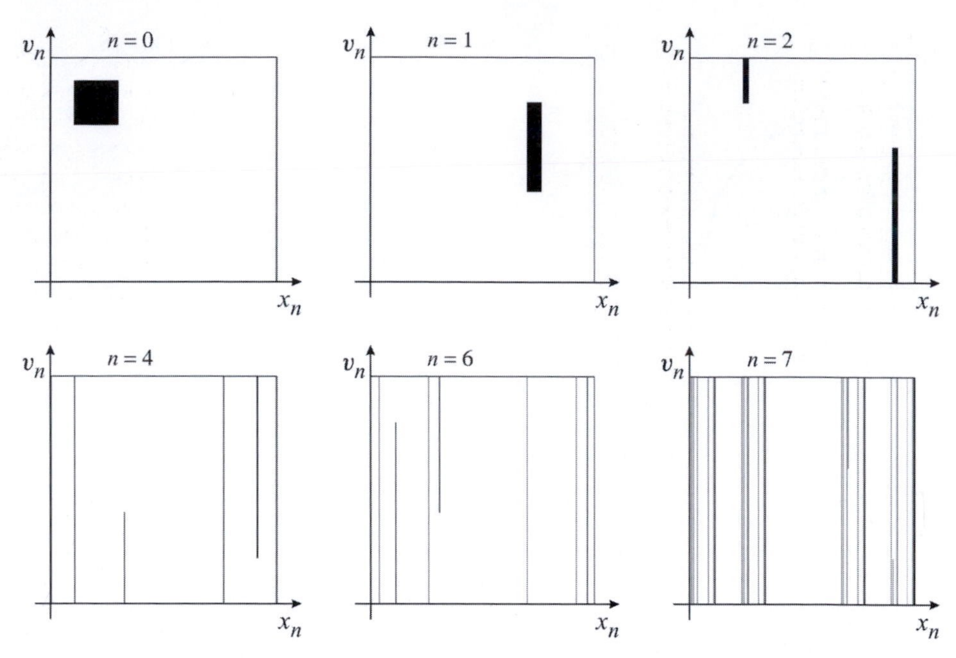

Fig. 5.6. Time evolution of a phase space domain. In the numerical process a grid of size 0.0001 is created in the range $0.1 \leq x_0 \leq 0.3$; $0.7 \leq v_0 \leq 0.9$, and from each grid-point a trajectory is started (a total of four million). The images of each point are plotted up to the seventh iterate ($c = 1/3$).

An important property of chaotic systems is that the dynamics (time evolution) and the geometrical structure (a stationary phase space object) are closely related. This illustrates that chaos does not represent complete disorder (i.e. it is not the same as noise); if it did, the points would fill the full unit square instead of tracing out a pattern. The initial condition is not on the above-mentioned object of phase space; the trajectory nevertheless approaches it very closely within a few iterates.

5.1.3 The chaotic attractor

The subset of phase space which the trajectory converges to is an attractor, namely a chaotic attractor. In order to illustrate this, we could use many different initial conditions and show that they all converge to the same set, on which the character of the motion, and thus the Lyapunov exponent, is the same. Instead of individual trajectories, however, it is better to explore the dynamics of a phase space volume filled with a large number of points in a numerical simulation.

The phase space domain converges to the same set as a single trajectory (Fig. 5.6). This experiment can be repeated from different initial positions with different phase space volumes, and the same asymptotic

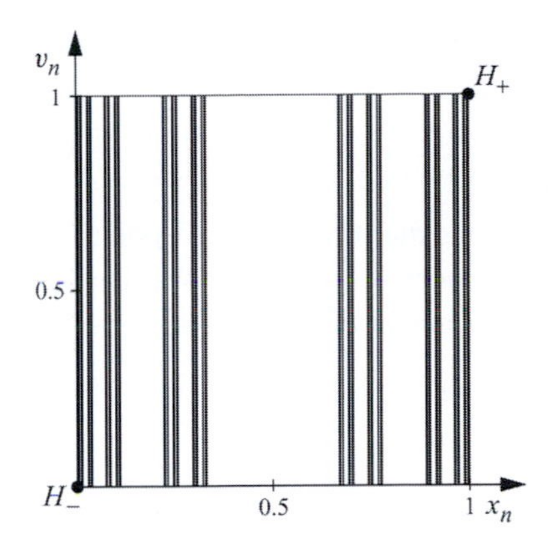

Fig. 5.7. The baker attractor ($c = 1/3$). Numerically, the plot has been created by performing many ($\approx 10^5$) iterates of an arbitrary initial point and omitting the first few dozens of images because they are usually not close enough to the chaotic attractor.

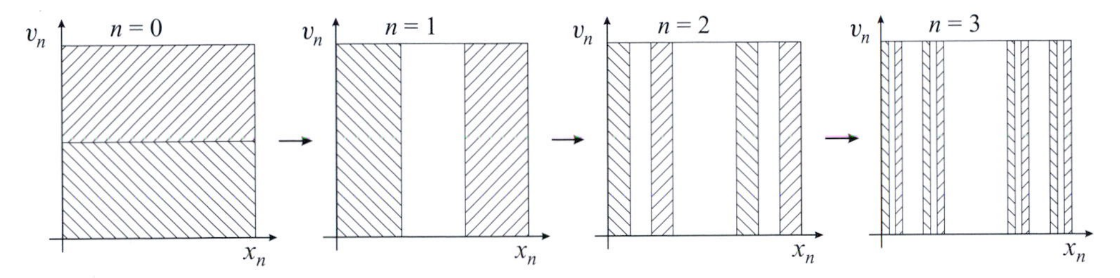

Fig. 5.8. Change of the phase space volume of the entire unit square throughout three iterations. Areas originating from the top and from the bottom half are distinguished by different types of shading. As the phase space volume shrinks onto the chaotic attractor, the emergence of a fractal structure is discovered ($c = 1/3$).

structure is obtained in *all* cases. It can therefore be stated that this structure is an attracting set in phase space, i.e. an attractor. The basin of attraction is the entire unit square.

The chaotic attractor of the baker map has thus been identified as shown in Fig. 5.7. Note that the attractor of this simple model map has a structure similar to that of the local patterns of the chaotic attractors of the physical systems in Section 1.2.1.

The phase space volume tends to zero because of dissipation, in agreement with the fact that a chaotic attractor has no area; it is a *set of measure zero*, a fractal (see Section 2.2). The structure and development of this fractal are easier to understand by following the motion of the entire unit square over a few iterations (Fig. 5.8). The successive widths of the resulting columns are $c, c^2, c^3, \ldots$, whereas their number increases as powers of two. The phase space evolution is exactly the same process as that used in Section 2.2.2 to construct Cantor filaments (see Fig. 2.9); the contraction rate, r, is now the parameter, c, of the baker map. The chaotic attractor therefore is a set of Cantor filaments. According to relation (2.10), its fractal dimension is

$$D_0 = 1 + \frac{\ln 2}{\ln(1/c)}. \tag{5.11}$$

The construction of Fig. 5.8 also shows that the two fixed points are, in every step, parts (corners) of the outermost columns. We therefore conclude that the hyperbolic points $H_{\pm}$ are *on* the chaotic attractor.

Problem 5.1 By means of numerical simulation, determine the chaotic attractor of the baker map for different values of parameter c.

5.1.4 The chaotic attractor and unstable manifolds

Hyperbolic points always possess stable and unstable manifolds (see Sections 4.3 and 4.5) which follow the directions of the eigenvectors in the immediate vicinity of the fixed points. The eigenvector of fixed points $H_{\pm}$ belonging to eigenvalue Λ_+ is $\mathbf{u}_+ = (0, 1)$; the unstable manifolds begin therefore with a vertical segment. The question is, what does the entire unstable manifold look like? First, we determine the longest connected piece of the manifold emanating from a fixed point, called the *basic branch*. For the fixed points H_- and H_+, these are the segments $(0, 0) - (0, 1)$ and $(1, 0) - (1, 1)$, respectively, since in the inverted map all points on these segments move towards the corresponding fixed points, along the segment. The construction of the entire manifold is based on the observation that an initial segment of the unstable manifold is mapped onto a longer segment of the manifold. In the baker map the length of a vertical segment is multiplied by a factor of two in each iteration (it can, however, only fit into the unit square if it is simultaneously 'folded'). Using this, the unstable manifold can be constructed for arbitrary lengths.

Figure 5.9 exhibits the beginning of the construction. The segment containing the basic branch of the origin above the line $v_n = 1/2$ does not remain on the vertical axis, but is mapped onto a vertical line with co-ordinate $x_n = 1 - c = 2/3$, since, according to B_+, (5.2), the images of points with $x_n = 0$ are the points with $x_{n+1} = 1 - c$. The segment of the basic branch of the fixed point $(1, 1)$ below the line $v_n = 1/2$ is mapped onto a vertical line with co-ordinate $x_n = c = 1/3$ (according to B_-, the images of points with $x_n = 1$ are the points with $x_{n+1} = c$).

In Fig. 5.10, further iterations are carried out and longer segments of the unstable manifolds of the two fixed points are drawn. In each iteration the unstable curves are split in two (they are not connected here

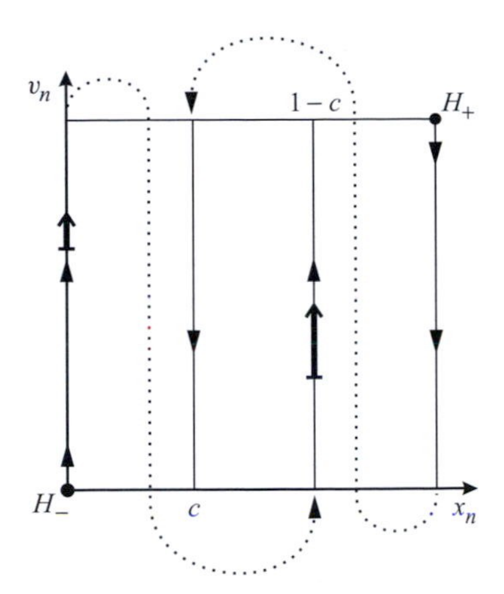

Fig. 5.9. Unstable manifolds of the fixed points H_+ and H_- (arrows pointing downwards and upwards, respectively), obtained in the first two steps of the construction: the basic branch (the left and right edges of the square) and its first iterate ($c = 1/3$). Dotted lines indicate how the manifold segments are joined to each other. The arrow with a tail, and its image, illustrate stretching along the unstable manifold.

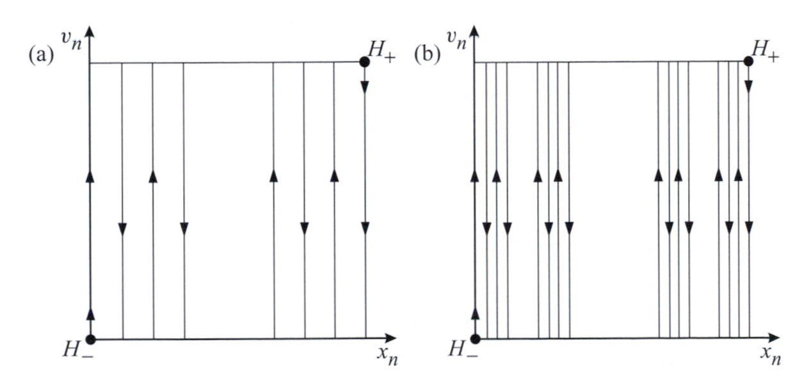

Fig. 5.10. Continuation of the unstable manifolds of H_+ and H_- (arrows pointing downwards and upwards, respectively) over further iterations. Images of the basic branches are shown after (a) two and (b) three iterations. The smallest distance between the vertical segments is (a) c^2, (b) c^3.

with dotted lines, but the arrows show unambiguously which fixed point they belong to). Continuing the procedure, longer and longer segments of both unstable manifolds become visible. Each of the unstable manifolds is a discontinuous line of infinite length 'squeezed' into the unit square.

Comparing the construction of the unstable manifolds with Fig. 5.8, which shows the shrinking of the entire phase space volume, we make the interesting observation that pieces of the unstable manifold coincide with the edges of the phase space volume columns. Since these columns converge to the attractor (without their edges moving, by merely producing further and further edges via splitting), we have to conclude that the unstable manifolds of the fixed points belong to the attractors. Moreover, the unstable manifolds appear to be *identical* to the chaotic attractor.

All this is in accordance with our previous observations (Section 3.3): unstable manifolds lead into attractors. Now we have a new situation where the hyperbolic point is on the chaotic attractor, and, since the manifold has to lead into this same attractor (there exists no other attractor within the unit square), all this is possible only if the entire unstable manifold is part of the chaotic attractor itself.

5.1.5 Two-cycles

Let us investigate whether there exist cycles in the baker map. The calculation of the two-cycles is similar to that of the fixed points, but the point remains invariant after two iterations only: $B^2(x^*, v^*) \equiv B(B(x^*, v^*)) = (x^*, v^*)$. It may occur that both points of a two-cycle are in the bottom half, both are in the top half or one is in the top and the other in the bottom half of the unit square. This corresponds, respectively, to applying the form B_- or B_+ twice, or B_- and B_+ successively. In the first two cases, the resulting points are the fixed points $(0, 0)$ and $(1, 1)$, since fixed points are simultaneously two-, three-, etc. cycles as well. In the third case, we obtain $(x^*, v^*) = B_+B_-(x^*, v^*) = (c^2x^* - c + 1, 4v^* - 1)$. Thus, the coordinates of one two-cycle point are $x^* = 1/(1+c)$ and $v^* = 1/3$. This point[3] is $P_1 \equiv (1/(1+c), 1/3)$. Since this is in the bottom half-square, the other cycle point can be determined by applying map B_- to P_1 once, and is found to be $P_2 \equiv (c/(1+c), 2/3)$. It is easy to verify that iterating this again (applying B_+) leads back to P_1:

$$\cdots \xrightarrow{B_+} P_1 \xrightarrow{B_-} P_2 \xrightarrow{B_+} P_1 \xrightarrow{B_-} P_2 \xrightarrow{B_+} \cdots . \quad (5.12)$$

We have thus found three two-cycles. Two of them are the known fixed points H_+ and H_-, and there exists only one non-trivial two-cycle: the couple P_1, P_2.

Problem 5.2 Write the map $(x_{n+2}, v_{n+2}) = B^2(x_n, v_n)$ in a form similar to (5.2). Using this, show that P_1 and P_2 form a two-cycle.

[3] By applying the maps in reversed order (first B_+, afterwards B_-), the other point (P_2) of the two-cycle is obtained.

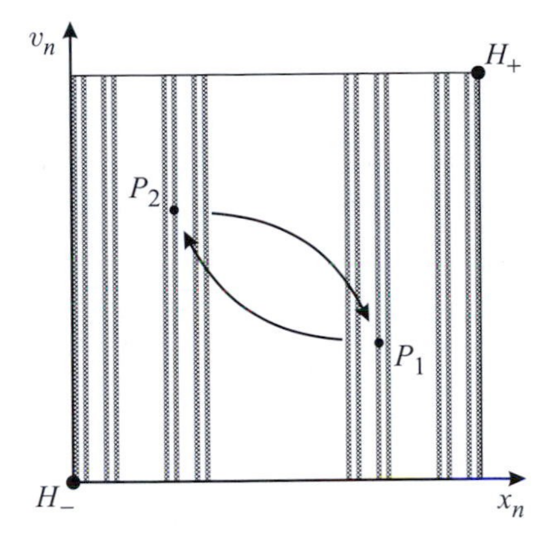

Fig. 5.11. The two-cycle P_1, P_2 and the baker attractor. The two-cycle is also hyperbolic and is part of the attractor.

Figure 5.11 suggests that the two-cycle is on the chaotic attractor. We can check this by plotting points P_1 and P_2 in figures of higher resolutions. However high the resolution of the attractor is, we would always find the points of the two-cycle to be on the attractor.

Both points of the two-cycle behave as hyperbolic fixed points, each possessing its individual stable and unstable manifolds. We have seen that the direction and strength of stretching and contraction are uniform throughout the baker map. The eigenvalues of the fixed points of the twice-iterated map $B^2(x_n, v_n)$ are those of the once-iterated map, squared: $\Lambda_+^2 = 4$, $\Lambda_-^2 = c^2 < 1$. The square roots of these, i.e. the eigenvalues corresponding to one iteration, are the same for both cycle elements as for the fixed points.

Problem 5.3 Construct the unstable manifolds of the two-cycle P_1, P_2 in a manner analogous to that applied for the fixed points.

We emphasise that the unstable manifolds of the two-cycle do not coincide anywhere with the unstable manifolds of the fixed points. This is only possible if the manifolds discussed so far are very close to each other. This is the reason why any one of the manifolds approximates the attractor very closely.

5.1.6 Higher-order cycles

Similar to the method of Section 5.1.5, other periodic orbits can also be determined. Contrary to two-cycles, there usually exist several independent higher-order cycles of the same length. We can, for example, find two independent three-cycles, consisting of six cycle points altogether.

Problem 5.4 Determine analytically the three-cycles of the baker map.

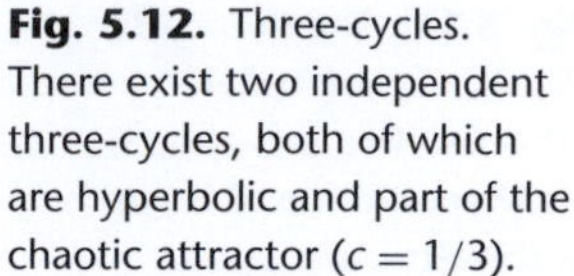

Fig. 5.12. Three-cycles. There exist two independent three-cycles, both of which are hyperbolic and part of the chaotic attractor ($c = 1/3$).

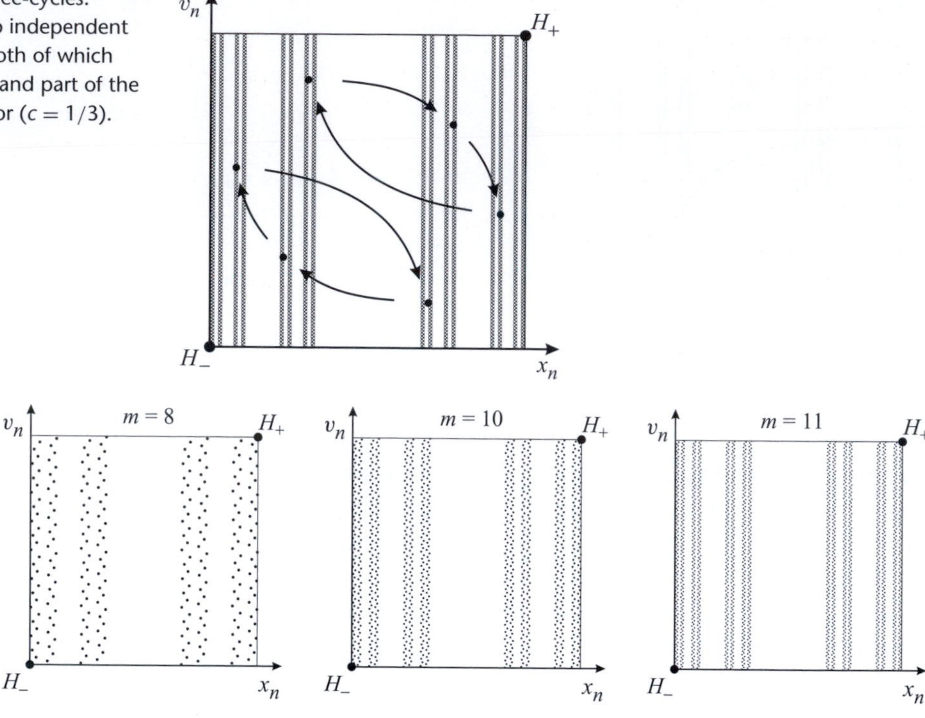

Fig. 5.13. Higher-order cycles: all points of the cycles of length $m = 8$, 10 and 11. The cycle points of cycles of increasing length approximate the entire chaotic attractor more and more accurately ($c = 1/3$).

Figure 5.12 exhibits the three-cycles and the arrows indicate the motion of the cycle points under the mapping. The figure suggests, and it can be checked in higher resolutions, that all the cycle points lie on the chaotic attractor. Both three-cycles are hyperbolic, and the eigenvalues of the cycle points for one iteration are Λ_+ and Λ_-, since the stability matrix, L, is the same everywhere.

The elements of an m-cycle are the fixed points of the m-fold iterated map, i.e. the roots of the equation $(x^*, v^*) = B^m(x^*, v^*)$. Figure 5.13 exhibits all the points of a few higher-order cycles. There are visibly more and more points that belong to longer cycles, and they can all be checked to be on the chaotic attractor. Consequently, the points of higher- and higher-order cycles represent the chaotic attractor itself more and more accurately. In other words, the cycle points lie *densely* on the chaotic attractor.

For each cycle point, the same statement holds as for the fixed points and the two-cycles: their eigenvalues for one iteration are Λ_- and Λ_+. Thus, all the cycles are hyperbolic. Consequently, the hyperbolic cycles form the skeleton of the chaotic attractor.

We will find it useful to deduce a relationship between the cycle length and the number of cycle points. There exist two fixed points. There is one two-cycle, and this implies two cycle points. Together with the two fixed points, the total number of the elements of two-cycles is four. We have found two three-cycles with six cycle points, and, on adding the two fixed points, we have a total of eight elements for the three-cycles. As far as the four-cycles are concerned, there exist three independent ones with 12 cycle points. Including the two-cycle and the fixed points, the total number of cycle points for the four-cycles is 16. The number of five-cycles is five, which amounts to 32 points together with the fixed points. These suggest that the total number, N_m, of cycle points belonging to the m-cycles is 2^m.[4] The number of fixed points of map B^m can also be obtained by direct calculation.

Problem 5.5 Determine the mth iterate of the baker map for the velocity variable v_n. (This can be achieved without taking variable x_n into consideration.) Determine the v^* co-ordinates of all the fixed points of $B^m(x_n, v_n)$.

The total number of elements of cycles of length m is thus given by

$$N_m = 2^m. \tag{5.13}$$

Accordingly, there exist cycles of *arbitrary length*; moreover, the number of all the points belonging to the cycles increases *exponentially*. The growth rate, h, defined by the relation $N_m \sim e^{hm}$ is called the *topological entropy* (for more details, see Section 5.4.1). The topological entropy of the baker map is therefore given by

$$h = \ln 2 = 0.693. \tag{5.14}$$

We arrive at the verification of our previous statement that chaos is an unstable state (for example, the state illustrated by a pencil standing on its point) appearing with infinite multiplicity. Thus, chaotic motion can be considered as a *random walk among unstable cycles*. The iterated point can temporarily approach one of the cycles (in Fig. 5.3, for example, a four-cycle and a three-cycle take shape around $n \approx 30$ and $n \approx 250$, respectively). Since, however, the cycles are unstable, the trajectory can only remain in this neighbourhood for a finite time, and sooner or later

[4] The same conclusion can be drawn by recognising that each column that appears in Fig. 5.8 showing the mth image of the entire phase space volume contains one (and only one) m-cycle point. The fixed points are always in the outermost columns, the two central columns of the diagram for $n = 2$ contain points P_1 and P_2 of the two-cycle, and in each of the six inner columns of the $n = 3$ case lies one three-cycle point (cf. Fig. 5.12).

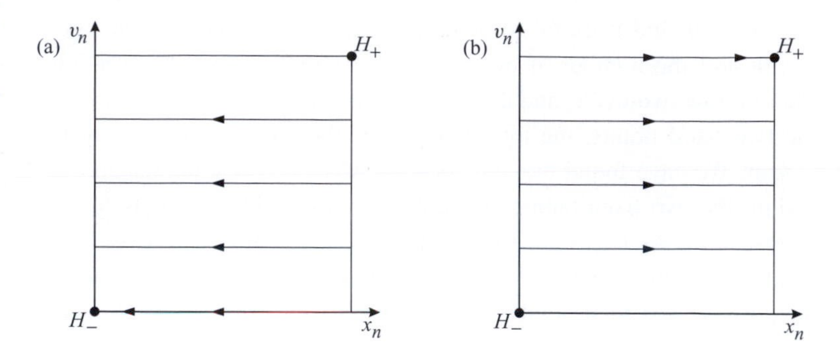

Fig. 5.14. A few branches of the stable manifold of (a) H_- and (b) H_+ in the unit square. Note that some branches coincide, which is a mark of the space-filling property of these manifolds ($c = 1/3$).

it approaches another cycle. This is the origin of the irregular nature of chaotic dynamics.

There is an unstable manifold of infinite length that emanates from each cycle point, and these are also part of the attractor. In summary, unstable manifolds of infinite length belonging to an infinity of cycle points all belong to the chaotic attractor.

5.1.7 Stable manifolds, homoclinic points and heteroclinic points

The hyperbolic cycle points also possess stable manifolds. The cycles can be approached along these manifolds, and, since they are dense on the attractor, the chaotic attractor itself is then approached. In the baker map the stable manifolds of the cycle points are straight line segments parallel to the x_n-axis (Fig. 5.14), which provide a dense foliation of the unit square, and are not of fractal nature. Together, all the stable manifolds form the basin of the chaotic attractor.

The intersections of all stable and unstable manifolds are points with interesting properties. They are usually divided into two groups. *Homoclinic points* are the intersections of the stable and unstable manifolds of the same cycle point. *Heteroclinic points*, on the other hand, are the intersections of the stable and unstable manifolds of different cycle points (Fig. 5.15).

Homoclinic points are, at the same time, on both the stable and the unstable manifold of the given cycle point. During iterations, they must therefore simultaneously deviate from and approach the cycle point. This is only possible if the homoclinic points never reach the cycle points, i.e. they only approach them. Stable and unstable manifolds are invariant curves that are mapped onto themselves. The images of the manifold intersections must therefore also be intersections. Homoclinic points are mapped onto other homoclinic points. In general, an *infinity* of homoclinic points is formed along the manifolds of a single cycle

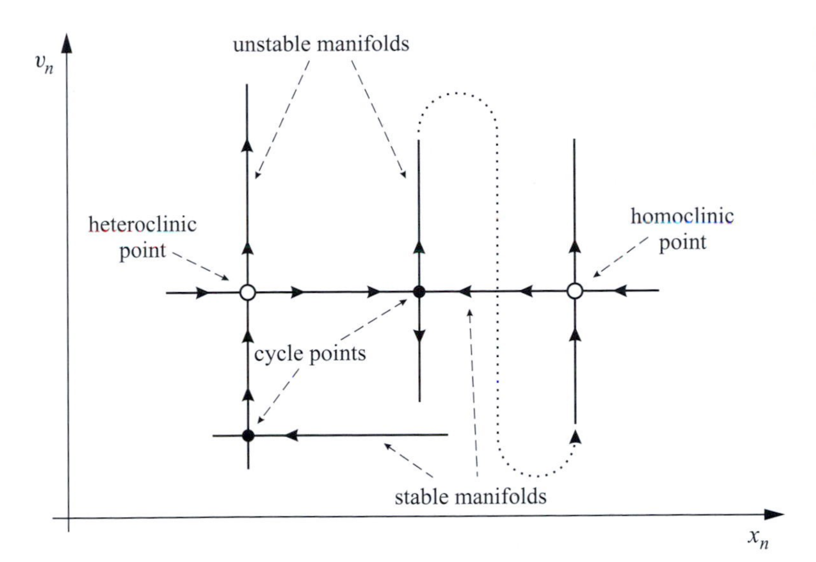

Fig. 5.15. Homoclinic and heteroclinic points. Black dots and empty circles mark cycle points and homoclinic and heteroclinic points, respectively. A dotted line indicates the continuation of the unstable manifold of the cycle point.

point. Heteroclinic points fulfil a similarly complicated requirement, since – as we have seen – the unstable manifold of any cycle point is nearly identical to that of another one. A typical heteroclinic point can be mapped only onto another heteroclinic point in the course of iterations. Homoclinic and heteroclinic points must therefore belong to *aperiodic, chaotic* motion.

Since the unstable manifolds are part of the attractor, the homoclinic and heteroclinic points, being subsets of these manifolds, also all lie on the chaotic attractor, and, in addition, they are very close to each other. If we choose a point on the attractor at random, then points of all three types will be found in an arbitrarily small neighbourhood of the chosen point. Chaotic motion, understood as wandering among cycle points, is realised through the presence of homoclinic and heteroclinic points.[5]

5.1.8 Asymmetric baker map

The fact that the stability characteristics of the two fixed points and of any periodic orbit are identical is a consequence of the symmetry of the baker map (5.2). It is, however, easy to give an asymmetric extension, where both the expansion and the contraction parameters depend on which half-square the point is in. The form of this map is given by

$$(x_{n+1}, v_{n+1}) = (c_1 x_n, a_1 v_n), \quad \text{for} \quad v_n \leq b, \tag{5.15}$$

$$(x_{n+1}, v_{n+1}) = (1 + c_2(x_n - 1), 1 + a_2(v_n - 1)), \quad \text{for} \quad v_n > b. \tag{5.16}$$

[5] The relation between periodic points vs. homoclinic and heteroclinic points is similar to that of rational and irrational numbers.

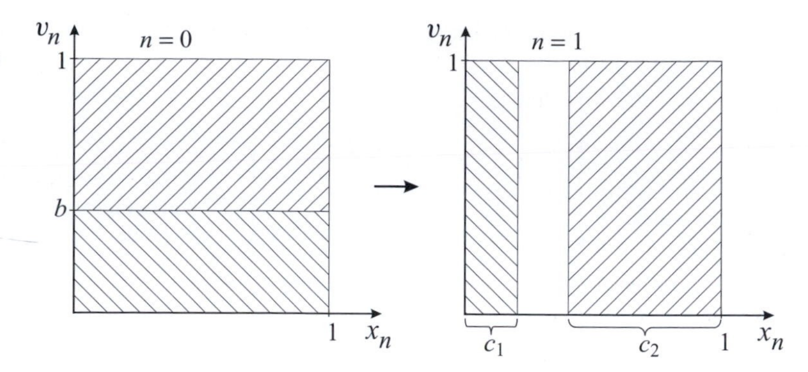

Fig. 5.16. Action of an asymmetric baker map on the unit square ($c_1 = 0.2$, $c_2 = 0.6$, $b = 0.4$).

Fig. 5.17. The asymmetric baker attractor ($c_1 = 0.2$, $c_2 = 0.6$, $b = 0.4$).

Here, $a_1 = 1/b$, $a_2 = 1/(1 - b)$, c_1, c_2 and $b < 1$ are positive parameters and $c_1 + c_2 < 1$. The mapping acts as follows: the region $v_n < b$ is compressed along the x-axis by a factor c_1, while it is expanded by a factor $a_1 = 1/b$ along the v_n-axis (Fig. 5.16). In the region $v_n > b$, the contraction and stretching factors are c_2 and $a_2 = 1/(1 - b)$, respectively. With $c_1 = c_2$, $b = 1/2$, the symmetric baker map is recovered.

The more the ratio c_1/c_2 differs from unity, the more conspicuous the asymmetry of the new baker attractor (Fig. 5.17). The directions of the stable and unstable manifolds are unchanged; therefore, we present no plots of the manifolds.

We emphasise, however, that the asymmetry is present even in the special case when $c_1 = c_2$, as long as $b \neq 1/2$: a different number of particles go to the left and right sides of the chaotic attractor. Then the asymmetry does not arise in the geometrical structure of the attractor, but rather in the probability distribution related to it (see Section 5.4.4).

Problem 5.6 Derive the equation for the fractal dimension of the asymmetric baker map with parameter $c_1 \neq c_2$.

Problem 5.7 What relation must parameter b of the asymmetric baker map fulfil between itself and the other two parameters so that the Jacobian of the map is the same constant everywhere?

Problem 5.8 Under what condition will the Jacobian of the top half-square be the reciprocal of that in the bottom half? Is such a map globally area contracting?

5.2 Kicked oscillators

5.2.1 General properties

The stroboscopic map of periodically kicked harmonic oscillators was derived in Section 4.4 (see (4.31)) in the dimensionless form

$$x_{n+1} = E v_n, \quad v_{n+1} = -E x_n + I(x_{n+1}). \tag{5.17}$$

Here, $E \equiv e^{-(\alpha/2)T} < 1$ is a damping parameter measuring the power of dissipation, and the phase space is the entire plane (x_n, v_n).

It is worth transforming map (5.17) into an equivalent form. This is useful because, if E decreases (friction increases), the picture representing the motion in the phase plane (x_n, v_n) becomes compressed along the x_n-axis. By introducing a new variable and notation via

$$E v_n \to v_n, \quad E I(x_n) \equiv f(x_n), \tag{5.18}$$

respectively, the picture size becomes independent of E, provided the values of function f are of order unity. From (5.17),

$$x_{n+1} = v_n, \quad v_{n+1} = -E^2 x_n + f(v_n). \tag{5.19}$$

In what follows we use this form of the kicked oscillator map and denote it by K as follows:

$$(x_{n+1}, v_{n+1}) = K(x_n, v_n) \equiv (v_n, -E^2 x_n + f(v_n)). \tag{5.20}$$

In components:

$$x_{n+1} = K_1(v_n) \equiv v_n, \quad v_{n+1} = K_2(x_n, v_n) \equiv -E^2 x_n + f(v_n). \tag{5.21}$$

The Jacobian of the map is (see Section 4.6) $J = E^2$.

The inverse of K, $(x_{n+1}, v_{n+1}) = K^{-1}(x_n, v_n)$, is obtained from (5.19) for *any* (x_n, v_n) as follows:

$$x_{n+1} = -\frac{1}{E^2} v_n + \frac{1}{E^2} f(x_n), \quad v_{n+1} = x_n. \tag{5.22}$$

It can easily be checked that $(x_n, v_n) = K^{-1}K(x_n, v_n)$.

In the subsequent sections we analyse cases driven by amplitude functions of the following form:

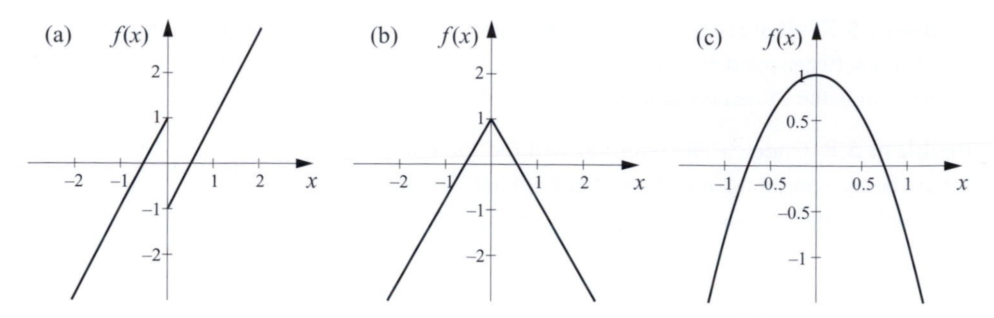

Fig. 5.18. (a) Sawtooth, (b) roof and (c) parabola amplitude functions. (The parameter a is equal to 1.95, 1.77 and 1.8, respectively.)

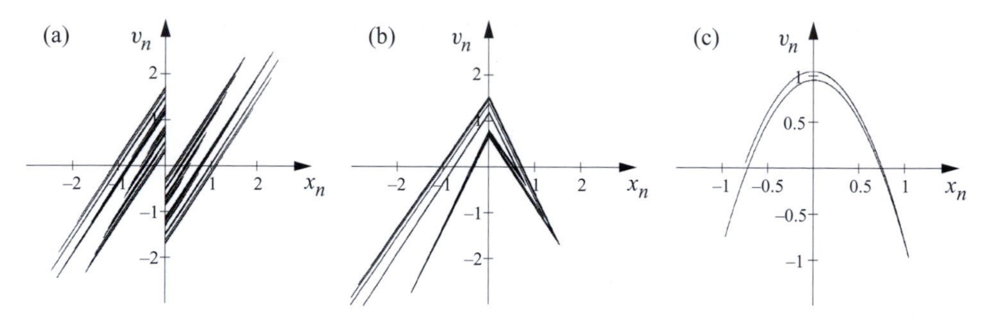

Fig. 5.19. Chaotic attractors of the oscillator kicked with (a) a sawtooth, (b) a roof and (c) a parabola amplitude. Parameter a is identical in each case to that set in Fig. 5.18, and the respective values of E are 0.8, 0.7 and 0.25.

- sawtooth amplitude:

$$f(x) = ax - \operatorname{sgn}(x), \tag{5.23}$$

- roof amplitude:

$$f(x) = 1 - a|x|, \tag{5.24}$$

- parabola amplitude:

$$f(x) = 1 - ax^2. \tag{5.25}$$

These are presented in Fig. 5.18. Since all three functions are non-linear (even though the first two are piecewise linear), the quantity $a > 0$ is called the *non-linearity parameter*. The stroboscopic dynamics of kicked oscillators is characterised by two dimensionless numbers, the damping and the non-linearity parameters E and a, respectively.

The motion of kicked oscillators is chaotic over a wide range of the parameters. It is interesting to observe that the chaotic attractor appears to be the union of infinitely many copies of the graph of $f(x)$, as illustrated in Fig. 5.19. (Note that the co-ordinates in Figs. 5.18 and 5.19 are not the same.)

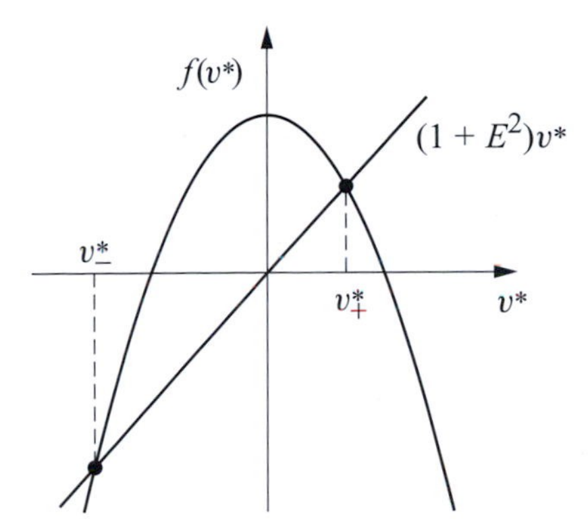

Fig. 5.20. Fixed points are obtained from the intersections of the graph of $f(v^*)$ and the straight line $(1 + E^2)v^*$ (schematic diagram).

Our aim in this section is not to determine the chaos parameters (topological entropy, Lyapunov exponents, fractal dimensions), but rather to explore the typical geometrical structures of chaotic attractors and their basins.

5.2.2 Fixed points and their stability

We have seen in the example of the baker map that the unstable manifolds of the fixed points are intimately related to the chaotic attractor. Therefore, we determine here the possible fixed points and investigate when they can be unstable (with a general amplitude $f(x)$). The equation $K(x^*, v^*) = (x^*, v^*)$ for the fixed point co-ordinates yields

$$x^* = v^*, \quad v^* = -E^2 x^* + f^*. \tag{5.26}$$

Here, and in the following, we use the notation $f^* \equiv f(v^*)$ and $f'^* \equiv f'(v^*)$ (the prime denotes the derivative). According to (5.26), a condition for the existence of fixed points is that the equation

$$(1 + E^2)v^* = f^* \tag{5.27}$$

has solutions (Fig. 5.20). We investigate functions f that possess two branches, as shown in Fig. 5.18; we therefore usually obtain two fixed points. As in the baker map, these will be denoted by H_+ and H_-.

The stability of a fixed point (see Section 4.5) follows from the dynamics in its vicinity. Map (5.19) linearised around any of the two (x^*, v^*) is governed by the stability matrix

$$L = \begin{pmatrix} 0 & 1 \\ -E^2 & f'^* \end{pmatrix}, \tag{5.28}$$

whose eigenvalues are given by

$$\Lambda_\pm = \frac{f'^* \pm \sqrt{f'^{*2} - 4E^2}}{2}. \tag{5.29}$$

Henceforth we follow the convention of denoting the eigenvalue with the greater absolute value by Λ_+. In (5.29), therefore, the positive sign belongs to Λ_+ if f'^* is positive; otherwise the negative sign should be taken.

When determining the stability of the fixed points, we make use of the fact (see (4.41)) that the product of the eigenvalues equals the Jacobian:

$$\Lambda_+\Lambda_- = E^2 < 1. \tag{5.30}$$

We assume that the eigenvalues are real; in other words, that the discriminant of (5.29) is positive: $|f'^*| > 2E$. If $|\Lambda_+| > 1$, then, according to (5.30), $|\Lambda_-| < 1$, and the fixed point is *hyperbolic*. If f'^* is positive – if the fixed point lies in the domain where the kicking amplitude is an increasing function – the condition for hyperbolicity is obtained via rearranging (5.29): $f'^* > 1 + E^2$. If, on the other hand, f'^* is negative, then $\Lambda_+ = \left(f'^* - \sqrt{f'^{*2} - 4E^2}\right)/2 < -1$ and, according to (5.30), Λ_- is negative as well,[6] and the fixed point is hyperbolic if $f'^* < -(1 + E^2)$. In summary: hyperbolic points exist if

$$|f'^*| > 1 + E^2. \tag{5.31}$$

This implies that the kicking amplitude must change sufficiently rapidly around the fixed point in order for this latter to be unstable.

The eigenvectors of the hyperbolic point, from the solution of equation $\Lambda_\pm \mathbf{u}_\pm = L\mathbf{u}_\pm$, are given by

$$\mathbf{u}_\pm = (1, \Lambda_\pm). \tag{5.32}$$

The eigenvectors yield the local directions of the manifolds. Accordingly, the eigenvalues Λ_+ and Λ_- are now the slopes of the *basic branches* of the unstable and stable manifolds emanating from a fixed point, respectively. The local forms, $v_\pm(x)$, of the basic branches around fixed points are therefore given by

$$v_n \equiv v_\pm(x_n) = \Lambda_\pm(x_n - x^*) + x^*. \tag{5.33}$$

Here we have taken into account that $x^* \equiv v^*$. The top (bottom) sign belongs to the unstable (stable) manifold (see Fig. 5.21).

Problem 5.9 Under what condition will the fixed point of a kicked oscillator be a node attractor or a spiral fixed point (for an arbitrary function f)?

[6] The consequence of negative eigenvalues is that the point jumps to the opposite side of the fixed point in each iteration.

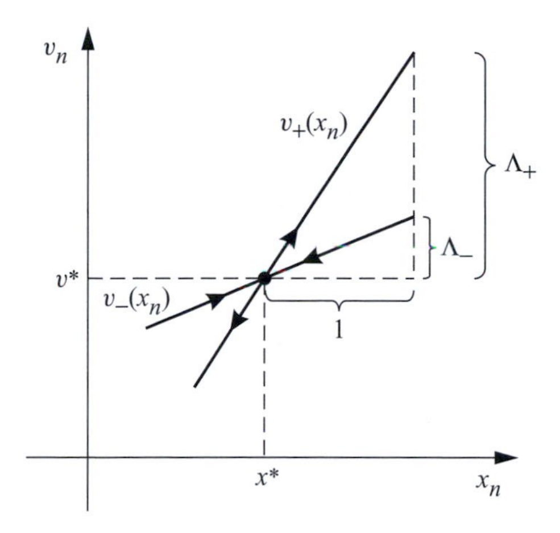

Fig. 5.21. Local form of the stable (v_-) and unstable (v_+) manifolds of a hyperbolic point in kicked oscillators. The slopes are given by the eigenvalues.

Problem 5.10 Analyse the stability of the fixed points of an oscillator kicked with the linear amplitude function $f(x) = 1 - ax$ ($a > 0$) and give the equations of the stable and unstable manifolds.

Problem 5.11 Analyse the behaviour of a kicked oscillator for a kicking amplitude $f \equiv 1 =$ constant.

Problem 5.12 Derive the equations that determine the two-cycles of a kicked oscillator with an arbitrary $f(x)$.

5.2.3 Sawtooth attractor

The stroboscopic map of an oscillator kicked with a sawtooth amplitude, (5.23), is the sawtooth map:

$$x_{n+1} = v_n, \quad v_{n+1} = -E^2 x_n + a v_n - \text{sgn}(v_n). \tag{5.34}$$

The fixed point co-ordinates are given by

$$x_\pm^* = v_\pm^* = \frac{\pm 1}{a - (1 + E^2)}, \tag{5.35}$$

where the upper and lower signs belong to $H_+ = (x_+^*, v_+^*)$ and $H_- = (x_-^*, v_-^*)$, respectively.

It can be seen from Fig. 5.20 that the straight line $(1 + E^2)v^*$ can intersect the half-lines $av^* - 1$, $v^* > 0$ and $av^* + 1$, $v^* < 0$ only if its slope is less than a. Therefore there exist fixed points only if $a > 1 + E^2$. This is equivalent to condition (5.31) of hyperbolicity. The fixed points of an oscillator kicked with the sawtooth amplitude are therefore always unstable. The stability matrix, (5.28), is, in our case, independent of the position; consequently, all higher-order cycles are also hyperbolic. In addition, their eigenvectors are identical, all given by (5.32).

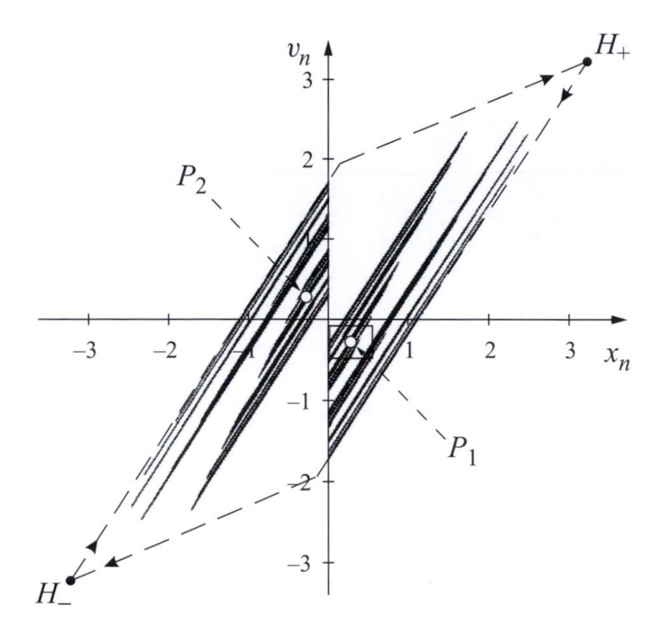

Fig. 5.22. The sawtooth attractor ($a = 1.95$, $E = 0.8$). The attractor has been determined via numerically following a single trajectory and omitting the first few steps, until the plot stopped changing at the given resolution. The two-cycle marked by white dots (P_1, P_2) are on the attractor. The dashed lines indicate the basic branches of the manifolds of $H_{\pm}$.

Accordingly, regular attractors cannot be present in this system. The chaotic attractor obtained via numerical simulation is presented in Fig. 5.22, which shows that the fixed points, $H_{\pm}$, lie *outside* of the attractor. Readers may check for themselves, however, that there exists an infinity of higher-order cycles and that these are all part of the chaotic attractor.

Problem 5.13 Demonstrate the self-similarity of the sawtooth attractor by consecutive magnifications of the rectangle marked around P_1 in Fig. 5.22.

Problem 5.14 Follow numerically the motion of a phase space domain towards the sawtooth attractor.

Note that the sawtooth attractor is very similar to the baker attractor (Fig. 5.7), as it resembles a 'cut-in-two' and distorted version of the latter. The parallelogram formed by the basic branches of the manifolds of the fixed points surrounds the chaotic attractor.[7] This parallelogram corresponds to the unit square of the baker map, but it is 'wider', since its corners do not pertain to the attractor.

Problem 5.15 Determine numerically the chaotic attractor of a baker map whose expansion factor is only 1.8 instead of 2 as given in (5.2).

The chaotic attractor of Fig. 5.22 appears to be a much more disordered set of lines than those of the baker attractor. Nevertheless, this

[7] In fact, to obtain this parallelogram, the basic branch of the unstable manifolds has to be extended beyond the v_n-axis (cf. Fig. 5.25).

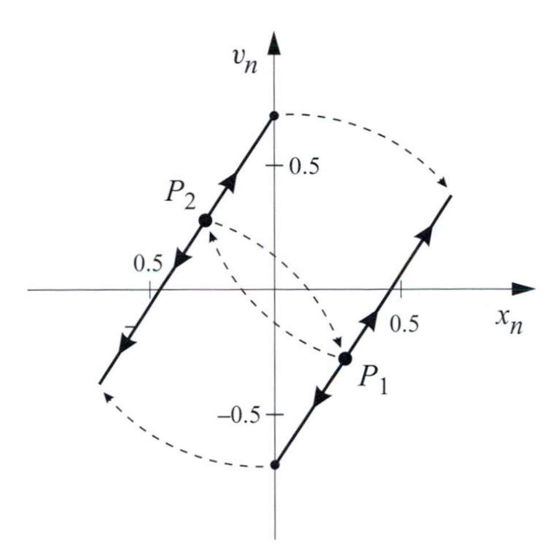

Fig. 5.23. Basic branches of the unstable manifold of the two-cycle (P_1, P_2). The dashed lines indicate the mapping of the end-points falling on the v_n-axis, and of the cycle points ($a = 1.95$, $E = 0.8$).

object can also be constructed *exactly*. This is possible because map (5.34) is linear within each half-plane (it only changes when crossing the x_n-axis.) Consequently, the basic branch of the manifolds of any cycle point lying off the x_n-axis is a straight line within an *extended* neighbourhood of the cycle point. The end-points of the basic branches follow from the mapping rule. As the lowest-order cycle on the attractor is a two-cycle, it is helpful to construct the unstable manifold of this cycle.

Problem 5.16 Determine the two-cycle $P_i(x_i^*, v_i^*)$ $(i = 1,\ 2)$ of the sawtooth map. Demonstrate that $x_i^* = -v_i^*$.

The equation, $v_{i+}(x)$, of the basic branch of the unstable manifold of the cycle point $P_i = (x_i^*, v_i^*)$ $(i = 1,\ 2)$ can be expressed in terms of the direction of the eigenvector, $\mathbf{u}_+ = (1, \Lambda_+)$, and the cycle co-ordinates as follows:

$$v_{i+}(x) = \Lambda_+(x - x_i^*) - x_i^*. \tag{5.36}$$

Here, Λ_+ is given by (5.29) with $f'^* = a$, and we have taken into account that $x_i^* = -v_i^*$.

These basic branches are plotted in Fig. 5.23. One end-point of each branch falls on the v_n-axis. This is because all points $(x_n,\ v_n = -0)$ that approach the x_n-axis from below are mapped into $(x_{n+1} = 0,\ v_{n+1} = 1 - E^2 x_n)$ according to rule (5.34), while the points $(x_n,\ v_n = +0)$ are mapped into $(x_{n+1} = 0,\ v_{n+1} = -1 - E^2 x_n)$. For not too large values, x_n, these fall on the upper and lower v_n-axes, respectively. The points of the two-cycle change positions in each iteration, along with the segments of the basic branches between the x_n and v_n axes. Accordingly, the image of the end-point on the v_n-axis must coincide with the end-point of the other basic branch off the v_n-axis, as indicated in Fig. 5.23.

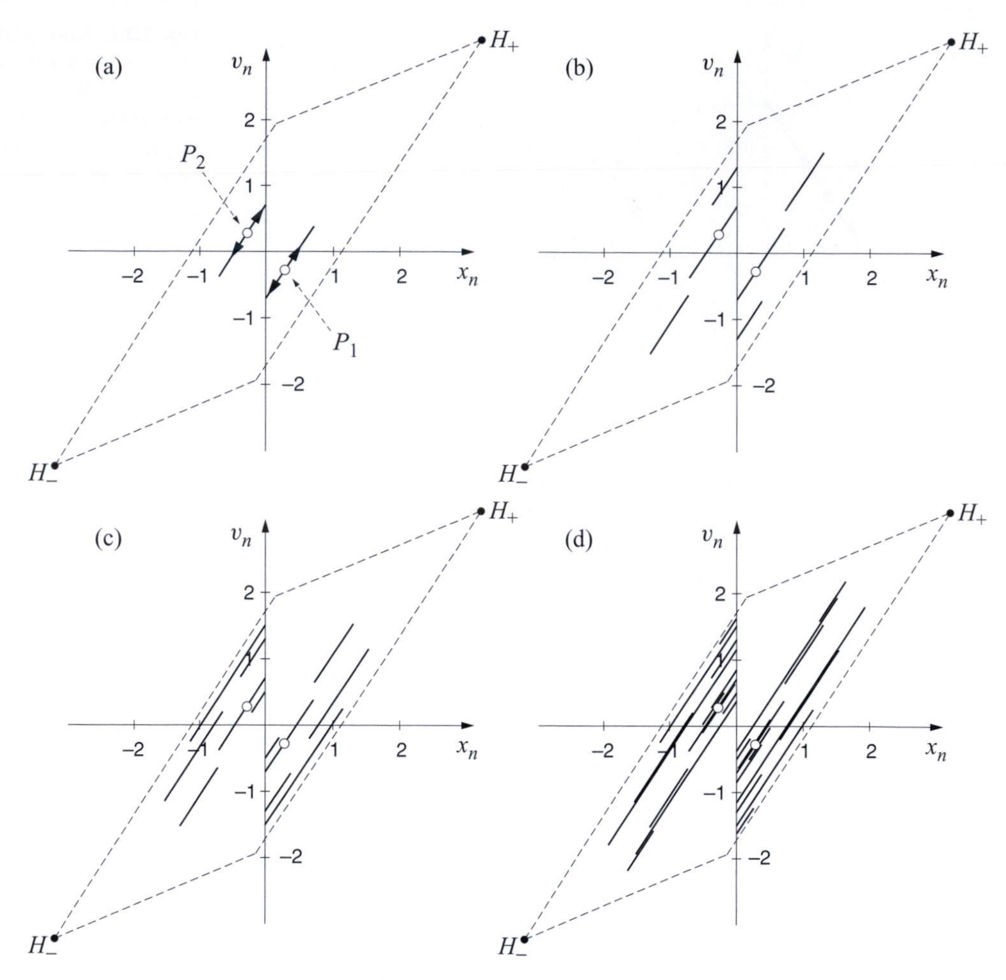

Fig. 5.24. (a) Basic branches of the unstable manifold and their images in (b) two, (c) four and (d) six iterations. Longer and longer segments of the two-cycle's unstable manifold are generated, providing better and better approximants to the chaotic attractor ($a = 1.95$, $E = 0.8$).

Problem 5.17 Determine the end-point co-ordinates of the basic branches of the two-cycle's unstable manifold. Demonstrate that the image of the end-point on the v_n-axis is, in fact, on the straight line of the other basic branch.

The segment of the basic branch which is not in the quadrant containing the cycle points is mapped outside the basic branch (see Fig. 5.24(b)), since all segments along the unstable direction are stretched by a factor of $\Lambda_+ > 1$. Thus, repeated iterations generate longer and longer segments of the entire manifold.

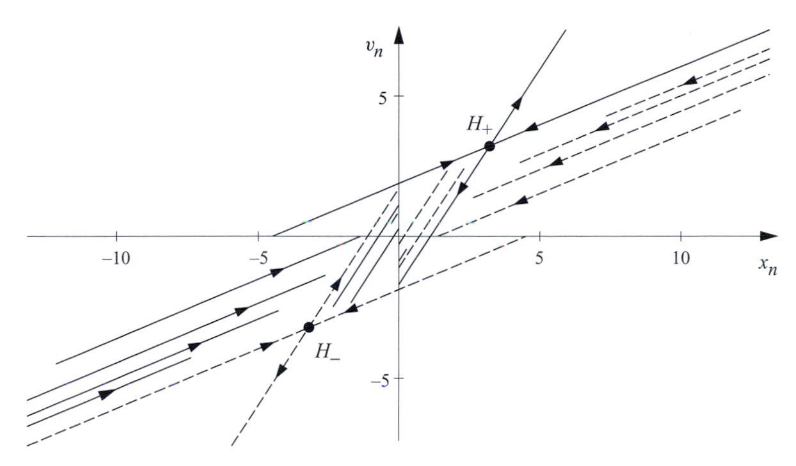

Fig. 5.25. Stable and unstable manifolds of the fixed points. The curves belonging to H_+ and H_- are plotted with continuous and dashed lines, respectively, and arrows mark the directions of contraction and stretching. One branch of each unstable manifold leads into the attractor; the other leads to states with infinite velocity (which can be considered as simple attractors). The stable manifolds also extend beyond the region shown ($a = 1.95$, $E = 0.8$).

Figure 5.24 presents the result obtained after a few iterations. Surprisingly, the shape of the chaotic attractor is also satisfactorily traced out. The chaotic attractor appears to be *identical* to the unstable manifold of the two-cycle. Of course, to construct the unstable manifolds to infinite length, an infinite number of iterations would be required. During this construction, no point would leave the parallelogram surrounding the attractor even once. This fact alone illustrates the rather complex pattern of the full unstable manifold and the fractal structure of the chaotic attractor. As in the case of the baker map, there exist cycles of arbitrary length on the attractor.

The unstable manifolds of $H_\pm$ can be constructed similarly (see Fig. 5.25). The fixed points are not part of the chaotic attractor, but the unstable manifold branches that emanate towards the attractor lead into it. In the other direction the manifold runs out to infinity: the oscillator can take up an arbitrary amount of energy from the kicks, and form (5.23) of the kicking amplitude allows it to develop arbitrarily large velocities both in the positive and negative directions.

Problem 5.18 Determine the end-points of the unstable basic branches of the fixed points, $H_\pm$.

Next, we investigate the *stable* manifolds of the fixed points (and cycle points). They consist of segments of slope Λ_-, as follows from the eigenvectors $\mathbf{u}_- = (1, \Lambda_-)$. The equations of the stable basic branches can be determined in a similar way to (5.36). Further segments of the

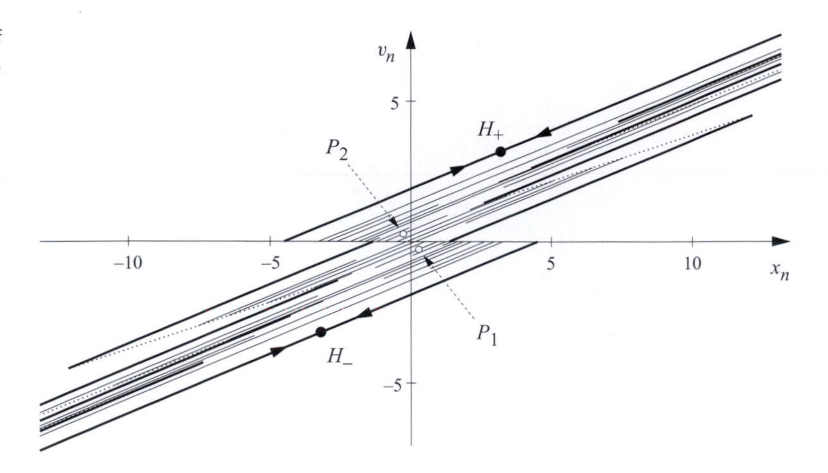

Fig. 5.26. Stable manifold of the two-cycle (thin lines). The stable manifolds of the fixed points surrounding them are marked with thick lines, and the straight segments connecting the points of discontinuity, which are also parts of the boundary, are denoted by dotted lines ($a = 1.95$, $E = 0.8$).

stable manifolds can then be obtained by applying the *inverted* map, (5.22). During this, the stable manifold behaves as an unstable manifold with stretching factor $1/\Lambda_-$. The stable manifold can therefore be constructed by means of the inverted map in the same manner as the unstable manifold by means of the original map (Fig. 5.25).

In Fig. 5.26 we have plotted the stable manifolds of both the fixed points, $H_\pm$, and of the two-cycle. Contrary to the stable manifolds of the fixed points, the two-cycle's stable manifold fills a connected domain, namely the area *surrounded* by the fixed point's stable manifolds and by the lines connecting the end-points of the discontinuities.

If we determined the stable manifold of any periodic orbit in a similar way, it would also lie within this area and would also fill it. The union of the stable manifolds of the cycle points is therefore – as opposed to the unstable ones – *not* of fractal structure. The area filled by them is simply the *basin* of the chaotic attractor (see Fig. 5.27). This

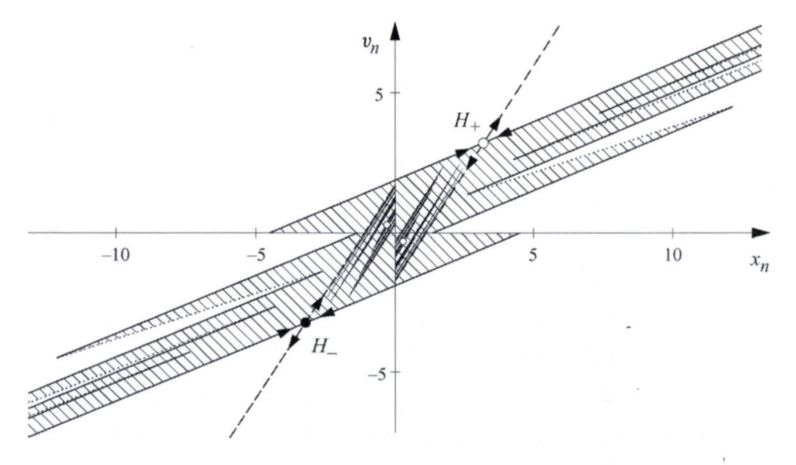

Fig. 5.27. Phase portrait of the sawtooth map. The basin of attraction of the chaotic attractor is shaded. It becomes thinner and thinner for increasing values of $|x_n|$ ($a = 1.95$, $E = 0.8$).

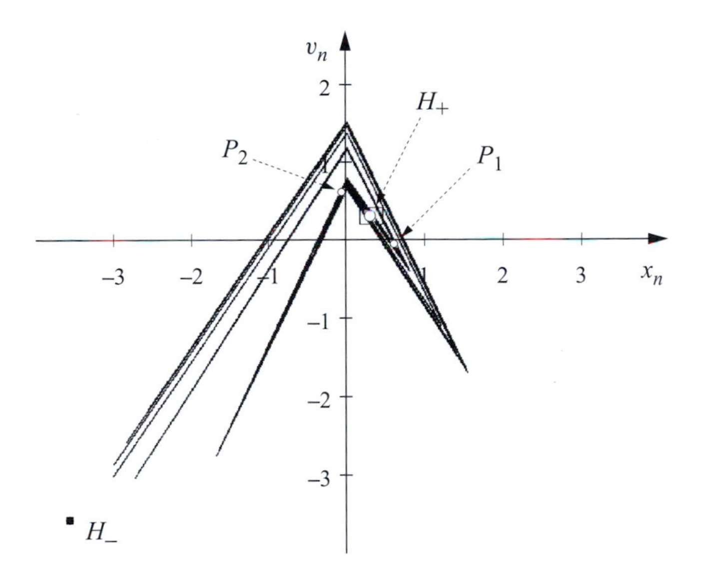

Fig. 5.28. The roof attractor ($a = 1.77$, $E = 0.7$) has been determined by numerically generating a single long trajectory. The hyperbolic points, H_+ and H_-, are marked by large black and white dots, respectively. Two smaller white dots indicate the two-cycle.

is in harmony with what we have seen for regular motion (see Section 3.3.1), since the stable manifolds of H_+ and H_- do indeed form the basin boundary between the chaotic attractor and the attractors at $\pm$ infinity.

5.2.4 Roof attractor

The stroboscopic map of an oscillator kicked with a roof amplitude, (5.24), is the roof map:

$$x_{n+1} = v_n, \quad v_{n+1} = -E^2 x_n + 1 - a|v_n|. \tag{5.37}$$

The fixed point co-ordinates are given by

$$x_\pm^* = v_\pm^* = \frac{1}{1 + E^2 \pm a}. \tag{5.38}$$

Both fixed points are unstable for $a > 1 + E^2$. Over a wide range of the parameters where a chaotic attractor exists, we can see (Fig. 5.28) that the fixed point H_+ is *on* the attractor. The fixed point H_- is, however, not part of the attractor. Contrary to the sawtooth map, the two fixed points are no longer mirror images of each other, and they play different roles.

Problem 5.19 Demonstrate the self-similarity of the roof attractor by consecutive magnifications of the rectangle around H_+.

The derivative of the roof amplitude function, (5.24), is either $f' = -a =$ constant or $f' = a =$ constant, depending on whether x is

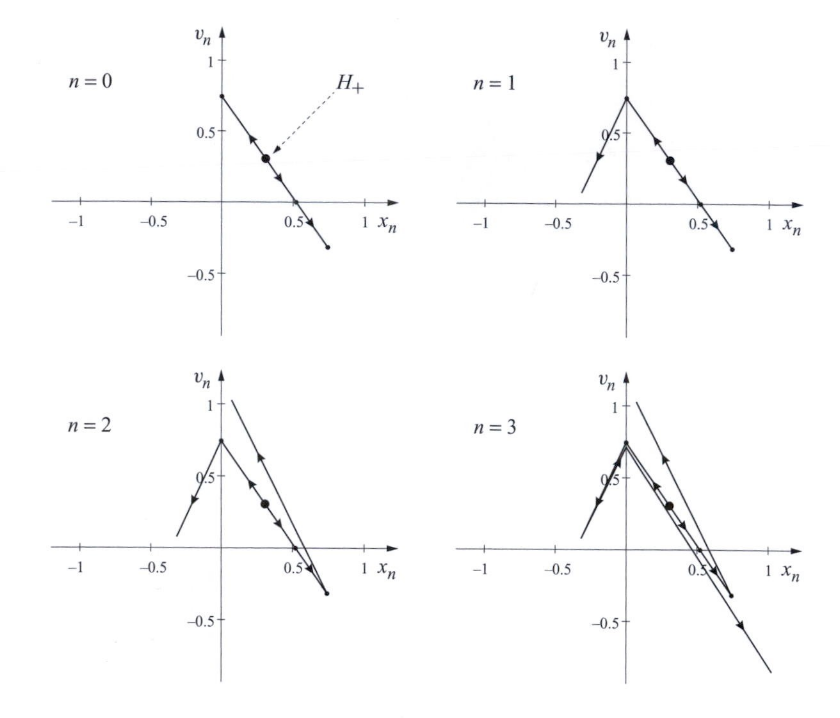

Fig. 5.29. Unstable basic branch of H_+ and its first three images. The unstable manifold is continuous, but broken ($a = 1.77$, $E = 0.7$).

positive or negative. The eigenvalues of the fixed points can be obtained by substituting $f'^* = \pm a$ into equation (5.29). For H_+ lying on the attractor, $\Lambda_\pm < 0$, while for H_- lying off the attractor, $\Lambda_\pm > 0$. Contrary to the baker and sawtooth maps, the eigenvalues and eigenvectors are now *not identical* for the two fixed points.

The map is piecewise linear; therefore, the stable and unstable manifolds can again be constructed exactly. Since H_+ is on the chaotic attractor, the attractor can be well approximated by its unstable manifold.[8] Again, one end-point of the unstable basic branch of H_+ lies on the v_n-axis. As with the sawtooth attractor, its image is the other end-point of the basic branch.

Problem 5.20 Determine the end-points of the unstable basic branch of H_+. Demonstrate that the image of the end-point lying on the v_n-axis is, in fact, on the straight line of the basic branch.

Figures 5.29 and 5.30 exhibit the first three and the next six images of the basic branch, respectively. The essential difference with respect to the cases analysed so far is that the unstable manifold has no discontinuities

[8] Higher-order cycles of infinite number are again on the attractor.

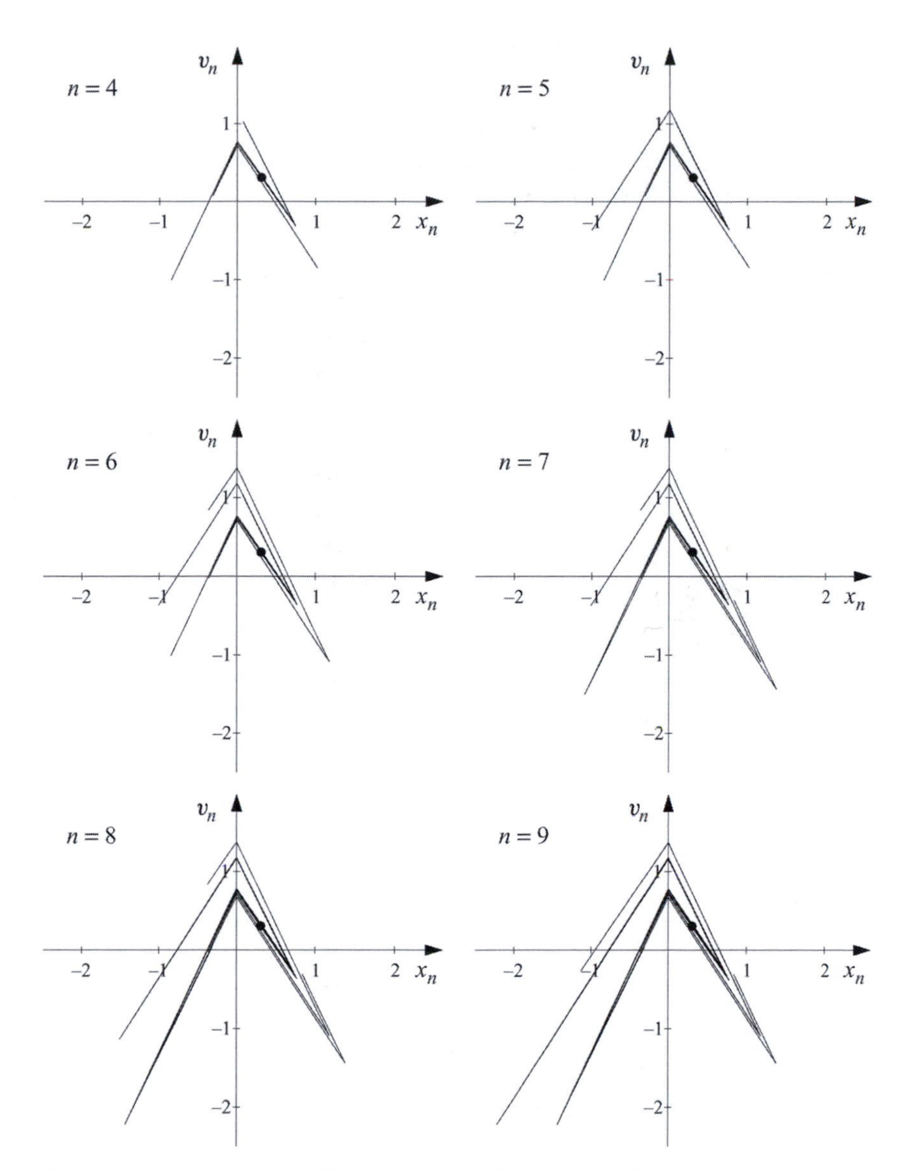

Fig. 5.30. Further images of the unstable basic branch of H_+ (a continuation of the sequence in Fig. 5.29). The chaotic attractor presented in Fig. 5.28 becomes more and more clearly visible.

and remains *continuous* over its full length; i.e., the manifold inherits the continuity of the amplitude function. After an infinite number of iterations the infinitely long unstable manifold traces out the chaotic attractor, but the result of the ninth iterate presented in Fig. 5.30 already provides a rather good approximant.

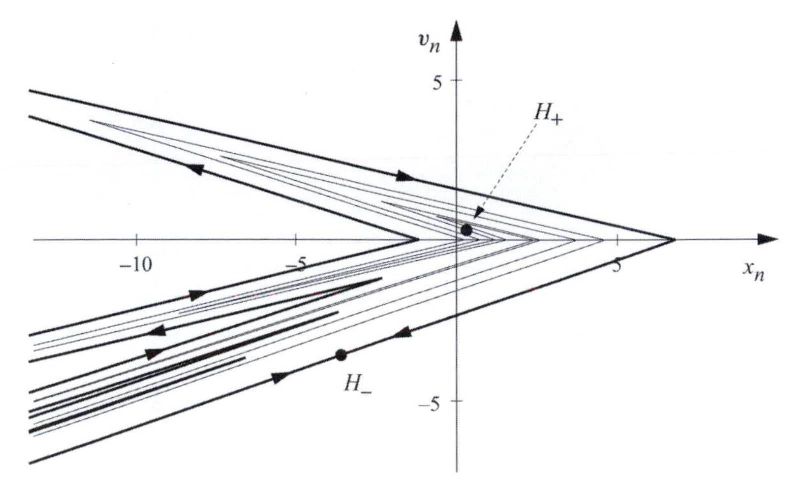

Fig. 5.31. Stable manifolds of the fixed points. The thick line marks the stable manifold of H_-, the basin boundary, while the thin lines represent (part of) the stable manifold of H_+ ($a = 1.77$, $E = 0.7$).

Problem 5.21 Based on map (5.37), mark on Fig. 5.29 the images of the end-points and break-points. (Following the motion of these points also provides a better understanding of Fig. 5.30.)

Even though the values of f' – and therefore the eigenvalues of the stability matrix L (see (5.28)) – are identical in each half-plane, the stable and unstable directions of the fixed point and of the higher-order cycle points are not identical, even within one half-plane. The reason for this is that the eigenvalues of a higher-order cycle point provide the total stretching and contraction rates over a complete cycle. For an m-cycle, the matrix of the linearised, m-fold iterated map can be calculated by multiplying the matrices, $L(x_i^*, v_i^*)$ $(i = 1, \ldots, m)$, evaluated in the cycle points. The eigenvalues and eigenvectors of the product naturally differ from those of any $L(x_i^*, v_i^*)$ and also depend on how many cycle points lie in one and in the other half-plane.

Problem 5.22 Determine the two-cycle of the roof attractor. Find also its stability eigenvalues and its stable and unstable directions.

Let us now investigate the stable manifolds. The fixed point, H_-, is off the attractor; therefore, its stable manifold provides the basin boundary. This manifold can be constructed by means of the inverted map (5.22) (see Fig. 5.31). Since H_+, on the other hand, is part of the chaotic attractor, its stable manifold remains within the basin of attraction, completely filling the latter (cf. Fig. 5.31).

One branch of the unstable manifold of H_- leads again into the chaotic attractor; the other goes towards increasingly negative values of (x_n, v_n) (see Fig. 5.32), as in Fig. 5.25. For large displacements, the oscillator always receives a negative momentum according to the

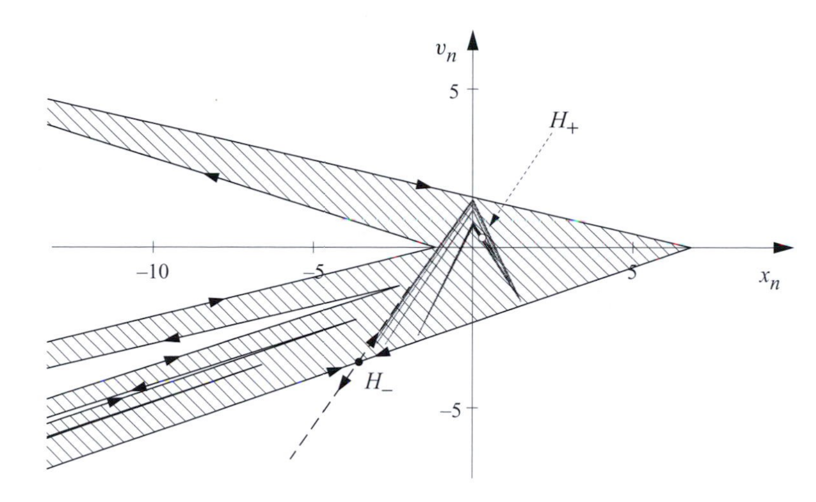

Fig. 5.32. Phase portrait of the roof map. The basin boundary is provided by the stable manifold of H_-. The fixed points, the attractor and the unstable manifold of H_- are also indicated ($a = 1.77$, $E = 0.7$).

amplitude function (5.24); therefore, it cannot reach arbitrarily large values of x_n.

5.2.5 Parabola attractor

The stroboscopic map of an oscillator kicked with a parabola amplitude, (5.25), is the parabola map:

$$x_{n+1} = v_n, \quad v_{n+1} = -E^2 x_n + 1 - a v_n^2. \tag{5.39}$$

The fixed point co-ordinates are now given by

$$x_\pm^* = v_\pm^* = \frac{-(1+E^2) \pm \sqrt{(1+E^2)^2 + 4a}}{2a}. \tag{5.40}$$

The derivative of the parabola amplitude function is $f' = -2ax$. This, contrary to our previous examples, depends on position; consequently, the matrix, L, of the linearised map is *different* in each point. The eigenvalues of the fixed points can be obtained via substitution into (5.29), and, as in the previous case, the eigenvalues of H_+ (H_-) are negative (positive). Fixed point H_- is always hyperbolic; H_+ is hyperbolic only if $a > 3(1+E^2)^2/4$. Over a wide parameter range there exists a chaotic attractor (see Figs. 5.33 and 5.34); fixed point H_+ is on the attractor, but H_- is not.

By means of the unstable manifold of H_+, we can again obtain the chaotic attractor. Since, however, the map is non-linear in every point, the manifold is bent; the eigenvectors of the stability matrix, L, yield the stable and unstable directions only *locally*. Equation (5.33) determines the tangent of the manifold in such a case. Consequently, when constructing the manifold numerically (see Fig. P.18 in the 'Solutions

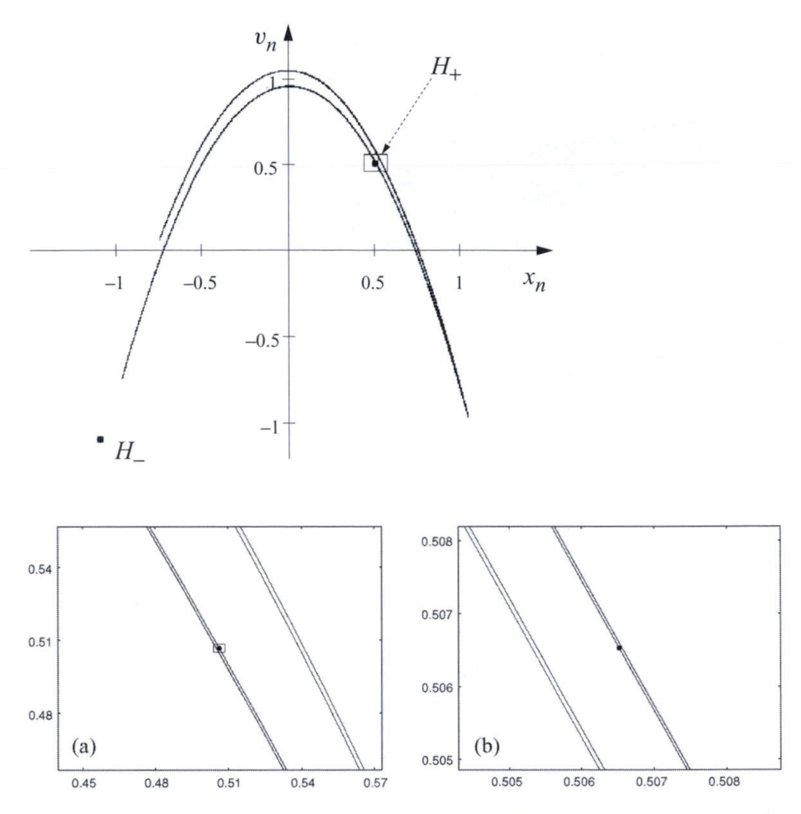

Fig. 5.33. The parabola attractor ($a = 1.8$, $E = 0.25$) determined from a long trajectory. The fixed points, H_+ and H_-, are marked with black dots.

Fig. 5.34. Magnification of the rectangle in Fig. 5.33 by a factor of (a) 30 and (b) 900. Even though at the original resolution the attractor appears to consist of two lines only, the magnified picture, (b), demonstrates the self-similarity characteristic of fractals.

to the problems' section), we take a short segment and iterate that. The unstable manifold proves to be not only continuous, but also a curve *without any break-points*.

Problem 5.23 Demonstrate by means of a numerical simulation how a short piece of the unstable manifold around H_+ changes under iterations.

The stable manifolds of the fixed points can be determined in a similar way by means of the inverted map. In Fig. 5.35 the basin of attraction is shaded, and the stable manifold of H_- forms the boundary. One branch of the unstable manifold of H_- leads into the attractor; the other one leads into negative infinity. As in the previous example, the stable manifold of point H_+ (on the attractor) fills the entire basin of attraction. The stable manifold that forms the boundary is continuous, but the branch emanating to the right only bends back again at a large distance from the origin (the first 'arm' does so at a distance of approximately −30 000 units);

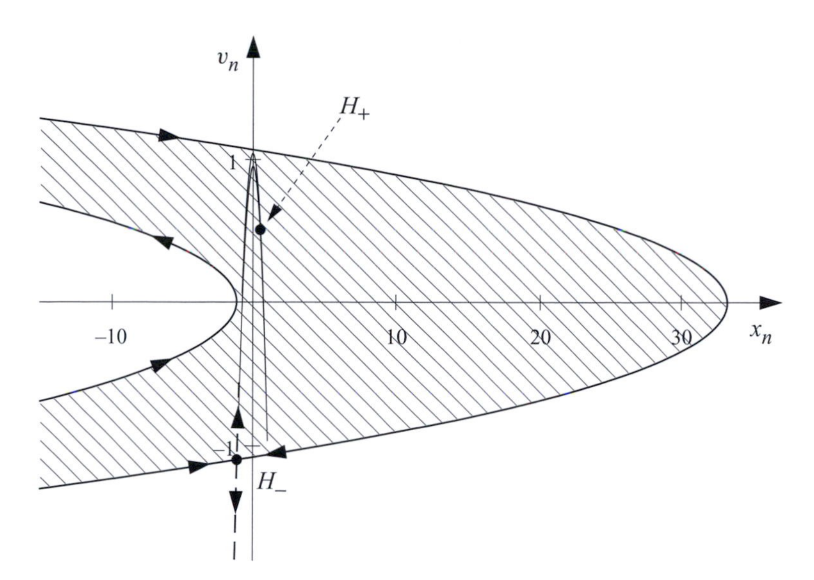

Fig. 5.35. Phase portrait of the parabola map. The basin boundary of the chaotic attractor is the stable manifold of fixed point H_- ($a = 1.8$, $E = 0.25$).

therefore, we have not plotted it. The topology is nevertheless similar to that of the roof map.

In summary, we have gradually deviated from the baker map through the sawtooth, roof and parabola maps towards typical chaotic systems (see also Problems 5.24 and 5.25). Meanwhile, the opportunity of using an analytical approach has been lost, and so this must be replaced in all respects by a *numerical simulation*.

Problem 5.24 By means of computer simulation, determine the chaotic attractor of an oscillator kicked with the bell amplitude function $f(x) = a(e^{-4x^2} - 1) + 1$ with parameters $a = 2,\ E = 0.7$.

Problem 5.25 By means of computer simulation, determine the chaotic attractor of an oscillator kicked with the sinusoidal amplitude function $f(x) = a \sin x$. Apply periodic boundary conditions; i.e., always shift x_n between $-\pi$ and π. Let the parameters be $a = 5,\ E = 0.7$.

Box 5.1 Hénon-type maps

In 1976 the French astronomer Michel Hénon introduced the map later named after him, which transforms a point (x_n, y_n) of the plane into

$$\begin{aligned} x_{n+1} &= 1 - a x_n^2 + b y_n, \\ y_{n+1} &= x_n. \end{aligned} \tag{5.41}$$

Hénon proved that this is the most general form of quadratic maps, which yields a nice chaotic attractor with the standard parameters $a = 1.4,\ b = 0.3$ (Fig. 5.36(a)).

It is worth comparing this with the parabola map. Introducing the new variable

$$v = 1 - ax^2 + by, \tag{5.42}$$

(5.41) becomes

$$\begin{aligned} x_{n+1} &= v_n, \\ v_{n+1} &= 1 - ax_{n+1}^2 + bx_n. \end{aligned} \tag{5.43}$$

This is identical to (5.39), but only if $E^2 = -b$. The original Hénon attractor does *not* correspond therefore to the dynamics of kicked oscillators, or indeed to any physical system, since its Jacobian is negative. For any negative b, however, it is equivalent to the stroboscopic map of an oscillator kicked with a parabola amplitude. The case discussed in Section 5.2.5 corresponds to the choice $b = -0.063$.

A piecewise linear version of the Hénon map was given by the French mathematician René Lozi in the form

$$\begin{aligned} x_{n+1} &= 1 - a|x_n| + by_n, \\ y_{n+1} &= x_n, \end{aligned} \tag{5.44}$$

where the usual b values are positive (Fig. 5.36(b)).

In general, a Hénon-type map,

$$\begin{aligned} x_{n+1} &= f(x_n) + by_n, \\ y_{n+1} &= x_n, \end{aligned} \tag{5.45}$$

is equivalent, with the choice $b = -E^2 < 0$, to the stroboscopic dynamics of an oscillator kicked with amplitude $f(x)$ (Fig. 5.36(c)). The mathematical attractors belonging to positive b values differ significantly from the physical attractors belonging to the opposites of b (they are not even necessarily both chaotic).

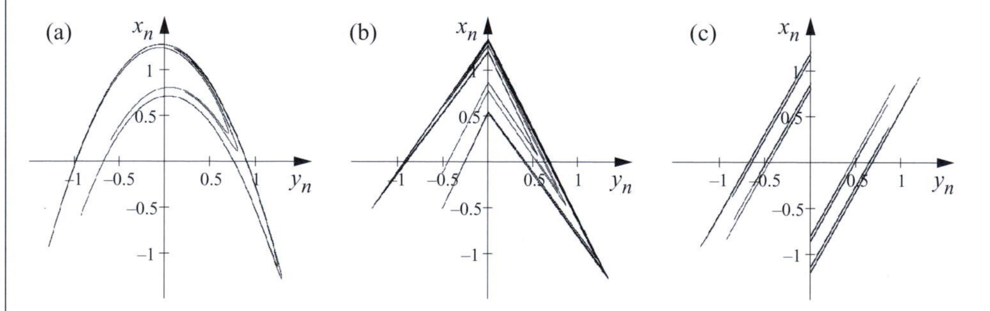

Fig. 5.36. Hénon-type chaotic attractors. (a) Hénon attractor (in (5.41), $a = 1.4$, $b = 0.3$). b) Lozi attractor (in (5.44), $a = 1.7$, $b = 0.5$). (c) Hénon-type sawtooth attractor (attractor of the map (5.45) generated with a function $f(x)$ of the form of (5.23); $a = 1.6$, $b = 0.3$).

5.2.6 The limit of extremely strong dissipation: one-dimensional maps

A strongly damped motion of kicked oscillators arises if $E = e^{-(\alpha/2)T} \ll 1$, which can occur not only because of strong friction, but

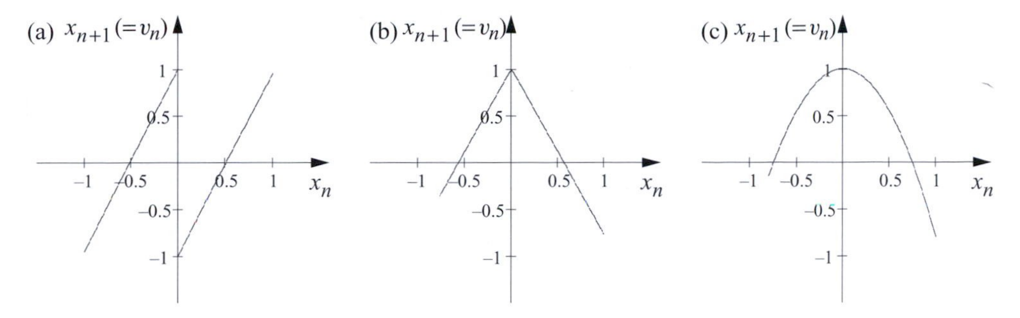

Fig. 5.37. Chaotic attractors for extremely strong dissipation ($E = 0$). (a) Sawtooth amplitude, (5.23); the one-dimensional map, (5.46), is $x_{n+1} = ax - \text{sign}(x_n)$, called a Bernoulli map ($a = 1.95$). (b) Roof amplitude, (5.24); the one-dimensional map is $x_{n+1} = 1 - a|x_n|$, called a tent map ($a = 1.77$). (c) Parabola amplitude, (5.25); the one-dimensional map is $x_{n+1} = 1 - ax_n^2$, called a logistic map ($a = 1.8$).

also because of rare kicking. In the limit $E \to 0$ of extremely strong dissipation, no chaotic motion can develop with bounded kicking amplitudes, I, since the body practically comes to a halt before the next kick. If, however, the magnitude of the momentum transfer is increased at the same rate as E decreases, then the motion never ceases between kicks. The change to new variables, (5.18), corresponds precisely to this limit since the momentum transfer is $I = f/E$, with a finite f for any E. Thus, in the form (5.19) chaos can be present even in the limit $E \to 0$, when

$$x_{n+1} = v_n, \quad v_{n+1} = f(x_{n+1}). \tag{5.46}$$

By means of the substitution $x_{n+1} \to x_n$, we obtain

$$x_{n+1} = f(x_n). \tag{5.47}$$

This is called a *one-dimensional map*, since the value of one of the co-ordinates (for example the position) is uniquely determined solely by the value of the same co-ordinate taken one step earlier. Moreover, the chaotic attractor is, according to (5.46), a segment of a simple curve in the (x_n, v_n)-plane, i.e. a piece of the graph $v = f(x)$ (see Fig. 5.37). The one-dimensional map, (5.47), describes the discrete-time motion on this segment.

Note that for f's with a local extremum (as in all the examples analysed so far) one-dimensional dynamics *loses* invertibility; x_n is no longer a single-valued function of x_{n+1}, and the equation $x_n = f^{-1}(x_{n+1})$ may have two solutions (see Box 4.1). For any dissipation parameter $E \neq 0$, the presence of the term $(-E^2 x_n)$ assures in (5.19) the existence of a unique inverse given by (5.22). From the point of view of invertibility, the cases $E \ll 1$ and $E \to 0$ are therefore not identical. One-dimensional maps represent a *non-physical limit*, when dynamics is non-invertible. Furthermore, fractality is lost, since in the plane (x_n, v_n) the chaotic attractor degenerates into a single curve segment (see Fig. 5.37). In spite of this, one-dimensional maps, and the logistic map among them, are useful mathematical tools in understanding certain concepts (such as, for example, the time evolution of probability distributions (see Section 5.4.4) and universality (see Section 5.3).

5.3 Parameter dependence: the period-doubling cascade

Whether chaotic attractors exist in a non-linear system depends on the parameters. In the example of kicked oscillators, one reaches the chaotic regime by increasing the non-linearity parameter, a, and this happens via bifurcations (see Section 3.3.2). Since such bifurcations are characteristic not only of kicked oscillators, but also of many other systems, we turn to a general notation.

Denote the tunable parameter by μ and assume that, for small values of μ, the system possesses only one fixed point attractor in the map (i.e., an attracting limit cycle in the flow). One of the most typical roads towards chaos is the *period-doubling* bifurcation sequence, or period-doubling cascade.[9] By increasing parameter μ, we find that, at a certain value μ_1, the fixed point becomes unstable and a two-cycle takes its role over as an attractor. (This is similar to the pitchfork bifurcation seen in Section 3.3.2, but the bifurcation now describes how limit cycles, instead of equilibrium states, lose their stability.) The period of the attractor is doubled. This attractor, however, is only stable up to another value μ_2, where an attracting four-cycle is born. At the value μ_n, a cycle of period 2^n appears, but it remains stable only up to a value μ_{n+1}, and so on (Fig. 5.38). An important feature of period-doubling bifurcations is that higher- and higher-order cycles are stable over *shorter and shorter* intervals of μ. Therefore, with a finite change in the parameter, the cascade reaches an *accumulation point*, where formally a cycle of length 2^∞ appears. The parameter, μ_∞, of the accumulation point is thus a finite value. For values of μ greater than this, the system is capable of exhibiting chaotic behaviour, since an infinity of unstable periodic orbits has been created by then, suitable for forming the skeleton of a chaotic attractor.

[9] If there exist several fixed point attractors, then each of these can pass through its own cascade independent of that of the others.

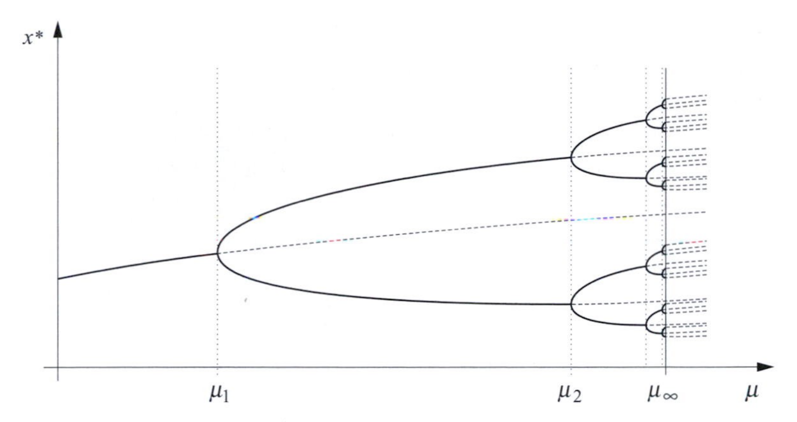

Fig. 5.38. Schematic diagram of a period-doubling cascade. The vertical axis is one of the co-ordinates (x^*) of the attractor's points. Dashed lines refer to the unstable limit cycles left in place of the simple attractors. Dotted vertical lines mark bifurcation points. An infinity of bifurcations occurs before the accumulation point, μ_∞, and an infinity of unstable states is born. Chaos is present in the region $\mu > \mu_\infty$.

In the vicinity of the accumulation point μ_∞, *all* period-doubling cascades behave as *geometric sequences*: the parameter, μ_n, of the nth bifurcation point can be expressed as

$$\mu_n = \mu_\infty - A\left(\frac{1}{\delta}\right)^n, \tag{5.48}$$

where in all systems described by smooth (differentiable) maps the constant δ is the same number:

$$\delta = 4.669. \tag{5.49}$$

This quotient of the cascade does not depend on the details of the system; it is a *universal* property (the factor A, on the contrary, is not). Furthermore, a similar rule applies to the width of the 'forks' appearing in the cascade: these quantities go to zero as negative powers of the number

$$\alpha = 2.503. \tag{5.50}$$

Since $\delta > \alpha$, when approaching the accumulation point, the forks become more and more open. The numbers α and δ are called Feigenbaum constants. We have now uncovered an important new property of chaotic behaviour: the transition to chaos may possess *universal* features.

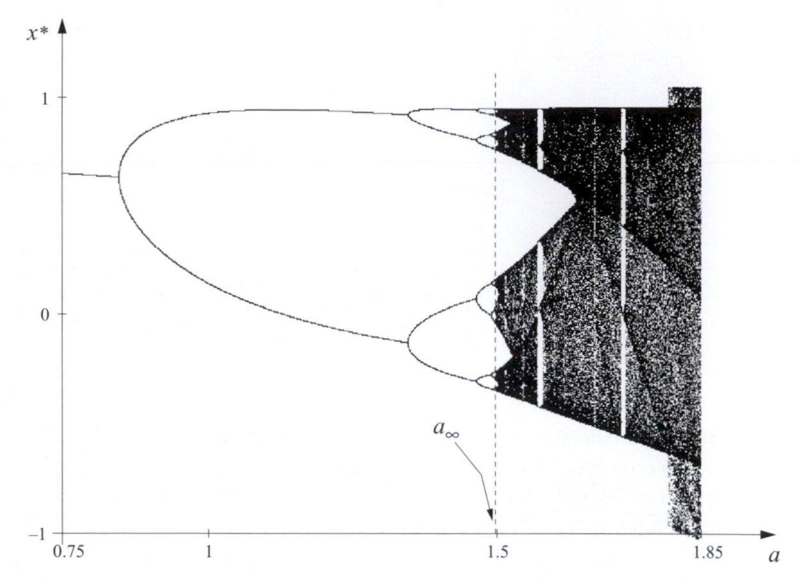

Fig. 5.39. Bifurcation diagram of an oscillator kicked with a parabola amplitude, of the parabola map (5.39) at $E = 0.25$ in terms of the non-linearity parameter $\mu \equiv a$. The accumulation point is $a_\infty \approx 1.499$. For $a > a_\infty$, the parameters belonging to a chaotic attractor are those where the values of x^* cover an entire interval. The diagram is obtained by omitting the first 2000 iterates to screen out transient effects.

The character of the cascade depends only on the form of the mapping function around its extremum.[10] For smooth maps this is typically well approximated by a quadratic function. The cascade of an oscillator kicked with a parabola amplitude (see Figs. 5.39 and 5.40) therefore proceeds according to the Feigenbaum constants. In other cases, however, where the mapping is not quadratic around its extremum points, the cascades are different. In maps with break-points or discontinuities at the extrema, as, for example, for the roof map, chaos appears abruptly when the fixed point attractor loses its stability.

Problem 5.26 Determine the bifurcation diagram of the oscillator kicked with a roof amplitude function ($E = 0.7$).

A kind of mirrored version of the period-doubling cascade can be observed in the chaotic region $\mu > \mu_\infty$. Here, sufficiently far from the accumulation point, the chaotic attractor appears as a single connected set with an unstable fixed point on it (Figs. 5.33 and 5.40(c)). By decreasing the parameter μ, the attractor splits into two pieces (see Fig. 5.40(b)). The iterated points appear on these pieces in a strictly alternating manner.

[10] In order to determine the stability of a limit cycle attractor of length 2^n, the 2^n-fold iterated map should be investigated, whose contraction rate is the 2^nth power of the original Jacobian. Thus, in the vicinity of the accumulation point, $n \gg 1$, as far as the stability of the attractor is concerned, *all* systems behave as if their Jacobians were zero. This behaviour is therefore determined by the one-dimensional map arising in the limit $J \to 0$ of the original map. It can be shown that the new attractors appear around the local extrema of the many-times-iterated map.

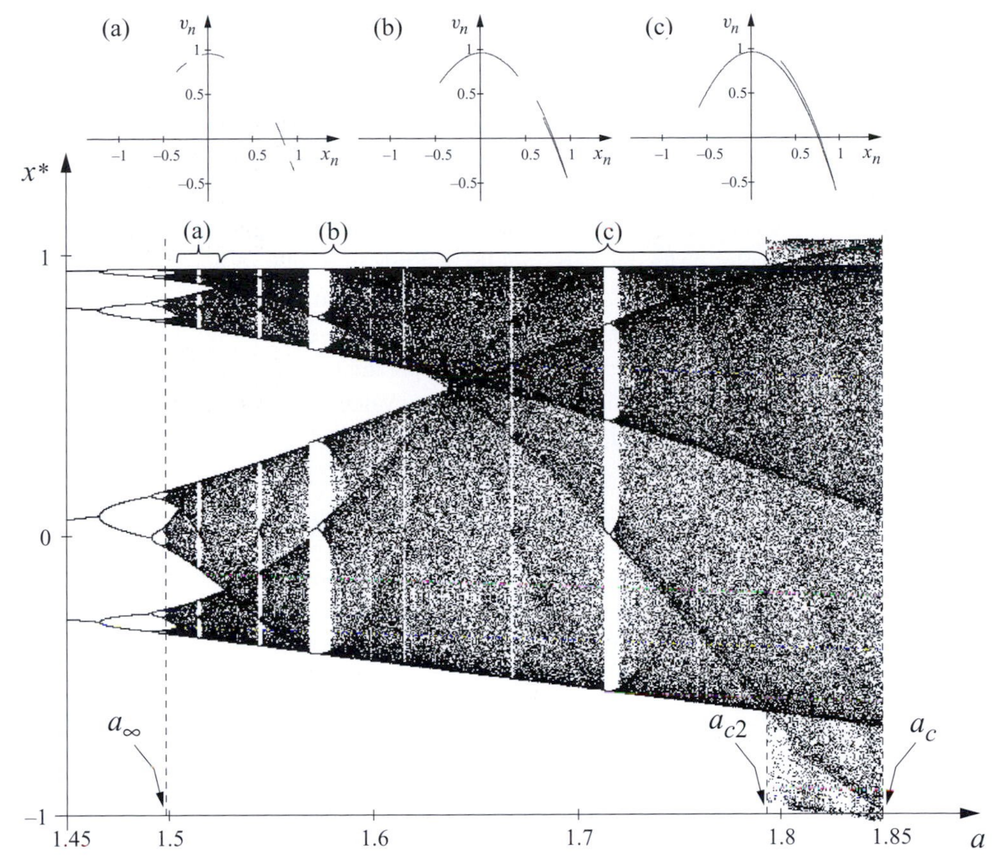

Fig. 5.40. The chaotic region of Fig. 5.39. In intervals (a)–(c) a typical form of the attractor on the (x_n, v_n)-plane is also shown. At $a = a_{c2} = 1.793$ the bifurcation diagram widens abruptly. In the range $a > a_{c2} = 1.793$ the chaotic attractors are of the type presented in Fig. 5.33. For $a > a_c = 1.859$ the kicks are so strong that the iterations always run out into infinity. Overall in the chaotic region there exist periodic windows where the attractor is a periodic cycle.

The motion is still irregular on both pieces: it can be considered to be a chaotic wandering around a two-cycle. The two-piece attractor does no longer contains the unstable fixed point. The shortest unstable orbit that belongs to the attractor is a two-cycle. By further decreasing the parameter, the attractor suddenly falls into four pieces (see Fig. 5.40(a)), and from then on the two-cycle is not part of it.

Above the accumulation point an increase in parameter μ typically implies an increase in both topological entropy and positive Lyapunov exponent. In certain regions, however, these quantities might also decrease. It is characteristic of systems described by smooth (differentiable) maps that their chaotic attractors sometimes disappear (Fig. 5.40). In these regions of parameter μ, which are parts of the so-called *periodic windows*, the attractors are periodic cycles[11] (their length is never 2^n ($n = 1, 2, \ldots$), since the stability of such cycles was already lost in the course of the period-doubling cascade).

The question arises how typical the parameters μ^* belonging to chaotic attractors are. By considering any finite interval of μ that contains chaotic parameters μ^*, one finds that the total length of the set μ^* *is finite* since the size of the windows belonging to higher- and higher-order cycles decreases very rapidly. All this is similar to the construction of Figs. 2.11 and 2.12, and the pattern of the fat Cantor filaments presented there is, in fact, similar to the structure seen in Fig. 5.40 above the accumulation point. The chaotic parameters, μ^*, therefore form a *fat* fractal (see Section 2.2.3). A typical value of the fat fractal exponent, α, defined by relation (2.15) is, for μ^*, $\alpha = 0.45$.

5.4 General properties of chaotic motion

Based on the previous examples, we are now in a position to describe the general properties of driven dissipative chaotic systems with three-dimensional flows, represented by two-dimensional invertible maps. We also give the general definition of a few basic characteristics as measures of chaos. The map will be written in the form $(x_1', x_2') = M(x_1, x_2)$.

5.4.1 The measure of complexity: topological entropy

The *skeleton* of chaotic attractors is the set of the unstable periodic orbits. At the level of the stroboscopic map, these orbits are hyperbolic fixed points or cycles. Which is the simplest hyperbolic periodic orbit lying on the attractor depends on the parameters and can usually be determined by numerical methods only. Trajectories started around hyperbolic orbits off the attractor never return there, contrary to those started around hyperbolic orbits on the attractor. For large, extended chaotic attractors, a fixed point or a two-cycle is typically part of the attractor.[12] Independently of whichever cycle is first in the sequence of periodic orbits belonging to the attractor, the number of such cycles increases with their length. It is a fundamental property that the longer the cycles, the more cycles of the same length are found on the attractor. Their number increases, in general, *exponentially* with the length: the number N_m of cycle points of

[11] Chaos is, however, *always* present in the form of transient chaos; see Section 6.7.

[12] If an unstable cycle is not on the attractor, but is close to it, the stable manifold of the cycle cuts through the attractor, which therefore consists necessarily of several pieces (see Figs. 5.40 (a) and (b)).

the unstable cycles of length m (of period mT) on the chaotic attractor increases, for sufficiently large cycle lengths m, according to

$$\boxed{N_m \sim e^{hm}.} \tag{5.51}$$

The parameter h is called the topological entropy.[13] This leads to a possible definition of chaos: a system is chaotic if its topological entropy is positive, i.e. if $h > 0$.

The chaotic attractor is of finite size, and the unstable periodic cycles are *dense* on it: one can find cycle points in arbitrarily small neighbourhoods of each point of the attractor. The proportion of the periodic cycles to those representing aperiodic motion (homoclinic and heteroclinic points) is the same as that of rational to irrational numbers. Thus, a point chosen at random on the attractor does not coincide with any of the cycle points: chaotic motion is, after all, a random walk among the unstable periodic orbits. Consequently, chaotic motion does not repeat itself: it cannot be decomposed into the sum of even an infinity of periodic motion with discrete frequencies.

A property of topological entropy that is easier to use in practice is that it is also the growth rate of the length of line segments of the phase space. A line segment of length L_0 initially lying in the basin of attraction is stretched more and more in the direction of the unstable manifold under iterations. Let L_n denote the length of the line segment after n iterations. Experience shows that, after a sufficiently large number of iterations ($n \gg 1$), this length increases exponentially, and the growth rate is given by just the topological entropy, according to the relation

$$\boxed{L_n \sim e^{hn}.} \tag{5.52}$$

Qualitatively speaking, in the course of its stretching the line segment approaches more and more unstable periodic orbits, and in n time steps each orbit of length n gives an approximately identical contribution to its growth. For a numerical determination see Appendix A.5.

Problem 5.27 On the basis of property (5.52), determine the topological entropies of the asymmetric baker attractor and the sawtooth attractor.

[13] This definition is based on the Boltzmann relation $S = k_B \ln N$ known from statistical physics, where N is the number of states, S is thermodynamical entropy and k_B is Boltzmann's constant. It can be seen that the equivalent of S is hm; h is thus some kind of entropy density.

5.4.2 The measure of unpredictability: the Lyapunov exponent

Dynamical instability

The existence of the densely lying hyperbolic cycles results in a rapid *deviation* of neighbouring orbits called *dynamical instability* or *sensitivity to the initial conditions*. In the neighbourhood of any point on the attractor, a periodic orbit exists which is unstable. We have seen that trajectories starting from the vicinity of hyperbolic points deviate at an exponential rate in time (see Section 3.1). Thus, for two nearby points around point $\mathbf{r} = (x_1, x_2)$ of the attractor, their distance increases according to powers of the repelling eigenvalue characterising the periodic orbit near $\mathbf{r}$. If the initial distance $\Delta r_0 \equiv [\Delta x_{1,0}^2 + \Delta x_{2,0}^2]^{1/2}$ in phase space is sufficiently small, the distance $\Delta r_n \equiv [\Delta x_{1,n}^2 + \Delta x_{2,n}^2]^{1/2}$ after $n \gg 1$ steps can be obtained from Δr_0 via multiplication by an exponential factor, i.e.

$$\boxed{\Delta r_n(\mathbf{r}) = \Delta r_0 \, e^{\lambda(\mathbf{r})n}.} \tag{5.53}$$

The quantity $\lambda(\mathbf{r})$ is called the *local Lyapunov exponent*. For relation (5.53) to hold, we must assume that the distance $\Delta r_n(\mathbf{r})$ after n iterations is still small compared with the total extension of the attractor. The factor $\exp(\lambda(\mathbf{r})n)$ is thus the repelling eigenvalue of the map M^n linearised around one of the n-cycle points near point $\mathbf{r}$. The local Lyapunov exponent is therefore the logarithm of the repelling eigenvalue of the n-fold ($n \gg 1$) iterated map taken for one iterate, i.e. the logarithm of the nth root of the repelling eigenvalue of M^n. If n is sufficiently large, then $\lambda(\mathbf{r})$ obtained in this manner only depends on the position in phase space.

The local Lyapunov exponent, $\lambda(\mathbf{r})$, is a positive number, as all periodic orbits on the chaotic attractor are hyperbolic. Since such a positive exponent can be assigned to *any* point of the attractor, typical pairs of points on the attractor deviate with some *average Lyapunov exponent*, $\bar{\lambda}$:

$$\boxed{\Delta r_n = \Delta r_0 \, e^{\bar{\lambda} n}.} \tag{5.54}$$

In chaotic systems, the average Lyapunov exponent is positive: $\bar{\lambda} > 0$. This is another property suitable for defining chaos. The exact manner in which the average is to be taken will be discussed in Section 5.4.5. The average Lyapunov exponent does not, of course, depend on the position in phase space, rather it is a number characteristic of the chaotic attractor (for numerical methods, see Appendix A.5).

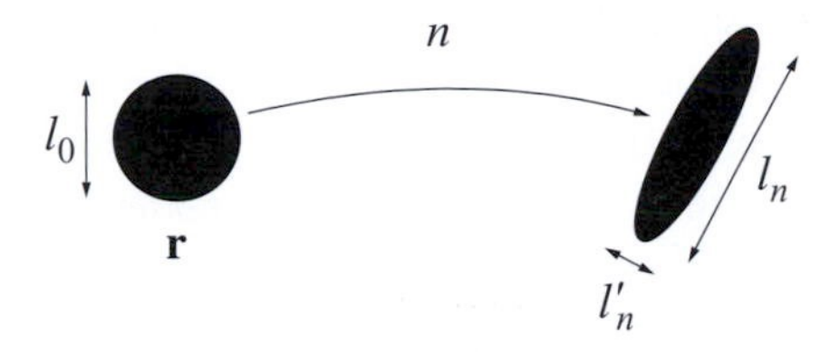

Fig. 5.41. An initial phase space element is stretched in one direction and contracted in another direction over a few iterations, *n*. The total area decreases.

Another consequence of local hyperbolicity is that any two points chosen in such a special way that they fall exactly on the stable manifold approach each other. In this case, in the neighbourhood of point **r**,

$$\Delta r_n(\mathbf{r}) = \Delta r_0 \, e^{\lambda'(\mathbf{r})\,n}, \tag{5.55}$$

where $\lambda'(\mathbf{r})$ is negative. The average of these exponents is the average *negative Lyapunov exponent*, $\bar{\lambda}'$.

These statements can be summarised in an illustrative manner as follows: a small disc of diameter l_0 around point **r** of the attractor is mapped into a stretched object in n steps, whose length and thickness (see Fig. 5.41 for the initial phase) are given by

$$l_n(\mathbf{r}) = l_0 \, e^{\lambda(\mathbf{r})\,n}, \quad l'_n(\mathbf{r}) = l_0 \, e^{\lambda'(\mathbf{r})\,n}. \tag{5.56}$$

As the area in n steps is approximately $l_n l'_n$, and we know that the area is multiplied by the Jacobian $J(\mathbf{r}) < 1$ in each step, (5.56) yields

$$\lambda(\mathbf{r}) + \lambda'(\mathbf{r}) = \ln J(\mathbf{r}). \tag{5.57}$$

For typical points, therefore,

$$\bar{\lambda} + \bar{\lambda}' = \overline{\ln J}, \tag{5.58}$$

where $\overline{\ln J}$ is the mean value of the logarithm of the Jacobian on the attractor. In a dissipative system this average is negative (the area decreases); therefore, we see that the contraction experienced in the stable direction is, on average, stronger than the stretching in the unstable direction, i.e. $|\,\bar{\lambda}'\,| > \bar{\lambda}$.

It is worth noting what happens to the points and point pairs starting outside the chaotic attractor. Inside the basin of attraction, such points move towards the attractor, since it attracts all trajectories. When they get so close to one of the points of the attractor that the linear approximation becomes applicable, their distance from the attractor is multiplied in each step by the factor $e^{\bar{\lambda}'} < 1$, on average (while they also move along the attractor, obviously). Therefore, the approach to the chaotic attractor (as to any other attractor) occurs at an *exponential* rate. The average negative

Lyapunov exponent, $\bar{\lambda}'$, therefore also characterises the rate at which points converge to the attractor. As the approach is of an exponential character, the points are, in practice, on the attractor within a few steps, of the order of $1/|\bar{\lambda}'|$. From then on the distance between neighbouring points increases rapidly at the rate given by the positive average Lyapunov exponent. In the period of time preceding the arrival on the attractor, however, the distance between neighbouring points can also decrease, as can be seen in the first few steps of Fig. 5.4.

Prediction time

A motion is considered to be predictable if, starting from a well defined initial condition, the state arising after a long time is also well defined. In practice, the definition of a state is never perfect; the initial condition can only be given with some error. The motion is predictable if the uncertainty arising from the initial error is still relatively small after a long time.

The dynamics on chaotic attractors is not of this kind. The initial distance, Δr_0, between two nearby points can also be considered as the uncertainty in defining a state in phase space. Measuring the distance in the units of some characteristic phase space distance (for example the extension of the chaotic attractor), the dimensionless Δr_0 is exactly the relative error in defining the state. Due to the dynamical instability on the attractor, the relative error, Δr_n, seen after n steps is given by expression (5.54). If the error reaches the order of magnitude of the quantity to be determined, i.e. if the relative error increases to unity (to 100%), then the prediction is no longer reliable. The time over which this does not occur yet is, from (5.54), the *prediction time*, given by

$$\boxed{t_{\mathrm{p}} = \frac{1}{\bar{\lambda}} \ln \frac{1}{\Delta r_0}} \tag{5.59}$$

(measured in units of the driving period, T). This time is principally determined by the average Lyapunov exponent, or, more accurately, by its reciprocal. Since in chaotic systems the Lyapunov exponent is usually of order unity, the prediction time is typically that of a few iterations only. Even though t_{p} also depends on the initial error, it does so logarithmically, which is a rather weak dependence. The average Lyapunov exponent can therefore be considered as the measure of unpredictability: the greater the $\bar{\lambda}$, the shorter the time for which the behaviour of the system can be predicted.

The significance of this result can truly be appreciated by examining the predictability of regular systems. In such systems the error either decreases with time or, if it increases, it does so, generally, at a linear rate at most. Considering this worst case, the relation $\Delta r_n = \Delta r_0(1 + \tilde{\lambda} n)$

replaces (5.54), where $\tilde{\lambda}$ is a parameter of order unity characteristic of the non-chaotic error growth. With a small initial error Δr_0, the prediction time is now given by

$$t_{\mathrm{p}}^{(\text{non-chaotic})} = \frac{1}{\tilde{\lambda}}\left(\frac{1}{\Delta r_0} - 1\right) \approx \frac{1}{\tilde{\lambda}}\frac{1}{\Delta r_0}, \tag{5.60}$$

i.e. it is inversely proportional to the initial uncertainty.

As an example, let $\Delta r_0 = 10^{-6}$ and take $\bar{\lambda} = \tilde{\lambda} = 1$. The prediction times are

$$t_{\mathrm{p}} = 6 \ln 10 \approx 14, \quad t_{\mathrm{p}}^{(\text{non-chaotic})} = 10^6. \tag{5.61}$$

The long-term prediction of the state is *impossible* on the attractor of a chaotic system.

It is also worth considering what happens if the accuracy of the measurement is improved. Suppose that we are able to decrease the error

Box 5.2 The trap of the 'butterfly effect'

The American meteorologist Edward Lorenz (see Box 5.6) gave a lecture in 1972 with the title: 'Predictability: does the flap of a butterfly's wings in Brazil set off a tornado in Texas?'. The term 'butterfly effect' went into general use due to J. Gleick's popular book *Chaos: Making a New Science*, which implicitly suggested that the answer to Lorenz's question is affirmative. Outside scientific literature this is often interpreted as if modern sciences would claim that everything is related to everything, and we could not therefore be sure of anything. In *contrast* to this, the analysis of chaotic systems shows that unpredictability is limited, i.e. it only holds *on* the chaotic attractor. As far as motion before reaching the attractor is concerned, we know *for sure* that it converges in phase space to a very small (but extended) set of zero volume: the attractor. Nearby orbits do not diverge before reaching the attractor, since they do not enter the vicinity of hyperbolic cycles. In addition, *from a statistical point of view*, the motion developing on the attractor can be described with *perfect accuracy* (see Section 5.4.4). Even by accepting that the term 'butterfly effect' can be used as a synonym of sensitivity to the initial condition or unpredictability, we have to emphasise that this effect can always be observed on chaotic attractors (chaotic sets) only. The question addressed in Lorenz's lecture can therefore only be answered if one can decide whether the initial trajectory starting from Brazil and the one flicked by the flap of the wing are on the attractor to which the Texan tornado pertains. This would be unlikely since there is practically no interaction between the air masses of the southern and northern hemispheres (not to mention that in our terminology weather cannot even possess a chaotic attractor since it not a simple system, being made up of a multitude of components). For the sake of historical faithfulness, we add that Lorenz also noted in *The Essence of Chaos* that if the flutter of a butterfly can generate a tornado then it can also prevent it from occurring. The frequency of extreme events does therefore not increase due to a butterfly; the parallel illustrates the *random* character of the phenomenon. With these metaphors, Lorenz intended to illustrate the difficulties of weather forecasting, and did not give an answer to the question raised in the lecture's title.

by three orders of magnitude.[14] The non-chaotic prediction time would become 1000 times longer; in the chaotic case, however, the change is not proportional to the improvement of the accuracy. In fact, t_p does increase, but by very little: by $\ln 1000/\bar{\lambda} \approx 7/\bar{\lambda} \approx$ seven iterates. Thus, the prediction time becomes 21 time units (as opposed to 10^9). In a chaotic system, a significant improvement of the prediction time is *impossible*.

5.4.3 The measure of order in phase space: fractal dimension

In the course of iterating a phase space domain its area shrinks, since $J < 1$, and the images move closer and closer to the chaotic attractor. In each step the domain is stretched along the *unstable* direction, i.e. parallel to the unstable manifold constituting the attractor, and is contracted in the direction across the unstable manifold. Simultaneously, the domain is *folded* again and again, since its length increases while being confined to a bounded region. Since the thin filaments cannot intersect each other, a structure similar to a Cantor set develops, perpendicular to the unstable direction. The domain takes on a more and more filamentary structure, and finally it adopts the shape of the chaotic attractor (see, for example, Fig. 5.6).

The structure of a chaotic attractor is therefore that of a direct product: in the unstable (stable) direction it is a continuous line (a fractal similar to a Cantor set). All chaotic attractors thus have the structure of distorted Cantor filaments (see Section 2.2.2). Their dimension can always be written as the sum of two partial dimensions:

$$\boxed{D_0 = 1 + D_0^{(2)},} \tag{5.62}$$

where the 1 is the partial dimension along the unstable direction, and $D_0^{(2)}$ is the partial dimension along the stable direction, with $0 < D_0^{(2)} < 1$. The fractal dimension, D_0, measures how much more complicated the phase space structure is than for regular motion. In addition, its existence demonstrates that an *ordered* phase space structure is associated with motion which appears to be irregular in time.

5.4.4 Natural distribution

After overviewing the three essential characteristics of chaotic motion, we now discuss in detail a quantity that provides their synthesis and also generalisation: the *probability distribution* developing on the chaotic

[14] Such sudden improvements were very rare in the history of science and were always consequences of significant discoveries.

attractor. The use of this quantity is unavoidable, since – as we have seen – after a prediction time of merely a few time units, the motion on the attractor can only be described with a 100% error. The observer finds the motion to be random. Long-term behaviour can therefore be characterised only by giving the *probability* that the state of the system comes to a small neighbourhood of one point or another of the attractor.

In order to interpret this accurately, it is useful to investigate the dynamics of a phase space domain in a probabilistic sense before reaching the attractor. The points constituting the phase space domain will henceforth be called *particles* in order to facilitate a probabilistic interpretation.

If initially particles are uniformly distributed, the density, $P_0(\mathbf{r})$, of the probability that there is a particle within the phase space domain of initial area A_0 is $P_0 = 1/A_0$, and $P_0(\mathbf{r}) \equiv 0$ outside this domain. In one iteration of the map, the area decreases; therefore, the probability is greater than at the beginning, since the number of particles is conserved and the distribution is normalised to unity. The particle density defines some distribution, $P_1(\mathbf{r})$, above the new shape of the phase space domain as support. Furthermore, the distribution is *no longer homogeneous*, since, in general, neither stretching nor contraction is homogeneous (the local Lyapunov exponents are position-dependent). In the course of further iterations the area continues to decrease, and the probability density keeps increasing. As a consequence of stretching, the probability distribution, $P_n(\mathbf{r})$, remains smooth along the unstable direction. Due to the repeated folding, however, this is not so along the stable direction.

Continuing the iterations, the distributions, P_n, converge for $n \to \infty$ to a limit distribution P^*, as follows:

$$P_n(x_1, x_2) \to P^*(x_1, x_2), \tag{5.63}$$

which is non-zero only on the chaotic attractor, a set of zero area. This is the *natural distribution*[15] of the chaotic attractor, a stationary, time-independent distribution. The natural distribution yields the probability that a point moving on the attractor for a long time visits a certain region of the attractor. The natural distribution is the only proper tool for the long-term characterisation of chaotic systems.

As a particular example, consider first the time evolution of the probabilities in the symmetric baker map, (5.2). It is worth starting from the initial distribution, $P_0 \equiv 1$, which corresponds to uniformly filling the entire unit square with particles. In the first step the phase space

[15] In mathematical terminology, the natural measure.

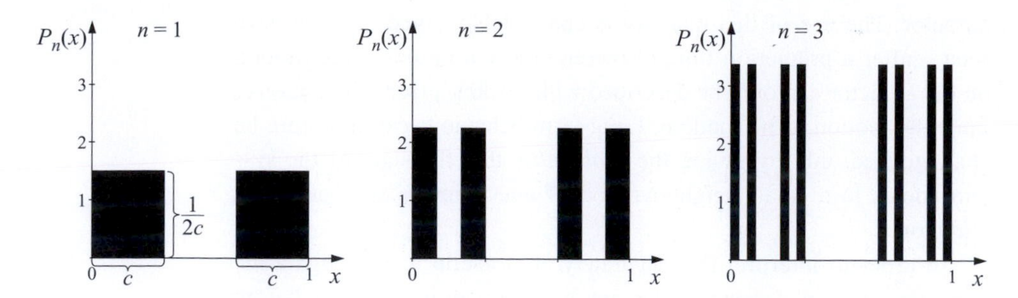

Fig. 5.42. Time evolution of the probability in the symmetric baker map over the first three steps, starting from a uniform distribution, $P_0 \equiv 1$, on the entire phase space ($c = 1/3$). (The initial distribution is not presented.) Since the distribution is independent of the velocities, only the x-dependence is displayed. The sequence converges to the natural distribution, P^*. The total area of the columns remains equal to unity throughout.

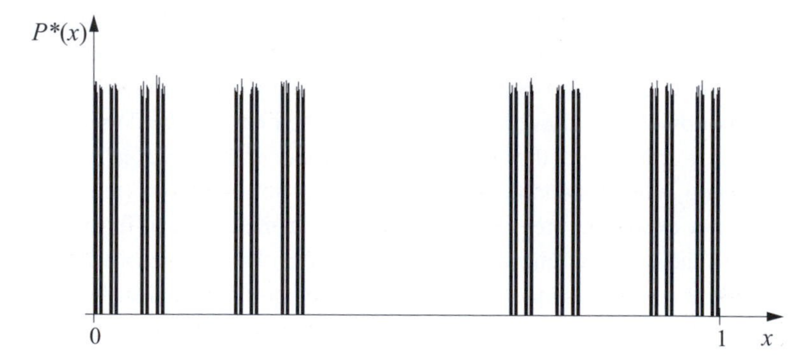

Fig. 5.43. Natural distribution of the symmetric baker attractor obtained from numerical simulation. A total of 300 000 iterations has been carried out from a single initial point on the attractor, and a resolution of $\varepsilon = (1/3)^6$ has been used in x to represent recurrences.

volume is mapped into two rectangles of width c and unit length (see Fig. 5.8). The number of particles is unchanged; the probability of finding them is $1/2$ on both sides, and the distribution function therefore takes on the value $P_1 = (2c)^{-1} < 1$ on the rectangles. In the next step, $P_2 = (2c)^{-2} > P_1$, and this value belongs to four rectangles each of width c^2. The distribution is uniform in the direction of the unstable manifold (along the v_n-axis) (Fig. 5.42).

In the limit $n \to \infty$, the sequence converges to the natural distribution, P^*, whose support is formed by a Cantor filament. A point moving on the chaotic attractor of the symmetric baker map can therefore occur on any of the Cantor filaments with equal probability. The natural distribution can also be obtained via a long-term numerical simulation of a single trajectory (see Fig. 5.43).

Problem 5.28 Show that in the symmetric baker map a linear initial distribution $P_0 = 1 + \gamma(x - 1/2)$ (independent of velocity) converges to the same natural distribution, P^*, as the homogeneous initial distribution.

Problem 5.29 Show that in the symmetric baker map a linear initial distribution $P_0 = 1 + \gamma(v - 1/2)$ (independent of position) converges to the same natural distribution, P^*, as the homogeneous initial distribution.

It is important to emphasise that the natural distribution is *independent* of the initial distribution. If the particles are not uniformly distributed in the initial domain, or if the domain is chosen in a different way, the initial differences disappear upon approaching the attractor. The limit distribution is determined by stretching along the unstable manifold and contraction along the stable manifold. The natural distribution is therefore a distribution characteristic of the attractor only.

Based on the preservation of the particle number, a general relation can be found between probability distributions P_n and P_{n+1}. The integral of the probability distribution, P_n, over a small area is proportional to the number of particles in that area. In the nth step the number $P_n\, dx_1\, dx_2$ of the particles in the area element $dx_1\, dx_2$ around point (x_1, x_2) has to be equal to the particle number $P_{n+1}\, dA'$ in the area dA' around the image point $(x_1', x_2') = M(x_1, x_2)$; i.e., $P_n\, dx_1 dx_2 = P_{(n+1)} dA'$. Since the area dA' is a factor of the Jacobian $J(x_1, x_2)$ ($J > 0$) smaller than the original, the probability distribution at the $(n + 1)$st iterate follows from that at the nth iterate as follows:

$$\boxed{P_{n+1}(x_1', x_2') = \frac{P_n(x_1, x_2)}{J(x_1, x_2)}.} \tag{5.64}$$

This equation, called the Frobenius–Perron equation, provides the time evolution of distributions. Since time is discrete, it can also be considered as a map acting on probability distributions. The sequence of distributions, P_n, with arbitrary P_0 (within the basin of attraction) converges to the natural distribution, P^*, of the chaotic attractor, a solution of (5.64) with $P_{n+1} \equiv P_n \equiv P^*$. The natural distribution, P^*, is therefore a *fixed point* of (5.64), a *simple* attractor in the space of probability distributions. Since the natural distribution is unique and is the solution of an exactly known equation, in a probabilistic sense chaos can be described with perfect accuracy.

Problem 5.30 By means of equation (5.64), determine the distribution $P_1(x, v)$ obtained in one step from a general normalised initial distribution $P_0 \equiv g(x, v) > 0$ defined over the entire phase space of the symmetric baker map.

Problem 5.31 Applying the same arguments as in the derivation of equation (5.64), determine the relation between probability distributions P_n and P_{n+1} for a one-dimensional map $x_{n+1} = f(x_n)$, where the function $f(x)$ has two branches. What is the natural distribution in the extremely strong dissipation limit ($E \to 0$) of (a) the sawtooth, (b) the roof and (c) the parabola maps at parameter $a = 2$?

Despite the simple form of (5.64), the natural distribution is usually rather complicated. Its support is a fractal of measure zero; therefore, in practice, the natural distribution cannot be determined with arbitrary accuracy. Numerically, the procedure is to cover the attractor by small squares of size ε, measure for each square how many points of a long trajectory (obtained, for example, in 10^6 iterations) fall in the box, and divide this number by the total number of points (see Fig. 5.43). In this manner we determine the integral of the natural distribution over that box to a good approximation. The average of a certain quantity taken with this distribution only slightly deviates from the average taken with the exact distribution (the difference is of the order of ε).

Natural distributions show a new facet of chaotic attractors. On a fractal support embedded in two dimensions a usually highly inhomogeneous distribution appears (in a third dimension) whose local maxima belong to the most often visited subsets of the attractor. A numerical determination of the natural distribution proves in itself the chaoticity of a system. Figure 5.44 presents the natural distributions developing on the attractors of oscillators kicked with different amplitudes. Plate XII and the back cover exhibit other views of these distributions in colour.

Problem 5.32 Determine via numerical simulation the natural distributions on the chaotic attractors of the oscillators kicked with a bell-shaped and a sinusoidal amplitude (see Problems 5.24 and 5.25).

From the evolution of the distributions, P_n, or directly examining P^*, the natural distribution proves to be a *fractal distribution* (see Section 2.3). This implies that on plotting the distribution at finer and finer resolutions (boxes of smaller and smaller size ε), more detail becomes visible and more information is gained. Accordingly, a finite information dimension, D_1, can be assigned to the natural distribution (for its numerical determination see Appendix A.5). As seen, D_1 is the fractal dimension of the regions that are *typical* with respect to the distribution. Since the natural distribution is smooth along the unstable direction, the total information dimension, D_1, can be decomposed as follows:

$$\boxed{D_1 = 1 + D_1^{(2)},} \tag{5.65}$$

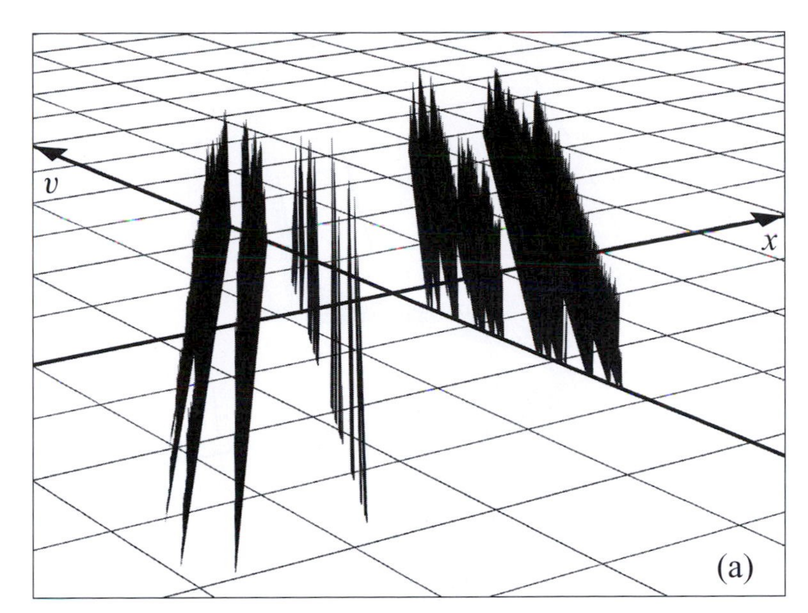

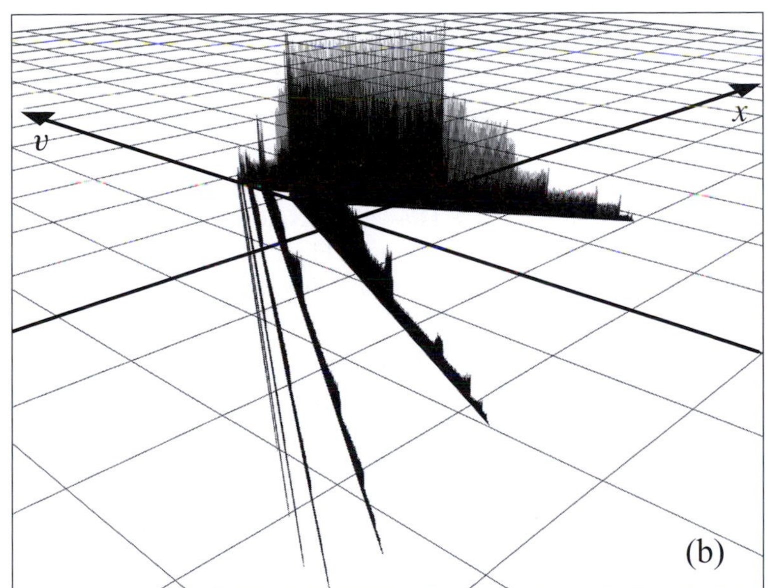

Fig. 5.44. Natural distributions of kicked oscillators. (a) Sawtooth attractor (see Fig. 5.22). (b) Roof attractor (see Fig. 5.28). (c) Parabola attractor (see Fig. 5.33). The apparent sudden jumps in distributions (b) and (c) are caused by the sharp bends of the unstable manifold, constituting the attractor, at certain locations. Along the manifold, however, the distribution changes smoothly. The grid is an aid seeing the perspective, it does not indicate units.

where $D_1^{(2)}$ is the partial information dimension in the stable direction. The chaotic attractor contains regions that are non-typical with respect to the natural distribution; the information dimension, D_1, is therefore less than the fractal dimension, D_0, of the attractor, i.e. $D_1^{(2)} \leq D_0^{(2)}$.

This is well illustrated by the example of the asymmetric baker map. Starting from a homogeneous distribution on the unit square and again investigating only a section parallel to x, we see that in one step the total

Fig. 5.44. (c).

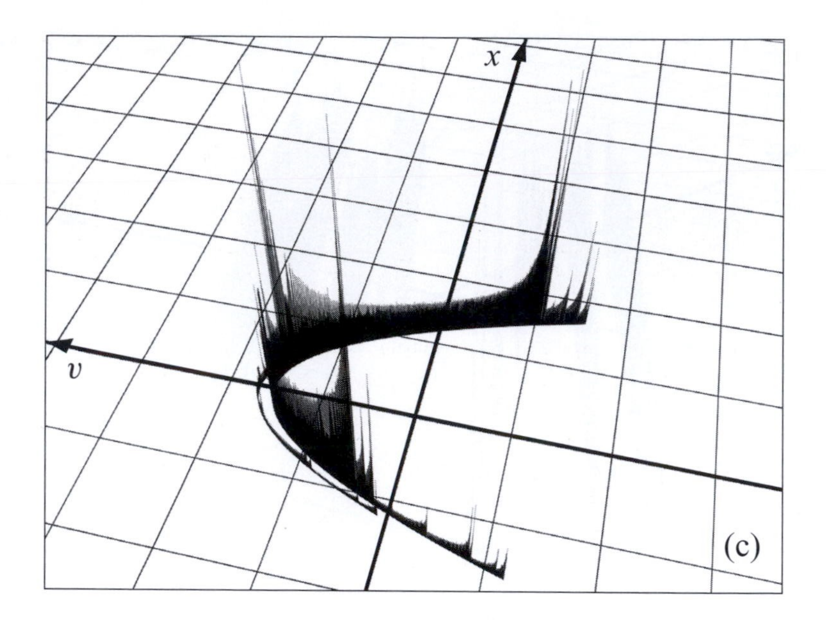

probability of the interval of width c_1 on the left is b, that of the interval of width c_2 on the right is $(1-b)$, and so on (Fig. 5.45).

In order to determine the information dimension in a simple way, we assume that $c_1 = c_2 \equiv c$; i.e., that the contraction rates in the x-direction are identical. After n steps, the probabilities of the intervals of length c^n take on the values $p_m = b^m(1-b)^{n-m}$. Note that this is exactly the construction of the fractal distribution of Fig. 2.14 (with the contraction parameter c replacing $1/3$). The information dimension, according to definition (2.18), is

$$D_1^{(2)} = \frac{b \ln b + (1-b)\ln(1-b)}{\ln c} < D_0^{(2)} = \frac{-\ln 2}{\ln c}. \tag{5.66}$$

For example, for $b = c = 1/3$, $D_1^{(2)} = 0.579 < D_0^{(2)} = 0.631$.

Problem 5.33 Determine the information dimension of the natural distribution of the asymmetric baker map with $c_1 \neq c_2$.

5.4.5 The Lyapunov exponent as an average

A pair of points moving on the chaotic attractor is subjected to the effect of different local Lyapunov exponents: the average Lyapunov exponent is therefore the time-average of the local exponents. Since, however, this same dynamics generates the natural distribution, the average can also be considered as an average taken with respect to the natural distribution. Consequently, the average Lyapunov exponent, $\bar{\lambda}$, is obtained as the mean

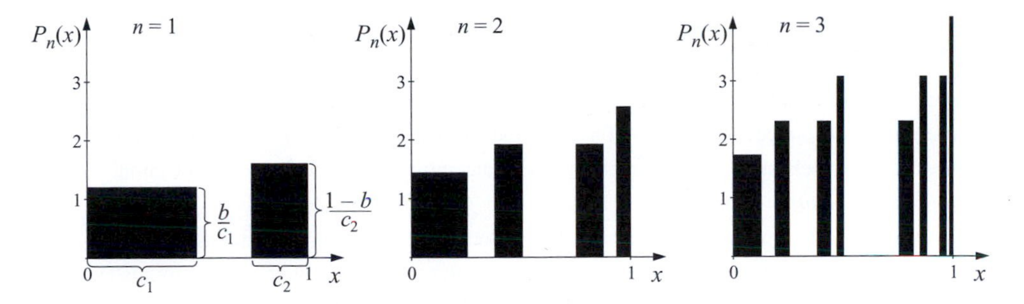

Fig. 5.45. Convergence of a uniform initial distribution of the asymmetric baker map to the natural distribution (only the x-dependence is shown) in the first three steps with $c_1 = 0.5$, $c_2 = 0.25$, $b = 0.6$.

value of the local Lyapunov exponents, $\lambda(\mathbf{r})$ (see (5.56)), with respect to the time-independent natural distribution, i.e.

$$\bar{\lambda} = \int dx_1\, dx_2\, \lambda(x_1, x_2)\, P^*(x_1, x_2). \tag{5.67}$$

This also leads to a relation between topological entropy and the average Lyapunov exponent. According to (5.52), topological entropy is the growth rate of the logarithm of the total length resulting from different local line segments. On the other hand, the local Lyapunov exponent is proportional to the logarithm of the length of a line segment (see (5.56)), and $\bar{\lambda}$ is the mean value of these exponents. The mean of a logarithm (a convex function) is generally less than the logarithm of the mean. Therefore, the topological entropy and the average Lyapunov exponent usually differ, and the latter cannot be greater than the topological entropy:

$$\boxed{h \geq \bar{\lambda}.} \tag{5.68}$$

The left-hand side of (5.68) represents the stretching rate (h) of a line in phase space (see (5.52)), which can also be viewed as the filamentary support of a probability distribution which is converging towards and spreading along the attractor. The right-hand side of (5.68) is the stretching rate, $\bar{\lambda}$, of regions typically populated with respect to the natural distribution along the line. (Remember, any distribution converges asymptotically to the natural one, as discussed in Section 5.4.4.) Thus, $\bar{\lambda}$ can be interpreted as the growth rate of the *average* lengths obtained by weighting with the natural distribution. Inequality (5.68) implies that the stretching of a line is in general faster than that of its typical regions. The difference is due to atypically populated (i.e. very rare) regions which do not contribute to the average. Equality only holds in (5.68) if the values of the local Lyapunov exponents, the stretching rates, are the same everywhere, in which case the natural distribution is a homogeneous distribution on the attractor.

Box 5.3 Determinism and chaos

It is wort emphasising that chaotic systems are *deterministic*. In a mathematical sense there exists a *unique* end-point to each trajectory, since the equation of motion (the Newtonian equation) is an ordinary (i.e. not noisy, in other words not stochastic) differential equation. Predictability is therefore present for initial conditions specified with *infinite* accuracy.

From a *practical* point of view, however, the interesting issue is the behaviour with initial conditions known with finite accuracy. The phenomenon of chaos demonstrates that for a practically deterministic (i.e. predictable) behaviour of real processes – in which initial conditions are inevitably known with some finite errors only – it is not sufficient that the system be deterministic in an idealised sense (for exactly known initial conditions). Determinism in a practical sense is present in non-chaotic systems only.

We must be sure that initial errors can usually not be neglected; their time evolution might be the *essence* of the phenomenon. Chaotic motion is an *error amplifier*. Beyond the short prediction time, chaotic motion appears to be *random*, as if it originated from a stochastic equation of motion. It should be emphasised that the random behaviour is not a consequence of the interaction with an environment composed of many particles (as, for example, for Brownian motion) but a property of the deterministic *inherent dynamics* of a few variables. Referring to this random behaviour of novel origin, chaos is often called *deterministic chaos*. Chaotic processes thus shed new light on the role played by chance in the description of Nature: the origin of random behaviour is not only the complexity of many-component systems, but also, surprisingly, the inherent non-linear dynamics of simple systems.

The negative average Lyapunov exponent is the mean value of the local exponents, $\lambda'(\mathbf{r})$ (see (5.56)):

$$\bar{\lambda}' = \int dx_1\, dx_2\, \lambda'(x_1, x_2)\, P^*(x_1, x_2). \tag{5.69}$$

Similarly, the average of any other quantity can also be calculated by means of the natural distribution. For example, the average of the logarithm of the local Jacobian, $J(x_1, x_2)$, is the following integral:

$$\overline{\ln J} = \int dx_1\, dx_2\, \ln J(x_1, x_2)\, P^*(x_1, x_2). \tag{5.70}$$

In view of (5.57), the above relations provide the proof of (5.58).

The support of the natural distribution is a fractal; therefore the integrals must always be carried out with finite resolutions, i.e. over a distribution determined on boxes of size ε. For the asymmetric baker map, the first construction step of the natural distribution (Fig. 5.45, $n = 1$) can readily be considered as a distribution determined with a finite resolution. Small segments mapped into the left domain of width c_1 become stretched by a factor of $1/b$ in the v-direction, while those mapped into the right are stretched by a factor of $1/(1 - b)$. The local Lyapunov exponent is therefore $\lambda = -\ln b$ on the left side and $\lambda = -\ln(1-b)$ on the right side. The average taken with respect to the piecewise constant distribution (with probabilities b and $1-b$) is therefore

$$\bar{\lambda} = -b\ln b - (1-b)\ln(1-b). \tag{5.71}$$

This process should be repeated with better and better approximants of the natural distribution, but the baker map is so simple that the result does not change. For $b \neq 1/2$, $\bar{\lambda} < h = \ln 2$. For $b = 1/2$ we recover the Lyapunov exponent, $\bar{\lambda} = \ln\ 2(= h)$, of the symmetric baker map.

Problem 5.34 By using the nth approximant of the natural distribution of the asymmetric baker map, show that the average Lyapunov exponent is indeed given by (5.71).

Furthermore, the local negative Lyapunov exponents in one step are $\ln c_1$ and $\ln c_2$, which yields an average Lyapunov exponent given by

$$\bar{\lambda}' = b\ln c_1 + (1-b)\ln c_2. \tag{5.72}$$

For $c_1 = c_2 = c$, $\bar{\lambda}' = \ln c$.

The average Lyapunov exponents are the weighted averages of all the local exponents, and the weights to be applied are determined by the natural distribution.

Problem 5.35 Determine the quantity $\overline{\ln J}$ on the asymmetric baker attractor.

Problem 5.36 Determine the average Lyapunov exponents of the saw-tooth attractor.

5.4.6 Link between dynamics and geometry: the Kaplan–Yorke relation

Both the information dimension and the average Lyapunov exponents are determined by the natural distribution. We can therefore expect to find an explicit relation between them. This rule, called the Kaplan–Yorke relation is valid for chaotic attractors of general two-dimensional invertible maps, and can be obtained from a simple argument. A phase space volume element of radius l_0 lying on the attractor is distorted in $n \gg 1$ steps into a filament of average length $l_0 e^{\bar{\lambda}n}$ and width $l_0 e^{\bar{\lambda}'n}$, i.e. of average area $A_n \approx l_0^2 e^{(\bar{\lambda}+\bar{\lambda}')n}$, folded in a fractal fashion (Fig. 5.46). The time evolution of the phase space volume itself determines a typical small length scale, which is, in the nth step, $l_0 e^{\bar{\lambda}'n}$. After $n \gg 1$ iterations, this can be considered as the size of the boxes needed

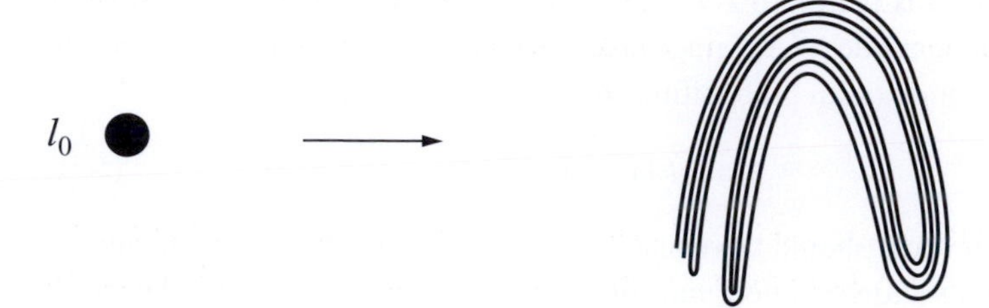

Fig. 5.46. A phase space volume element on the attractor is stretched (contracted) by a factor of $e^{\bar{\lambda}n}$ ($e^{\bar{\lambda}'n}$) along the unstable (stable) direction over $n \gg 1$ iterations, and, simultaneously, it becomes folded several times.

to cover the fractal attractor traced out by this time, i.e. as the resolution ε used in Section 2.1.2. The number of boxes of size $\varepsilon_n = l_0 e^{\bar{\lambda}' n}$ needed to cover the fractal is therefore given by

$$N(\varepsilon_n) = \frac{A_n}{\varepsilon_n^2} \sim e^{(\bar{\lambda}-\bar{\lambda}')n}. \tag{5.73}$$

This number is the power $(\bar{\lambda}/\bar{\lambda}' - 1)$ of resolution $\varepsilon_n \sim e^{\bar{\lambda}' n}$. The negative of this exponent can be interpreted as a fractal dimension defined by either formula (2.3) or (2.16). Having considered here a *typical* behaviour, (2.16) applies, and this dimension is the *information dimension*, D_1, of the natural distribution. Thus, we obtain the *Kaplan–Yorke relation*:

$$\boxed{D_1 = 1 + \frac{\bar{\lambda}}{|\bar{\lambda}'|}.} \tag{5.74}$$

Even though, in general, neither the information dimension nor the average Lyapunov exponent can be determined exactly, this explicit rule between them always holds. Equation (5.74) is one of the most important relations for attractors: it establishes a link between *dynamics* and *geometry*, and indicates that, in chaos, unpredictability and fractal structure are inseparable.

The partial dimension, $D_1^{(2)} \equiv D_1 - 1$ (see (5.65)), of the stable direction is thus given by the ratio of the average Lyapunov exponents. According to relation (5.58),

$$D_1^{(2)} = -\frac{\bar{\lambda}}{\bar{\lambda}'} = \frac{\bar{\lambda}}{\bar{\lambda} - \overline{\ln J}}, \tag{5.75}$$

which shows that the partial dimension is small for small positive Lyapunov exponents, and is close to unity for large ones. At a fixed $\overline{\ln J} < 0$, the more chaotic the system, the larger the dimension.

The difference between the fractal dimension, D_0, and the information dimension is generally a small percentage; the relation

$$D_0 \gtrsim 1 + \frac{\bar{\lambda}}{|\bar{\lambda}'|} \tag{5.76}$$

can therefore be used to estimate the fractal dimension of the attractor. The values of the two dimensions are identical only if the local Lyapunov exponents are independent of position and the natural distribution on the attractor is homogeneous.

Problem 5.37 Determine the fractal and information dimensions of the sawtooth attractor.

Box 5.4 What use is numerical simulation?

The law of exponential error growth raises the following question: When applied to chaotic systems, are numerical methods reliable at all? We give the answer in two stages.

As long as the moving point does not reach the vicinity of the chaotic attractor, there is no exponential amplification of the errors (the many unstable orbits all lie on the attractor). Computer simulations with sufficiently small numerical errors therefore correctly describe the approach towards the attractor. It also follows from this that a numerical process can provide the *shape* of the attractor accurately, since this is a consequence of the attraction along the stable direction.

In the course of the motion on the attractor, the concept of a single trajectory indeed becomes meaningless beyond the prediction time. A time series obtained via long-term running can, however, be considered as a *typical* trajectory. Statistical characteristics of the dynamics can thus be accurately obtained from long-term simulations, and averages and variances can be determined. By monitoring a single point on the attractor for a long time, the *natural distribution* can be determined with *arbitrary* accuracy. The same is obtained by simulating many points (an ensemble of particles) for a shorter time interval. Following pairs of nearby points and averaging over a sufficiently large number of pairs yields the average Lyapunov exponent. Numerical processes are therefore well suited for accurately determining the phase space structures and the statistical properties related to chaotic dynamics.

5.5 Summary of the properties of dissipative chaos

Chaotic behaviour is possible in systems with three-dimensional flows (two-dimensional invertible maps), and it generally occurs in sufficiently non-linear cases. The characteristics of such systems are essentially determined by their unstable periodic motions. Hyperbolic cycles that are not on a chaotic attractor play the role of a water-shed divide. Their stable manifolds form the basin boundary between the attractors

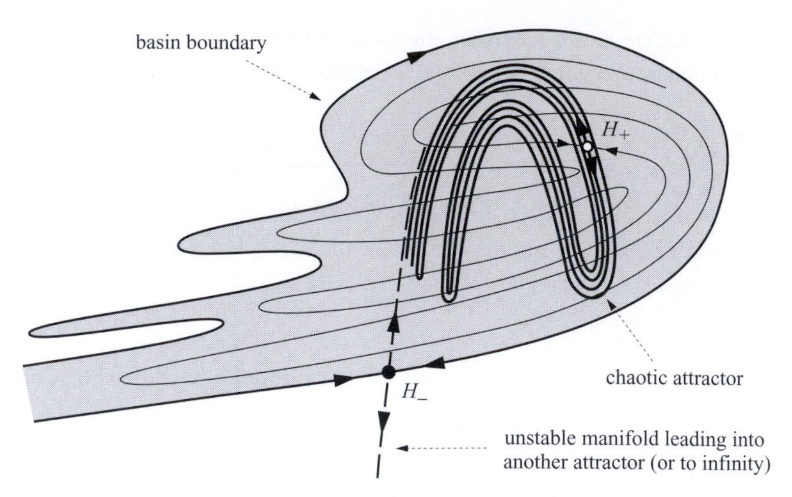

Fig. 5.47. Schematic phase portrait of a general invertible two-dimensional map.

(see Fig. 5.47). One branch of their unstable manifolds leads into one, the other leads into another attractor.

The chaotic attractor is an attracting invariant set on which a never-recurring sustained motion takes place. An infinity of unstable periodic orbits accumulate on the chaotic attractor, each of which possesses an unstable manifold. (From a practical point of view, the unstable manifold of any single cycle (of low period) can be considered to be the chaotic attractor.) This shows that chaos is, in fact, a sustained instability. The presence of hyperbolic cycles alone is, however, not sufficient to explain chaotic behaviour. This latter is the consequence of the development of homoclinic and heteroclinic points as intersections between the stable and unstable manifolds of cycles. If one is formed, there has to exist an infinity of them (see Box 6.3), since these can be mapped only into points like themselves. In the course of their motion they have to approach towards and move away from cycle points, which is only possible with a never exactly recurring time evolution. Homoclinic and heteroclinic points are therefore the elements that hold the cycle points together to form a chaotic attractor.

The properties of a chaotic attractor can be formulated at several levels. It is true to a good approximation (the difference is hardly perceptible numerically) that a chaotic attractor is

- the union of all the unstable (hyperbolic) periodic orbits lying on the attractor, or
- the unstable manifold of a single hyperbolic periodic orbit lying on the attractor.

A more precise definition of a chaotic attractor is

- the union of all the hyperbolic periodic orbits and of all the homoclinic and heteroclinic points formed among the manifolds of these points, or
- the union of the unstable manifolds of all the unstable periodic orbits lying on the attractor.

The main properties of chaotic motion: irregularity, unpredictability and ordered but complex phase space structure, are different manifestations of the same phenomenon. Characteristic numbers can be assigned to each of them as follows.

- Irregularity: positive topological entropy, $h > 0$,
- Unpredictability: positive average Lyapunov exponent, $\bar{\lambda} > 0$,
- Phase space order: fractal or information dimensions less than the dimension of the phase space, $2 > D_1 > 1$.

These properties reflect different aspects of chaos, manifested in one-, two- and many-particle behaviour, respectively (by particle we still mean a phase space point). Dissipative chaos is characterised by the simultaneous fulfilment of the above three inequalities.

The underlying reason for these properties is the existence of a natural distribution that determines the statistical properties of long-term motion. Consequently, the characteristic numbers of the individual properties are not independent (see, for example, equations (5.68) and (5.74)). These numbers also indicate the intensity of chaos: the larger their values, the more chaotic the system.

A phenomenon analogous to sensitivity to the initial conditions can also be observed in the parameter space: tiny changes of the parameters can lead to strong alterations in the nature of chaos or even to the disappearance of the chaotic attractor. This is a determinative feature of chaos, as well as, for example, the universal character of the bifurcation cascade leading to it. Any of these properties may be of help in deciding whether a system is chaotic or not.

Problem 5.38 In order for a chaotic attractor to exist it is sufficient that its information dimension be less than 2. In very exceptional cases, the attractor can be space-filling: its fractal dimension can be two. This is the case for the asymmetric baker map with parameters $c_1 = 1 - b$, $c_2 = b$ (the globally area preserving version of Problem 5.8). Determine the attractor via numerical simulation. Compute the quantity $-\overline{\ln J}$, the average phase space contraction rate, and show that it is always positive in this case. Determine the information dimension, $D_1^{(2)}$, up to leading order in terms of the asymmetry parameter $\delta = 1/2 - b$.

Box 5.5 Ball bouncing on a vibrating plate

One of the simplest examples of dissipative chaotic motion that can also be demonstrated in practice is that of a small ball bouncing on a harmonically oscillating plate, for example a vibrating loudspeaker diaphragm. The vibration of the plate is periodic, but the driving – which is, from the point of view of the ball, the collision with the plate – is not necessarily so. The origin of chaotic behaviour is that the flight time of the ball is usually not identical to the period of the vibration; collisions therefore occur at different phases each time. It is helpful to determine the state of the ball at the instants of the collisions, for example right after bouncing back, and to monitor the motion in the form of a map relating these states (which, in this case, is not a stroboscopic map). The form of the map can be derived because the motion between collisions is the well known vertical projection.

Let the velocity of the plate be $V(t) = V_0 \cos(\Omega t)$. By neglecting air drag, a ball which starts moving upwards with velocity v_n at time t_n will have velocity $\tilde{v}_n = v_n - g\Delta t_n$ immediately before the next bounce on the plate at time $t_{n+1} = t_n + \Delta t_n$. The most important dissipative effect is the energy loss during collisions. This is taken into account via a collision parameter, $k < 1$: the magnitude of the velocity relative to the plate after a collision is k times that of the impact value. The relative velocity at the instant before the $(n+1)$st collision is $\tilde{v}_n - V(t_{n+1})$; the relative velocity after the bounce is therefore $k(-\tilde{v}_n + V(t_{n+1}))$. Consequently, the velocity, v_{n+1}, right after the $(n+1)$st collision is $-k\tilde{v}_n + (1+k)V(t_{n+1})$. The time of flight, Δt_n, is determined by the fact that the distance between the positions of the plate at instants t_n and t_{n+1} is the same as the displacement $v_n\Delta t_n - g(\Delta t_n)^2/2$ of the ball.

The complete map is therefore given by

$$\left.\begin{aligned} t_{n+1} &= t_n + \Delta t_n, \\ v_{n+1} &= -k(v_n - g\Delta t_n) + (1+k)V_0 \cos(\Omega t_{n+1}), \\ \frac{V_0}{\Omega}&(\sin(\Omega t_{n+1}) - \sin(\Omega t_n)) = v_n \Delta t_n - \tfrac{g}{2}(\Delta t_n)^2. \end{aligned}\right\} \tag{5.77}$$

Measuring the velocity in units of V_0, and using the phase $\phi = \Omega t$ of the vibration instead of time, the dimensionless map is given by

$$\left.\begin{aligned} \phi_{n+1} &= \phi_n + \Delta\phi_n, \\ v_{n+1} &= -k\left(v_n - \tfrac{2\Delta\phi_n}{q}\right) + (1+k)\cos(\phi_{n+1}), \end{aligned}\right\} \tag{5.78}$$

where the phase difference, $\Delta\phi_n$, can be obtained from solving the following equation numerically:

$$\sin(\phi_n + \Delta\phi_n) - \sin(\phi_n) = v_n\Delta\phi_n - \frac{(\Delta\phi_n)^2}{q}.$$

Here, $q \equiv 2V_0\Omega/g$ is the dimensionless driving frequency. We suggest that the reader explores the details of the dynamics individually.

We demonstrate the chaoticity of the system by presenting the chaotic attractor of the explicit map obtained in the limit $q \gg 1$, corresponding to very fast vibrations of the plate (Fig. 5.48).

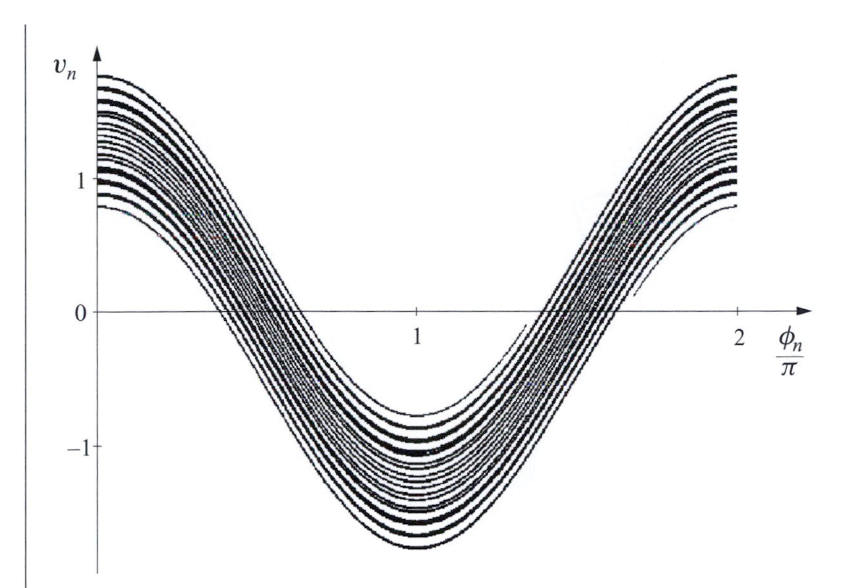

Fig. 5.48. Bouncing ball: chaotic attractor of the map $\phi_{n+1} = \phi_n + qv_n$ $(\Delta\phi_n = qv_n)$, $v_{n+1} = kv_n + (1 + k)\cos(\phi_{n+1})$ obtained in the limit $q \gg 1$ $(k = 0.3,\ q = 20)$.

5.6 Continuous-time systems

The occurrence of chaos is generally due not to a sudden external intervention, but to non-linearity. The reason we started with kicked systems was because their maps can be derived exactly. Here we investigate periodically driven dissipative systems with smooth time dependence. In such cases the stroboscopic maps can only be determined numerically; the discrete-time behaviour generated by them is, however, of a character very similar to that seen in the previous examples.

In relation to such systems, we explore the connection between the structures of three-dimensional flows and two-dimensional maps. All unstable periodic orbits are of hyperbolic character in the entire phase space. This implies that there is a (non-closed) surface, the stable manifold, pertaining to the helical trajectory of an unstable limit cycle, and trajectories from each point of this surface run into the limit cycle (Fig. 5.49). Close to the limit cycle, all trajectories on the surface of the stable manifold approach the limit cycle at a rate of $e^{\lambda'_f t}$, where $\lambda'_f < 0$ is the continuous-time negative eigenvalue of the limit cycle. (For a one-cycle, this corresponds to the eigenvalue λ_- discussed in Section 3.1.1.)

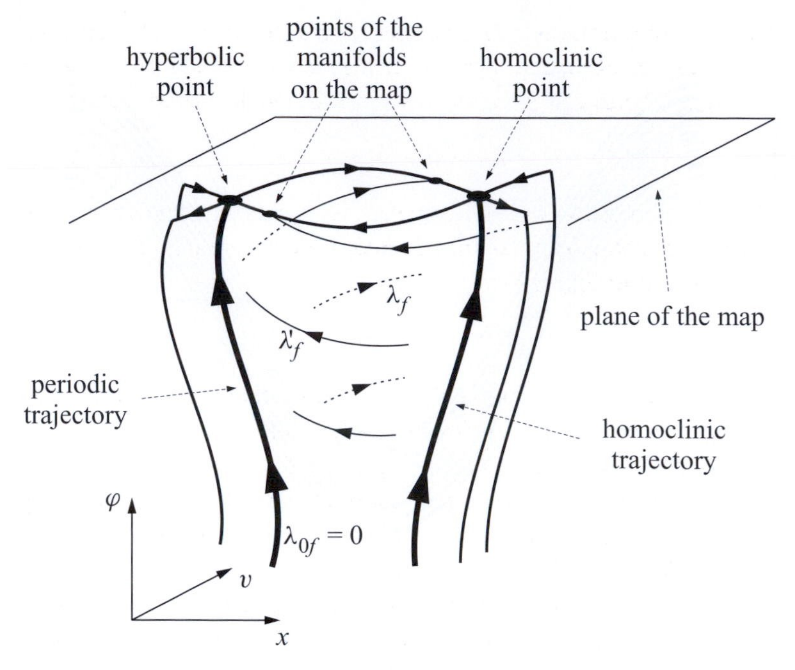

Fig. 5.49. Stable and unstable manifolds of a hyperbolic limit cycle of the flow. The intersection of the surfaces of the stable and unstable manifolds is a homoclinic trajectory. We have also marked a few trajectories within the surfaces of the manifolds.

Similarly, another (non-closed) surface also exists, the unstable manifold, in which a purely exponential deviation occurs according to the rule $e^{\lambda_f t}$, where $\lambda_f > 0$ is the positive eigenvalue of the limit cycle (analogous to the instability exponent λ_+ of Section 3.1.1). Since the exponential time dependence is also valid for pairs of points, the eigenvalues λ_f and λ'_f can be considered as local Lyapunov exponents.

The phase space distance between two nearby points along the trajectory of the limit cycle, will obviously not change in time. We can say that the local Lyapunov exponent, λ_{0f}, characteristic of the motion *along* a trajectory in the flow is zero: $\lambda_{0f} \equiv 0$.

The intersections between the stroboscopic planes taken with the period T of the driving force and the manifolds of the flow (see Fig. 5.49) determine the curves of the manifolds appearing in the stroboscopic map. (Continuous trajectories moving on the surface of a manifold generate points jumping along the manifold of the map.) The intersection lines of the surfaces of the three-dimensional manifolds form non-periodic homoclinic trajectories which generate the homoclinic points of the map. The earlier observation that homoclinic points are mapped into homoclinic points corresponds to following the motion along a single three-dimensional homoclinic trajectory (a homoclinic point and its image are a phase difference of 2π away along the trajectory).

Since all trajectories cross the plane of the map in every period T, the eigenvalues of a cycle point are in a simple relation with the eigenvalues characteristic of the flow. For a fixed point, $\Lambda_+ = e^{\lambda_f T}$ and $\Lambda_- = e^{\lambda'_f T}$.

Consequently, if the hyperbolic cycles form a chaotic attractor, the average Lyapunov exponents, $\bar{\lambda}$ and $\bar{\lambda}'$, of the map are related to the average Lyapunov exponents, $\bar{\lambda}_f > 0$ and $\bar{\lambda}'_f < 0$, of the flow as follows:

$$\bar{\lambda} = \bar{\lambda}_f T, \quad \bar{\lambda}' = \bar{\lambda}'_f T. \tag{5.79}$$

The sum of the flow's average Lyapunov exponents yields the opposite of the average phase space contraction rate, $\bar{\sigma}$, on the attractor (see (3.65)); therefore, according to (5.57),

$$\bar{\sigma} T = -\overline{\ln J}. \tag{5.80}$$

The average of the logarithm of the Jacobian is thus proportional to the average phase space contraction rate of the flow.

Structures developing in the map are shifted and slightly distorted when following three-dimensional trajectories (and return periodically every T). Such a smooth transformation *cannot* change the fractal properties of an object. Consequently, dimensions D_0 and D_1 *do not depend* on the phase in which the snapshot is taken. In the flow, the trajectories are smooth curves; the fractal dimension, D_{0f}, of the flow's chaotic attractor is therefore simply greater by 1 than that observed on the stroboscopic map:

$$\boxed{D_{0f} = D_0 + 1.} \tag{5.81}$$

Similarly, on the basis of the Kaplan–Yorke relation, (5.74), the information dimension of the flow's chaotic attractor is found to be

$$\boxed{D_{1f} = 2 + \frac{\bar{\lambda}_f}{|\bar{\lambda}'_f|}.} \tag{5.82}$$

If we define the topological entropy, h_f, of the flow as the growth rate of the number of unstable periodic orbit elements in continuous time then it is simply the topological entropy of the map divided by T: $h = h_f T$. Thus, relation (5.68) is also valid in the form $h_f \geq \bar{\lambda}_f$.

Now we present a few chaotic systems with smooth time evolution. Their attractors will be displayed on stroboscopic maps, since this is the best tool to provide a clear picture.

5.6.1 Harmonic oscillator driven by a position-dependent amplitude

We investigate a harmonic oscillator subjected to a sinusoidal driving of period T, whose amplitude, f_0, is a given function, $f_0(x)$, of the instantaneous position (cf. (4.7)). In this system the effect examined in kicked oscillators is smoothed out in time. As f_0 has the dimension of

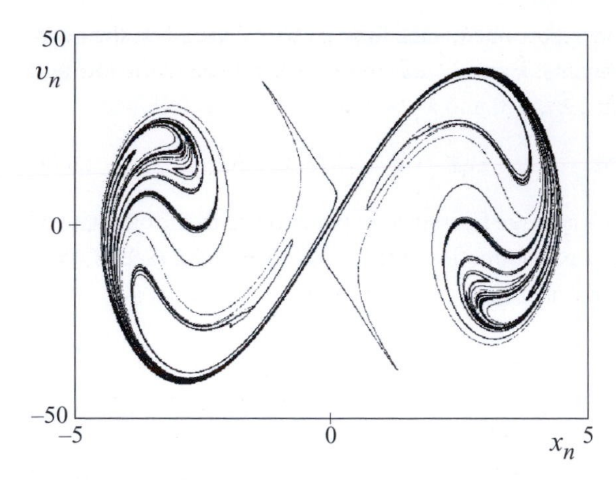

Fig. 5.50. Harmonic oscillator driven sinusoidally with amplitude $I(x) = a \sin x$ (equation (5.84)): a chaotic attractor on the stroboscopic map generated numerically with period T. The parameters are $\alpha T = 2.5,\ \omega_0 T = 6,\ a = 600$.

acceleration, we can write it in the form $f_0(x) = I(x/L)L/T^2$, where L is a characteristic length. The equation of motion is given by

$$\ddot{x} = -\alpha\dot{x} - \omega_0^2 x + \frac{L}{T^2} I\left(\frac{x}{L}\right)\cos\left(2\pi\frac{t}{T}\right). \tag{5.83}$$

Measuring time in units of the driving period, T, and distance in units of parameter L,[16] the dimensionless equation of motion is given by

$$\ddot{x} = -\alpha T\dot{x} - \omega_0^2 T^2 x + I(x)\cos(2\pi t). \tag{5.84}$$

The chaotic attractor obtained with the amplitude function $I(x) = a \sin x$ is presented in Fig. 5.50.

5.6.2 Non-linear oscillator driven with constant amplitude

With this model, the effect of deviation from a linear force law is investigated. We keep the driving amplitude at a constant value, $f_0 = IL/T^2$, where I is its dimensionless magnitude. The spring stiffens with elongation and the force per unit mass is written in the form $-(\omega_0^2 x + \varepsilon_0 x^3)$. The equation of motion is given by

$$\ddot{x} = -\alpha\dot{x} - \omega_0^2 x - \varepsilon_0 x^3 + \frac{L}{T^2} I \cos\left(2\pi\frac{t}{T}\right). \tag{5.85}$$

[16] For those who are not familiar with writing equations in dimensionless forms we suggest reading Appendix A.2.

Measuring time in units of the driving period, T, and distance in units of the parameter $L = \omega_0/\sqrt{\varepsilon_0}$, the dimensionless equation (see Appendix A.2.3) is given by

$$\ddot{x} = -\alpha T \dot{x} - \omega_0^2 T^2 (x + x^3) + I \cos(2\pi t). \tag{5.86}$$

The attractor of this system has been shown in Section 1.2.1 (Fig. 1.4) and in Plate I. with parameters $\alpha T = 0.6$, $\omega_0 T = 6$ and $I = 1800$. The corresponding natural distribution is presented in Plate XIII and on the front cover.

These examples illustrate that the chaotic attractor of an oscillator kicked with an amplitude depending smoothly on position basically does not differ from that of harmonically driven non-linear systems. It does not matter whether non-linearity arises from the amplitude or the spring force; the structure of distorted Cantor filaments always appears.

5.6.3 Pendulum with a harmonically moving suspension

In this example, the driving originates from the motion of the suspension. The point of suspension of a simple pendulum with a thin rod of length l (Fig. 5.51) is moved along a horizontal line according to a harmonic oscillation of amplitude A and period T. The displacement of the suspension point is $x_0(t) = A\cos(2\pi t/T)$. We derive the equation of motion in a reference frame fixed to the point of suspension. In such an accelerating system an inertial force is present, which is proportional to the acceleration:

$$a_0(t) = -\frac{d^2 x_0}{dt^2} = A\left(\frac{2\pi}{T}\right)^2 \cos\left(2\pi\frac{t}{T}\right). \tag{5.87}$$

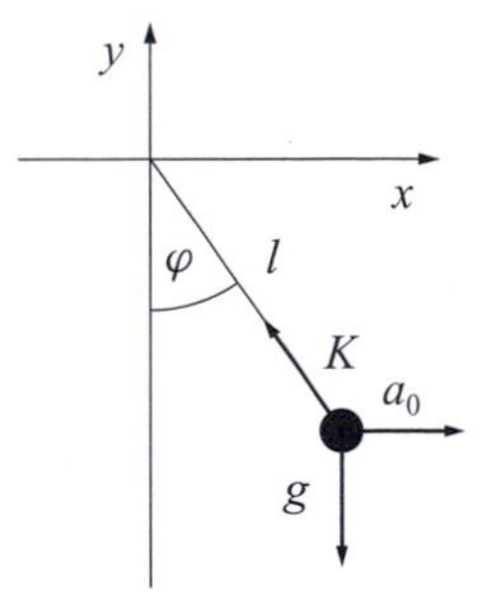

Fig. 5.51. Pendulum in a co-ordinate system fixed to its point of suspension, which oscillates horizontally. An inertial force proportional to acceleration, a_0, acts on the body. K is the force in the rod per unit mass.

By taking the algebraic sum of the horizontal and vertical force components, the equations valid in a vacuum are given by

$$\ddot{x} = -K\sin\varphi + a_0(t), \quad \ddot{y} = K\cos\varphi - g, \tag{5.88}$$

where K is the force exerted by the rod per unit mass and g is the gravitational acceleration (see Fig. 5.51). Using the relations $x = l\sin\varphi$, $y = -l\cos\varphi$, we switch to the angle variable, and find that $l\ddot{\varphi} = \ddot{x}\cos\varphi + \ddot{y}\sin\varphi$. Substituting equations (5.88), the force exerted by the rod is cancelled out. Assuming a bearing resistance (friction) of coefficient α proportional to the angular velocity, we obtain the equation of motion as follows:

$$\ddot{\varphi} = -\alpha\dot{\varphi} - \frac{g}{l}\sin\varphi + \frac{A}{l}\left(\frac{2\pi}{T}\right)^2 \cos\varphi\,\cos\left(2\pi\frac{t}{T}\right). \tag{5.89}$$

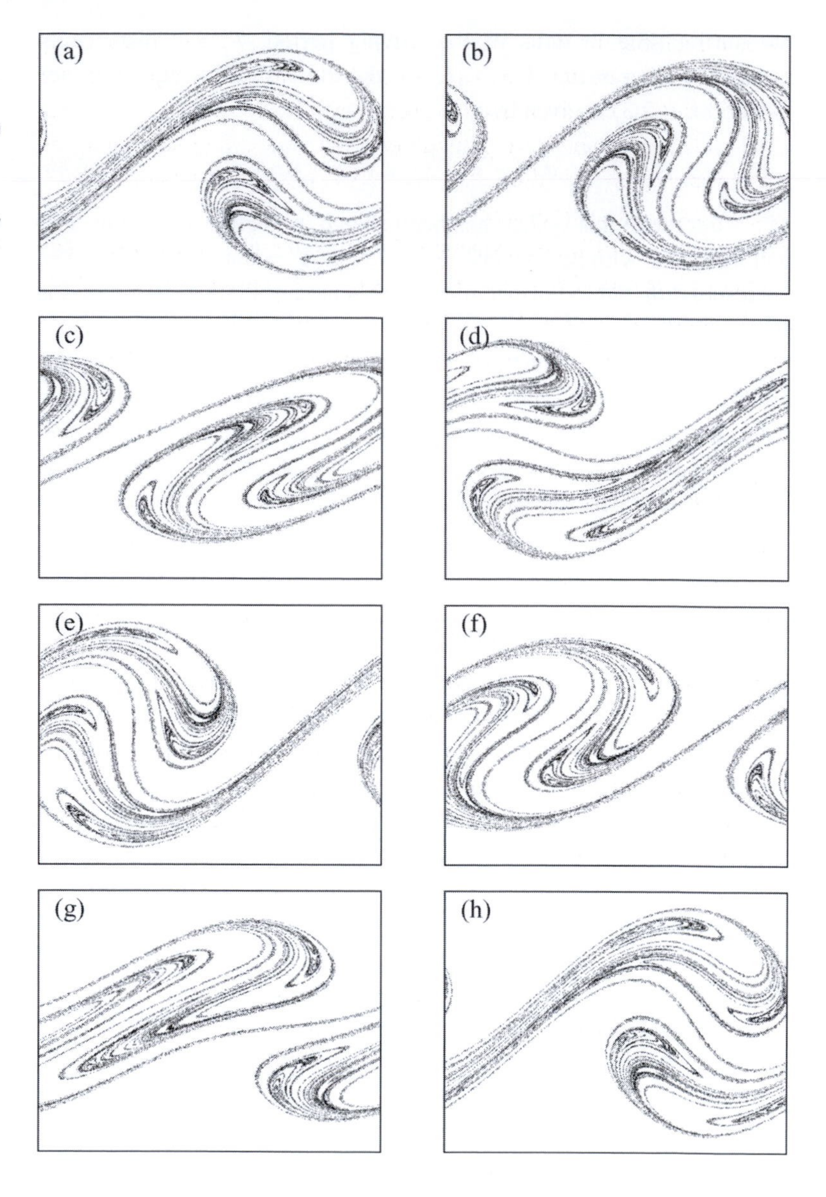

Fig. 5.52. Chaotic attractor of the driven pendulum on stroboscopic maps taken at different phases. In (a) (which is identical to Fig. 1.8) the initial phase is $\varphi_0 = 0$; in (b)–(h), it is $\varphi_0 = 2\pi/7, 4\pi/7, \ldots, 14\pi/7$, respectively. (The phase difference between (a) and (h) is exactly 2π; therefore these are identical.)

Measuring time in units of the driving period T, the dimensionless equation of motion (cf. Appendix A.2) is

$$\ddot{\varphi} = -\alpha T\dot{\varphi} - \frac{g}{l}T^2 \sin\varphi + \frac{A}{l}(2\pi)^2 \cos\varphi \ \cos(2\pi t). \tag{5.90}$$

The attractor of this system has been shown in Fig. 1.8 and in Plate II, with parameters $\alpha T = 0.2\pi$, $T\sqrt{g/l} = 2\pi/3$ and $A/l = 2$. The corresponding natural distribution is presented in Plate XIV.

We now augment our knowledge about the structure of the attractor by presenting its dependence on the driving phase. The flow trajectories are intersected at the instants $kT/7 + nT$ $(k = 0, \ldots, 7)$, corresponding to the respective initial phases $\varphi_0 = k2\pi/7$ $(k = 0, \ldots, 7)$. This is as if we looked at the chaotic attractor on different 'floors' within a single period. With a continuous changing of the initial phase, we could move with an 'elevator' upwards along the phase axis, obtaining thus the motion of the chaotic attractor in continuous time (which naturally repeats itself periodically with T). The sequence of figures in Fig. 5.52 exhibits a few frames of this film. The shape of the chaotic attractor naturally depends on the phase φ_0; its fractal dimension D_0 (and information dimension D_1) are, however, *independent* of it – as mentioned at the beginning of Section 5.6.

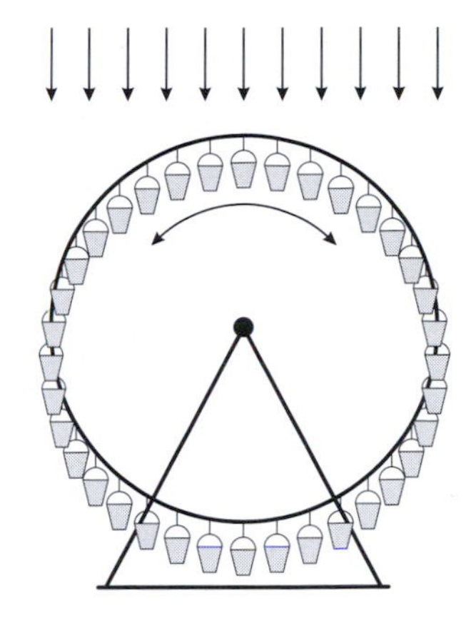

Fig. 5.53. Water-wheel. Buckets are distributed along the rim of the wheel mounted on a horizontal shaft, and rain falls onto them continuously. The wheel can rotate freely in both directions.

5.7 The water-wheel

Chaotic motion developing in dissipative systems can be a consequence not only of a time-periodic external driving, but also of any kind of energy input. In this general case, phase space is again at least three-dimensional, but neither variable must necessarily go to infinity (as the phase φ of the driving would do). In this case the chaotic attractor can appear within some bounded three-dimensional domain of phase space. A good example of this is the problem of a water-wheel, the equations for which are similar to that of the celebrated Lorenz model.

Consider a heavy circular wheel with buckets distributed homogeneously along its rim (Fig. 5.53). The wheel can pivot freely to the right or to the left about a shaft crossing its centre. Let a constant rain of spatially homogeneous distribution fall onto this apparatus and assume that some water is continuously leaking out of the buckets. The question is whether, despite the symmetric layout, the water-wheel may start rotating, and if it can, what will its motion be like in the long run?

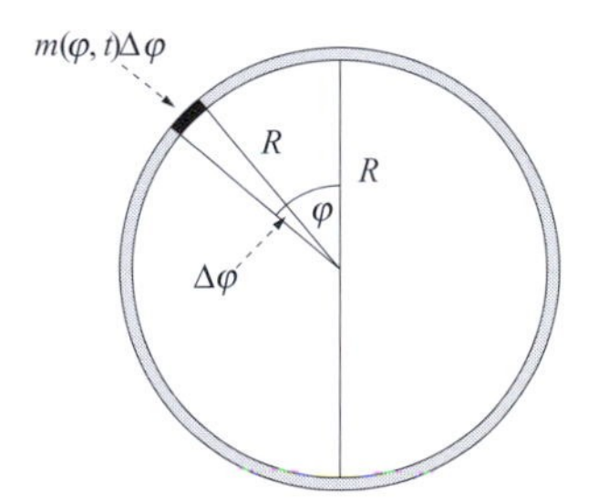

Fig. 5.54. Model of a water-wheel of radius R, in which the water distribution along the rim is considered to be continuous. Position along the rim is determined by angle φ; the water mass in an arc element, $\Delta\varphi$, is $m(\varphi, t)\Delta\varphi$.

5.7.1 Equation of motion

As a simple model of the water-wheel, assume that the buckets are very small, that there are many of them, and that their individual masses are negligible. Thus, the distribution of water along the rim can be considered to be continuous. Let the mass of water along a short arc element, $\Delta\varphi$, around angle φ fixed in space be denoted by $m(\varphi, t)\Delta\varphi$ (see Fig. 5.54). Due to the inflow and outflow, the mass distribution, m, might also depend on time.

In order to derive the wheel's equation of motion, we have to determine the gravitational torque of the water distribution about the axis.

The contribution from an arc element around angle φ is given by

$$m(\varphi, t)\ \Delta\varphi\ g\ R\ \sin\varphi, \tag{5.91}$$

where g is the gravitational acceleration and R is the radius of the wheel. The gravitational torque of the wheel vanishes because of the symmetry of its shape. Thus the resultant torque is the sum, or rather the integral, of the elementary contributions (5.91):

$$M = \int_0^{2\pi} d\varphi\, m(\varphi, t)\ g\ R\ \sin\varphi. \tag{5.92}$$

Note that the resultant torque is non-zero only if the mass distribution, $m(\varphi, t)$, is asymmetric with respect to the angle φ.

A damping torque is also present due to friction, and is assumed to be proportional to the instantaneous angular velocity, ω. The equation of motion relates the applied torques to the angular acceleration:

$$\Theta\dot{\omega} = M - \alpha\omega. \tag{5.93}$$

Here, Θ is the moment of inertia of the system with respect to the axis and $\alpha > 0$ is the friction coefficient. For the sake of simplicity, we assume that the wheel is much heavier than the total weight of the water in the buckets; the moment of inertia Θ is therefore constant in time.

In order to calculate the torque, N, the mass distribution, $m(\varphi, t)$, of the water must be known. Assume that it can be written in the form

$$m(\varphi, t) = A(t)\ \sin\varphi + B(t)\ \cos\varphi. \tag{5.94}$$

By this, we have only chosen a simple form for the angular dependence, and the amplitudes A and B remain unknown. The idea is to derive differential equations for $A(t)$ and $B(t)$. If the solution of the complete set of equations yields $A(t) \neq 0$, then the mass distribution is asymmetric. If this is not the case, the resultant torque is always zero, and only a simple damped motion can develop (if the initial angular velocity is non-zero).

Evaluating integral (5.92) using (5.94) (the integral of $\sin^2\varphi$ over the entire period is π), the total torque is given by

$$M = gR\pi A. \tag{5.95}$$

The equation of motion is therefore given by

$$\dot{\omega} = \frac{gR\pi}{\Theta}A - \frac{\alpha}{\Theta}\omega. \tag{5.96}$$

Since function $A(t)$ is not yet known, in order to obtain a closed set of equations, another relation is needed, which will be provided by the conservation of the water mass.

The change of the mass per unit time, $\varphi \cdot \partial m(\varphi, t)/\partial t$, is due (i) to the precipitation falling onto the arc element $\Delta\varphi$, (ii) to leaking out, and (iii) to the 'turning in' and 'turning out' of a certain amount of water because of the rotation of the wheel. (Remember that the arc element, $\Delta\varphi$, is fixed in space.)

The precipitation is assumed to be uniformly distributed in space. Let the mass of water falling onto a horizontal line segment of length R per unit time be denoted by $q > 0$. This is a measure of the rain intensity. Since the horizontal projection of an arc element of length $R\Delta\varphi$ is $R \cos\varphi \, \Delta\varphi$, the amount of water falling onto the arc element per unit time is $q \cos\varphi \, \Delta\varphi$.[17] The amount of leaking water (which does not flow into any of the other buckets) per unit time is assumed to be proportional to the instantaneous mass, m, and the proportionality constant is denoted by $\kappa > 0$. The loss of water per unit time is thus $-\kappa \, m(\varphi, t) \, \Delta\varphi$. On a given arc element, the water mass turning in and out per unit time is $\omega \, m(\varphi, t)$ and $-\omega \, m(\varphi + \Delta\varphi, t)$, respectively. For very small $\Delta\varphi$, their sum can be written as $-\omega \, \Delta\varphi \, \partial m(\varphi, t)/\partial\varphi$.

Adding these and dividing by $\Delta\varphi$, we obtain the equation of mass conservation or the continuity equation of the water-wheel:[18]

$$\frac{\partial m(\varphi, t)}{\partial t} = q \cos\varphi - \kappa m(\varphi, t) - \omega \frac{\partial m(\varphi, t)}{\partial \varphi}. \tag{5.97}$$

By substituting the mass distribution, (5.94), we obtain

$$\begin{aligned} &\dot{A} \sin\varphi + \dot{B} \cos\varphi \\ &= q \cos\varphi - \kappa A \sin\varphi - \kappa B \cos\varphi - \omega A \cos\varphi + \omega B \sin\varphi. \end{aligned}$$

Whereas in equation (5.96) only the amplitude, A, of the anti-symmetric part appears, here both amplitudes are important. The equation can hold for all angles φ only if the $\sin\varphi$ and $\cos\varphi$ terms vanish separately. This leads to

$$\begin{aligned} \dot{A} &= -\kappa A + \omega B, \\ \dot{B} &= q - \kappa B - \omega A. \end{aligned} \tag{5.98}$$

Equations (5.98) and (5.96) form a closed, autonomous set of first-order differential equations for variables A, B and ω. The set of equations is

[17] The precipitation distribution along the lower semi-circle should be determined separately, and it depends on details like, for example, the degree of shadowing due to the upper buckets. We keep here the form $q \cos\varphi$ because the actual form does not essentially change the final result. If, for example, shadowing is complete, and $q = 0$ along the lower semi-circle, the parameter q in (5.98) should be replaced by $q/2$.

[18] The general form of the continuity equation is $\partial\varrho/\partial t + \mathrm{div}(\varrho \mathbf{v}) = Q$, where ϱ is the density of the conserved quantity, $\mathbf{v}$ is the flow velocity and Q is the source strength. In our system, $\varrho \to m$, $\mathbf{v} \to \omega$, $Q \to q \cos\varphi - \kappa m$, and divergence is replaced by partial differentiation with respect to φ.

non-linear, and contains the simplest type of quadratic non-linearity via the terms ωB and ωA. This set of equations uniquely determines the rotation dynamics of the water-wheel. A sustained rotation implies the existence of a solution in which ω does not tend to zero.

It is remarkable that these results remain unchanged if the assumed mass distribution, (5.94), is replaced by the form $m(\varphi, t) = A(t)\sin\varphi + B(t)\cos\varphi + B_0(t) + \sum_{n=2}^{\infty}(A_n(t)\sin(n\varphi) + B_n(t)\cos(n\varphi))$. This is because only the term $A(t)\sin\varphi$ contributes to integral (5.92), and the torque is given by (5.95) in the presence of this general distribution also. The continuity equation (5.97), yields, apart from (5.98), the relations $\dot{A}_n = -\kappa A_n + n\omega B_n$, $\dot{B}_n = -\kappa B_n - n\omega A_n$, $n \neq 1$, for the coefficients of the different harmonics. The coefficients A_n and B_n, $n \neq 1$, are therefore *decoupled* from equations (5.96) and (5.98). They do *not* affect A, B and the angular velocity, i.e. whether, or how fast, the wheel rotates.[19]

The phase space of the essence of the water-wheel dynamics is thus spanned by the variables A, B and ω. The system is dissipative, which is shown by the positive sign of the phase space contraction rate, σ (see (3.55)): $\sigma \equiv -(\partial\dot{A}/\partial A + \partial\dot{B}/\partial B + \partial\dot{\omega}/\partial\omega) = 2\kappa + \alpha/\Theta$.[20] The water-wheel therefore possesses *attractors*.

5.7.2 Fixed points and their stability

In a fixed point of the flow, $\dot{A} = \dot{B} = \dot{\omega} = 0$ holds. Therefore, the equations for a fixed point (ω^*, A^*, B^*) are, from (5.96) and (5.98), given by

$$\left.\begin{aligned} gR\pi A^* - \alpha\omega^* &= 0,\\ -\kappa A^* + \omega^* B^* &= 0,\\ q - \kappa B^* - \omega^* A^* &= 0. \end{aligned}\right\} \tag{5.99}$$

One of the possible solutions is the trivial fixed point $\omega^* = 0$, $A^* = 0$, $B^* = q/\kappa$. The angular velocity being zero, this is a state of rest of the wheel. In accordance with our expectations, the condition for equilibrium is that the term proportional to $\sin\varphi$ is not present in the mass distribution.

There may, however, exist other solutions. With $\omega^* \neq 0$, after eliminating A^* and B^*, we obtain

$$\omega^* = \pm\sqrt{\frac{qgR\pi}{\alpha} - \kappa^2}. \tag{5.100}$$

[19] This decoupling is valid even if the rain distribution is of the form $q_0 + q\cos\varphi + \sum_{n=2}^{\infty} q_n \cos(n\varphi)$. This implies that the water-wheel dynamics is governed by (5.96) and (5.98) even if the plane of the wheel is tilted and rain is falling onto a part of it only, as long as $q \neq 0$.

[20] The system is dissipative even in the frictionless case due to the loss of water (the term 2κ).

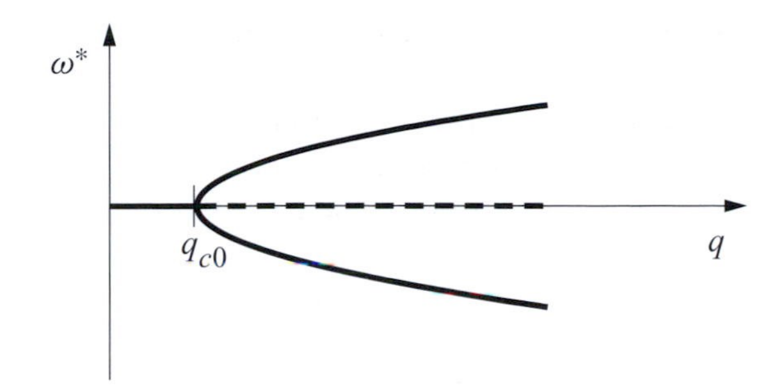

Fig. 5.55. Bifurcation diagram of the water-wheel. As long as the intensity of the rain is less than the critical value q_c, only one point attractor exists, which represents a standing state. This fixed point becomes unstable at $q = q_{c0}$, where two new point attractors are born corresponding to uniform rotations in different directions.

The condition for ω^* to be real is that the discriminant be positive, i.e.

$$q > q_{c0} = \frac{\kappa^2 \alpha}{gR\pi}. \tag{5.101}$$

The intensity of the rain must therefore be *greater* than a critical value, q_c, for the wheel to rotate. The $\pm$ signs then correspond to rotations in opposite directions, and there exist two non-trivial fixed points (see Fig. 5.55). The two other variables take on the following values:

$$A^* = \omega^* \frac{\alpha}{gR\pi}, \qquad B^* = \frac{\alpha\kappa}{gR\pi}. \tag{5.102}$$

The mass distribution is asymmetric, since $A^* \neq 0$, and A^* is proportional to the angular velocity.

Problem 5.39 Estimate the critical rain intensity if the parameters of the water-wheel are $R = 0.3$ m, $\kappa = 0.1$ s^{-1} and $\alpha = 1$ kg m^2 s^{-1} (this latter is realistic for a wheel with a mass of approximately 1 kg). What are the steady angular velocity and the values of A^*, B^* for a rain intensity $q = 2q_{c0}$ above the critical value?

If the rain is light, i.e. if $q < q_{c0}$ the fixed point $\omega^* = 0$, $A^* = 0$, $B^* = q/\kappa$ is stable (see Problem 5.40); it is an attractor. Consequently, if the wheel is set into rotation, sooner or later it comes to a halt. If, on the other hand, $q > q_c$ then the trivial fixed point is unstable (it becomes a hyperbolic point) and the two non-trivial fixed points appear as attractors (see Problem 5.41). The system therefore undergoes a pitchfork bifurcation (see Section 3.3.2) as the intensity of the rain increases. The phenomenon is analogous to the emergence of bistability presented in Fig. 3.14. In the range $q > q_{c0}$, small deviations that break the symmetry of the mass distribution become magnified and lead to the appearance of a finite A^*. The water-wheel is then in uniform rotation to the right or to the left.

It is worth writing the equations in dimensionless forms. Let us introduce the new dimensionless variables x, y, z via the relations $\omega = \kappa x$, $A = B^* y$, $B = -B^* z + q/\kappa$, and measure time in units of $1/\kappa$. Thus, (5.96) and (5.98) become

$$\left.\begin{aligned} \dot{x} &= -\sigma(x - y), \\ \dot{y} &= rx - y - xz, \\ \dot{z} &= -z + xy, \end{aligned}\right\} \tag{5.103}$$

where only two dimensionless parameters remain:

$$\sigma \equiv \frac{\alpha}{\Theta \kappa} \quad \text{and} \quad r \equiv \frac{qgR\pi}{\alpha \kappa^2}. \tag{5.104}$$

The first is the ratio of the decay times related to leakage and friction,[21] while the second is the dimensionless rain intensity. In the phase space spanned by the new variables x, y and z, the trivial fixed point is the origin, while the non-trivial fixed points are given by

$$x^* = y^* = \pm\sqrt{r - 1}, z^* = r - 1, \tag{5.105}$$

which exist for $r > 1$ only.

Problem 5.40 Show that in the range $r < 1$ ($q < q_{c0}$) the trivial fixed point is stable.

Problem 5.41 Determine the stability matrix of the non-trivial fixed points. In what range of parameter r are these fixed points stable? (Stability is lost when the real parts of two complex eigenvalues vanish.)

Box 5.6 The Lorenz model

Edward Lorenz discovered in 1963 that in the model (since named after him) given by

$$\left.\begin{aligned} \dot{x} &= \sigma(y - x), \\ \dot{y} &= rx - y - xz, \\ \dot{z} &= -bz + xy, \end{aligned}\right\} \tag{5.106}$$

with parameters $r = 28$, $\sigma = 10$, $b = 8/3$, even tiny differences in the initial conditions lead to large deviations in the numerically simulated time evolution. This was the first example of the appearance of unpredictability (and of the underlying chaotic attractor) in an autonomous system with few degrees of freedom. The Lorenz model has since become a paradigm of continuous-time chaotic systems.

Equations (5.106) are related to the hydrodynamical equations of thermal convection, i.e. of the flow of a fluid layer heated from below, under certain conditions. This phenomenon is, in fact, fundamental in the

[21] The notation σ is used because of historical reasons; we emphasise that the σ of (5.103) is not identical to the phase space contraction rate, which is $\sigma + 2$.

study of atmospheric motion, since the ground warmed up by the Sun heats the air from below. As long as the temperature difference, ΔT, across the fluid is small, no flow is initiated; only thermal conduction occurs, since buoyancy is not large enough to overcome viscosity. Exceeding a critical temperature difference, ΔT_c, the fluid starts flowing. *Slightly* above the critical value, the flow pattern is that of counter-rotating cylinders whose axes lie in a horizontal plane. The diameter of these cylinders is, to a good approximation, identical to the thickness of the fluid layer. In this regime, (5.106) can indeed be derived from the hydrodynamical equations. Quantity x proves to be proportional to the intensity of the cylindrical flow. Variables y and z measure the magnitude of temperature deviations from the temperature profile of linear height dependence at wavelengths corresponding to the layer thickness and to its half, respectively. The parameter σ is the ratio of the kinematic viscosity to the coefficient of thermal diffusion, Prandtl's number (which is 0.72 in air and 7.1 in water). The value $b = 8/3$ results from the cylindrical flow geometry, and $r \equiv \Delta T/\Delta T_c$, with the restriction $r \approx 1$. As long as r is less than unity, all initial motion dies out. The Lorenz equation describes convection in the range $0 < r - 1 \ll 1$ only. In the parameter range where chaos appears ($r > 24.06$), however, the Lorenz model (5.106) is no longer to do with the problem of convection. (This, of course, does not contradict the statement that if the temperature difference (and thus r) increases, the hydrodynamical flows become increasingly complicated. This may be seen, for example, in the process of warming water. These flows are described by the complete hydrodynamical equations that cannot be reduced to a system of three ordinary differential equations in this range.)

Fortunately, there exist several physical problems where the Lorenz equations provide a faithful model of the phenomenon within the physically relevant parameter range. This is the case of the water-wheel, with the only difference that the value of b cannot be changed: the dynamics of the water-wheel is described by the Lorenz equations with the choice $b \equiv 1$, in a context basically different from the original convection problem.

5.7.3 Chaos of the water-wheel

By increasing the rain intensity (the dimensionless parameter r), the sustained rotation of the water-wheel does not always remain uniform. Above a certain value, r_c, the wheel rotates to the right for some time and then to the left, and the directions are alternated at irregular intervals: the motion is chaotic. Fixing the parameter σ, we observe (Fig. 5.56) that by increasing r, chaos appears *suddenly* (more detail will be given in Section 6.6), and not via the period-doubling cascade of Section 5.3. For sufficiently large r-values, the two non-trivial fixed points also become unstable and a single chaotic attractor exists. A trajectory running on the attractor winds around one of the unstable fixed points a few times and then switches over to the other unstable fixed point, and this is repeated with an unpredictable number of rounds on each side. Neither fixed point is part of the chaotic attractor; there are therefore two large holes visible around them. The shape of the flow's attractor resembles a butterfly (see Fig. 5.57).

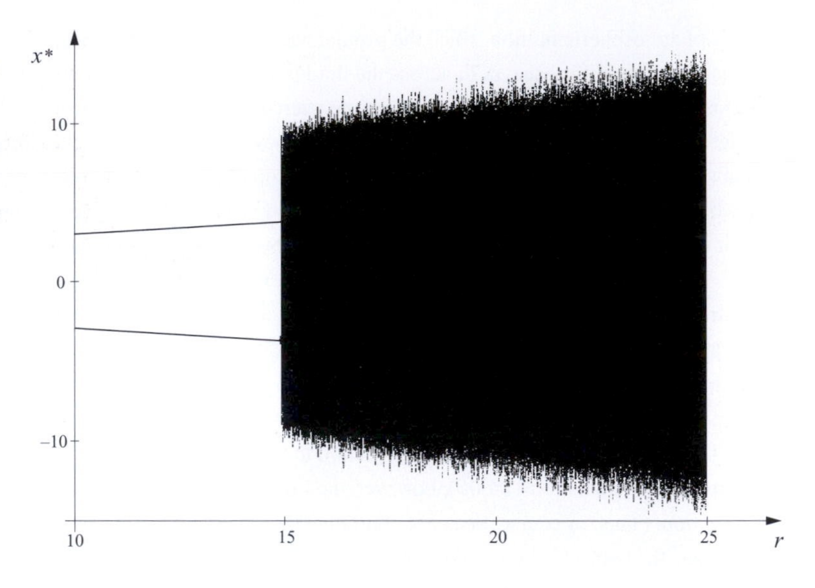

Fig. 5.56. Bifurcation diagram of the water-wheel at $\sigma = 10$. At each r, four trajectories are simulated and the x values are plotted in the time interval $500 < t < 600$. Chaos suddenly appears at the value $r = r_c \approx 15.0$.

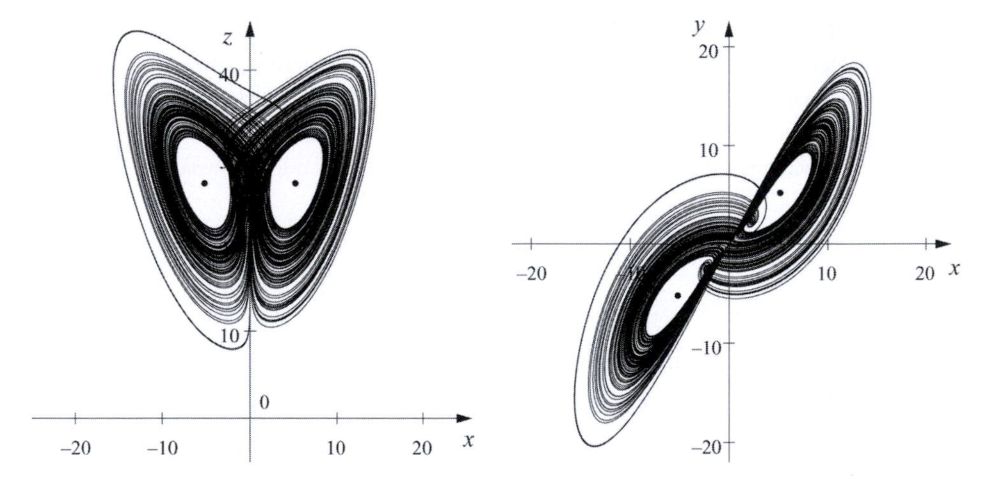

Fig. 5.57. The water-wheel attractor projected on the (x, z)- and (x, y)-planes ($\sigma = 10$, $r = 28$). The unstable fixed points of the flow are each marked by a black dot. The attractor is reminiscent of the Lorenz attractor (equation (5.106) with $b = 8/3$).

A stereoscopic picture provides a clearer view (Fig. 5.58), and we can perceive that the attractor is not a single, twisted, smooth surface but that each butterfly wing consists of many surfaces very close to each other, forming a fractal set.

In order to obtain a two-dimensional representation, we intersect the attractor with a plane and mark the intersections of trajectories coming from a given direction (for example from above). This defines what is called a *Poincaré map*, to be discussed in more detail in Section 7.1.

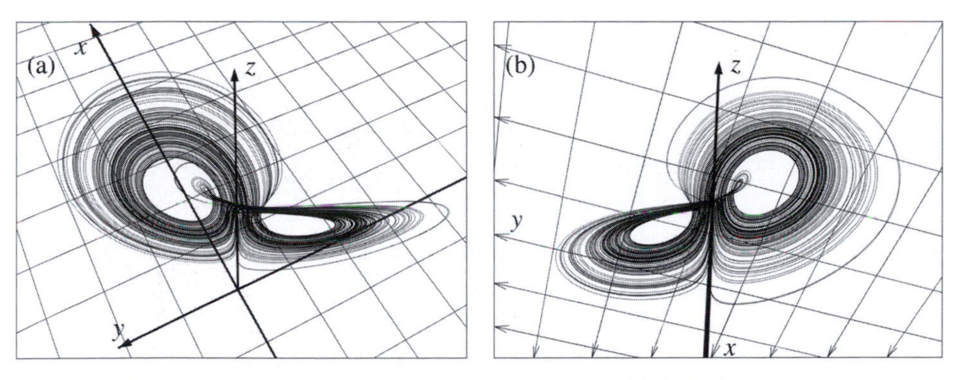

Fig. 5.58. Two stereoscopic views of the water-wheel attractor: (a) from above, (b) from below ($\sigma = 10$, $r = 28$). The thin arrows in (b) mark the direction of increase of x and y; the x- and y-axes are not visible.

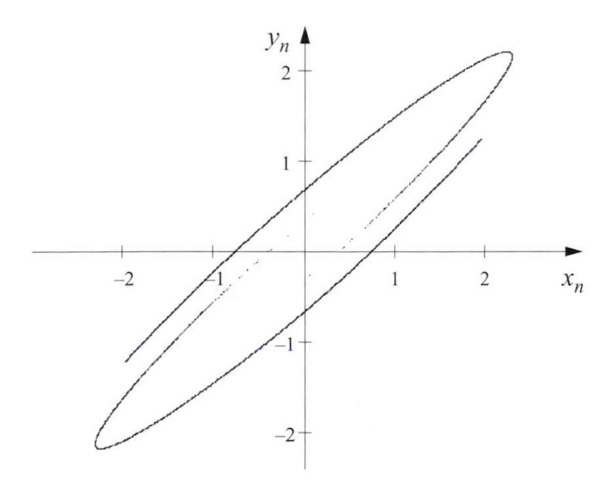

Fig. 5.59. Water-wheel attractor on the Poincaré section $z = r - 1$ ($\sigma = 10$, $r = 28$).

On the Poincaré plane the attractor appears to be two simple curves (see Fig. 5.59) since dissipation is strong (the volume shrinks by a factor of $e^{\sigma+2} = e^{12} \approx 1.6 \times 10^5$ per unit time).

Problem 5.42 Determine the nth maximum, z_n, of function $z(t)$ on the water-wheel attractor, and plot z_{n+1} vs. z_n.

The great advantage of applying a map is that the natural distribution can clearly be represented on it. By running a single long trajectory, we can determine how often it visits different points of the attractor. The distribution reflects the fractal structure better than the shape of the attractor alone. At certain points (Fig. 5.60) the distribution takes on very large values, while at others it is nearly constant, and these domains

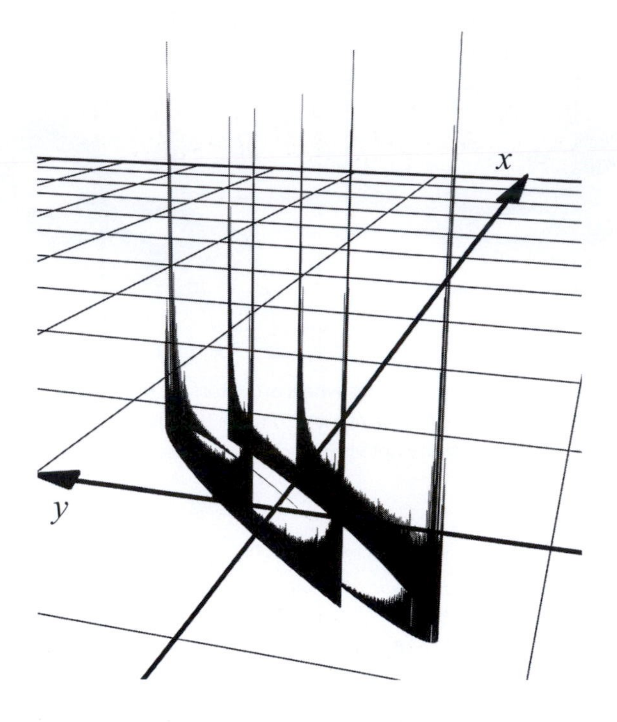

Fig. 5.60. Natural distribution of the water-wheel attractor on the Poincaré section $z = r - 1$ ($\sigma = 10$, $r = 28$).

alternate at all scales. Even though the physical process behind the map fundamentally differs from that of kicked oscillators, the character of the distribution is similar to that for the parabola attractor (see Fig. 5.44(c)). All this indicates that the structure of chaotic attractors does not essentially depend on whether the external energy supply is periodic in time or of some other type.

Problem 5.43 Determine the chaotic attractor of the system of equations

$$\dot{x} = -(y + z), \quad \dot{y} = x + ay, \quad \dot{z} = b + xz - cz,$$

which is called the Rössler model, via numerical simulation, with parameters $a = b = 0.2$, $c = 5.7$. The entire attractor belongs to positive values of the variables.

Chapter 6
Transient chaos in dissipative systems

Under certain circumstances chaotic behaviour is of finite duration only, i.e. the complexity and unpredictability of the motion can be observed over a finite time interval. Nevertheless, there also exists in these cases a set in phase space responsible for chaos, which is, however, non-attracting. This set is again a well defined fractal, although it is more rarefied than chaotic attractors. This type of chaos is called *transient* chaos, and the underlying non-attracting set in invertible systems is a *chaotic saddle*. The concept of transient chaos is more general than that of permanent chaos studied so far, and knowledge of it is essential for a proper interpretation of several chaos-related phenomena. The basic new feature here is the *finite lifetime* of chaos.

In dissipative systems transient chaos appears primarily in the dynamics of approaching the attractor(s). It is therefore also called the *chaotic transient*. The temporal duration of the chaotic behaviour varies even within a given system, depending on the initial conditions (see Fig. 6.1 and Section 1.2.2). Despite the significant differences in the individual lifetimes, an *average* lifetime can be defined. To this end, it is helpful to consider several types of motion (trajectories) instead of a single one: the study of particle ensembles is even more important in transient than in permanent chaos.

In phase space, chaotic behaviour is confined to the vicinity of the saddle. If we choose a region that overlaps the saddle but does not contain attractors, and start $N(0) \gg 1$ trajectories on it, these trajectories escape the pre-selected region sooner or later, and the long-lived ones appear to be chaotic (but after escape their complexity may cease). The number $N(t)$ $(N(n))$ of trajectories remaining in the region up to time t

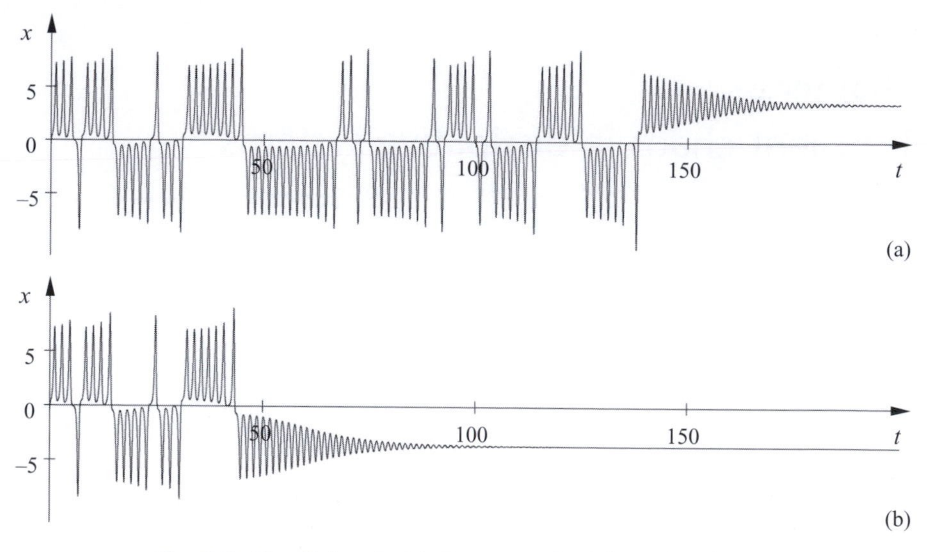

Fig. 6.1. Chaotic transient in the water-wheel dynamics. The parameter $r = 14$ ($\sigma = 10$) is now *outside* of the chaotic region ($r > r_c = 15$) determined in Section 5.7 for $\sigma = 10$. The initial conditions are (a) $x_0 = -0.648$, $y_0 = 0.591$, $z_0 = 13$ (rounded to three decimal places) and (b) the same, but co-ordinate x_0 is decreased by 10^{-7}. Initially, the signals are similar to that of the permanently chaotic water-wheel, but after some time they suddenly switch to a time-independent behaviour corresponding to the fixed points $x^* = y^* = \pm\sqrt{13}$, $z^* = 13$. Note the major changes resulting from a tiny difference in the initial condition: the lifetimes differ significantly, and, in addition, the different types of motion converge to different fixed points.

(or iteration steps n) is thus a monotonically decreasing function of t. After a sufficiently long time the decay is, in general, exponential[1] (see Fig. 6.2), i.e.

$$\boxed{N(t) \sim e^{-\kappa t}, \quad \text{or} \quad N(n) \sim e^{-\kappa n}.} \tag{6.1}$$

Coefficient κ is called the *escape rate* (in continuous or discrete time, respectively), whose reciprocal,

$$\boxed{\tau \equiv \frac{1}{\kappa},} \tag{6.2}$$

can be considered to be the average lifetime of chaos. The non-zero escape rate is thus a new, important chaos characteristic.

We start the study of transient chaos with an open version of our 'model' map, the baker map. Next, the chaotic transients of kicked

[1] Similar to the law of radioactive decay.

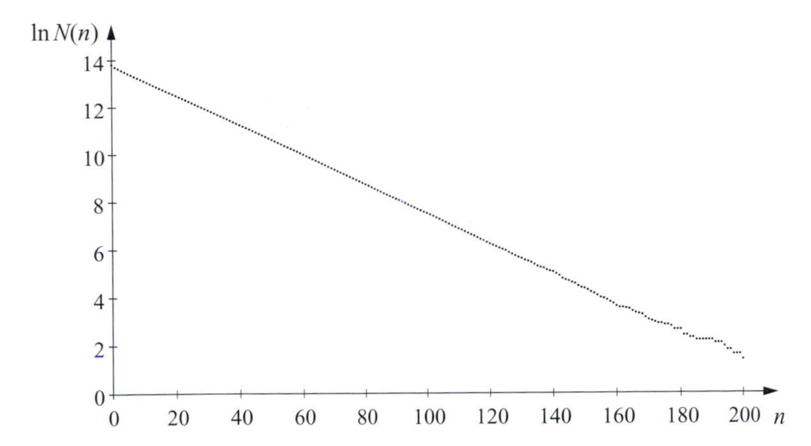

Fig. 6.2. Decay of the number $N(n)$ of non-escaped particles in the water-wheel dynamics ($\sigma = 10$, $r = 14$). Note that $N_0 = 10^6$ points were started in a square of size 2, centred at the origin of the Poincaré map, and were followed up to n iterates. The slope on this log-linear plot is -0.063, corresponding to an escape rate $\kappa = 0.063$ and an average lifetime $\tau = 15.9$.

oscillators are investigated. Based on these experiences, the most important transient chaos properties can be summarised. Emphasis is laid again on the natural distribution developing in such cases on chaotic saddles. Critical phase space configurations, called 'crises', are identified which mediate transitions from one kind of chaos (for example transient) to another (for example permanent). Finally we introduce the problem of fractal basin boundaries and discuss its relation to transient chaos.

6.1 The open baker map

A simple model of transient chaos is obtained by 'opening up' the baker map discussed in Section 5.1. In this case the rectangles obtained in one step from the two half-squares stretch beyond the unit square, thus making escape possible (see Fig. 6.3). Accordingly, in the mathematical definition, the stretching factor 2 in equation (5.2) is substituted by an arbitrary value, a, greater than 2. The open symmetric baker map is written as follows:

$$B(x_n, v_n) = \begin{cases} B_-(x_n, v_n) \equiv (cx_n, av_n), & \text{for } v_n \leq 1/2, \\ B_+(x_n, v_n) \equiv (1 + c(x_n - 1), 1 + a(v_n - 1)), & \text{for } v_n > 1/2. \end{cases} \quad (6.3)$$

The map is defined on the entire (x_n, v_n)-plane, but the interesting dynamics is bound to the unit square. The Jacobian is $J = ac$, and the

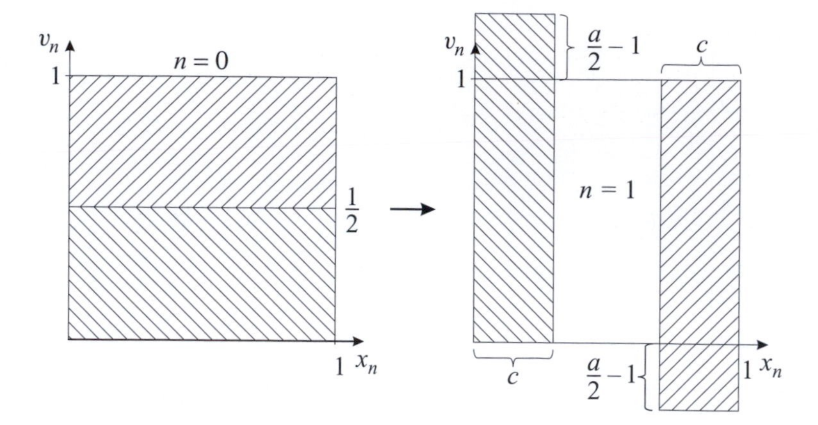

Fig. 6.3. Open (symmetric) baker map ($a > 2$) acting on the unit square.

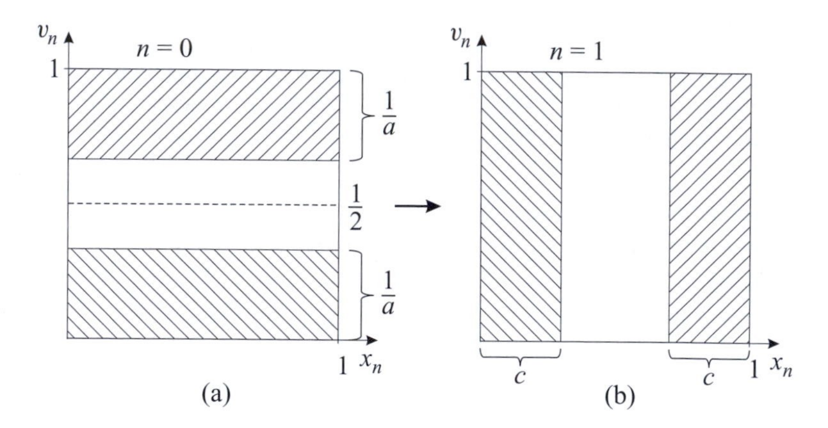

Fig. 6.4. (a) Points not escaping the unit square within one step and (b) their images after one iteration.

system is dissipative for $0 < c < 1/a$. For the sake of simplicity, we fix b to be $1/2$. The fixed points are again $H_- = (0, 0)$ and $H_+ = (1, 1)$. In the boxing analogy mentioned in Section 5.1.1, it is now possible to escape the 'ring', but points that have difficulties in finding their way out are exposed to the 'kicks' of the two fixed points for a long time, and perform chaotic motion while inside.

Let us find first the points that do not leave the unit square within one step. The half-squares are mapped into columns of height $a/2 > 1$, pieces of which lie outside the square (see Fig. 6.3). Points remaining inside must therefore start from two horizontal bands, each of height $1/a$ (see Fig. 6.4).

Points remaining inside for two steps must start from bands of height $1/a^2$ since two consecutive a-fold stretchings should lead to the unit length. Four such columns exist (see Fig. 6.5). The fate of points not escaping over four steps is shown in Fig. 6.6.

In general, points remaining inside for n steps start from 2^n horizontal bands, each of height $1/a^n$. Thus, out of $N(0)$ points distributed

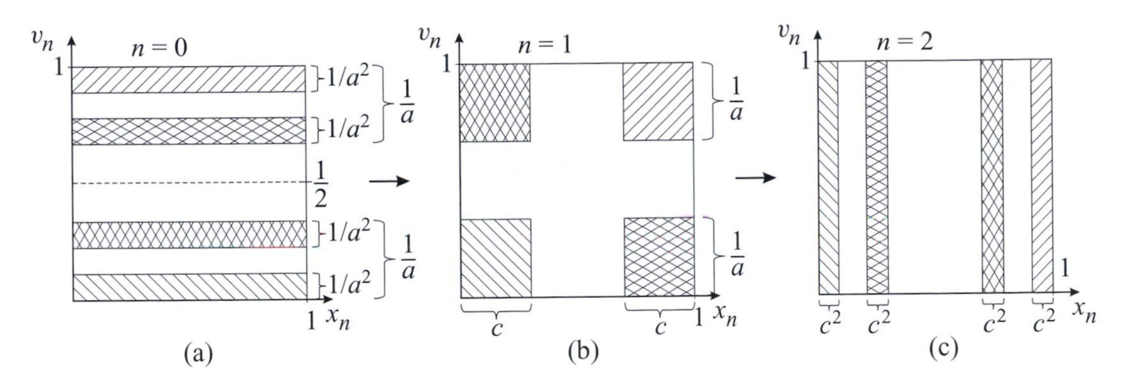

Fig. 6.5. (a) Points not escaping within two steps and their images after (b) one and (c) two iterations.

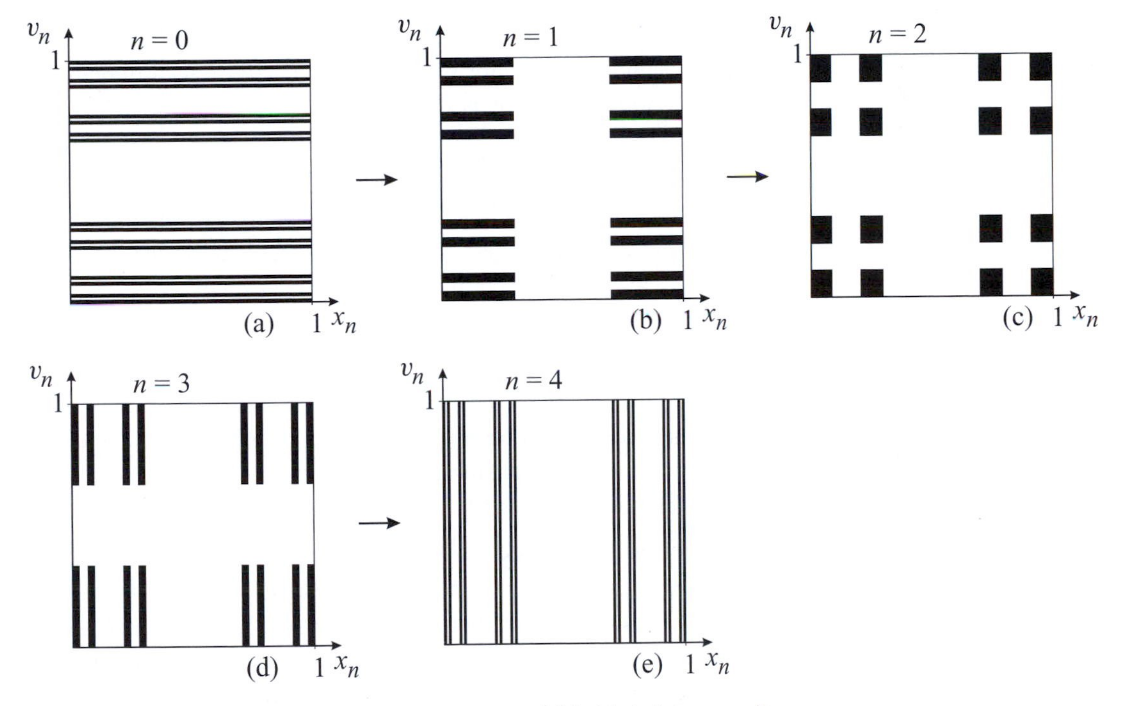

Fig. 6.6. (a) Points not escaping within four steps and (b)–(e) their images after the first four iterations.

uniformly over the unit square, $N(n) = N(0) \cdot (2/a)^n$ survive n iterates. According to (6.1),

$$e^{-\kappa n} = \left(\frac{2}{a}\right)^n, \tag{6.4}$$

which yields

$$\kappa = \ln \frac{a}{2} > 0. \tag{6.5}$$

Therefore the escape rate of the open baker map increases as the stretching factor, a, increases.

The distance of neighbouring point pairs increases – as follows from (6.3) – by a factor of $a > 1$ in each step. The positive Lyapunov exponent of the dynamics before escape is thus given by

$$\lambda = \ln a. \tag{6.6}$$

Note that this is greater than the value $\ln 2$ valid for the baker attractor: transiently chaotic systems are often more unstable than permanently chaotic ones. The negative Lyapunov exponent characteristic of contraction is $\lambda' = \ln c$ in this case also.

By extrapolating Fig. 6.6 to n steps, and taking the limit $n \to \infty$, we see that trajectories remaining inside for an arbitrary number of steps are rather exceptional. Their initial conditions form along the v_n-axis a set of measure zero, a Cantor set of parameter $1/a$ (see Figs. 2.7 and 6.6(a)). The dimension of this Cantor set, called the *partial* fractal dimension along the *unstable* direction, is given by

$$D_0^{(1)} = \frac{\ln 2}{\ln a} < 1. \tag{6.7}$$

Similarly, the end-points of the trajectories remaining inside for $n \gg 1$ steps form a Cantor set of parameter c along the x_n-axis (see Figs. 2.7 and 6.6(e)). The dimension of this Cantor set is given by

$$D_0^{(2)} = \frac{\ln 2}{\ln 1/c} < 1, \tag{6.8}$$

which is the partial fractal dimension along the stable direction. Points never leaving the unit square (neither forwards, nor backwards, in time, i.e. the direct product of the aforementioned two sets) form an asymmetric Cantor cloud (see Fig. 2.10) of dimension

$$D_0 = \frac{\ln 2}{\ln a} + \frac{\ln 2}{\ln (1/c)}. \tag{6.9}$$

This set is the chaotic saddle of the baker map (see Figs. 6.7 and 6.6(c)). The basic difference from (5.11) is that the partial fractal dimension along the unstable manifold is now less than unity.

The points of the saddle are mapped onto each other; the Lyapunov exponents, λ and λ', can thus be measured for an arbitrary length of time. These points are, however, exceptional. Typical transiently chaotic trajectories only come close to the saddle. Their divergence can therefore be observed on average over a lifetime τ only.

The chaotic set is called a saddle because around each of its points stretching takes place in one direction and compression takes place in another, just like around hyperbolic (saddle) points (see Section 3.1.1). Accordingly, we can talk of the stable and unstable manifolds of the

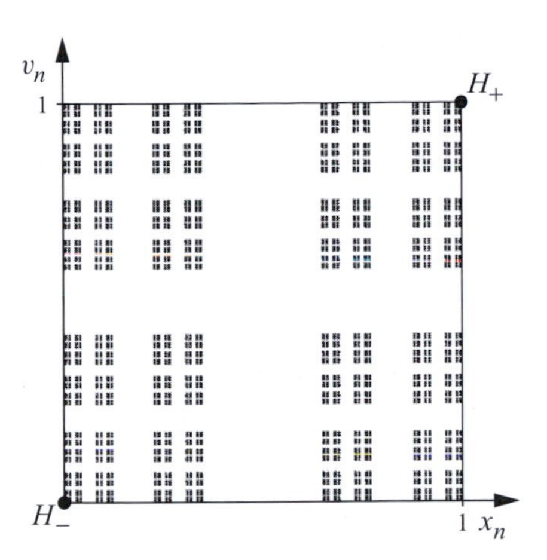

Fig. 6.7. Baker saddle: the chaotic saddle of map (6.3) with $a = 2.4$, $c = 0.35$.

entire saddle, which are Cantor filaments (see Section 1.2.2), whose fractal dimension $D_0^{(s)}$ and $D_0^{(u)}$ is 1 + the partial dimension along the unstable and stable directions, respectively:

$$D_0^{(s)} = 1 + D_0^{(1)} \quad \text{and} \quad D_0^{(u)} = 1 + D_0^{(2)}. \tag{6.10}$$

Since $D_0^{(1)} < 1$, $D_0^{(s)} < 2$, the stable manifold is thus *not space-filling*; the chaotic saddle therefore does not have a basin of attraction of finite area. This is why it appears to be non-attracting. Note also that $D_0 < D_0^{(u)}$; thus, the dimension of the saddle is smaller than that of its unstable manifold. It *cannot* therefore be approximated by the unstable manifold.

In order to obtain a deeper insight into the saddle's structure, it is important to recognise that the open baker map also possesses a two-cycle and higher-order cycles, and, moreover, that there is no upper bound on the length and number of the cycles. All the cycles are hyperbolically unstable.

Problem 6.1 Determine the two- and three-cycles of the open baker map (6.3).

Problem 6.2 How does the m-fold iterated open baker map act on the velocity variable v_n? Based on this, determine the total number of the fixed points of $B^m(x_n, v_n)$.

The number of all the elements of the unstable m-cycles in the open baker map is again

$$N_m = 2^m. \tag{6.11}$$

The topological entropy defined by $N_m \sim e^{hm}$ (see Section 5.4.1) is thus

$$h = \ln 2 = 0.693, \tag{6.12}$$

which is now *smaller* than the positive Lyapunov exponent.

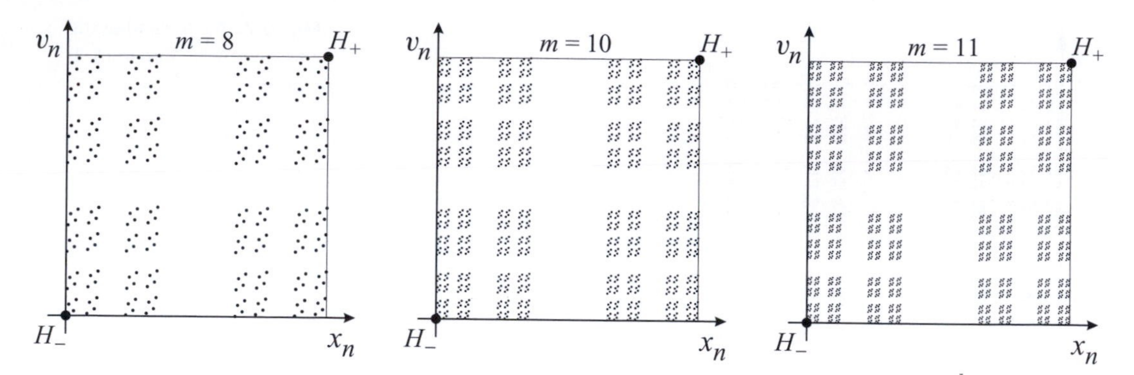

Fig. 6.8. Higher-order cycles of the open baker map (6.3): all points of the cycles of length $m = 8$, 10 and 11 ($a = 2.4$, $c = 0.35$), which gradually approximate the chaotic saddle.

Figure 6.8 illustrates how the number of cycle points increases with m. One can verify that all of these are on the chaotic saddle. The points of the increasingly higher-order cycles trace out the chaotic saddle more and more accurately. The hyperbolic cycles therefore provide a skeleton for the chaotic saddle.

An unstable manifold of infinite length emanates from each cycle point, and these manifolds prove to be all similar to each other. The unstable manifold of the chaotic saddle is therefore simply the *union* of the unstable manifolds of all the cycles. A similar statement holds for the stable manifolds. The chaotic saddle can, therefore, also be considered as the *intersection* of its own stable and unstable manifolds.

Problem 6.3 Construct the manifolds of the fixed points of the open baker map, starting from the basic branches emanating from these points.

The intersection points of the stable and unstable manifolds of the cycles again form homoclinic and heteroclinic points. Trajectories starting (exactly) in the homoclinic or heteroclinic points must therefore perform an aperiodic motion for an arbitrary length of time. All the homoclinic and heteroclinic points are *on* the chaotic saddle, very close to each other. Accordingly, the chaotic saddle can also be considered as the union of the homo- and heteroclinic points formed among the manifolds of all the hyperbolic cycles on it. Chaotic motion, interpreted as a random walk between the cycles, is thus realised via the presence of homoclinic and heteroclinic points again.

All these properties also hold for a generalised open baker map, in which the stretching and contraction rates are different in the two half-squares. As an extension of equations (5.15) and (5.16), such an

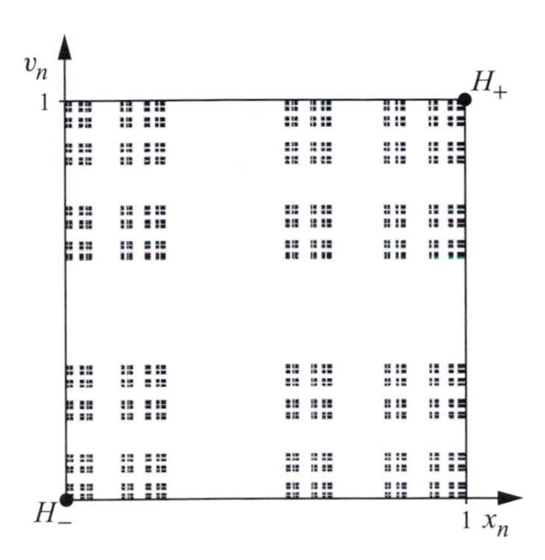

Fig. 6.9. Asymmetric baker saddle with $a_1 = 3$, $a_2 = 2.5$, $c_1 = 0.25$, $c_2 = 0.45$.

asymmetric open map can be defined as follows:

$$
\begin{aligned}
(x_{n+1}, v_{n+1}) &= (c_1 x_n, a_1 v_n), \quad \text{for} \quad v_n \le b, \\
(x_{n+1}, v_{n+1}) &= (1 + c_2(x_n - 1), 1 + a_2(v_n - 1)), \quad \text{for} \quad v_n > b,
\end{aligned}
\tag{6.13}
$$

where $0 < b < 1$, $a_1 b$, $a_2(1 - b) > 1$ and, c_1 and c_2 are positive parameters for which $c_1 + c_2 < 1$ holds. The stretching factors a_1, a_2 are now independent of b; for the sake of simplicity, we fix b to be $1/2$. The saddle of this map is an asymmetric Cantor cloud (Fig. 6.9).

Problem 6.4 Derive the equations for the partial fractal dimensions of the baker saddle of (6.13).

Problem 6.5 Another variant of the open asymmetric baker map is given by

$$
\begin{aligned}
(x_{n+1}, v_{n+1}) &= (c_1 x_n, a_1 v_n), \quad \text{for } v_n \le 1/2, \\
(x_{n+1}, v_{n+1}) &= (1 - c_2 x_n, a_2(1 - v_n)), \quad \text{for } v_n > 1/2.
\end{aligned}
\tag{6.14}
$$

Compute the fixed points, the two-cycle and the partial fractal dimensions of the saddle.

6.2 Kicked oscillators

The dynamics, (5.19), of oscillators kicked with the amplitude functions discussed in Section 5.2 is, for sufficiently large values of the non-linearity parameter a, always transiently chaotic. Qualitatively this is so because the momentum transfer of kicks is then so large that the motion cannot be confined to finite x_n and v_n values for arbitrarily long times; it converges ultimately to an attractor at infinity.

In Fig. 6.10 we present the saddles for values of the non-linearity parameter slightly higher than those belonging to the attractors discussed

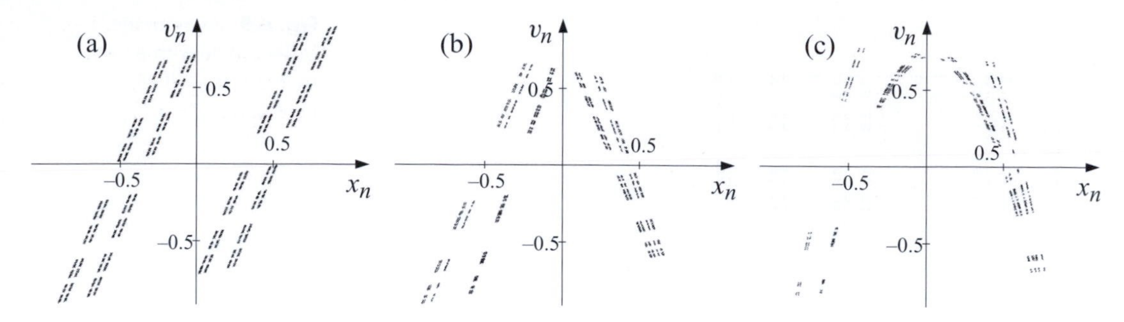

Fig. 6.10. The (a) sawtooth, (b) roof and (c) parabola saddles at respective parameters $a = 2.6$, $a = 2.6$ and $a = 3.2$ ($E = 0.7$).

in Section 5.2. Although the structures are more complicated than that of the baker saddle, in a co-ordinate system defined by the local stable and unstable directions, each saddle is the direct product of the two Cantor sets, i.e. topologically they are asymmetric Cantor clouds.

It is remarkable that with the parabola amplitude one can easily find chaotic saddles even for large E values (weak friction) where chaotic attractors do not exist for any a. This illustrates that transient chaos is more robust than permanent chaos.

Finally, we call the reader's attention to the interesting fact that, for large values of the non-linearity parameter, the saddle is confined to smaller and smaller regions; it is more and more unstable, and becomes increasingly similar to a baker saddle (see Fig. 6.11). The dynamics is then basically controlled by the eigenvalues of the hyperbolic points, H_- and H_+.

It can be shown that, for $a \gg 1$, the kicked oscillator dynamics can be approximated by a baker map. The baker map, which could not so far be related to a physical system, emerges in the strongly non-linear limit of the kicked oscillators (and of some other physical systems, too) producing transient chaos.

Problem 6.6 Choosing fixed point H_- as the origin and H_+ as point $(1, 1)$ of the co-ordinate system, show that a roof map (5.37) tends to a baker map in the limit $a \gg 1$. Give the parameters of the chaotic saddle in this limit.

Problem 6.7 To what kind of baker map does the general kicked oscillator map, (5.19), tend for large non-linearity parameter, if function f has two branches that become steeper with increasing values of a? (It is worth introducing new variables such that $H_- = (0, 0)$ and $H_+ = (1, 1)$.)

Fig. 6.11. Parabola saddle for a relatively large value of the non-linearity parameter ($a = 5.2$, $E = 0.7$).

Problem 6.8 Under what condition on function $f(x)$ does the one-dimensional map (5.47) (the extremely dissipative, $E \to 0$ limit of kicked oscillators) generate transient chaos? What is the equivalent of the saddle now? Assume that $f(x)$ is an even function.

Box 6.1 How do we determine the saddle and its manifolds?

It is much more difficult to generate chaotic saddles numerically than for attractors since following a single trajectory over a long time does not yield the desired result. Numerous different methods have therefore been developed, from which we present the simplest and most efficient possibility (all the saddles of the book have been obtained by means of this method).

We start $N(0) \gg 1$ trajectories distributed uniformly over some region R of the phase space. If the number of non-escaped points decreases very rapidly, or the decay is not exponential, then R does not contain the saddle; we therefore choose new regions until an exponential decay is found. Once this is so, the escape rate and its reciprocal (the average lifetime, τ) can be estimated. (In this case R already overlaps with the saddle, but it does not necessarily contain the entire saddle.) Next, we choose an iteration number, n_0, corresponding to a multiple of the lifetime, and we follow the time evolution of each point of R up to time n_0. We keep only those trajectories that do not escape R in n_0 steps (their number is approximately $N_0 e^{-n_0/\tau}$). If n_0/τ is sufficiently large (but not so large that only a few points remain inside) then we can be sure that trajectories with this long lifetime become close to the saddle in the course of the motion. This necessarily implies that their initial conditions were in the immediate vicinity of the stable manifold of the saddle (or of the saddle itself). Simultaneously, the end-points must be close to the unstable manifold of the saddle since most points still inside after n_0 steps are already in the process of leaving the region. The mid-points of these trajectories (with $n \approx n_0/2$) are then certainly in the vicinity of the saddle. This we have seen in Figs. 6.6(a), (c) and (e) for a specific example. It applies in general that the initial, mid- and end-points of trajectories with lifetimes of at least

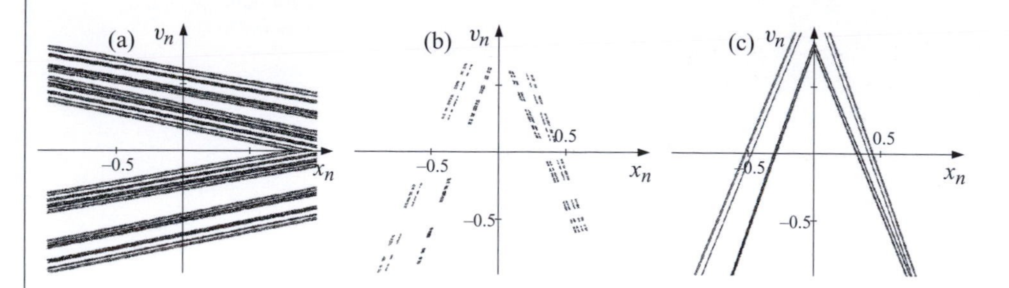

Fig. 6.12. Generating the invariant sets of transient chaos in the roof map. Region R is a square of size 2×2 centred at the origin; $N_0 = 10^7$, $\kappa = 0.257$ ($a = 2.6$, $E = 0.7$). Shown are the points of trajectories with a minimum lifetime $n_0 = 16$ at iteration numbers (a) $n = 0$, (b) $n = 8$ and (c) $n = 16$, well approximating the stable manifold, the saddle and the unstable manifold, respectively.

n_0 trace out, respectively, the stable manifold, the saddle and the unstable manifold within region R to a good approximation. In practice, it is often sufficient to choose an n_0 that is approximately four to six times the lifetime, τ. (After this, it is easy to choose an R that contains the complete saddle.) Figure 6.12 illustrates the algorithm applied to the roof saddle.

6.3 General properties of chaotic transients

6.3.1 Chaos characteristics

The most important new feature of chaotic transients is their average lifetime, τ, defined in (6.2), and its reciprocal, the escape rate. The other characteristics are essentially the same as those introduced for chaotic attractors; we therefore only refer to them briefly, and only provide formulae that generalise previous relationships.

A measure of the complexity of the dynamics on and around saddles is the *topological entropy*, i.e. the growth rate towards infinity of the number of cycle points vs. the cycle length, as expressed by (5.51). The same quantity can also be read off from the stretching process of a line segment of initial length L_0. Let L'_n denote the length of the nth image of the line segment measured within some fixed region around the saddle.

This fulfils, for $n \gg 1$,

$$\boxed{L'_n \sim e^{hn}.} \tag{6.15}$$

The distinction between L'_n and the full length, L_n, is necessary because the latter increases faster (in the open baker map, (6.3), as a^n, while

$L'_n \sim 2^n$). The topological entropy is thus the growth rate of length within the 'interesting' region around the saddle (for a numerical determination, see Appendix A.5).

Problem 6.9 Determine the topological entropy of the asymmetric baker maps given by equations (6.13) and (6.14).

Local Lyapunov exponents are again defined by relations (5.53) and (5.54), with the restriction that both trajectories remain close to the saddle.

The chaotic saddle has a direct product structure: its fractal dimension (see Section 2.2.2) is the sum of the two partial dimensions:

$$\boxed{D_0 = D_0^{(1)} + D_0^{(2)},} \tag{6.16}$$

where $D_0^{(1)}$ and $D_0^{(2)}$ are the partial dimensions along the unstable and stable directions, respectively, and $0 < D_0^{(j)} < 1$, $j = 1, 2$.

6.3.2 Natural distribution

Long-time dynamics is characterised by a probability distribution on chaotic saddles, which yields the probability that trajectories are in the vicinity of one point or another of the saddle. In a mathematical sense, this time-independent distribution characterises the dynamics of points never leaving the saddle. Since the chaotic set is non-attracting, the distribution cannot be constructed in a way similar to that seen for attractors (Section 5.4.4). Its support is a Cantor cloud, an object that is discontinuous in any direction. The simplest way to construct the distribution is to follow numerous points initiated in some region around the saddle for a sufficiently long time n_0. The points of trajectories that do not escape the region necessarily become close to the saddle at 'half-time' (see Box 6.1); therefore the distribution should be determined from these points. In a numerical algorithm we cover the saddle with small squares of size ε and for each box we record how many mid-points ($n = n_0/2$) of the non-escaped trajectories over iterations $n_0 \gg 1$ fall into the box, and we divide this value by the total number of mid-points. In this way, the integral of the natural distribution, P^*, over each box is determined to a good approximation. The natural distribution characteristic of saddles is independent of the initial distribution of the points used to construct it (just as for chaotic attractors; see Section 5.4.4).

Figure 6.13 shows the natural distributions on the chaotic saddles of the asymmetric open baker map and of the oscillators kicked with

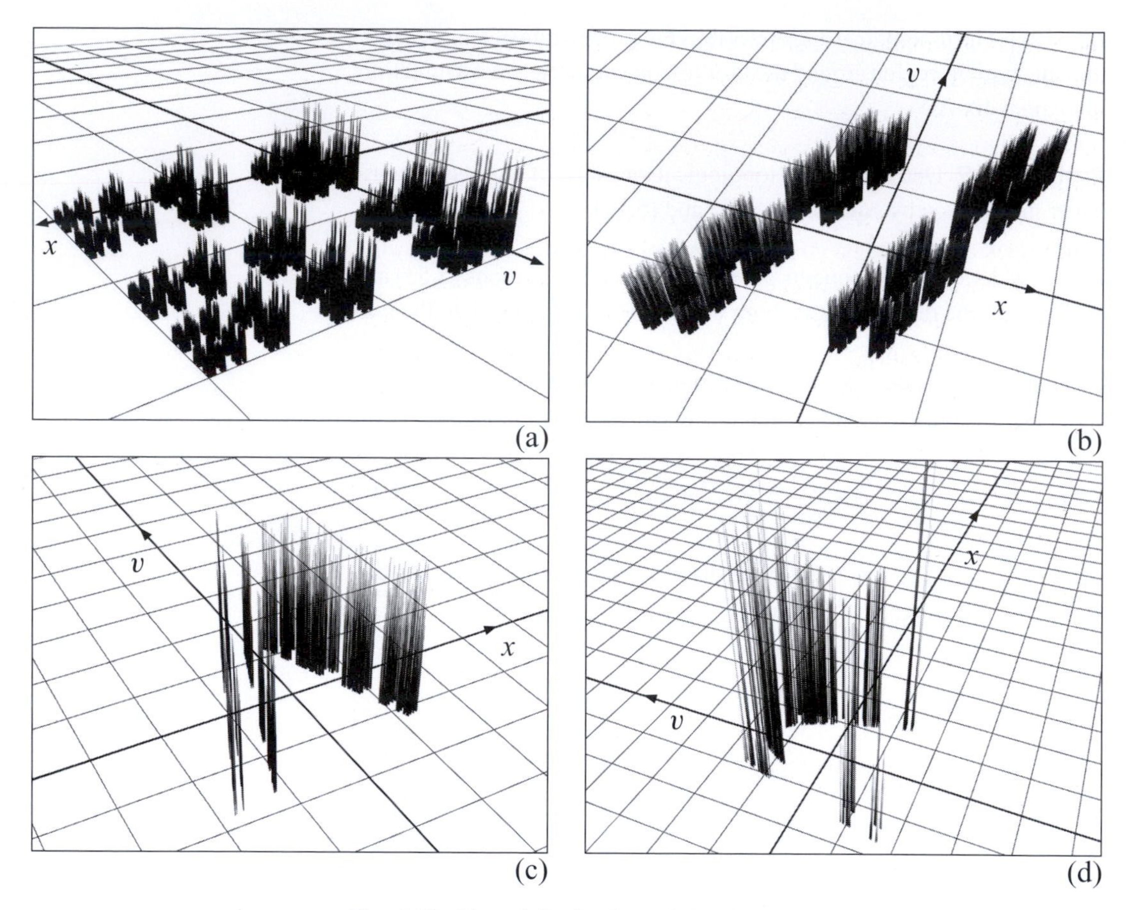

Fig. 6.13. Natural distributions of chaotic transients. (a) Asymmetric baker saddle (see Fig. 6.9). (The distribution on a symmetric baker saddle would be homogeneous in the sense that it would contain columns of the same height only.) (b) Sawtooth saddle (see Fig. 6.10(a)). (c) Roof saddle (see Fig. 6.10(b)). (d) Parabola saddle (see Fig. 6.10(c)). The resolution is $\varepsilon = 1/1000$ in all cases. These distributions can also be viewed as variants 'cut through', as if with a cookie cutter, of the distributions of chaotic attractors (cf. Fig. 5.44).

different amplitude functions. Some of these distributions, from different views, are presented in Plates XVII–XIX.

The natural distribution, P^*, of chaotic saddles is also a *fractal distribution* (see Section 2.3), which provides more and more detailed information as the resolution is refined. Accordingly, the natural distribution possesses an information dimension, D_1. Since the distribution is also of a direct product nature, it is fractal along both the unstable and the stable direction. The full information dimension is decomposed,

$$\boxed{D_1 = D_1^{(1)} + D_1^{(2)},} \tag{6.17}$$

into partial information dimensions $D_1^{(j)}$ ($j = 1(2)$ for the unstable (stable) direction). The saddle also contains very rarely visited, and thus atypical, regions; consequently, the information dimension, D_1 can only be smaller than the fractal dimension, D_0, which naturally holds for the partial dimensions also: $D_1^{(j)} \leq D_0^{(j)}$, $j = 1, 2$.

Since the local Lyapunov exponents have been defined as the divergence rates of particle pairs staying around the saddle for a long time, the average Lyapunov exponents are averages taken with respect to the natural distribution, P^*: Eqs. (5.67) and (5.69) also hold for transient chaos. In principle, the positive Lyapunov exponent, $\bar{\lambda}$, is the divergence rate of point pairs remaining on the saddle forever. In practice, however, it can also be obtained to a good approximation from pairs moving in the vicinity of the saddle for a long time (see Appendix A.5). The prediction time, given by (5.59), $t_p \sim 1/\bar{\lambda}$, thus refers to the motion both on and around the saddle.

Problem 6.10 Determine the average Lyapunov exponent in the asymmetric open baker proof maps (6.13) and (6.14).

Problem 6.11 Determine numerically the natural distribution on the asymmetric baker saddle of maps (6.13) and (6.14) at a given set of parameters a_1, a_2, c_1, c_2 ($b = 1/2$). Show that the information dimension is the same for each case. Determine analytically the partial information dimensions.

The topological entropy of transient chaos differs from the positive Lyapunov exponent, even in the simplest case, when all the local Lyapunov exponents are the same. Since this λ characterises the full stretching, the length of a line segment is proportional to $e^{\lambda n}$ after n steps. The length of segments, L'_n, remaining in a fixed region around the saddle is a factor $e^{-\kappa n}$ of the full length. The topological entropy, h, is the stretching rate of L'_n; i.e., from (6.15), $e^{hn} \approx e^{(\lambda-\kappa)n}$. For $n \gg 1$ this implies

$$h = \lambda - \kappa. \tag{6.18}$$

This relation holds for the open baker map (6.3).

In general, however, the local Lyapunov exponents are not identical, and the relation between the topological entropy and the average Lyapunov exponent is given by

$$\boxed{h \geq \bar{\lambda} - \kappa.} \tag{6.19}$$

This implies that the stretching rate (h) of a phase space line within a region around the saddle is larger than the stretching rate, $\lambda - \kappa$, of the typical segments within the same region. Equation (6.19) holds, for

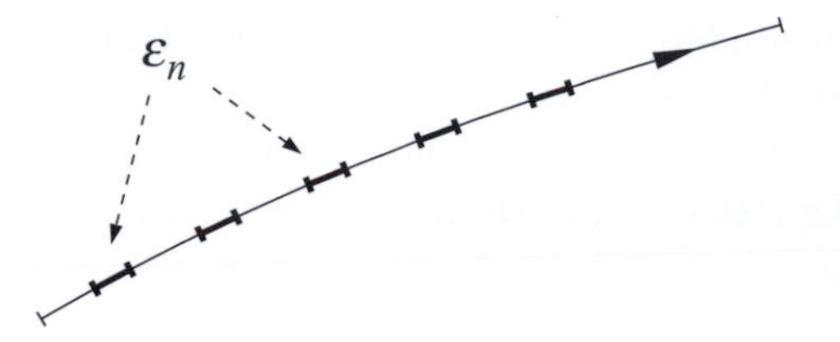

Fig. 6.14. Points that do not escape an unstable manifold segment of unit length in $n \gg 1$ steps form short intervals of average length ε_n. For a particular example, see the v_n-axis of Fig. 6.6(a).

example, for the asymmetric open baker map (6.13). (For the permanent case, see (5.68).)

For a single hyperbolic point the escape rate coincides with the instability exponent, $\lambda = \ln|\Lambda_+|$. The topological entropy is, of course, zero, from (6.18) so $\kappa = \lambda$ holds. For chaotic saddles, h is strictly positive. From (6.19) we thus obtain the important inequality:

$$\boxed{\kappa \leq \bar{\lambda}.} \tag{6.20}$$

A saddle is thus less unstable globally than locally. It is the set of the homoclinic and heteroclinic points among the infinitely many hyperbolic points that makes the saddle *more stable* than its separate elements. Even if a trajectory has escaped the neighbourhood of a hyperbolic point, it will be trapped by other hyperbolic cycles for a while due to the interwoven fractal topology of the stable manifolds. (Perfect global stability is realised in the limit $\kappa = 0$ only, when the saddle turns into an attractor.)

6.3.3 Link between dynamics and geometry: the Kantz–Grassberger relation

Since the chaotic saddle is characterised by two partial dimensions, the link between geometry and dynamics is now reflected in two relations: one valid along the unstable manifold, the other along the stable manifold. The qualitative argument used in this section is an extension of one leading to the Kaplan–Yorke relation; see Section 5.4.6.

Let us start with the partial dimension along the unstable manifold. Consider a segment of unit length of the unstable manifold that intersects the stable manifold. We distribute $N(0) \gg 1$ points uniformly on this segment. After $n \gg 1$ iterations the majority of these have already escaped the segment. The initial points of trajectories not escaping the segment up to n steps (see Fig. 6.14) are on several small intervals, which stretch to the full segment of unit length in n steps.

Their average length, ε_n, can therefore be estimated from the relation $\varepsilon_n \exp(\bar{\lambda} n) = 1$, since the mean stretching rate per iteration is $\exp(\bar{\lambda})$, where $\bar{\lambda}$ is the average Lyapunov exponent. The typical interval length is thus given by

$$\varepsilon_n = e^{-\bar{\lambda}n}. \tag{6.21}$$

For $n \gg 1$, the small intervals converge towards a fractal; their number, $N^*(\varepsilon_n)$, therefore scales as

$$N^*(\varepsilon_n) \sim \varepsilon_n{}^{-D_1^{(1)}}. \tag{6.22}$$

Since these are typical intervals, it is not the fractal dimension, but rather the *information dimension*, of the natural distribution that appears in the exponent (see (2.16)). Moreover, the total length, $\varepsilon_n N^*(\varepsilon_n)$, of the small intervals is, according to definition (6.1), proportional to the number of points that have not yet escaped in n steps: $e^{-\kappa n} \approx \varepsilon_n N^*(\varepsilon_n)$. Using expression (6.20) for the interval length, we obtain

$$e^{-\kappa n} \sim \left(e^{-\bar{\lambda}n}\right)^{\left(1-D_1^{(1)}\right)}, \tag{6.23}$$

from which

$$\boxed{D_1^{(1)} = 1 - \frac{\kappa}{\bar{\lambda}} = 1 - \frac{1/\bar{\lambda}}{\tau}.} \tag{6.24}$$

This relation, the *Kantz–Grassberger relation*, is perhaps the most important rule of transiently chaotic systems. It states that the dimension observed along the unstable direction increasingly deviates from unity the larger the ratio of the escape rate (a characteristic of the global instability of the saddle) to the average Lyapunov exponent (a characteristic of local instability on the saddle). For a single hyperbolic point ($\kappa = \lambda$) the partial dimension along the unstable direction is zero according to the fact that the natural distribution of a hyperbolic point is localised to that point, a zero-dimensional object. Since atypical intervals are left out of the argument, $N^*(\epsilon_n)$ is a lower bound on the number of all intervals along the unstable manifold, and the partial fractal dimension is larger than the information dimension: $D_0^{(1)} > D_1^{(1)}$. In practice, however, the difference is often very small, and

$$D_0^{(1)} \gtrsim 1 - \frac{\kappa}{\bar{\lambda}} \tag{6.25}$$

holds.

Let us now turn to the relation between the stable partial dimension and the Lyapunov exponents. A phase space volume element of radius l_0 placed in the vicinity of a chaotic saddle becomes, in $n \gg 1$ steps, a ribbon folded in a fractal-like manner (Fig. 6.15), with average length $l_0 e^{\bar{\lambda}n}$ and width $l_0 e^{\bar{\lambda}' n}$.

In accordance with escape, the ribbon extends at several locations beyond the narrowest strip (the height of which is chosen to be unity)

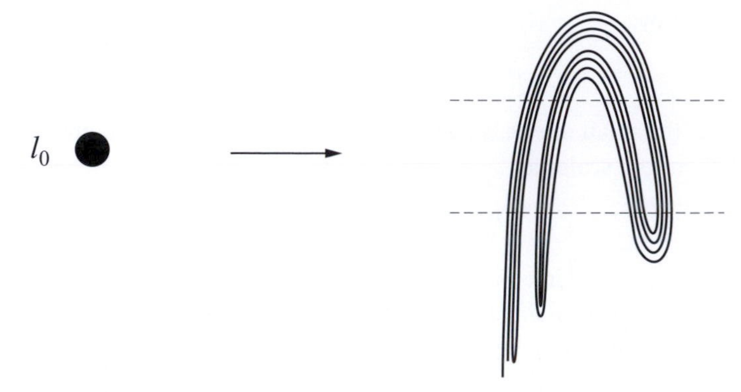

Fig. 6.15. A phase space volume element is in the vicinity of a chaotic saddle stretched (contracted) by a factor of $e^{\bar{\lambda}n}$ ($e^{\bar{\lambda}'n}$) along the unstable (stable) direction over $n \gg 1$ iterations. Simultaneously, it becomes folded several times and extends beyond the narrowest strip (dashed lines) containing the saddle.

containing the saddle. Only a fraction $\exp(-\kappa n)$ of the entire length remains inside the strip (see Fig. 6.15), i.e. length $l_0 e^{(\bar{\lambda}-\kappa)n}$. Since the height is unity, this equals the number $N^*(\varepsilon_n)$ of typical ribbon pieces, lying parallel to the unstable direction, within the strip. The average width of the ribbons is $\varepsilon_n = l_0 e^{\bar{\lambda}'n}$. Intersecting them along the stable direction, we obtain intervals of size ε_n. According to the definition of the partial dimensions, the number of typical intervals of size ε_n required for full coverage is $N^*(\varepsilon_n) \sim \varepsilon_n{}^{-D_1^{(2)}}$ (see (2.16)). Since this equals the number of ribbon pieces, which is proportional to $e^{(\bar{\lambda}-\kappa)n}$, we obtain

$$\varepsilon_n{}^{-D_1^{(2)}} \sim e^{|\bar{\lambda}'|nD_1^{(2)}} \sim e^{(\bar{\lambda}-\kappa)n}, \tag{6.26}$$

from which

$$D_1^{(2)} = \frac{\bar{\lambda}-\kappa}{|\bar{\lambda}'|}. \tag{6.27}$$

The partial dimension along the stable direction is thus the ratio of the difference of the positive Lyapunov exponent and the escape rate to the absolute value of the negative Lyapunov exponent. Substituting (6.24) into (6.27), we obtain

$$\boxed{D_1^{(2)} = \frac{\bar{\lambda}}{|\bar{\lambda}'|} D_1^{(1)}.} \tag{6.28}$$

This implies that the products of the information dimensions and the average Lyapunov exponents are the same in *both* stability directions.

The information dimension of the chaotic saddle is the sum of the partial dimensions (see (6.17)):

$$\boxed{D_1 = \left(1 - \frac{\kappa}{\bar{\lambda}}\right)\left(1 + \frac{\bar{\lambda}}{|\bar{\lambda}'|}\right).} \tag{6.29}$$

The dimension of the saddle thus depends not only on the average Lyapunov exponents, but also on the escape rate. Its numerical value generally hardly differs from that of the fractal dimensions.

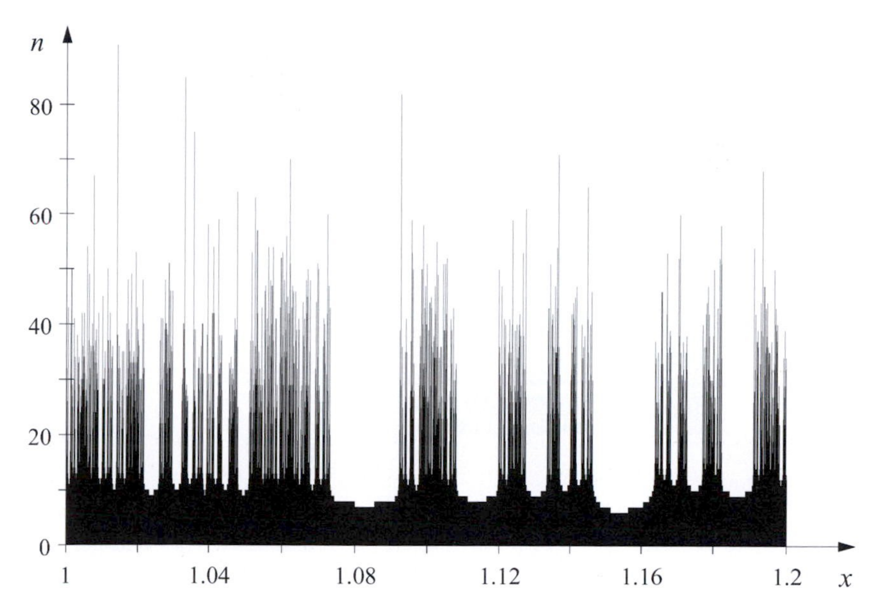

Fig. 6.16. Fractal lifetime distribution: the number, n, of iteration steps occurring in a square of size 2.8 centred at the origin (and containing the saddle of the roof map) by points started from the segment $v = 0$, $1 \leq x \leq 1.2$, vs. the initial x co-ordinate ($a = 2.3$, $E = 0.7$).

The respective information dimensions of the unstable and stable manifolds are given by

$$D_1^{(u)} = 1 + D_1^{(2)} = 1 + \frac{\bar{\lambda} - \kappa}{|\bar{\lambda}'|} \quad \text{and} \quad D_1^{(s)} = 1 + D_1^{(1)} = 2 - \frac{\kappa}{\bar{\lambda}}, \tag{6.30}$$

while, for the fractal dimensions, Eq. (6.10) holds.

The Kaplan–Yorke relation, (5.74), valid for chaotic attractors, is recovered in the limit $\kappa \to 0$.

6.3.4 Fractal lifetime distribution

If we select a region that contains (part of) the saddle and initiate points along some curve in phase space, we find that trajectories remain in the pre-selected region over a very wide range of times (Fig. 6.16). This is because small differences in initial conditions lead to significant variations in lifetimes.

Very large values can only belong to points where the selected curve intersects the *stable* manifold of the saddle. It is therefore a fractal set of points where the lifetime distribution takes on infinite values: the singularities along the segment sit on a fractal set of points of dimension $D_0^{(s)} - 1 = D_0^{(1)} < 1$.

The lifetime distribution, despite its complexity, is not in contradiction with the overall exponential decay. If we ask how many initial points remain in the region for time n, we find that their number decays

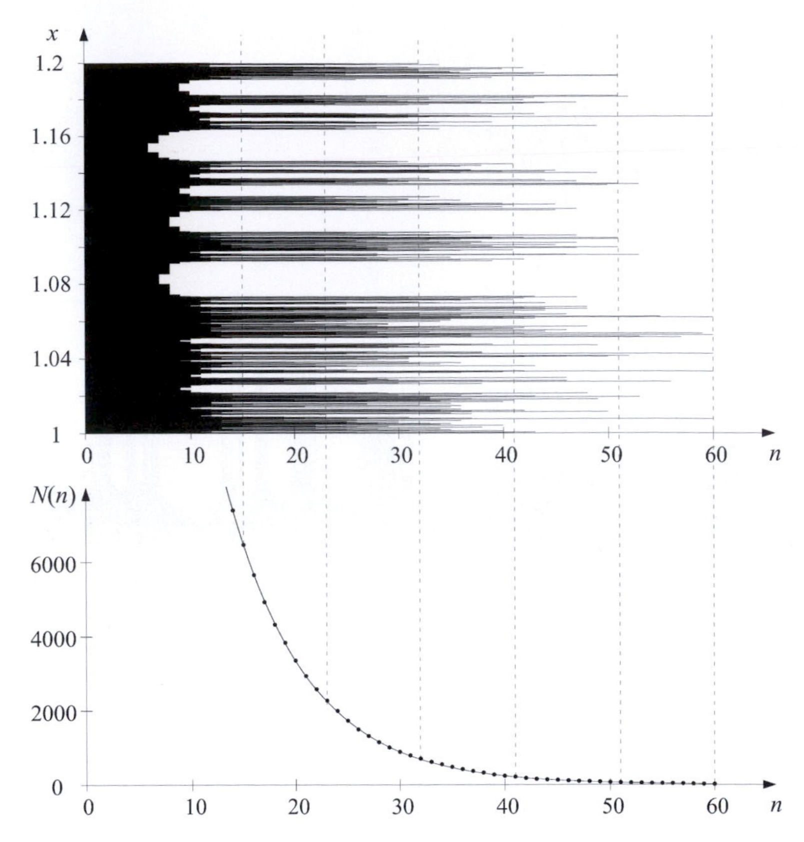

Fig. 6.17. Occurrence number, $N(n)$, of lifetimes, n, of Fig. 6.16, projected on the n-axis. The continuous curve is $N(15)\exp(-\kappa(n-15))$ with $N(15) = 6464$ and $\kappa = 0.132$, determined from an independent numerical measurement.

according to $\exp(-\kappa n)$ (see Fig. 6.17). This is so, irrespective of the curve on which the lifetime distribution is taken. It is true that, in general, the escape rate, κ, is *independent* of the region in which the statistics is investigated (provided it overlaps the stable manifold).

Box 6.2 Significance of the unstable manifold

In transient chaos, the unstable manifold does not coincide with the chaotic set, i.e. the saddle. The unstable manifold is nevertheless of great importance since trajectories of long lifetime ultimately leave the saddle along this manifold. This feature can best be illustrated by the fate of a droplet in phase space. Those exceptional points of the droplet that are on the stable manifold converge necessarily to the points of the saddle; the others, however, escape. Points initially far from the stable manifold escape rapidly, while those close to it stay around for a long time. The droplet is a connected phase space domain; its shape becomes more and more complicated in time due to the different fates of nearby points, and, after a sufficiently long time, the shape of the droplet *traces out* the *unstable* manifold of the saddle. Therefore, one can also say that the unstable manifold is the *main asymptotic transport route* in phase space.

The effect is similar to what has been observed for a single hyperbolic point (Fig. 3.26(a)), but the shape of the droplet is much more complicated now (see Fig. 6.18). The unstable manifold thus acts as a kind of

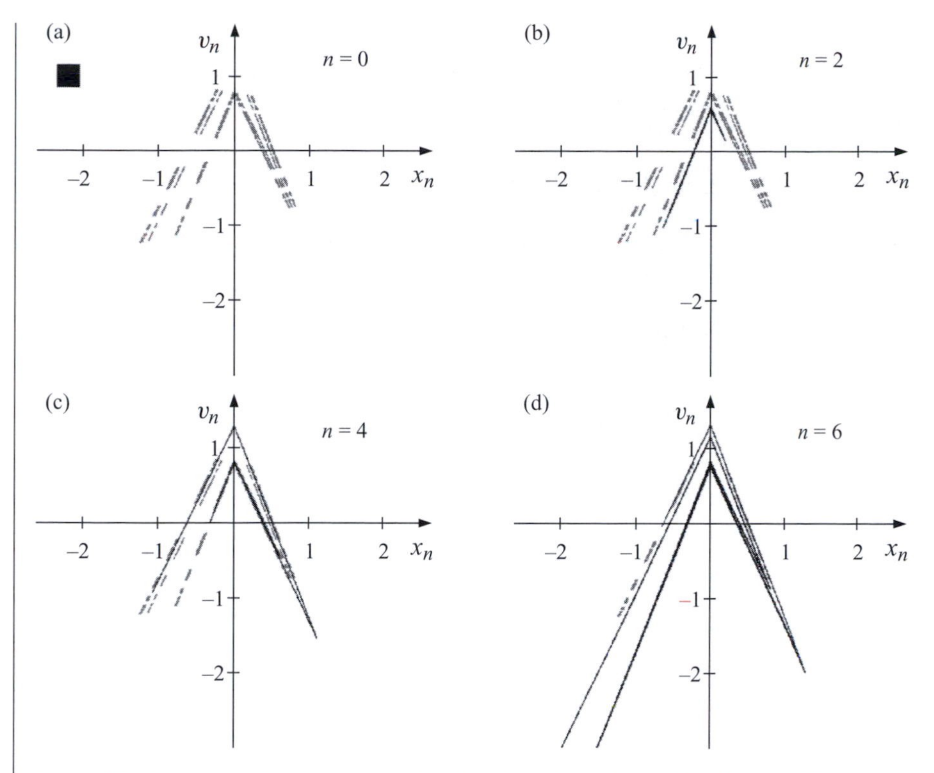

Fig. 6.18. Spreading of a phase space droplet (black) from the vicinity of the saddle (grey). Time evolution of a square of size 0.3, centred initially, (a), at $(-2.2, 1)$ in the roof map ($a = 0.7$, $E = 2.3$). Panels (b), (c) and (d) show the droplet after two, four and six iterations, respectively. While flowing away towards infinity, the droplet traces out the unstable manifold. Note, however, that due to escape practically no point of the droplet stays in the region shown for ever. A plot corresponding to $n \gg 1$ would therefore contain the saddle on its own.

attractor for droplets overlapping with the saddle's stable manifold: observed over a finite time, the droplets contract onto the manifold. The fact that this is not a real attractor becomes evident, however, after a long time when the overwhelming majority of the points of the droplet reach – along the unstable manifold – the real attractor(s).

6.4 Summary of the properties of transient chaos

Transient chaos is a kind of *metastable*, decaying state, whose average lifetime is finite. It precedes the convergence to the attractors, whether they are simple or chaotic. In the latter case, the long-term behaviour is also characterised by a positive Lyapunov exponent, but the value typically differs from that for the transients.

Underlying transient chaos (in invertible systems) is a chaotic saddle, which is, to a good approximation

- the union of all the unstable (hyperbolic) periodic orbits lying on it (or the set of the intersections of the stable and unstable manifolds of a single periodic orbit on the saddle).

With a more precise definition, the chaotic saddle is

- the union of all the hyperbolic periodic orbits on the saddle and of all the homoclinic and heteroclinic points formed among their manifolds (or the common part of the stable and unstable manifolds of all the infinitely many periodic orbits on the saddle).

Accordingly, the chaotic saddle can also be constructed by taking the intersection points of the stable and unstable manifolds of some periodic orbit (see Box 6.3). Thus, it is true in general that the vicinity of a saddle is characterised by a kind of 'tartan' pattern, which is traced out by the branches of the stable and unstable manifolds (Fig. 6.19). The manifolds are not equivalent: along the stable manifold, infinite lifetimes can be observed, whereas the unstable manifold marks the routes leading away from the saddle. The inverse of a saddle is, of course, also a saddle, but when reversing the direction of time, the arrows on the manifolds are also reversed.

Dynamics on chaotic saddles, just like on chaotic attractors, is a random walk among the infinitely many simple (periodic) but unstable cycles. The most important properties of transient chaos are again

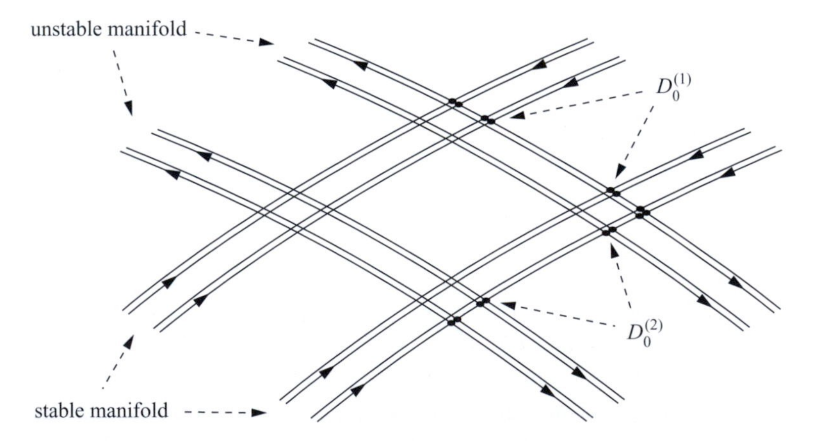

Fig. 6.19. 'Tartan' pattern of transient chaos showing the schematics of the saddle's manifolds. Partial fractal dimensions, $D_0^{(1)}$ and $D_0^{(2)}$, along the unstable and stable manifolds, respectively, are marked. Due to the Cantor filament structure, the fractal dimensions of the stable and unstable manifolds are $D_0^{(s)} = 1 + D_0^{(1)}$ and $D_0^{(u)} = 1 + D_0^{(2)}$, respectively.

Box 6.3 The horseshoe map

The observation that the existence of a single homoclinic point implies the chaoticity of a motion is due to the American mathematician, Steve Smale. He showed that the dynamics around a hyperbolic fixed point (cycle point) can then be approximated by – as we call it today – an open baker map. In order to demonstrate this, let us consider a fixed point and the first homoclinic point of its manifolds (Fig. 6.20). A square around the fixed point is stretched so much after a certain number of steps that its image – which resembles a horseshoe – extends beyond the homoclinic point (Fig. 6.20(a)). Let this number of steps be denoted by l. When the same square is iterated backwards, after a number of steps, m, we obtain a rectangular object stretched along the stable manifold, which also contains the homoclinic point (Fig. 6.20(b)). If the map is applied to this stretched rectangle $k = l + m$ times, the image of rectangle takes the shape of a horseshoe (Fig. 6.20(c)).

The application of the k-fold iterated original map on the stretched rectangle is therefore called a horseshoe map. Note that, within the rectangle, the horseshoe map has the same effect as the baker map: the rectangle is mapped into two columns. After one more iteration it contains four columns, then eight, and, after n steps, 2^n columns. The topological entropy is thus $h = \ln 2$ in this case also (the number of unstable cycles is therefore also infinite; see (5.51)). A similar statement holds in the vicinity of heteroclinic intersection points.

The argument is valid for all invertible maps, i.e. for area preserving ones also: the chaotic behaviour of a system follows from the existence of a single homoclinic or heteroclinic point. Note, however, that this ensures transient chaos only. For permanent chaotic behaviour, it is also necessary that the homoclinic and heteroclinic points are dense along the unstable direction.

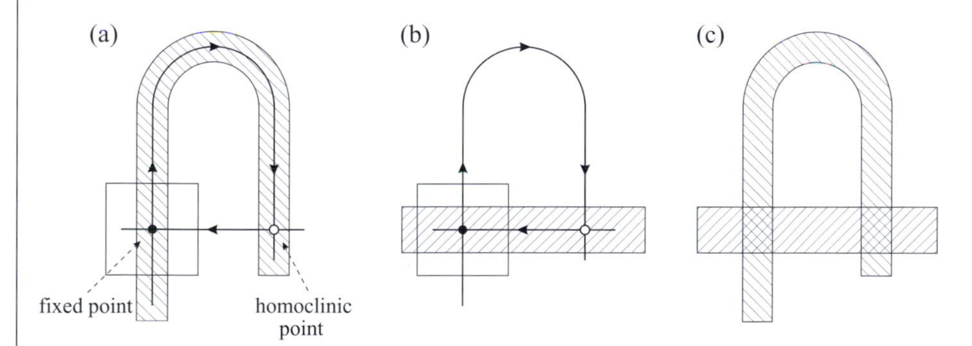

Fig. 6.20. Schematic construction of an invertible system's horseshoe map. (a) The image of the square around the fixed point surrounds the unstable manifold in the form of a horseshoe after l steps. (b) The mth pre-image of the square. (c) The map iterated forward $l + m$ times transforms the stretched rectangle into the horseshoe, via an action similar to that of a baker map.

irregularity, unpredictability and ordered but complex phase space structure. In order to give a general definition of chaos, which also covers transient chaos, one has to specify where to define the characteristic quantities. Chaotic dynamics is

- irregular: the topological entropy in some part of the phase space (not neccessarily an attractor) is positive, i.e. $h > 0$;
- ordered: there exists at least one invariant set, the chaotic set (which might also be a saddle), with a fractal natural distribution, i.e. $2 > D_1 > 0$;
- unpredictable: the average Lyapunov exponent on the chaotic set(s) is positive, i.e. $\bar{\lambda} > 0$.

Transient chaos sheds new light on what has been discovered about chaotic attractors (see Section 5.4.5). In the stable direction (i.e. on a section along some branch of the stable manifold), all the chaotic sets (attractors or saddles) of dissipative systems are fractals: $D_1^{(2)} < 1$. It is in the unstable direction that considerable differences show up. The attractor is a chaotic set that is continuous along the unstable manifold, with a smooth natural distribution in this direction: $D_0^{(1)} = D_1^{(1)} = 1$. Consequently, its stable manifold is actually space filling: $D_0^{(s)} = D_1^{(s)} = 2$ (cf. (6.30)), and its basin of attraction is finite. Since escape is impossible, the entire unstable manifold is identical to the chaotic set in this case. Chaotic saddles do not share these properties; they are discontinuous along the unstable direction also and they appear as the direct products of two Cantor sets.

The existence of chaotic transients indicates that even sets with basins of attraction of zero area (volume) may become *observable* via studying the finite-time dynamics. In general, we learn about transient chaos not via the everlasting dynamics on the saddle, but via trajectories coming close to the saddle and thus having finite, long lifetimes.

6.5 Parameter dependence: crisis

Let us investigate now how chaotic transients emerge when leaving the parameter region of permanent chaos. We have seen in Fig. 5.47 that the chaotic attractor is surrounded by the stable manifold of a hyperbolic point situated outside of the attractor, which is simultaneously its basin boundary. When some parameter, μ, is changed (for the kicked oscillator, for example, when the non-linearity parameter, a, is increased) the attractor becomes more extended and it becomes closer and closer to the basin boundary. At a certain parameter value, μ_c, the attractor touches its own basin boundary (Fig. 6.21). This situation is critical for further development; it is referred to as *crisis*. For the parabola attractor (at $E = 0.25$), the crisis value is (see Fig. 5.40) $a_c = 1.859$.

In a crisis configuration, not only a few 'extremal' points of the attractor touch the boundary. If a single such point (a heteroclinic point

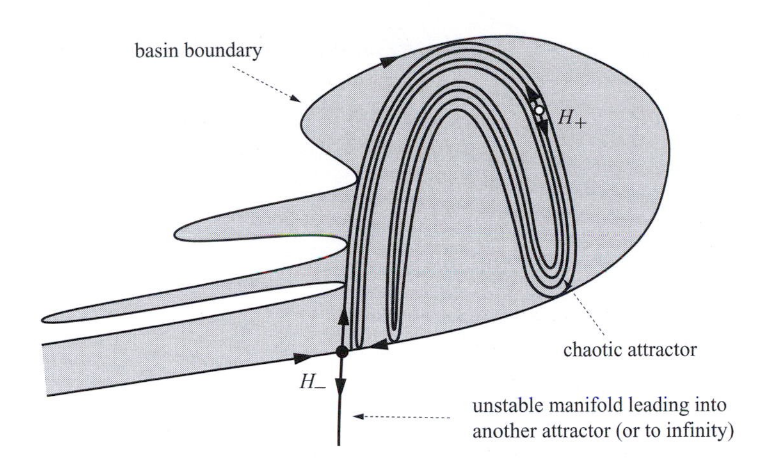

Fig. 6.21. Crisis situation: $\mu = \mu_c$ (schematic diagram). The attractor reaches the basin boundary and the basin boundary touches the attractor. Contact occurs at an infinity of points simultaneously (only a few are shown).

between H_- and H_+) exists, then all of its images and pre-images must also share this property (see Box 6.3). Therefore, an *infinite* number of points simultaneously become common points of the attractor and the basin boundary. Moreover, fixed point H_- is included into the attractor. A crisis situation therefore generally corresponds to the fully developed, most extended state of an attractor, a configuration which is, however, almost unstable.

Problem 6.12 Show that the attractors of the symmetric and asymmetric baker maps of Section 5.1 are in crisis.

For parameters beyond crisis ($\mu > \mu_c$), *no chaotic attractors exist*. This is so because, on the one hand, the unstable manifolds that constituted the attractor before extend beyond the former basin boundary, and, on the other hand, the latter curve, the stable manifold of H_-, 'bites into' the former attractor at several (in fact at an infinity of) locations (Fig. 6.22). All of these observations imply that, in the vicinity of the former attractor, arbitrary long motion is impossible and escape takes place. Nevertheless, an infinity of unstable periodic orbits is still present; the motion is therefore chaotic, but has a finite lifetime.

As a particular example, we present the crisis of the roof map (Fig. 6.23). The phase portraits at parameter values $\mu \equiv a$ slightly and greatly beyond crisis are shown in Figs. 6.24 and 6.25, respectively. In Fig. 6.25 it is clearly visible that the primary structure of the saddle is determined by the manifolds biting into each other.

Problem 6.13 Derive the crisis value, a_c for the sawtooth map from the equations of the manifolds around the fixed points.

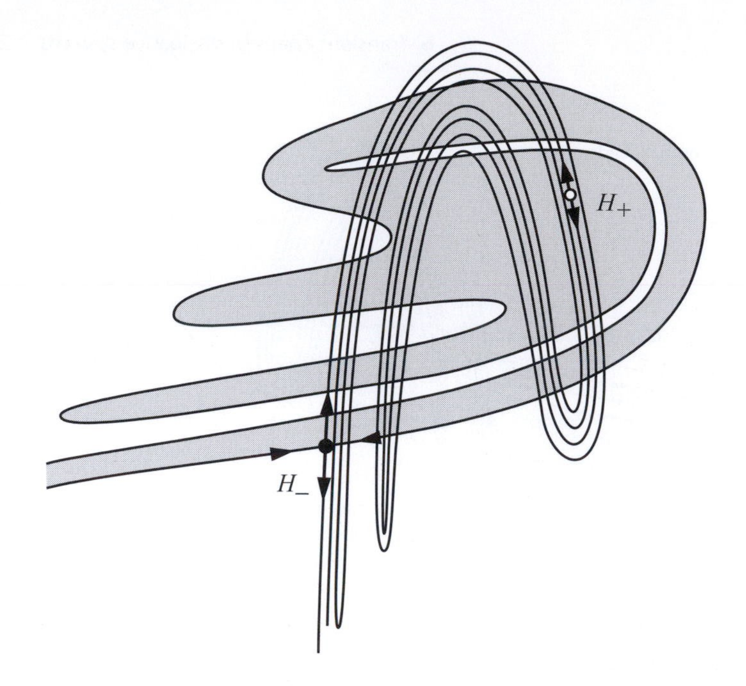

Fig. 6.22. A state beyond crisis: $\mu > \mu_c$ (schematic diagram) in terms of the stable manifold of H_- and of the unstable manifold of H_+. Shading indicates the area surrounded by the stable manifold: it no longer corresponds to a basin of attraction.

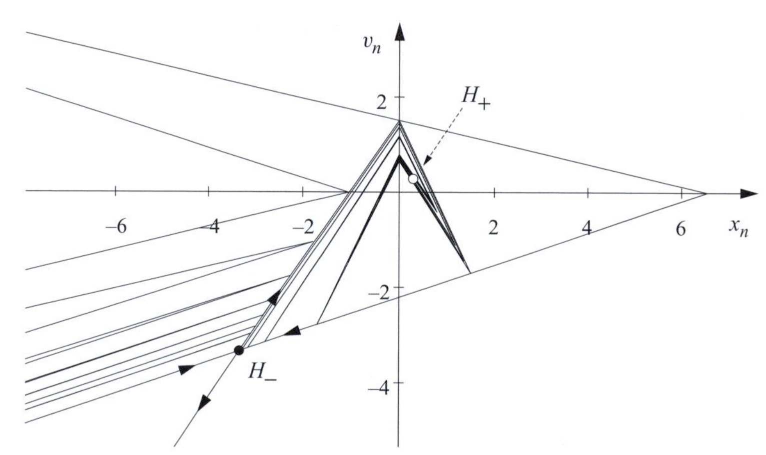

Fig. 6.23. Phase portrait of the roof map in crisis ($a_c = 1.7898$, $E = 0.7$). Around H_-, its own stable manifold accumulates and touches the unstable manifold in an infinity of points.

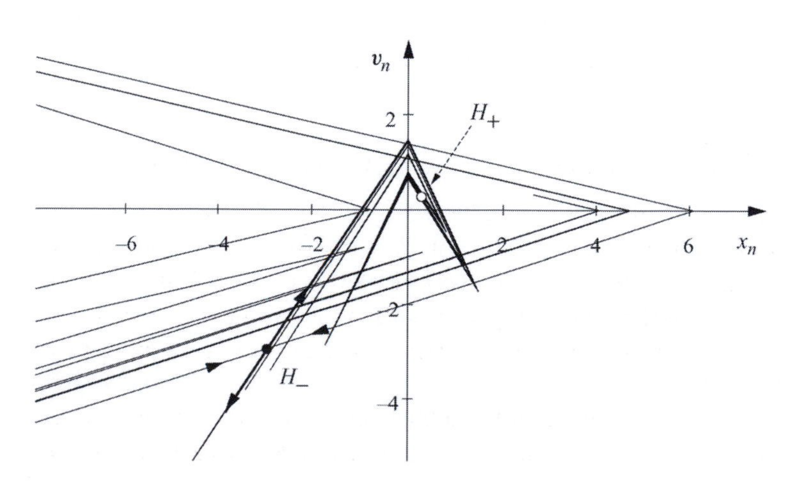

Fig. 6.24. Phase portrait of the roof map immediately after crisis ($a_c < a = 1.83$, $E = 0.7$). The unstable manifold of H_+ stretches beyond the area bounded by the stable manifold of H_-, which 'bites' into the unstable manifold of H_+ in a sequence of increasingly narrow, and ever deeper, bands (while it also becomes more and more folded).

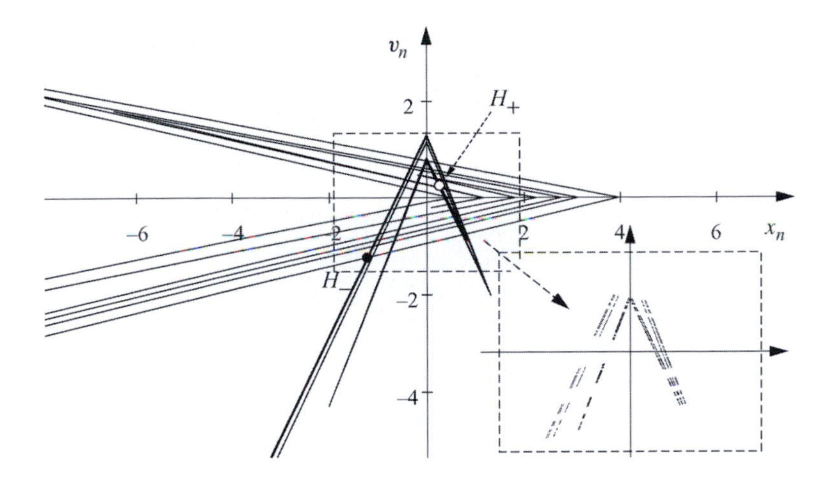

Fig. 6.25. Phase portrait of the roof map greatly beyond crisis ($a = 2.3$, $E = 0.7$). The inset shows the saddle.

Problem 6.14 Determine the phase portrait of the sawtooth map in crisis for $E = 0.8$.

Problem 6.15 Derive the equation determining the crisis value, a_c for the roof map.

Problem 6.16 What condition must function $f(x)$ of a one-dimensional map (5.47) fulfil (in the realm of functions with a single maximum) in order to be in crisis?

When departing from crisis with parameter μ within the transiently chaotic region, the holes along the unstable manifold increase, escape becomes easier and easier and, the lifetime, $\tau(\mu)$, of transient chaos decreases. Experience shows that near crisis this is proportional to a negative power of the parameter's deviation from the crisis value:

$$\tau(\mu) \sim (\mu - \mu_c)^{-\gamma}, \qquad \mu > \mu_c, \tag{6.31}$$

with γ as a positive exponent. Crisis is thus a kind of bifurcation, whose vicinity, when approaching crisis from above (decreasing μ), is indicated by a rapid increase in the lifetime.

6.6 Transient chaos in water-wheel dynamics

Permanent chaos has been observed in the dynamics of the water-wheel for parameters $\mu \equiv r > 15.0$ only (at $\sigma = 10$; see Fig. 5.56). If, conscious of the existence of transient chaos, we investigate the range $r < 15.0$ without excluding the approach towards the attractor, we find chaos to be present down to surprisingly small values of parameter r (see Fig. 6.26).

Numerically, the chaotic saddle seems to appear around $r = 10$, where it is still of very small extension and rather unstable. It becomes

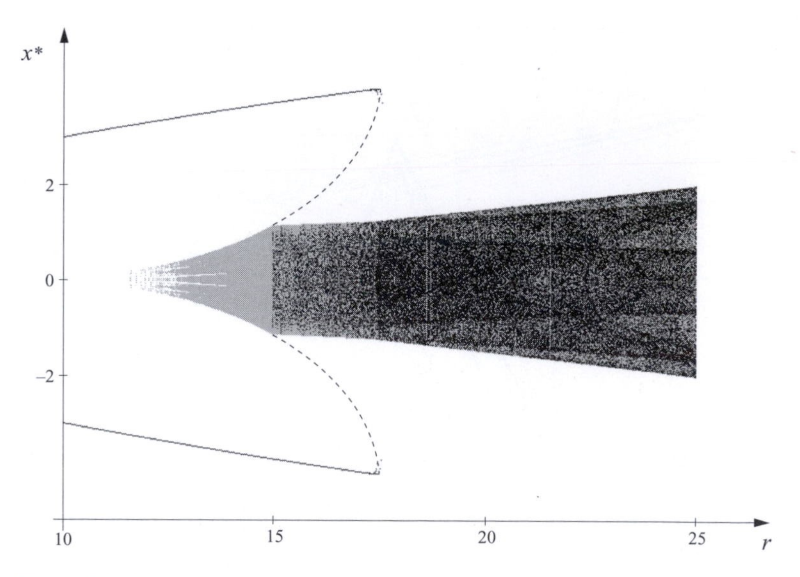

Fig. 6.26. Bifurcation diagram of the water-wheel indicating the presence of transient chaos: the projection of the chaotic saddle on the x-axis is represented by grey shading. Trajectories with lifetimes of at least 40 iterates within the band $|x_n| < 2$ of the Poincaré map have been considered for $r < r_c = 15.0$, and their x-values between the 5th and 20th iterate have been plotted. Note that the diagram also differs from Fig. 5.56 in the region of permanent chaos (black shading) since that has been generated in continuous time. In the present representation, the non-trivial fixed points are clearly visible (black lines) up to the point $r = 17.5$, where they lose stability. In the region $15 < r < 17.5$, two unstable limit cycles exist (represented by dashed lines) outside of the chaotic attractor.

more extended and easier to observe as parameter r increases. The water-wheel may therefore exhibit chaotic dynamics in relatively weak rain, but for a finite time only. The chaotic saddle co-exists here with the two non-trivial fixed point attractors. Due to the strong dissipation, the saddle has practically no extension in the stable direction; it seems to be the union of two Cantor sets sitting on slightly bent curves (Fig. 6.27). The natural distribution appears therefore in the shape of two curtains.

The critical value, $r_c = 15.0$, is approached from below in a sequence of increasingly long chaotic transients. Permanent chaos emerges in the water-wheel dynamics for increasing r via a *crisis* bifurcation (just like in the kicked oscillator dynamics with decreasing non-linearity parameter a).

The period-doubling cascade discussed in Section 5.3 and the crisis just investigated are two very common forms of the transitions towards (permanent) chaos.

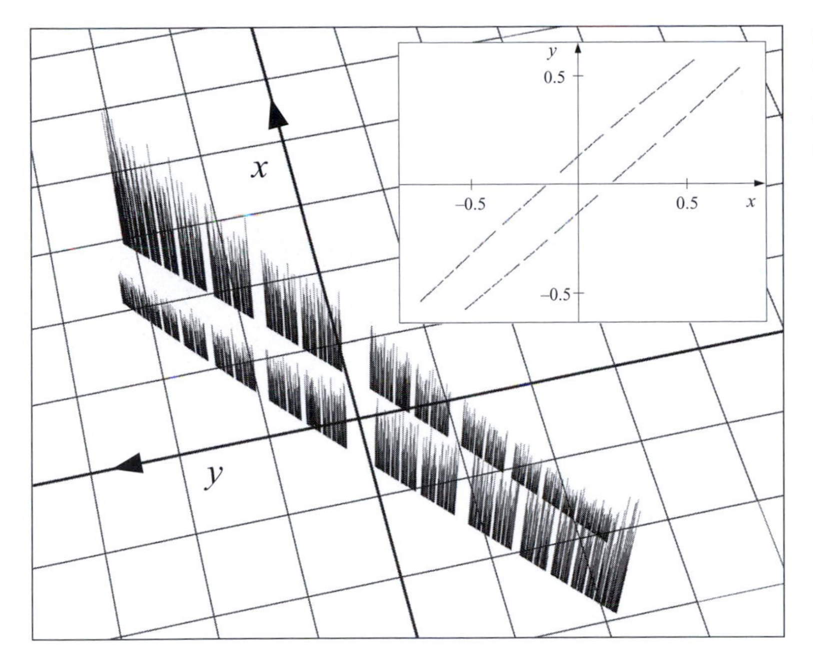

Fig. 6.27. Water-wheel saddle (inset) and the natural distribution on it in the Poincaré map $z = r - 1$ ($\sigma = 10$, $r = 14$). The resolution is $\varepsilon = 1/4000$

6.7 Other types of crises, periodic windows

In the structure of chaotic attractors, drastic changes, such as sudden enlargements, may occur when a parameter, μ, is changed. The critical configurations where these changes take place are also called crises. In contrast to the external or boundary crises investigated in Sections 6.5 and 6.6, when attractors touch their own basins, these are called internal crises (see Table 6.1). In the example of the parabola map ($\mu \equiv a$), a jump in attractor size can be observed in Fig. 5.40 at the parameter value $a_{c2} = 1.793$. For parameter values less than a_{c2}, the chaotic attractor is of small size; its neighbourhood, however, is, not 'empty': a chaotic saddle can be found there (Fig. 6.28).

Consequently, some trajectories may wander for a long time performing chaotic transients before settling on the attractor. For increasing parameter μ, the saddle becomes more extended, the lifetime of the transients increases and finally, at μ_{c2}, a part of the saddle and the small attractor suddenly merge to form an enlarged attractor, via an internal crisis.

Problem 6.17 Determine the saddle surrounding the small-size roof attractor for $a = 1.62$ preceding the crisis value $a_{c2} = 1.632$ of attractor enlargement (see Problem 5.26 and Fig. P.21 in the 'Solutions to the problems' section) ($E = 0.7$).

We have seen in Section 5.3 that parameter intervals exist, called periodic windows, a part of which does not contain any chaotic

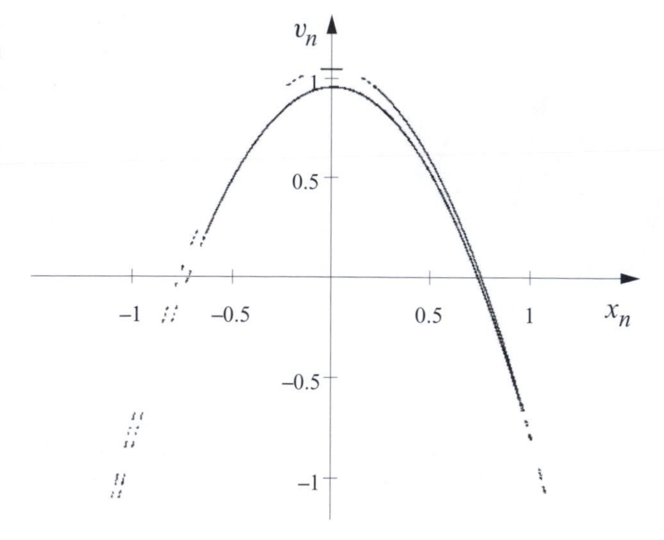

Fig. 6.28. Chaotic saddle around the small-size parabola attractor for $a = 1.79 < a_{c2} = 1.793$ ($E = 0.25$).

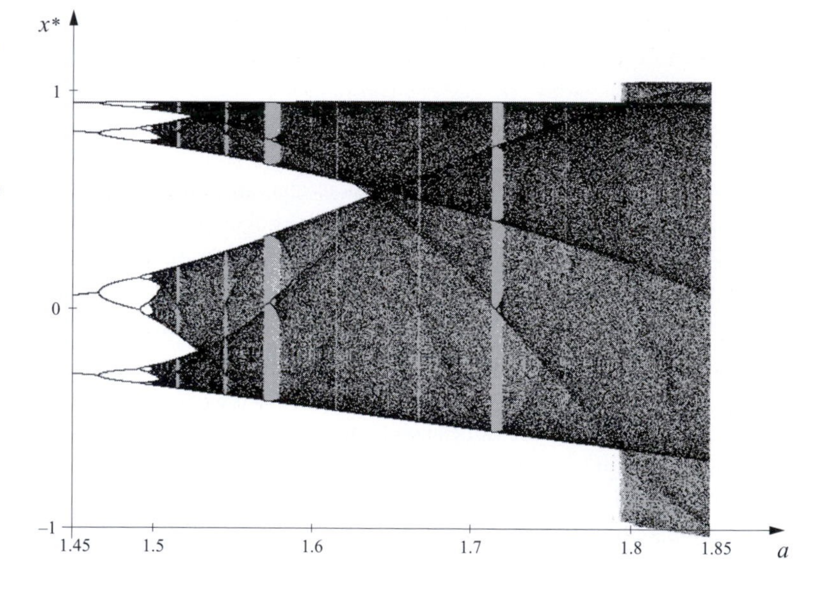

Fig. 6.29. Bifurcation diagram of the parabola map (5.39) obtained by omitting the first 50 iterates only (cf. Fig. 5.40). In this way the chaotic saddles filling the periodic windows also appear in the diagram, projected on the x-axis (grey regions).

attractors.[2] The chaotic parameter range above the accumulation point, μ_∞, has been found to be a nested set of chaotic and non-chaotic parameter values. Taking transient chaos into account, the situation becomes considerably simpler: the periodic windows might be empty of chaotic attractors, but not of chaotic saddles (Fig. 6.29). The infinity of unstable

[2] The other parts of the periodic windows contain multi-piece, small-size chaotic attractors (see e.g. Fig. 5.40) since the periodic orbits undergo a period-doubling cascade within each window. Periodic windows end with internal crises.

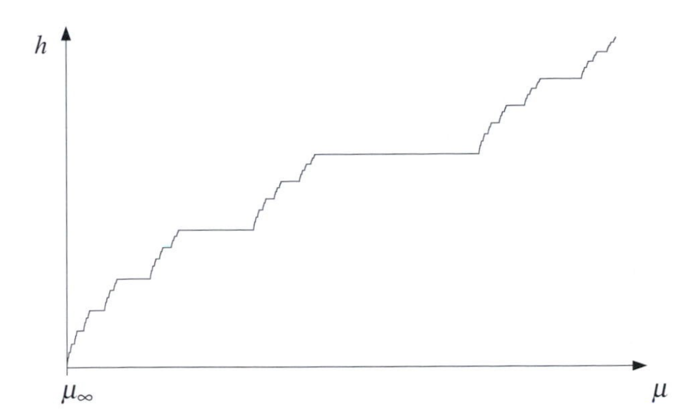

Fig. 6.30. Devil's staircase structure of topological entropy in the range $\mu > \mu_\infty$ (schematic diagram). In contrast, the average Lyapunov exponent of the chaotic set (attractors and saddles) is expected to be a smooth, monotonically increasing, function of *a*.

cycles developed on the route towards the accumulation point, μ_∞, cannot suddenly disappear from the system. Consequently, chaotic transients are present *in each periodic window*.

By accepting that the concept of chaos includes its transient variant also, the statement that the union of the chaotic parameter values, μ^*, constitutes a fat fractal (Section 5.3) must be revised: chaos is present on *continuous parameter intervals* above the accumulation point, μ_∞. Moreover, in certain cases, for example for kicked oscillators, the intervals may be of infinite length.

The average Lyapunov exponent determined over the chaotic set (attractor or saddle) is in general, for $\mu > \mu_\infty$, a monotonically increasing function. A similar statement holds for the topological entropy, whose typical shape is that of a *devil's staircase*. This can be considered as the integral of a homogeneous distribution over a Cantor set: it consists of long constant intervals, with sudden jumps on a fractal set (Fig. 6.30). Each jump in the topological entropy devil's staircase corresponds to a crisis, where new unstable orbits emerge.

6.8 Fractal basin boundaries

6.8.1 The phenomenon

For the kicked oscillators studied so far, the basin boundary (the stable manifold of fixed point H_-) between the chaotic attractor and an attracting fixed point at infinity is a simple curve, just like the boundary between the simple attractors in Section 3.3.1 (see Fig. 3.17). In view of this, it may seem surprising that in systems where chaos may occur at all (see Section 4.8), the basin boundary between even *simple* point attractors is often very complex, i.e. of fractal nature (Fig. 6.31).

This implies that the basins of attraction become interwoven and become very close to each other in the form of narrow bands. Particles that

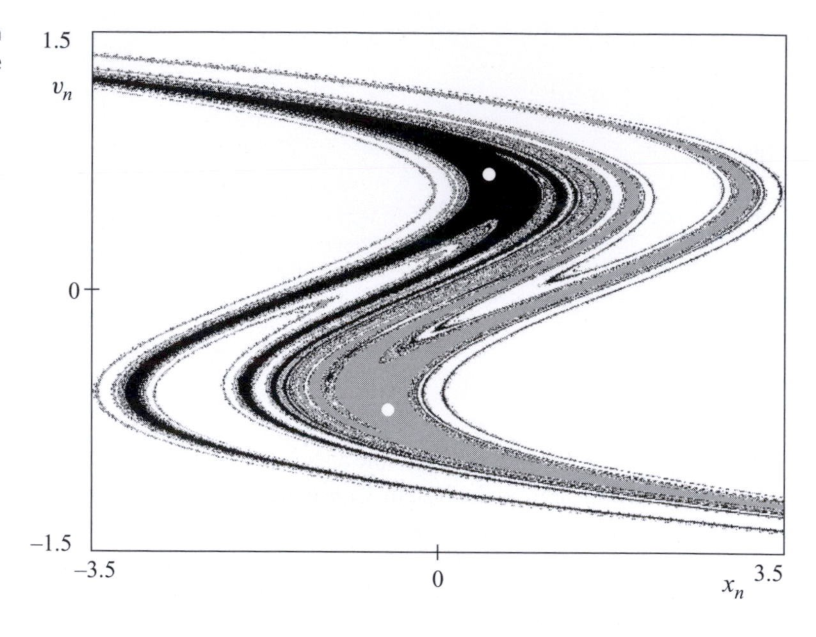

Fig. 6.31. Basins of attraction (black and grey regions) of the two fixed point attractors of a harmonic oscillator kicked with the amplitude function $f(x) = ax(1 - x^2)$ ($a = 2.8$, $E = 0.7$). The attractors are marked by white dots.

start out from this part of phase space hesitate for a long time: it is difficult for them to 'decide' which attractor to converge to. In the meantime, they move in a seemingly random manner without coming particularly close to either attractor. Consequently, there must exist a chaotic saddle between the attractors. It is the 'hesitation' that is a transiently chaotic behaviour.

The basins of attraction touch each other in an intricate way. The role of water-shed divide between different attractors, as already seen in simple systems (Sections 3.1–3.3), is played by stable manifolds. Fractal basin boundaries are thus *always* stable manifolds of chaotic saddles, and their appearance is therefore always accompanied by chaotic transients.

Problem 6.18 Determine the basins of attraction of the two attractors co-existing in the two-fold iterated roof map at $a = 1.55$, $E = 0.7$.

When changing a system parameter, there might occur sudden changes in the structure of the basin boundary: the latter might go over from a smooth into a fractal curve. Such *basin boundary metamorphoses* are themselves crises. In the course of such events, a horseshoe is formed between the originally smooth manifolds of a simple hyperbolic orbit on the boundary. With the creation of a chaotic saddle formed in this way, the basin boundary becomes a fractal, i.e. the stable manifold of the new saddle. When increasing the non-linearity parameter, a, of the kicked oscillators, it depends on the dissipation parameter, E, whether a basin boundary metamorphosis takes place before the boundary crisis is

Table 6.1. *Comparison of different crises.*

Type of crisis	Phenomenon
Boundary crisis	chaotic attractor ceases to exist: it is converted into a chaotic saddle
Internal crisis	chaotic attractor suddenly enlarges: a saddle is merged with a small size attractor
Basin boundary metamorphosis	the basin boundary changes its fractal character: a hyperbolic point on the boundary becomes part of a chaotic saddle

reached and the chaotic attractor disappears. Table 6.1 provides a comparison of the main types of crises discussed.

Problem 6.19 The parabola attractor approaches crisis for $E = 0.32$ at $a_c = 1.843$. By determining the basin boundary between the parabola attractor and the attractor at infinity for $a = 1.74$ and $a = 1.80$ ($E = 0.32$), demonstrate that a basin boundary metamorphosis occurs at around $a_c = 1.77$. Show numerically that the fixed point H_- does not have homoclinic intersections below a_c.

6.8.2 The uncertainty exponent

The existence of a fractal basin boundary implies that the outcome of motion started from its vicinity is difficult to predict. Namely, the error in determining the initial condition corresponds to a box of finite size in phase space, which may overlap both basins of attraction. When initiating the motion, it is therefore often impossible to predict which attractor the trajectory will converge to. A quantity measuring the degree of this uncertainty is the *uncertainty exponent*.

Let us cover the phase space with square boxes of size ε. A box is considered to be *certain* if any number of points started from it converges to the same attractor. Boxes without this property are called *uncertain*, and their number is denoted by $N(\varepsilon)$. The ratio of the number of uncertain boxes to that of all the boxes, $N_0(\varepsilon)$, is $f(\epsilon) \equiv N(\varepsilon)/N_0(\varepsilon)$. For sufficiently fine resolution, the ratio, f, scales as a power of the resolution:

$$f(\varepsilon) \sim \varepsilon^{\alpha}, \tag{6.32}$$

where the positive exponent, α, is the uncertainty exponent. For smooth basin boundaries, $\alpha = 1$, while, for fractal ones, $\alpha < 1$. In the latter case, by increasing the resolution by a factor of, say, 100, the number of accurately predictable boxes increases by a smaller factor only (for $\alpha = 1/2$, for example, by 10). The smaller the value of α, the more hopeless it is to improve the accuracy: for $\alpha = 0.2$, a decrease of f by a factor of 10 would require an increase in resolution by 10^5!

The uncertainty exponent can be expressed in terms of the characteristics of the chaotic saddle whose stable manifold is the basin boundary. The number of uncertain boxes varies with an exponent corresponding to the fractal dimension of the stable manifold, i.e. with $-D_0^{(s)}$. Since the number of all boxes in a plane scales as ε^{-2},

$$\boxed{\alpha = 2 - D_0^{(s)} = 1 - D_0^{(1)}.} \tag{6.33}$$

Applying the Kantz–Grassberger relation, (6.25), we obtain

$$\alpha \approx \frac{\kappa}{\bar{\lambda}}. \tag{6.34}$$

Thus, uncertainty is significant if the stable manifold of the saddle is almost space-filling, i.e. if the average chaotic lifetime is long.

6.8.3 Continuous-time systems: magnetic and driven pendula

The equation of motion of the magnetic pendulum presented in Section 1.2 is of simplest form if the thread is assumed to be long compared with the amplitude of swinging. Without magnets, the pendulum oscillates in this case with natural frequency $\omega_0 = \sqrt{g/l}$ around the origin, practically in a horizontal plane. Let magnet i be located at distance d_i under point (x_i, y_i) of the plane of motion. Assuming a point-like interaction of Coulomb type between the magnets, the force exerted by magnet i on the swinging body is of magnitude γ_i / D_i^2, where γ_i is a measure of the magnet's strength (positive for attraction), and

$$D_i = \sqrt{(x_i - x)^2 + (y_i - y)^2 + d_i^2} \tag{6.35}$$

is the distance between the body and the magnet. The x- and y-components of the force are obtained by multiplying the magnitude by the factors $(x_i - x)/D_i$ and $(y_i - y)/D_i$, respectively. Taking into account an additional damping term linearly proportional to the velocity, the equation of motion of the magnetic pendulum in the presence of N magnets is given by

$$\ddot{x} = -\omega_0^2 x - \alpha \dot{x} + \sum_{i=1}^{i=N} \frac{\gamma_i (x_i - x)}{D_i^3}, \tag{6.36}$$

Table 6.2. *Parameters used in Figs. 1.10 and 1.11 and Plates III–VI.*

In all cases, $\gamma_i = 1$.

	α	w_0	d
Fig. 1.10	0.2	0.5	0.3
Fig. 1.11	0.05	0.5	0.3
Plate III	0.3	0.5	0.3
Plates IV, V	0.2	1.0	0.3
Plate VI	0.3	2.0	1.0

$$\ddot{y} = -\omega_0^2 y - \alpha \dot{y} + \sum_{i=1}^{i=N} \frac{\gamma_i (y_i - y)}{D_i^3}, \tag{6.37}$$

$\alpha > 0$. For solely attractive interactions, the number of the attractors is the same as the number of the magnets (for weak magnets there may be an additional one, around the origin), but in a repelling case the number of attractors depends essentially on the system parameters.

Each attractor is necessarily a point attractor, as all motion stops due to damping. If we colour in the initial position (x_0, y_0) of the pendulum released without any initial velocity ($\mathbf{v}_0 = 0$) according to the attractor above which it stops, the basins of attraction become visible. We obtain in this way a *real space* image of the basin boundary generally present in phase space, or, more precisely, an intersection of the basins of attraction with the plane $\mathbf{v} = 0$. Parameters used in figures and plates exhibiting fractal boundaries of magnetic pendulums are given in Table 6.2.

The equation of motion of the driven pendulum, the other example in Section 1.2.2, has been given by equation (5.90). The parameters used in Figs. 1.13 and 1.14 and in Plate VII are $\alpha T = 0.1\pi$, $T\sqrt{g/l} = \pi/5$ and $A/l = 0.8$. The filamentary structure of the basin boundary in this example only appears in phase space.

Problem 6.20 Determine the basins of attraction of the non-trivial fixed points of the water-wheel for $r = 14$, $\sigma = 10$.

Problem 6.21 The dimensionless equation of motion of a ship subjected to periodic waves (the driven version of Problem 3.12) is $\ddot{x} = Ax(x+1) - \alpha\dot{x} + f_0 \cos(2\pi t)$. Determine numerically the basins of attraction of the ship's stable oscillation and of its capsized state (escape from the potential well) for parameters $\alpha = 0.6$, $f_0 = 3.8$ and $A = 55$. Plot the unstable manifold of the chaotic saddle.

Box 6.4 Other aspects of chaotic transients

Chaotic transients occur in numerous, some of them everyday, phenomena. The small, double pendulum-like, battery-driven machines designed to demonstrate chaos are usually seen in shop-windows performing periodic motion only. The reason is that, in general, the system, reaches a transiently chaotic state with a lifetime of a few minutes only, due to a gradual discharge of the battery (for several other examples, see Chapter 9).

In the course of *transport processes* (diffusion, electric or thermal conduction, etc.), particles move chaotically. Particles that do not leave a sample of finite length for a long time wander in the neighbourhood of a chaotic saddle. The corresponding transport coefficient, for example the diffusion coefficient, is found to be proportional to the escape rate; therefore, the longer the lifetime of chaos, the slower the transport.

If we select a sub-region of a chaotic attractor and track the trajectories starting from it until they leave the region, we find transiently chaotic behaviour. Points never escaping the region (neither forwards nor backwards in time) form a chaotic saddle. The periodic orbits of the saddle are a sub-set of all the cycle points located in the selected region of the attractor. Transient chaos thus helps to explore the dynamics on *sub-sets of chaotic attractors*, and it is therefore also related to the *control of chaos* (see Box 9.4).

It is interesting to note that weak external noise may transform a chaotic saddle into an effective attractor. This is so because chaotic saddles and nearby attractors may merge into a single extended attractor in the presence of noise. This phenomenon can be observed even in systems with simple deterministic attractors: then weak noise converts transient chaos into permanent chaos (this is called *noise-induced chaos*).

Chapter 7
Chaos in conservative systems

A special but important class of dynamics is provided by systems in which friction is negligible, or, more generally, where dissipative effects play no role. In this case the direction of time is not specific, the process described by a differential equation is reversible: forward and backward time behaviour is similar. Think of, for example, a planet: one cannot decide whether its motion recorded on a film takes place in direct or in reversed time. In frictionless systems phase space volume is preserved, and attractors cannot exist. In such conservative systems, the manifestation of chaos is of a different nature than in dissipative cases. In this chapter we investigate persistent conservative chaos where escape is impossible, and defer the problem of transient conservative chaos to Chapter 8. We start with the area preserving baker map and the stroboscopic map of a kicked rotator. Next, the dynamics of continuous-time, non-driven frictionless systems is considered. On the basis of these examples, we summarize the general properties of conservative chaos, including one of the most important relationships, the KAM theorem. The structure of chaotic bands characteristic of conservative systems is discussed and compared with that of chaotic attractors. Finally, we present how conservative chaos of increasing strength manifests itself and we discuss the consequences.

7.1 Phase space of conservative systems

Since dissipation is unavoidable in typical macroscopic dynamics, we consider the frictionless case only after studying the dissipative one. Examples will be taken from the realm of motion taking place in a vacuum. Note, however, that the dynamics of spacecraft or planets in the Solar System, or of charged particles in the electric and magnetic fields of accelerators or fusion equipment, are all conservative processes to a very good approximation. The advection dynamics of particles in flows can be shown to be of the same kind to a first approximation. The main fields of application for conservative chaos are space research/astronomy, plasma physics and hydrodynamics/environmental physics.

Friction is always accompanied by phase space contraction. A basic property of frictionless systems is that their phase space volume does not change in time. This is why they are called *conservative*.[1] The phase space contraction rate, (3.55), of conservative systems is by definition zero:

$$\sigma \equiv 0. \tag{7.1}$$

An important consequence of this is that there cannot be any sub-sets of phase space towards which a volume element could converge.[2] In conservative systems attractors (repellors) do not exist; the motion does not forget the initial state: its character *depends on the initial condition* even after a long time. This is why in conservative systems chaos typically appears in co-existence with regular motion (as shown in Figs. 1.17 and 1.20 and Plate VIII).

This kind of chaos is naturally present in driven one-dimensional motion with a vanishing friction coefficient ($\alpha = 0$). Conservative chaos appears, however, in an additional class: in non-driven systems of at least two spatial dimensions. In such systems it is of great importance that the *total energy* is conserved. The position of a body in the *plane* is specified by two position co-ordinates (x, y), to which two velocity components, (v_x, v_y), belong. The Newtonian equation determines the two acceleration components in terms of the position. This system of two second-order autonomous equations can be rewritten as a system of four first-order equations. Due to the conservation of energy, one of the four variables (for example, v_y) can be expressed in terms of the others, so only three independent first-order differential equations remain. We have seen in Section 4.8 that a necessary condition

[1] Conservative systems are frictionless limits of dissipative systems. This limit, however, might be very complicated (an increasing number of spiral attractors appear with diminishing basins); it is therefore useful to study the conservative case on its own.

[2] Such systems can be described efficiently by means of the canonical formalism of classical mechanics, but we do not assume in the following that the reader is familiar with this.

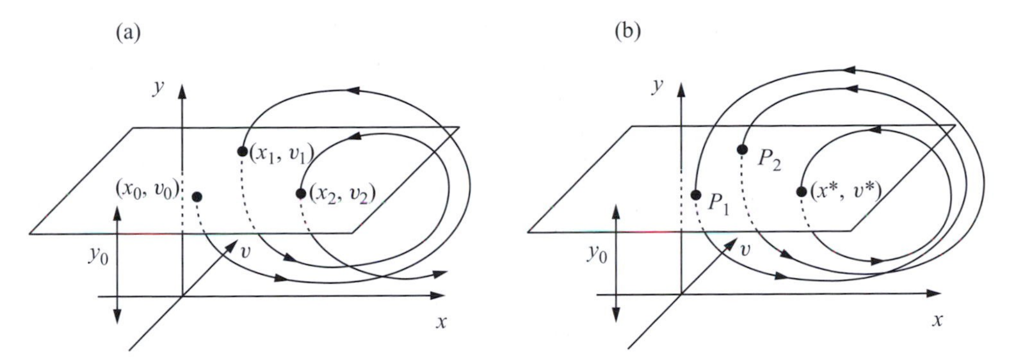

Fig. 7.1. Poincaré map (conservative planar motion of a single point mass). The section is taken here with the plane $y = y_0$. (a) Points (x_n, v_n) are the intersections with trajectories coming from above. (b) Periodic orbits consisting of one loop correspond to the fixed points (x^*, v^*) of the map; those with two loops correspond to two-cycles (P_1, P_2).

for chaos is an at least three-dimensional phase space. The same situation arises in the case of two bodies moving along a straight line. We therefore conclude that conservative chaos may occur in the motion of a *single body* in a *plane* or of *two bodies* along a *line*, even without driving. Henceforth we do not study cases more complicated than these.

In this new class it is also useful to monitor the three-dimensional phase space motion in a plane in the form of a map, i.e. in discrete time. This can be achieved by defining a *Poincaré map*: one of the position and velocity co-ordinates is recorded when the system happens to be in a certain position. This can be, for example, that co-ordinate y takes on a certain value y_0. Defining a Poincaré map corresponds to intersecting a continuous trajectory of the flow by a surface (see Fig. 7.1). This surface is called a Poincaré section (and the entire phase portrait is a Poincaré portrait). In order for the velocity value v_x to be unique, the intersections of trajectories coming from a certain direction, for example from above, are recorded. The successive intersections obey a map:

$$(x_{n+1}, v_{n+1}) = M(x_n, v_n). \tag{7.2}$$

(For the sake of simplicity, subscript x of the velocity component is omitted.) One sees from the definition that the fixed points of the map correspond to *periodic* orbits of the flow.

In contrast to stroboscopic maps, the intersections are now taken not at certain phases, but at certain *configurations*. The Poincaré section has to be chosen in such a way that typical trajectories can intersect the surface many times. The explicit form of the map depends on the choice. For properly chosen surfaces, however, the conclusions regarding the global dynamics (whether it is chaotic, or what the ratio of the areas belonging to chaotic and to regular trajectories is) are independent of the location of the surface.

The Poincaré map is, of course, invertible since it is derived from a differential equation (see Section 4.7). Due to the lack of friction, the forward and backward dynamics are of the same nature. Therefore, the inverse map, M^{-1}, is equivalent to the original one. Since the area contraction rate of an inverse map is the reciprocal of the original, the Jacobian must be unity:

$$\boxed{J \equiv 1.} \tag{7.3}$$

This ensures that there exist no attractors in the map either.

The Jacobian of driven conservative systems is also unity (cf. equation (5.80) with $\sigma = 0$). Thus, a common property of maps related to any kind of conservative systems is that they are *area preserving* (in suitably chosen co-ordinates at least). Relevant features of conservative chaos can thus be understood by studying area preserving maps in the plane. In view of (5.57), in such maps the local Lyapunov exponents are the same in absolute value but are each other's opposites:

$$\boxed{\lambda(\mathbf{r}) = -\lambda'(\mathbf{r}).} \tag{7.4}$$

The average Lyapunov exponent is therefore (-1) times the average negative exponent: $\bar{\lambda} = -\bar{\lambda}'$.

Problem 7.1 How does the stability of a periodic orbit depend on the stability matrix, L, of the corresponding fixed point of an area preserving two-dimensional map?

7.2 The area preserving baker map

First, let us consider the area preserving case of our simplest chaotic model, the baker map. It is worth considering immediately the asymmetric (closed) case, given by equations (5.15) and (5.16), which is area preserving if the height of the horizontal cutting line and the thickness of the lower rectangle after compression are equal, i.e. if $b = c_1 < 1$ and $c_2 = 1 - c_1$. The map is given by

$$(x_{n+1}, v_{n+1}) = \left(bx_n, \frac{v_n}{b}\right), \quad \text{for} \quad v_n \leq b, \tag{7.5}$$

$$(x_{n+1}, v_{n+1}) = \left(1 + (1-b)(x_n - 1), 1 + \frac{v_n - 1}{1-b}\right), \quad \text{for} \quad v_n > b. \tag{7.6}$$

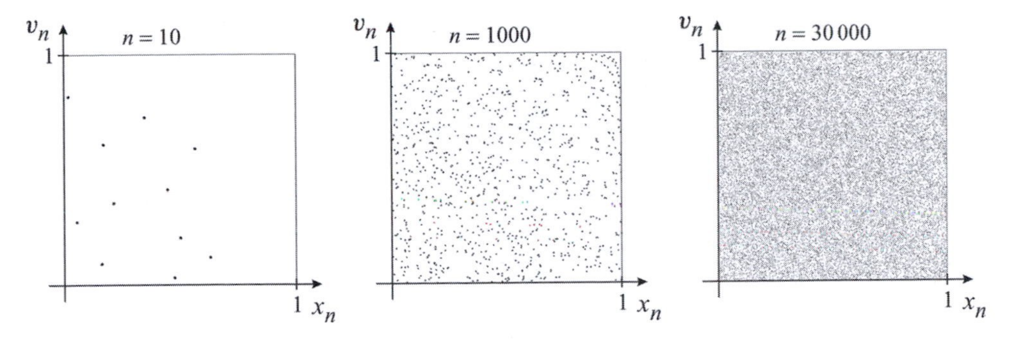

Fig. 7.2. The area preserving baker map: Points of a trajectory started with $x_0 = 1/2$, $v_0 = 2/\pi^2$ after $n = 10$, 1000 and 30 000 steps ($b = 1/3$).

The phase space is the unit square. If a trajectory is started from a point chosen at random inside this area, it visits the entire square (Fig. 7.2). Chaos is *more extended* than in the dissipative case. Figure 7.2 also illustrates that the distribution of the points is uniform after a sufficiently large number of iterations. The long-time chaotic behaviour in conservative systems is thus also characterised by a time-independent probability distribution. Its support is an extended area (not a fractal), and the density is *uniform*.

The chaoticity of the motion is also demonstrated by the fact that both the topological entropy and the average Lyapunov exponent – which are defined in the same way as in dissipative cases (see equations (5.51) and (5.54)) – are positive.

Problem 7.2 Determine the topological entropy of the area preserving baker map.

Problem 7.3 Determine the average Lyapunov exponents of the area preserving baker map. Derive the inverse map and show that it is equivalent to the original (and therefore that its Lyapunov exponents are the same).

The argument employed to unfold the relation between chaos and the unstable manifolds of hyperbolic periodic orbits can be applied in the same manner as in the closed dissipative case. We find that the chaotic region is the union of the hyperbolic cycles' unstable manifolds. Due to the equivalence of the forward and backward iterations, however, the stable manifolds now have a similar property.

It is also useful to follow the time evolution of the entire unit square. Since it is mapped onto *itself*, details can only be discovered by distinguishing points originating above and below the dividing line. As the number of iterations increases (Fig. 7.3), the two colours become dispersed in an increasing number of thin vertical bands (although not of the

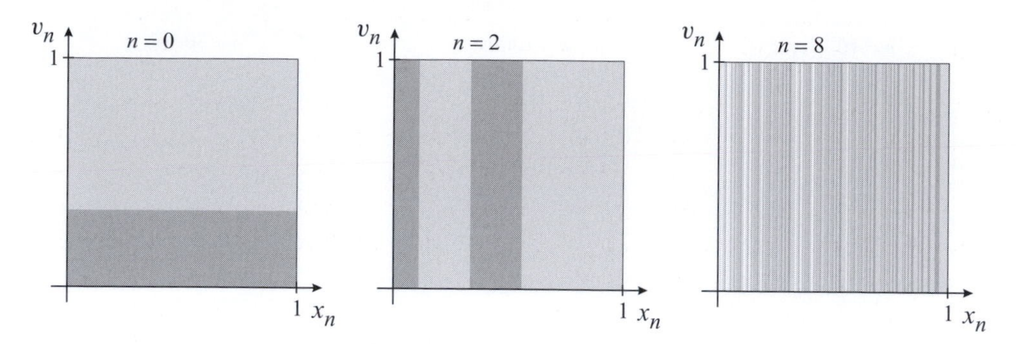

Fig. 7.3. Evolution of the entire phase space volume of the area preserving baker map in eight iterations ($b = 1/3$). The points starting below (above) the line $v_n = b$ are shaded dark (light) grey. After $n = 8$ steps, the two colours are already well mixed.

same width). After an infinite number of steps, *both* colours are present in an *arbitrarily close* vicinity of any point. A mixture of the two colours arises. This is exactly the mechanism that plays a role in the mixing of different materials (see Boxes 7.1 and 7.4 and Section 9.4).

Problem 7.4 A map similar to the area preserving baker map is called the 'cat' map, defined on the unit square by the rule $(x_{n+1}, v_{n+1}) = (2x_n + v_n, x_n + v_n)$ with both x_n and v_n periodic in (0, 1). The map is named after the cat head usually drawn into the unit square to illustrate the action of the map.

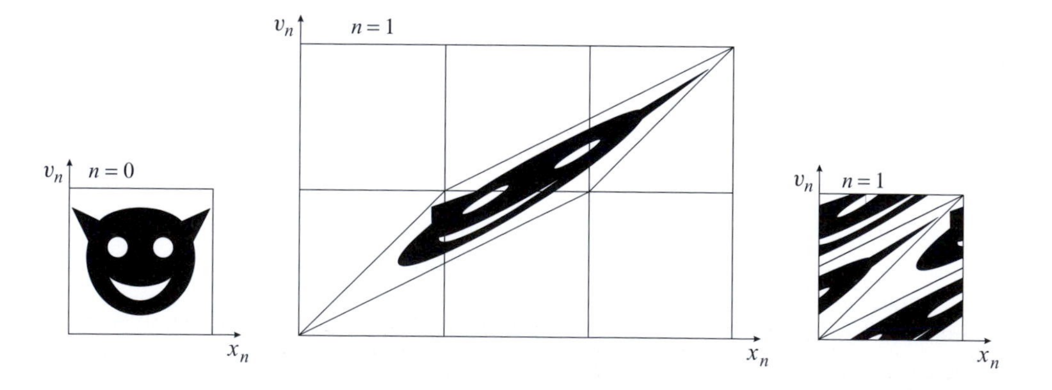

The middle and right panels show the image of the head in the entire plane and in the periodic representation, respectively. Determine the cat map's average Lyapunov exponent, its topological entropy and the directions of the unstable and stable manifolds.

Box 7.1 The origin of the baker map

The (symmetric) area preserving baker map appeared first in the works of the German mathematician, Eberhard Hopf, on ergodic theory in the 1930s. After defining the baker map, the author remarks that 'the repeated application of the map is something like the way puff pastry is made', and he proves with mathematical rigour the mixing property. The map's action on the unit square is indeed very similar to the process that the dough undergoes in the course of stretching. The stretched dough (whose volume is practically conserved) is folded. The so-formed two-layered piece is stretched again, then folded, and these steps are repeated (Fig. 7.4). A baker map given by (7.5) and (7.6) with $b = 1/2$ corresponds to a stretching process in which the stretched piece of dough is not folded, but cut into two identical pieces which are then put on top of each other (Fig. 7.4). (In the area preserving version of map (6.14) even the folding is similar to that of dough stretching.)

The name of the map, accepted after Hopf's studies, refers to the similarity with a baker's work. At the same time, this reminds us that chaotic processes result in continuous stretching, reordering and folding of phase space. As kneading and stretching dough homogenise the initial distribution of the ingredients (salt, sugar, flour, etc.), so do chaotic processes lead to good mixing in phase space.

The extension of the baker map to closed and open dissipative cases dates back to the 1980s. In the dough-kneading analogy, the effect of these maps mimics the action of a 'gluttonous' baker who eats up a certain (uniformly distributed) portion of the dough after each stretching step. No wonder that finally only infinitely thin 'salty sticks' remain: the fractal filaments of the chaotic attractor or of the unstable manifold of the saddle.

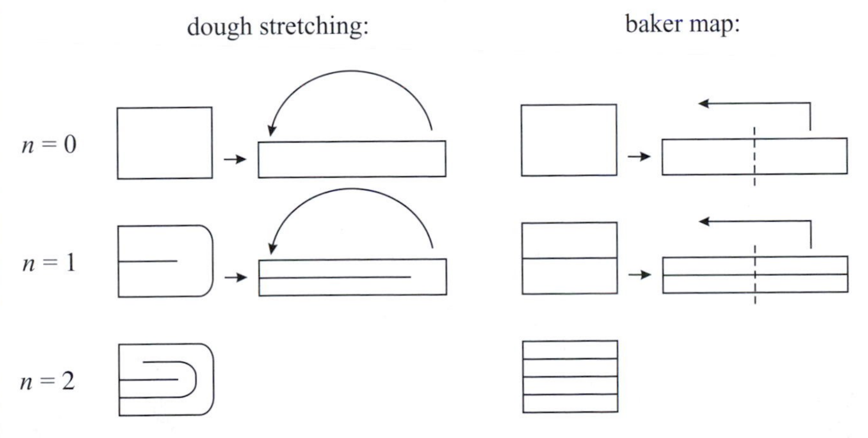

Fig. 7.4. Schematic representation of the traditional stretching (the piece of dough is shown from the side) and the stretching process corresponding to the baker map given by (7.5) and (7.6) ($b = 1/2$).

The area preserving baker map – although a good model from many points of view – does not illustrate the general property of conservative chaos (Section 1.2.3): that it co-exists in phase space with regular motion. This feature is evident, however, for example in the frictionless limit

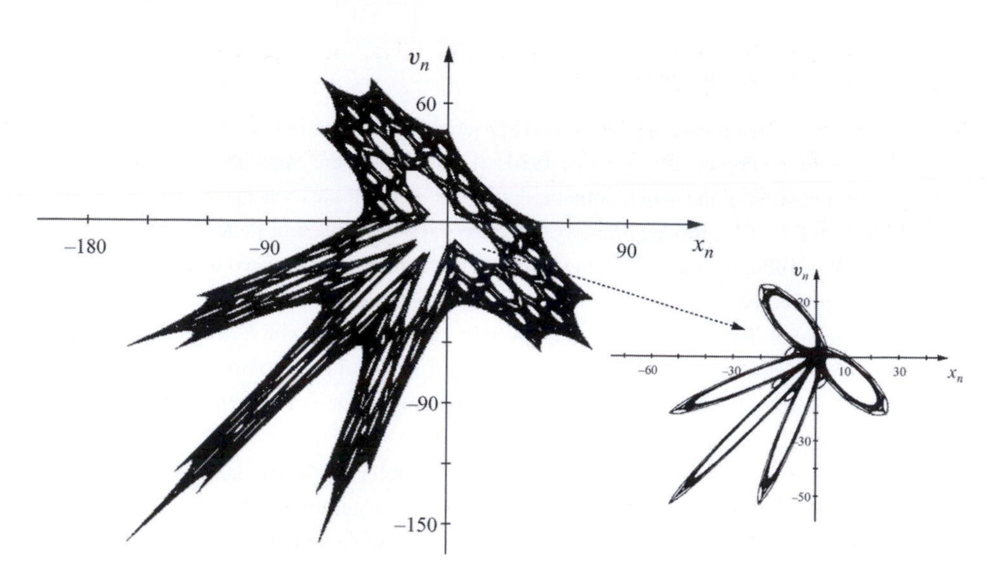

Fig. 7.5. Phase portrait of a typical area preserving map: two independent chaotic bands (black) of an undamped oscillator kicked with a roof amplitude function; see (5.37) ($a = 1.4$, $E = 1$). The chaotic bands have a large extension (see units), the white patches inside them belong to elliptic islands. (The point $H_+ = (0.294,\ 0.294)$ is so close to the origin that it is not marked.)

($E = 1$) of the kicked oscillator maps (Fig. 7.5). In the roof map (5.37), fixed point H_+ is stable for $a < 2$, and is necessarily elliptic. Around the elliptic point and the higher-order elliptic cycles the motion is regular. Between such elliptic islands *chaotic bands* appear, within which the iterated point moves at random. There is no crossing allowed from one band either to another band or to the elliptic islands, and vice versa. A chaotic band can be traced out by any trajectory whose initial condition falls into this band, since all such trajectories fill the band in the same way.

7.3 Kicked rotator – the standard map

7.3.1 The map

In conservative systems, variables are often angles that determine the position of a point along a closed phase space curve, for example an ellipse. We therefore present a kicked model with a periodic coordinate, which can be considered to be a prototype of conservative dynamics.

Consider a body rotating freely about a vertical axis: a rotator. The simplest way to realise this is to take a weightless rod of unit length in a horizontal plane, with one end fixed to a ball and the other end fixed to a rotating vertical shaft (Fig. 7.6). The instantaneous position of the ball is given in terms of its distance, x, along the unit circle's perimeter measured from a reference point (which is simply the deflection angle), and is considered to be periodic in 2π. The kicking amplitude is written as $uI(x)$, where u is a characteristic velocity and I is a dimensionless amplitude function.

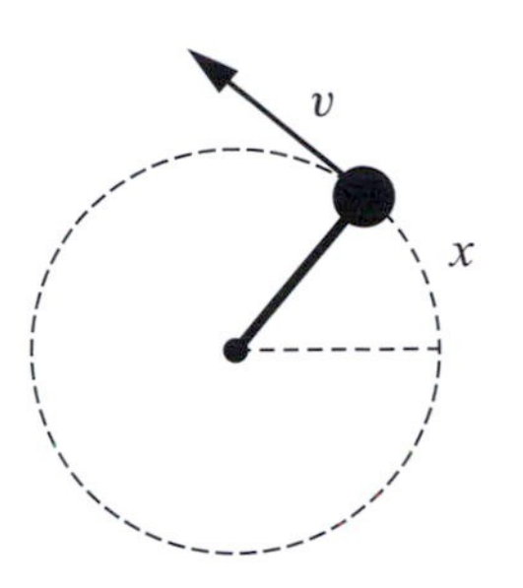

Fig. 7.6. Kicked rotator. A ball fixed to the end of a thin rod of unit length moves on a circular orbit, which is affected by instantaneous momentum transfers occurring with period T.

Since we are dealing with a free kicked motion, the result of Appendix A.1.5 applies:

$$x_{n+1} = x_n + v_n T, \qquad v_{n+1} = v_n + uI(x_{n+1}). \tag{7.7}$$

Variables x_n and x_{n+1} are angles; they represent the same position when shifted by 2π. Measuring velocity in units of u and distance in units of uT, the dimensionless form of the kicked rotator map is given by $(I \equiv f)$

$$x_{n+1} = x_n + v_n, \qquad v_{n+1} = v_n + f(x_{n+1}). \tag{7.8}$$

Problem 7.5 Derive the dimensionless map of the kicked rotator for the before-kick co-ordinates.

Problem 7.6 Derive the map of the kicked rotator in the presence of a drag linearly proportional to the velocity, both for the before-kick and the after-kick co-ordinates.

An important special case of the rotator map is obtained with a sinusoidal kicking amplitude function,

$$f(x) = a \sin x, \tag{7.9}$$

where a is a dimensionless positive constant called the non-linearity parameter. Map (7.8) is then given by

$$x_{n+1} = x_n + v_n, \qquad v_{n+1} = v_n + a \sin x_{n+1}. \tag{7.10}$$

This is the *standard map*, which plays an important role in understanding chaos in conservative systems.

Problem 7.7 Show that the standard map also describes the motion of a charged rotator placed into a homogeneous electric field switched on periodically for very short times.

Box 7.2 Connection between maps and differential equations

The example of the kicked rotator is well suited for exploring the connection between differential equations and maps. With a sinusoidal amplitude function (7.9), the dimensional stroboscopic map, (7.7), is as follows:

$$x_{n+1} = x_n + v_n T,$$
$$v_{n+1} = v_n + ua \sin x_{n+1}.$$

After rearrangement,

$$\frac{x_{n+1} - x_n}{T} = v_n,$$
$$\frac{v_{n+1} - v_n}{T} = \frac{ua}{T} \sin x_{n+1}.$$

In the limit $T \to 0$, the left-hand sides of the equations become the time derivative of x and v. If amplitude a is proportional to T, i.e. a/T is kept constant while taking the limit, this is equivalent to

$$\ddot{x} = \frac{ua}{T} \sin x,$$

which is the continuous-time equation of motion of a pendulum (see Problem 3.1 with $g/l \to ua/T$). Momentum transfers of decreasing strength and of increasing frequency indeed correspond to a continuously acting force. The point $x = 0$ provides the unstable equilibrium position of the pendulum (a pencil standing on its point).

The fact that the map belonging to finite values of T is chaotic is a warning concerning numerical solutions of differential equations: if the time step h (see Appendix A.3) of the numerical simulation is chosen to be too large, one may obtain erroneous results. This can make regular motion appear chaotic. More generally, the dynamics of a stroboscopic map of period T essentially differs from that of the differential equation obtained in the formal limit $T \to 0$ of the map. In a two-dimensional phase space, differential equations can describe regular motion only, while two-dimensional maps are usually chaotic.

7.3.2 Fixed points of the standard map and their stability

The condition for the existence of a fixed point (x^*, v^*) of (7.10) for $a \neq 0$ is as follows:

$$x^* = x^* + v^*, \quad v^* = v^* + a \sin x^*. \tag{7.11}$$

This yields $v^* = 0$ and $\sin x^* = 0$. A fixed point has to be a state of rest, and it may therefore only exist in positions where the body is not exposed to kicks. Two positions are therefore allowed, with co-ordinates $x_+^* = 0$ and $x_-^* = \pi$. (Obviously, the values shifted by 2π are formally also solutions, but these correspond to the same positions along the circle.)

The stability matrix, (4.36), of the fixed point $H_+ = (0, 0)$ is given by

$$L = \begin{pmatrix} 1 & 1 \\ a & 1+a \end{pmatrix}, \tag{7.12}$$

with eigenvalues

$$\Lambda_{\pm} = \frac{2 + a \pm \sqrt{(2+a)^2 - 4}}{2}, \tag{7.13}$$

which are positive for $a > 0$. As Λ_+ is greater than unity, the fixed point H_+ is always hyperbolic.

Problem 7.8 Determine the directions of the stable and unstable manifolds around H_+.

The stability matrix of the fixed point $H_- = (\pi, 0)$ is obtained from (7.12) by replacing a by $-a$. For $a < 4$, its eigenvalues are complex with unit absolute values. Thus, fixed point H_- is stable, elliptic for not too large values of a, but loses stability for $a > 4$ (see Table 4.2). For sufficiently strong kicks, even states with vanishing kick amplitudes ($x = 0$ and π) are unstable, because the momentum transfer changes sufficiently rapidly in their vicinity.

The points $P_1 = (0, \pi)$ and $P_2 = (\pi, \pi)$ represent a two-cycle of the standard map, which is stable for weak kicks. Since the velocity is not zero here, this cycle corresponds to a rotation of the rod (in the positive direction). Rotation in the opposite direction is described by the two-cycle $P_1' = (0, -\pi)$, $P_2' = (\pi, -\pi)$.

Problem 7.9 Show that $P_1 = (0, \pi)$, $P_2 = (\pi, \pi)$ is a two-cycle. For what values of a is it stable?

7.3.3 Discrete-time dynamics of the free rotator

When the rotator is not exposed to kicks ($a = 0$), it rotates with constant velocity: $v_n =$ constant. The position co-ordinate changes by the same amount in each step of map (7.10). How this simple dynamics manifests itself depends curiously on the precise value of the velocity. If, for example, the velocity is exactly π, we observe a two-cycle. Its position is arbitrary, and therefore orbits starting from any point along the line $v_n = \pi$ are two-cycles. (The cycle is marginal; the eigenvalues of the linearised dynamics are $\Lambda_{\pm} = 1$.) Similarly, the velocity values $2\pi/q$ (q being a natural number) result in q-cycles, and motion belonging to velocities of absolute values $v_n = 2\pi r$ ($r < 1$) is also periodic if r is rational (i.e. $r = p/q$, where p and q are natural numbers). If, however, r is irrational, the point does not return exactly to its initial position after any finite number of steps. The orbit, no matter what the initial co-ordinate x is, traces out the entire line segment $v_n = 2\pi r$ (see Fig. 7.7(a)).[3]

[3] Such trajectories represent quasi-periodic motions of the map (see Section 7.5.1).

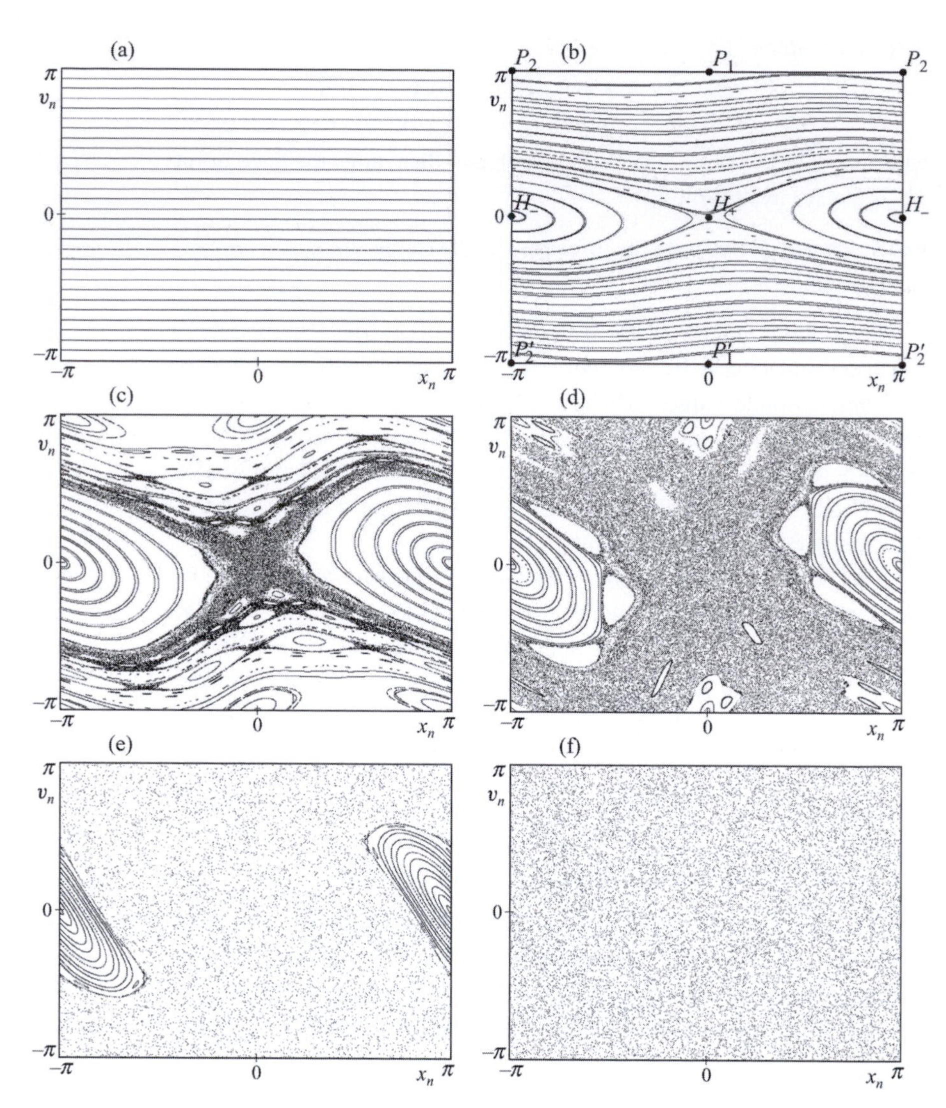

Fig. 7.7. Phase portraits of the kicked rotator for different kicking strengths. The parameter values are: (a) $a = 0$, (b) $a = 0.25$, (c) $a = 0.9$, (d) $a = 1.5$, (e) $a = 3$ and (f) $a = 10$. In all cases, the initial conditions are $x_0 = 0$ and $x_0 = -3.1$, with values of v_0 dividing the range $(-\pi, \pi)$ into 30 equal parts. The number of iterations is 5000. The closed loops and the curves crossing the entire interval $(-\pi \leq x_n < \pi)$ correspond to oscillations around stable cycles and to the turning over of the rotator, respectively.

7.3.4 Chaos in the kicked rotator

In the presence of kicks ($a \neq 0$), the only fixed points and two-cycles are those determined in Section 7.3.2. In accordance with elliptic behaviour, closed trajectories surround the fixed point H_- and the two-cycles as long as a is not too large. This implies that the rotator oscillates with some amplitude around the stable equilibrium state (similar to the way in which a compass needle oscillates around the north–south direction). Trajectories further away from the stable orbits are curves defined over the entire x-range, and correspond to the rotator's turning over (with non-uniform velocity). For small values of a (Fig. 7.7(b)), the manifolds of the unstable fixed point H_+ look like simple separatrices, which separate turnovers and oscillations.

For stronger kicks, however, it becomes visible that the 'separatrices' are not sharp curves. In their neighbourhoods, points of a single trajectory visit a whole two-dimensional region. This region is a chaotic band, in which points jump in a random fashion with uniform asymptotic distribution. Such bands exist also in other parts of the phase space, typically around separatrices of higher-order hyperbolic cycles (Fig. 7.7(c) and Plate XX).

When the non-linearity parameter is increased further, the chaotic bands widen. Oscillations around elliptic orbits are still present; some of the smooth trajectories corresponding to turnovers, however, are destroyed. They are replaced by a chain of tiny elliptic islands, with thin chaotic bands between them. Each chaotic band is bounded: the velocities cannot take on arbitrarily large values in the course of iterations.

At a certain critical value, a_c, however, even the last smooth curve corresponding to turning over disappears. This critical value is found numerically to be $a_c = 0.972$. For kicks stronger than this, the interesting situation arises that the velocity may take on arbitrary values. The total area of elliptic islands decreases as a increases (Figs. 7.7(d) and (e)). For sufficiently large values of a, all islands disappear, and the entire phase space becomes a single chaotic band (Fig. 7.7(f)). Then the rotator, starting from any initial condition chosen at random, moves chaotically. Even a single trajectory visits the entire phase space, and the distribution of the points becomes uniform. It is in this limiting case that the dynamics is similar to that of the area preserving baker map.

Box 7.3 Chaotic diffusion

For parameter values greater than a_c, nothing limits the velocity of the rotator. In every step it receives kicks of strength $a \sin x_{n+1}$, depending on the position co-ordinate, x_{n+1}. Since for large a the distribution of these values is uniform, the kicking strengths may take on arbitrary values within the interval $(-a, a)$. The velocity change corresponds therefore to a *random walk* along a straight line ($-\infty < v_n < \infty$). On starting a large number of points from an interval of length 2π on the v_n-axis, we find, accordingly, that the points spread out more and more along this axis (Fig. 7.8), just like a drop of ink in a fluid at rest. Chaotic dynamics thus generates a *diffusion process*.

The diffusion coefficient is easy to estimate. In a random walk with displacement r_i in step i (r_i is a real number) the total displacement in n steps is $R_n = \sum_{i=1}^{n} r_i$. If the r_i's can be considered as identical, independent variables of zero average, the average displacement is $\overline{R_n} = 0$, but the mean square displacement, $\overline{R_n^2} = \overline{(\sum_{i=1}^{n} r_i)^2}$, is non-zero (the average is taken over all the initial conditions). The average of the mixed terms in the full square is zero because of independence: $\overline{r_i r_j} = 0$ if $i \neq j$. Moreover, since the averages of the individual squares are equal, the mean squared displacement is proportional to the number of steps: $\overline{R_n^2} = \overline{r_i^2} n$. The usual definition of the diffusion coefficient, D, of a random walk is $\overline{R_n^2} = 2Dn$, which yields $D = \overline{r_i^2}/2$.

Applying this to the kicked rotator, where the random walk takes place along the velocity axis and the velocity jump is $r_i = a \sin x_{i+1}$, we obtain the diffusion coefficient

$$D = \tfrac{1}{2}\, \overline{a^2 \sin^2 x_{i+1}} = \tfrac{1}{4}\, a^2,$$

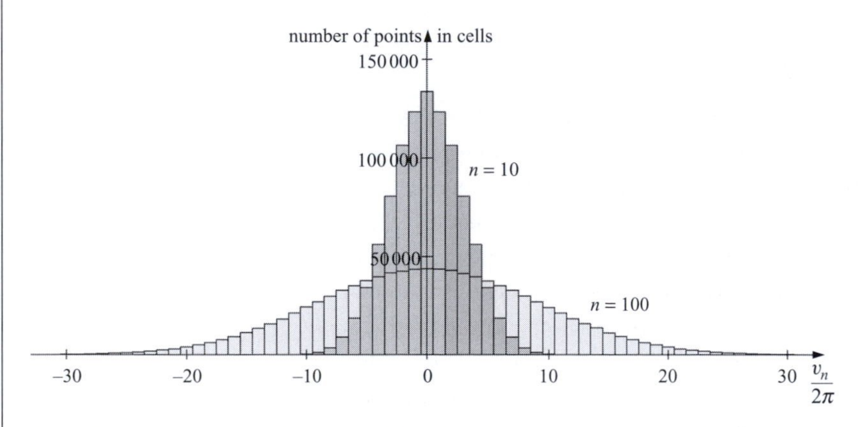

Fig. 7.8. Diffusion in the dynamics of the kicked rotator: distribution of 10^6 particles, started with a uniform distribution on the cell ($-\pi < x_0 < \pi, -\pi \leq v_0 \leq \pi$) along the v-axis after $n = 10$ and 100 steps ($a = 10$). The distribution is a bell-shaped one, typical of diffusion. The width of each column is 2π. The initial distribution at $n = 0$ (the middle column of height $10^6/(2\pi)$) is not displayed.

since the distribution of x is uniform, and the average of $\sin^2 x$ is $1/2$. This result is only valid in the limit of strong kicks ($a \gg 1$), because only in this case are the subsequent increments independent of each other.

The velocity $R_n \equiv v_n$ after n steps is the sum of the velocity increments; the mean square velocity is therefore

$$\tfrac{1}{2}\, \overline{v_n^2} = D\, n,$$

implying that the average kinetic energy increases linearly in time.

This shows that a motion of deterministic origin observed for a sufficiently long time can generate exactly the same process as external noise. In the case of Brownian motion, which is a prototype of diffusion, the force acting on the particle is noise-like and the equation of motion is stochastic, in contrast to our example, where kicking is described by a simple, completely *deterministic*, dynamics. Nevertheless, statistically they lead to the same result. Diffusion is *not* sensitive to the origin of the random behaviour.

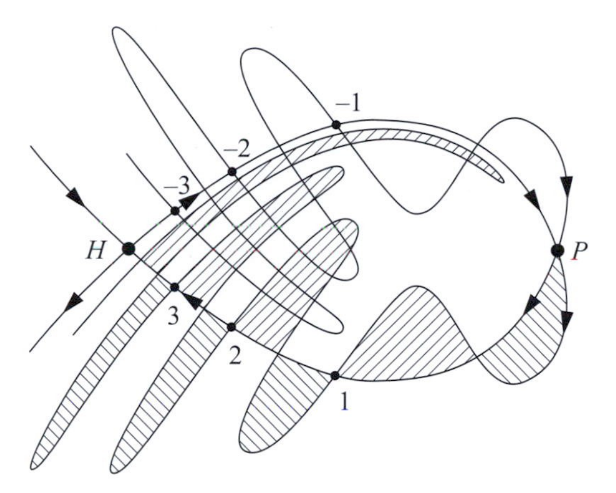

Fig. 7.9. Schematic diagram of the stable and unstable manifolds of an area preserving map's fixed point, H. The consecutive images (pre-images) of the first homoclinic point, P, are denoted by 1, 2, 3 (−1, −2, −3). Domains with identical shading are images of each other, and have therefore equal areas.

7.3.5 Structure of chaotic bands

Let us study now in more detail the chaotic band around fixed point H_+ when the velocity v_n cannot yet take on arbitrarily large values ($a < a_c$).

Figure 7.9 schematically represents the manifolds of a typical fixed point. These are only slightly bent in the vicinity of the fixed point, but further away they exhibit increasingly meandering and oscillating behaviour. Thus, they unavoidably intersect each other somewhere in a *homoclinic* point. Following the branches of the manifolds emanating from the fixed point, the first intersection point is the *first* homoclinic point, P. Whichever manifold is followed further, more and more homoclinic points are found (see Box 6.3). Where the manifold returns to the vicinity of the fixed point, its oscillations are already very large. The reason for this is that the homoclinic points are denser close to the fixed point (their distance is multiplied by $|\Lambda_-| < 1$ in each step). Due to area-preservation, the area enclosed by the manifolds between two consecutive homoclinic points is the same as that of their image one step later. The 'tongues' therefore become increasingly long and the manifolds meander more and more (generating newer and newer homoclinic points).

This structure can be observed in Fig. 7.10, which shows a segment of the manifolds of $H_+ = (0, 0)$ after an increasing number of numerical iterations (the segment of the stable manifold must, of course, be iterated backwards in order to obtain a longer piece). Following the manifolds of increasing length, we find that they visit the *entire* chaotic band. This holds for both manifolds. The width of the band is determined by how strongly the manifolds oscillate. The skeleton of a chaotic band is thus provided by the manifolds of the hyperbolic fixed points within the band. In the band there may also naturally exist many higher-order cycles. These possess manifolds of a similar nature, which also provide

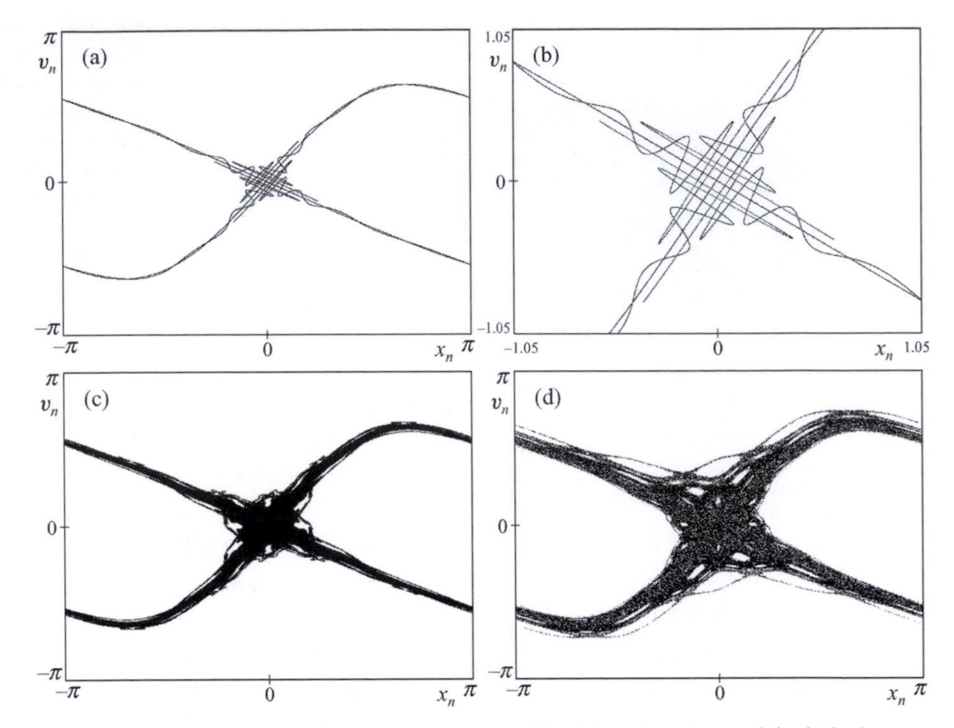

Fig. 7.10. Stable and unstable manifolds of fixed point H_+ of the kicked rotator ($a = 0.9$). (a) A square of size 0.00005, filled with 250 000 points and centred on the fixed point, has been iterated 18 times forwards and backwards. The images become so stretched along the manifolds that they return to the vicinity of the fixed point. (b) Blow-up of the region around the origin. The oscillation of the manifolds can be clearly observed. (c) Image of the manifolds after 50 iterations. (d) The entire chaotic band traced out numerically by a single trajectory.

a skeleton of the band by themselves. (Among manifolds of different cycles heteroclinic points are also generated; see Box 6.3.) A chaotic band is thus always the *union* of the stable and unstable manifolds of all the hyperbolic cycles it contains.

7.4 Autonomous conservative systems

7.4.1 Ball bouncing on a double slope

The simplest example of chaotic behaviour in non-driven conservative systems is provided by a ball bouncing elastically in a vacuum between two facing slopes of identical inclination, α. The description of the motion is based solely on the knowledge of oblique projection. Note that the velocity vector at the instants of the bounces should be recorded and a kind of Poincaré map should be constructed from this. The positions at

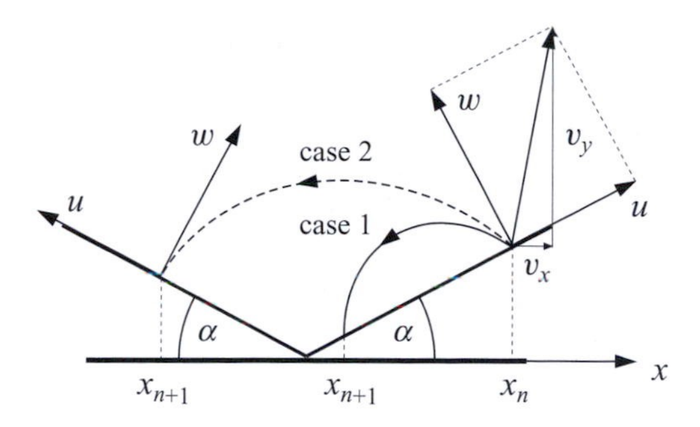

Fig. 7.11. Bouncing of a ball on a double slope. The velocity is given in terms of the perpendicular (w) and parallel (u) components. The rule connecting them depends on whether the ball remains on the same slope (case 1) or not (case 2).

the instants of the collisions then follow from the conservation of energy, while the paths between collisions are the well known parabola arcs.

Denoting the velocity components after the nth bounce by v_{xn} and v_{yn}, the displacement and velocity components evolve until the next collision according to

$$x(t) = x_n + v_{xn}t, \quad y(t) = y_n + v_{yn}t - \frac{g}{2}t^2, \tag{7.14}$$

$$v_x(t) = v_{xn}, \quad v_y(t) = v_{yn} - gt, \tag{7.15}$$

where t is the time elapsed since the bounce. Initially the ball is on the slope; its co-ordinates therefore fulfil $y_n/x_n = \tan\alpha$ (see Fig. 7.11). The dynamics basically differs according to whether the next bounce is on the same slope or not. In both cases, the impact velocity should be decomposed into components perpendicular and parallel to the slope, since these are the components in which the rule of elastic bounce is the simplest: the perpendicular component, w, changes sign, while the parallel one, u, remains unchanged.

Case 1

If the ball does not jump to the other slope, the time, t_n, elapsed until the next bounce follows from the condition that the impact point, $x_{n+1} \equiv x(t_n)$, $y_{n+1} \equiv y(t_n)$, lies on the original slope: $y_{n+1} = x_{n+1}\tan\alpha$. From (7.15),

$$gt_n = 2(v_{yn} - v_{xn}\tan\alpha). \tag{7.16}$$

Using this and (7.15), the velocity components, u and w, are obtained after the next impact as follows:

$$u_{n+1} = u_n - 2w_n\tan\alpha, \quad w_{n+1} = w_n. \tag{7.17}$$

This is the Poincaré map of bouncing on a single slope. It describes the steady increase of the parallel velocity; the perpendicular velocity does not change.

Problem 7.10 Derive map (7.17).

Case 2

Positioning the origin of the reference frame into the break-point between the two slopes, jumping over corresponds to the new co-ordinate x_{n+1} being negative. Since the ball hits the other slope, the new co-ordinates now fulfil $y_{n+1} = -x_{n+1} \tan\alpha$. Based on (7.15), this yields for the flight time, t_n, the quadratic equation:

$$\frac{g}{2}t_n^2 - (v_{xn}\tan\alpha + v_{yn})\,t_n - 2y_n = 0. \tag{7.18}$$

The solution is given by

$$gt_n = v_{xn}\tan\alpha + v_{yn} + \sqrt{D_n}, \tag{7.19}$$

where discriminant D_n is given by

$$D_n = (v_{xn}\tan\alpha + v_{yn})^2 + 4gy_n. \tag{7.20}$$

Co-ordinate y_n follows from the energy conservation:

$$gy_n + \frac{1}{2}\left(v_{xn}^2 + v_{yn}^2\right) = \mathcal{E}, \tag{7.21}$$

where $\mathcal{E}$ is the total energy per unit mass. The velocity components at the impact are obtained from (7.15):

$$u_{n+1} = -u_n + w_n\tan\alpha - w_{n+1}\tan\alpha, \tag{7.22a}$$

$$w_{n+1} = \sqrt{D_n}\cos\alpha. \tag{7.22b}$$

Problem 7.11 Derive map (7.22).

The mapping rule is made closed by taking the square of equation (7.22b) and applying (7.20), (7.21) and the transformation derived in the solution to Problem 7.10. Thus,

$$\begin{aligned} w_{n+1}^2 &= w_n^2(\cos^2 2\alpha - 2\cos^2\alpha) + u_n^2(\sin^2 2\alpha - 2\cos^2\alpha) \\ &\quad + u_n w_n \sin 4\alpha + 4\mathcal{E}\cos^2\alpha. \end{aligned} \tag{7.23}$$

We introduce the new variable, $z \equiv w^2$, which is proportional to the kinetic energy carried by the velocity component perpendicular to the slope. Thus, with (7.22), we have obtained the Poincaré map corresponding to jumping over to the other slope.

The entire map may be written in a dimensionless form by measuring the velocities in units of $\sqrt{2\mathcal{E}}$. Utilizing trigonometric identities, the dimensionless dynamics is given, in Case 1, by

$$u_{n+1} = u_n - 2\sqrt{z_n}\tan\alpha, \quad z_{n+1} = z_n, \tag{7.24}$$

and in Case 2 by

$$u_{n+1} = -u_n + \sqrt{z_n}\tan\alpha - \sqrt{z_{n+1}}\tan\alpha,$$
$$z_{n+1} = -z_n\left(1 + \frac{1}{2}\sin 4\alpha\tan\alpha\right) - u_n^2\frac{1}{2}\sin 4\alpha\,\mathrm{cotan}\,\alpha$$
$$+ u_n\sqrt{z_n}\sin 4\alpha + 2\cos^2\alpha. \tag{7.25}$$

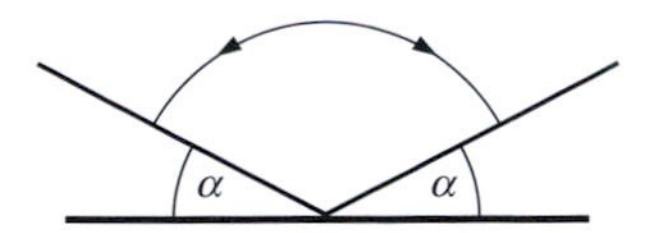

Fig. 7.12. Real space motion corresponding to the fixed point $(0, z^*)$ of the map.

Problem 7.12 Show that the ranges of the dimensionless variables are $| u_n | \leq 1$ and $0 \leq z_n \leq 1$, and that the constraint $z_n \leq 1 - u_n^2$ must hold in each step.

The mapping switches from (7.24) to (7.25) when the former loses its validity, i.e. when the new y_{n+1} computed from (7.24) becomes negative. Then the kinetic energy, $u_{n+1}^2 + w_{n+1}^2 = 1 - gy_{n+1}/\mathcal{E}$, measured in units of $\mathcal{E}$, would be greater than unity. Thus, map (7.24) is valid as long as the dimensionless kinetic energy expressed in terms of u_{n+1} and z_{n+1} is less than unity: $(u_n - 2\sqrt{z_n}\tan\alpha)^2 + z_n \leq 1$, i.e. $|u_n - 2\sqrt{z_n}\tan\alpha| \leq \sqrt{1 - z_n}$. Consequently, if the nth point is in the region[4]

$$u_n \geq 2\sqrt{z_n}\,\tan\alpha - \sqrt{1 - z_n}, \tag{7.26}$$

then (7.24) should be applied; otherwise (7.25) should be used. (The role of the boundary curve of inequality (7.26) is similar to that of the line $v_n = b$ for the baker map described by (7.5) and (7.6)).

The map can have a fixed point only if it appears in (7.25) (map (7.24) only describes bouncing downwards). The condition for this is that $u^* = 0$, i.e. that the impact is exactly perpendicular to the slope.

For the impact velocity $w^* = \sqrt{z^*}$, we obtain

$$z^* = \frac{\cos^2\alpha}{1 + (1/4)\sin(4\alpha)\tan\alpha} = \frac{\cos^2\alpha}{1 + \sin^2\alpha\cos(2\alpha)} = \frac{1}{2 - \cos(2\alpha)}. \tag{7.27}$$

In real space, the fixed point of the map corresponds to flying along a single symmetric arc between the two slopes (Fig. 7.12). Interestingly, the fixed point is stable only if the angle of the slopes is greater than 45°. For flat slopes the fixed point is hyperbolic.

Problem 7.13 Determine the map linearised around the fixed point.

Several higher-order cycles are, of course, also present in the map. Their stability depends sensitively on the inclination of the slopes. In addition to the case described in Section 1.2.3 (see Fig. 1.20, where $\alpha = 50°$), Fig. 7.13 presents a further example.[5]

[4] The condition $u_n - 2\sqrt{z_n}\tan\alpha \leq \sqrt{1 - z_n}$ is automatically fulfilled since $|u_n| \leq \sqrt{1 - z_n}$.

[5] We suggest that the reader explores the rich variety of the possible phase space structures by means of a computer program iterating the map.

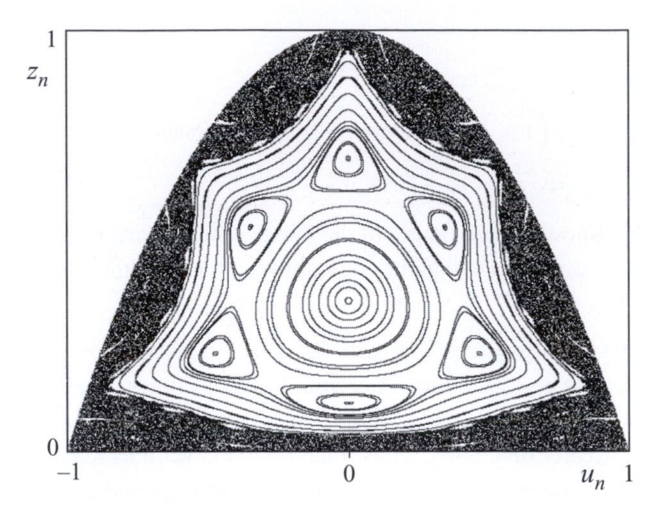

Fig. 7.13. Phase portrait of a ball bouncing on a double slope, in the plane $(u_n, z_n \equiv w_n^2)$, for $\alpha = 73°$ (1.274 radian). Trajectories begin from 30 different initial conditions.

Problem 7.14 Show that the dynamics is non-chaotic if the angle of each slope is 45°.

7.4.2 Spring pendulum

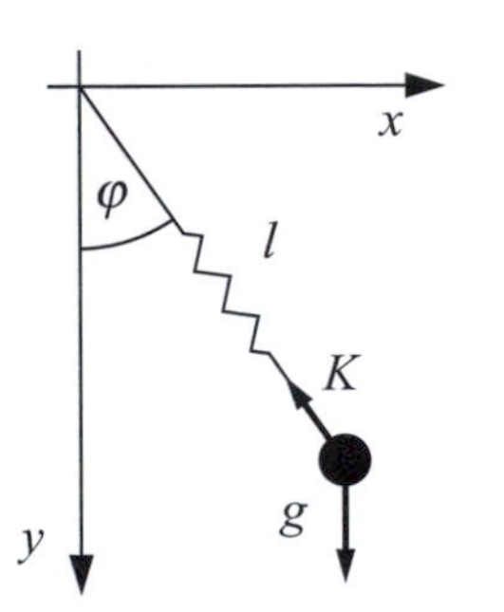

Fig. 7.14. Forces acting on the spring pendulum.

Consider a pendulum whose thread is a spring with rest length l_0 and natural frequency ω_0. This can also serve as a model of a body fixed to the end of a rubber thread; we assume that the thread never loosens. The spring pendulum can turn over.

Let the origin of the co-ordinate system be the point of suspension (Fig. 7.14). When the point of unit mass fixed to the end of the thread is at (x, y), the length of the spring is $l \equiv \sqrt{x^2 + y^2}$, and the force arising in it is $K = \omega_0^2(\sqrt{x^2 + y^2} - l_0)$. The horizontal and vertical components of this force are $K(-x/l)$ and $K(-y/l)$, respectively. Since gravity acts along the y-axis, the equations of motion are given by

$$\ddot{x} = -\omega_0^2 x \left(1 - \frac{l_0}{\sqrt{x^2 + y^2}}\right), \quad \ddot{y} = -\omega_0^2 y \left(1 - \frac{l_0}{\sqrt{x^2 + y^2}}\right) + g. \quad (7.28)$$

Another representation is obtained by using the spring's instantaneous length, l, and its angle, φ, as new variables ($x = l \sin\varphi$, $y = l \cos\varphi$):

$$\ddot{l} = l\dot{\varphi}^2 - \omega_0^2(l - l_0) + g\cos\varphi, \quad l\ddot{\varphi} = -2\dot{l}\dot{\varphi} - g\sin\varphi. \quad (7.29)$$

Measuring distance in units of the spring's rest length, l_0, and time in units of $1/\omega_0$ (see Appendix A.2.2), we obtain the dimensionless equations in rectangular coordinates as

$$\ddot{x} = -x \left(1 - \frac{1}{\sqrt{x^2 + y^2}}\right), \quad \ddot{y} = -y \left(1 - \frac{1}{\sqrt{x^2 + y^2}}\right) + q, \quad (7.30)$$

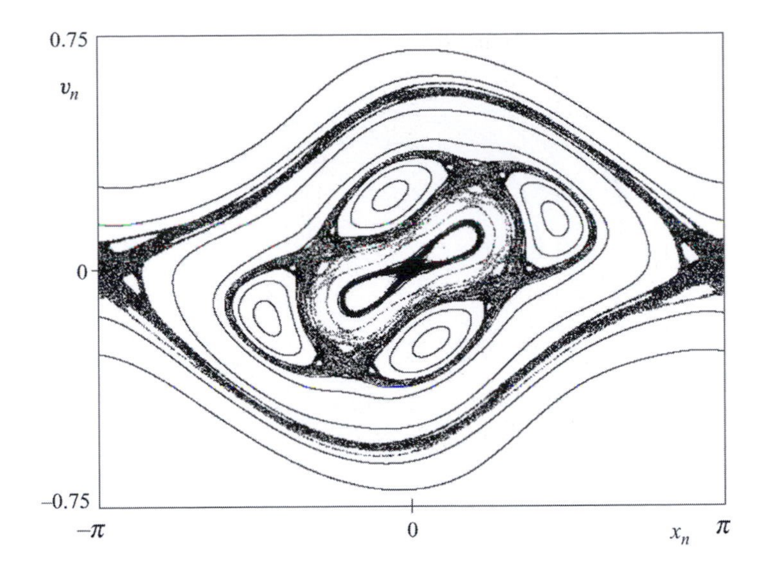

Fig. 7.15. Poincaré portrait, $x_n \equiv \varphi_n$, $v_n \equiv \dot{\varphi}_n$, of the spring pendulum obtained from eleven different initial conditions, with constant total energy $\mathcal{E} = \dot{l}^2/2 + l^2\dot{\varphi}^2/2 - ql\cos\varphi + (l-1)^2/2 = 0.195$; $q = 0.07$. The dynamics is that of a perturbed pendulum. In spite of the relatively small value of q, extended chaos already appears in phase space.

and

$$\ddot{l} = l\dot{\varphi}^2 - (l-1) + q\cos\varphi, \quad l\ddot{\varphi} = -2\dot{l}\dot{\varphi} - q\sin\varphi, \tag{7.31}$$

in polar co-ordinates. Here

$$q = \frac{g}{\omega_0^2 l_0} \tag{7.32}$$

is the only dimensionless parameter of the system: the square of the ratio of the spring's period to that of a pendulum of constant length.

A value for q of order unity indicates that the properties of a pendulum and of a spring are present in the system in about the same proportion. The coupling of these properties is highly non-linear; chaos is therefore at its strongest for such parameter values.

The Poincaré sections were constructed in two different ways. One was taken in the position when $l = l_0$, i.e. when the instantaneous length of the pendulum was equal to the rest length (with positive stretching velocity). The variables of the map are in this case the angular deflection, $x_n \equiv \varphi_n$, and the angular velocity, $v_n \equiv \dot{\varphi}_n$ (Fig. 7.15). In the other case, the map was defined by the condition $\varphi = 0$ ($\dot{\varphi} > 0$), and it was plotted in the plane of $x_n \equiv l_n$, $v_n \equiv \dot{l}_n$ (Fig. 7.16). Because of the choice $\varphi = 0$, this is equivalent to taking another Poincaré section, the section defined by $x = 0$, $\dot{x} > 0$ (on which $x_n \equiv y_n$, $v_n \equiv \dot{y}_n$) in the Cartesian representation (7.28). The phase portraits illustrate how elliptic and chaotic domains are embedded into each other.

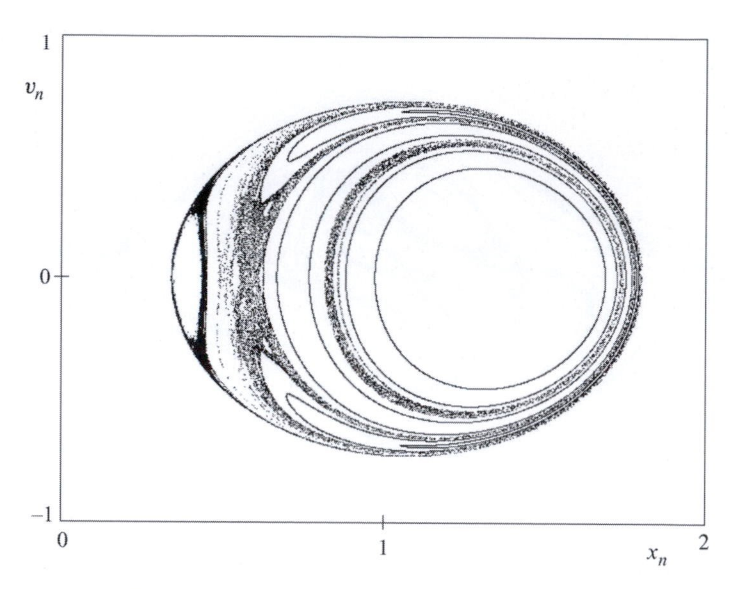

Fig. 7.16. Poincaré portrait, $x_n \equiv l_n$, $v_n \equiv \dot{l}_n$, of the spring pendulum with the same parameters and initial conditions as in Fig. 7.15. (The domains of different initial conditions, which succeed from the middle of Fig. 7.15 to its border, appear here in a different order.)

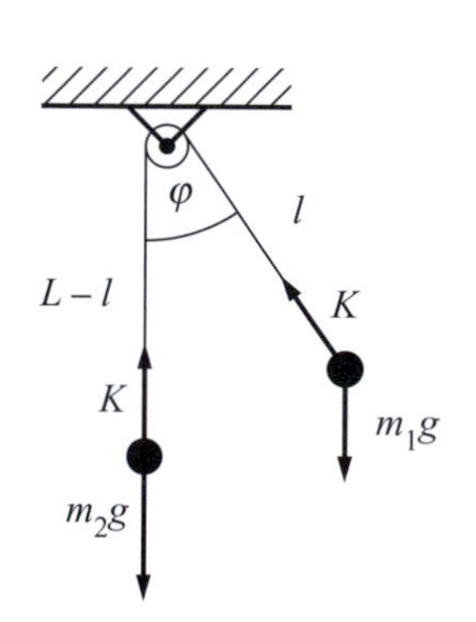

Fig. 7.17. Forces acting on the point masses at the ends of the thread (see also Fig. 1.15).

7.4.3 Body swinging on a pulley

Two point masses are fixed to the ends of an unstretchable thread of length L, wound around a fixed pulley with negligible radius and mass. The body of mass m_2 can only move vertically, while the other body, of mass m_1, can also swing in the vertical plane. The instantaneous position of the swinging point is given in terms of the angular deflection, φ, and the distance, l, measured from the pulley. Due to the constant length of the thread, the position of the other body is also uniquely determined.

When writing the equation of motion for m_1, it is worth using a frame co-rotating with this point. The instantaneous angular velocity is $\dot{\varphi}$; therefore, inertial forces also act on the body. The magnitude of the centrifugal force is $m_1 l \dot{\varphi}^2$, and it points outwards in the direction of the thread, along with component $m_1 g \cos\varphi$ of the weight (see Fig. 7.17). The acceleration in the direction of the thread is $\ddot{l}$, and therefore the radial component of the equation of motion is given by

$$m_1 \ddot{l} = m_1 g \cos\varphi + m_1 l \dot{\varphi}^2 - K, \qquad (7.33)$$

where K is the magnitude of the force in the thread. In the direction perpendicular to the thread, the other inertial force, the Coriolis force, acts in addition to the component $-m_1 g \sin\varphi$ of the weight. Since, in the co-rotating frame, m_1 moves radially with velocity $\dot{l}$, the magnitude

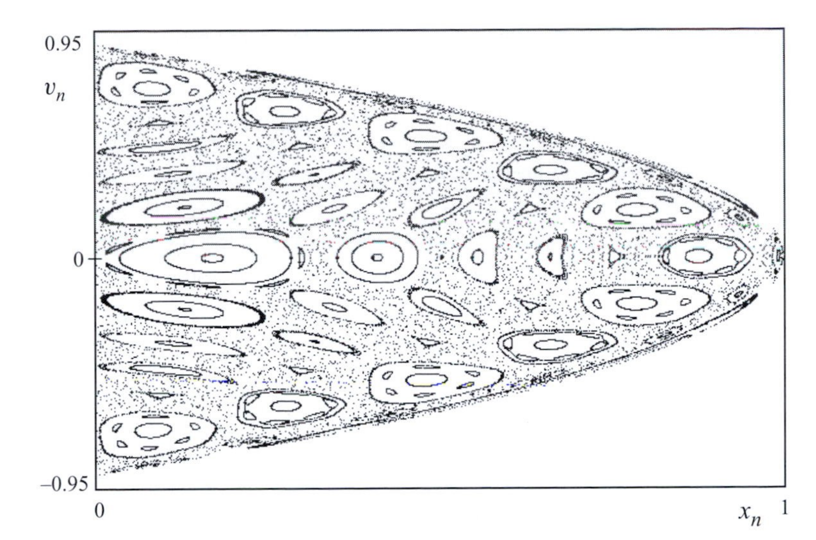

Fig. 7.18. Poincaré portrait, $x_n \equiv l_n$, $v_n \equiv \dot{l}_n$ ($\varphi = 0$, $\dot{\varphi} > 0$), of a body swinging on a pulley for mass ratio $\mu = 0.3$. The dimensionless total energy is $\mathcal{E} = \dot{l}^2/2 + \mu l^2\dot{\varphi}^2/2 - \mu l \cos\varphi - (1-\mu)(1-l) = -0.3$.

of the Coriolis force is $2m_1\dot{l}\dot{\varphi}$. For a positive angular velocity, $\dot{\varphi}$, it deflects the body to the *right* of the instantaneous velocity. Consequently, the equation of motion perpendicular to the thread is given by (see Fig. 7.17)

$$m_1 l\ddot{\varphi} = -m_1 g \sin\varphi - 2m_1\dot{l}\dot{\varphi}. \tag{7.34}$$

The equation for mass m_2 moving vertically is simply

$$m_2\ddot{l} = K - m_2 g, \tag{7.35}$$

since the thread is unstretchable and weightless, and the force in the thread is of equal magnitude at both ends.

Adding together equations (7.33) and (7.35) causes the forces K and $-K$ to cancel. Dividing this and (7.34) by the total mass, $m_1 + m_2$, we obtain the following equations:

$$\ddot{l} = \mu g \cos\varphi + \mu l\dot{\varphi}^2 - (1-\mu)g, \quad l\ddot{\varphi} = -g\sin\varphi - 2\dot{l}\dot{\varphi}, \tag{7.36}$$

where

$$\mu \equiv \frac{m_1}{m_1 + m_2}. \tag{7.37}$$

Measuring length in units of the total thread length, L, and time in units of $\sqrt{L/g}$ (which is proportional to the period of a pendulum of length L), we obtain the dimensionless equations

$$\ddot{l} = \mu\cos\varphi + \mu l\dot{\varphi}^2 - (1-\mu), \quad l\ddot{\varphi} = -\sin\varphi - 2\dot{l}\dot{\varphi}. \tag{7.38}$$

Figure 7.18 (and Plate VIII) shows a Poincaré portrait belonging to the mass ratio $\mu = 0.3$ (in Fig. 1.17, $\mu = 0.2$). It is clearly visible that elliptic and chaotic regions are embedded into each other.

Problem 7.15 The height of the vessel shown in the following figure can be expressed as follows:

$$V(x, y) = 0.5 - 10^{-2}(x^2 + 3y^2) + 10^{-4}(x^2 + 3y^2)^2 + 2 \times 10^{-3}x^2y^2,$$

where both the rectangular co-ordinates, x, y, and the height, V, is measured in centimeters (black and white bands indicate height differences of 0.5 cm). Show that the motion of a ball starting on the rim of height 5 cm with zero initial velocity is typically chaotic by plotting an orbit on the (x, y)-plane. Construct a Poincaré portrait. (For simplicity, consider the motion to be a motion in potential $V(x, y)$.)

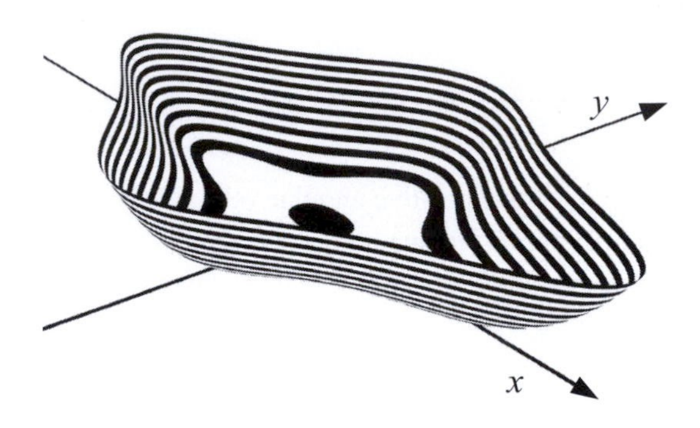

7.5 General properties of conservative chaos

7.5.1 The role of tori

The closed phase space curves displayed in Figs. 7.13, 7.15, 7.16 and 7.18 correspond in the three-dimensional flows to 'tube-like' smooth surfaces called *tori* (see Fig. 7.19(a)). These examples also illustrate that tori are typical formations of the phase spaces of conservative systems, even in most of the chaotic cases. One of the reasons for their ubiquity is that stable cycles are elliptic: oscillating motion develops in their close neighbourhood, but, due to the lack of friction, the point remains at a finite average distance from the periodic orbit. Flow trajectories then move on a surface surrounding the stable periodic orbit in the form of a tube (see Figs. 7.19 and 7.20). Tori may also be present far away from the stable orbits. A torus itself is always an *invariant surface*, i.e. a surface whose points all stay within the surface in the course of the entire motion.

The period of the motion along the central line of a torus is usually not equal to that winding around the torus surface (in Fig. 7.19(a) the motion in the vertical plane and that perpendicular to it are examples of this decomposition). The motion on tori is a combination of *two* types of periodic motion of different frequencies: Their ratio, the *winding number*

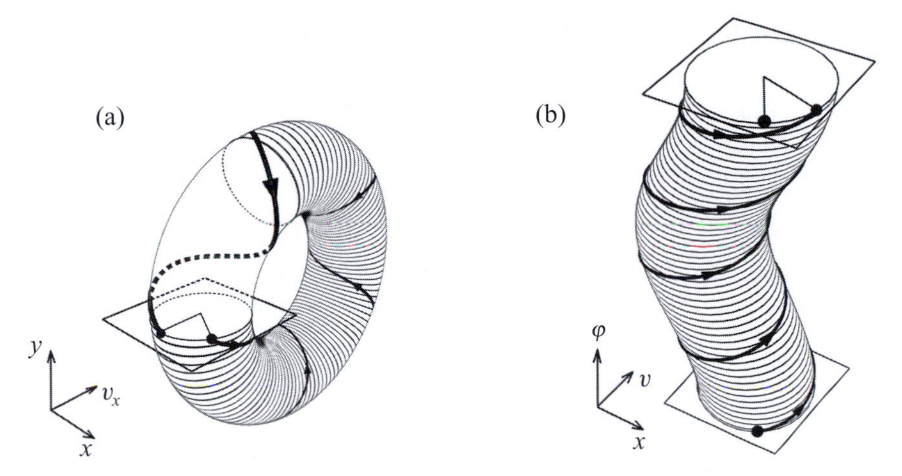

Fig. 7.19. Tori of conservative systems. (a) A torus in the phase space of an autonomous two-dimensional system. (b) A torus in the phase space of a driven one-dimensional system. The circular arc between the two black dots is the angular deviation, $\Delta\theta_1$, of a trajectory running on the torus after one iteration of (a) a Poincaré map and (b) a stroboscopic map.

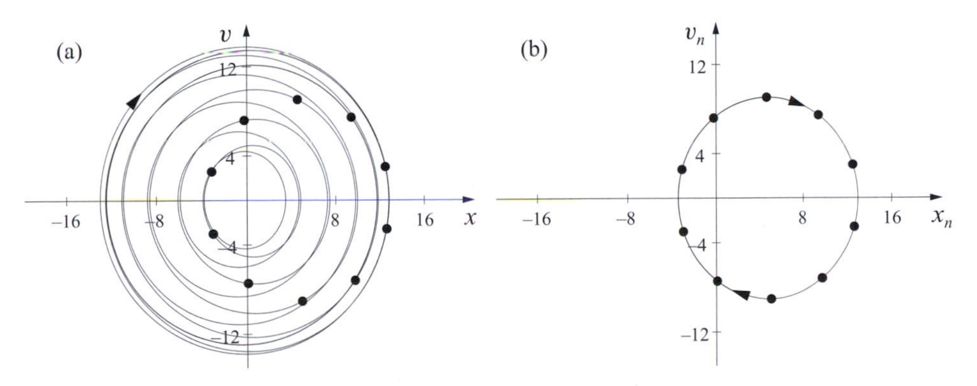

Fig. 7.20. Torus of a driven frictionless harmonic oscillator: $\ddot{x} = -\omega_0^2 x + f_0 \cos(2\pi t/T)$. The motion is represented (a) in the (x, v)-plane (the projection of a trajectory running on a torus as in Fig. 7.19(b)), and (b) in the stroboscopic map. The point turns in the map by the same angle in each step. The initial condition is the same as in Fig. 4.5; the parameters are $\Omega = 1, \omega_0 = 1.1$ and $\alpha = 0$; $E = 1$.

$\nu \equiv \omega_1/\omega_2$, is the same on the entire torus surface. The dynamics on a torus is, of course, never chaotic.

Tori are represented in maps (both stroboscopic and Poincaré) by *continuous curves*: the intersection of the torus surface and the plane of the map. Hereafter, continuous curves appearing in area preserving maps will *also* be called tori. Tori of maps are *invariant curves*, which are mapped onto themselves. For a trajectory that intersects the plane of

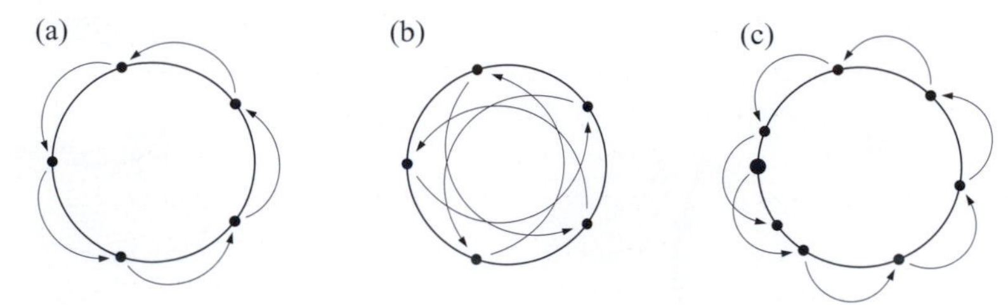

Fig. 7.21. Motion on tori of maps (schematic diagram). (a) Rational torus ($\nu = 1/5$); (b) rational torus ($\nu = 3/5$); (c) irrational torus ($\nu = \sqrt{2}/9$).

the map with period $T = 2\pi/\omega_2$, the angle of turn along the map's torus is given by

$$\Delta\theta_1 = \omega_1 T = 2\pi\nu = 2\pi\frac{\omega_1}{\omega_2}. \tag{7.39}$$

As far as maps are concerned, the winding number, ν, is the ratio of the angle of turn along the torus curve per iteration to the full angle (2π) (see also Fig. 7.19). This can clearly be seen in the example of the sinusoidally driven frictionless oscillator of natural frequency ω_0, where $T = 2\pi/\Omega$ is the driving period, $\omega_1 \equiv \omega_0$, and $\nu = \omega_0/\Omega$ (see Fig. 7.20).

Problem 7.16 Determine the winding number from the eigenvalues, $\Lambda_\pm$, of the dynamics linearised around elliptic fixed points of two-dimensional area preserving maps.

The motion on some of the tori may correspond to sliding over the entire co-ordinate space (in the examples of the kicked rotator and the spring pendulum it corresponds to turning over). In such cases, the curves representing the tori in maps are not closed loops (cf. Figs. 7.7(a) and (b) and 7.15).

The motion on tori is fundamentally different for rational than for irrational winding numbers. In the rational case, when the winding number can be written as $\nu = p/q$ (where p and q are natural numbers), the motion is *periodic* for any initial condition on the torus. Each discrete-time trajectory then consists of q points, and the order in which they are visited depends on p. If, for example, $p = 1$, the points follow one by one in a certain direction, while for $p = 3$ every third point is taken (see Figs. 7.21(a) and (b)). Trajectories beginning from different points on the torus perform the same periodic motion; their union traces out the closed curve of the map.

For an irrational winding number, when $\nu = \omega_1/\omega_2$ cannot be written as p/q, the motion is *not* periodic, and it never returns exactly to the initial position (see Fig. 7.21(c)). Such a motion is therefore called *quasi-periodic*. (A single trajectory is then sufficient to trace out the entire curve of the torus in the map.) There are, however, approximate returns. The value of the winding number can be read off from observing several close returns. Any point of the torus returns into a small neighbourhood of its initial position, on average after $1/\nu$ iterations.

According to the rational or irrational character of its winding number, a torus is henceforth called rational or irrational. It is always the higher frequency that is considered to be ω_2, and the winding number is thus at most unity. In fact, any point of the interval $(0, 1]$ can represent a winding number.

The role of irrational tori depends on *how irrational* their winding numbers are. Interestingly, the measure of a number's irrationality can be determined. This is possible because every positive irrational number, σ, less than unity can be expanded in a continued fraction, i.e. it can be written in the form

$$\sigma = \cfrac{1}{a_1 + \cfrac{1}{a_2 + \cfrac{1}{a_3 + \ldots}}} \equiv [a_1, a_2, a_3, \ldots], \tag{7.40}$$

where the a_i values are natural numbers. The expansion is unique, and is obtained by subtracting from $1/\sigma$ its integer part (a_1), then taking the reciprocal of the result and subtracting its integer part (a_2), and so on. If we stop after n steps in the expansion ($a_{n+1} = a_{n+2} = \ldots = 0$), we obtain a rational approximant to the irrational number. Obviously, the convergence to a given irrational number is slower the *smaller* the numbers, a_i, are. The numbers considered to be 'most irrational' are those that are the most difficult to approximate by rational numbers, i.e. whose continued fraction contains the most numeral ones.

Problem 7.17 Determine the first five elements of the continued fraction representation for the following numbers: $\pi - 3$, $e - 2$, $\sqrt{2} - 1$, $\sqrt{3} - 1$ and $(\sqrt{5} - 1)/2$. What is the deviation of these rational approximants from the respective numbers?

In this spirit, the 'most irrational' number is the *golden mean*, $g = (\sqrt{5} - 1)/2 = 0.618$. The expansion of g only contains '1' because it is the solution of the quadratic equation $g^{-1} = 1 + g$. Only a little less irrational are the so-called *noble* numbers: the continued fraction expansion of the kth noble number is $g_k = [k, 1, 1, 1, \ldots]$ $(k > 1)$.

Problem 7.18 Determine the first few noble numbers.

7.5.2 The KAM theorem

We now investigate what happens to a non-chaotic conservative system when it is perturbed by an external effect. Non-chaotic conservative

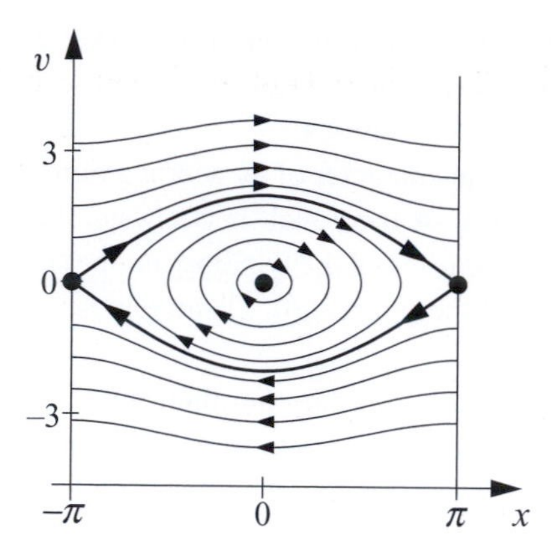

Fig. 7.22. Phase portrait of an integrable system (a simple pendulum) with a periodic position co-ordinate (the angle).

systems are called *integrable*. The phase spaces of these integrable systems are characterised by elliptic orbits surrounded by tori. A simple example is the pendulum (see Fig. 7.22). Even if a few hyperbolic orbits exist, their stable and unstable manifolds constitute a special torus that leads from one hyperbolic point to the other. This special torus is a separatrix between periodic or quasi-periodic motion of different kinds. Although the phase space of the unperturbed pendulum is two-dimensional, we might add a third direction along which nothing yet happens. The phase space surfaces are then shells of cylinders whose bases are the trajectories of the two-dimensional flow ($x \equiv \varphi,\ v \equiv \dot{\varphi}$). A weak perturbation, for example the pendulum thread becomes slightly elastic, modifies the behaviour in the (x, v)-plane and terminates the independence from the third co-ordinate (see Fig. 7.15).

The responses of rational and irrational tori to perturbative effects *differ* fundamentally. In the rational case (periodic dynamics), any point of the unperturbed map returns exactly to its starting position; perturbative effects can therefore *accumulate*. Along irrational tori (quasi-periodic dynamics), there is, however, no exact return, and perturbative effects might *average out*. We thus expect a kind of resonance to occur on every rational torus (and on nearby irrational ones): deviations from the unperturbed dynamics increase in time, and lead to the *destruction* of the torus. Rational tori are therefore also called *resonant* ones. No invariant curves similar to the original ones remain in the neighbourhood of destroyed tori, motion can no longer be quasi-periodic everywhere; these are the regions in which chaos appears. Thus, the stability of a torus against perturbations depends on whether the dynamics on it is sufficiently aperiodic, i.e. whether the torus is *sufficiently irrational*. The destruction of a few tori can clearly be seen in Fig. 7.7(b), where some smooth curves are replaced by a chain of short segments.

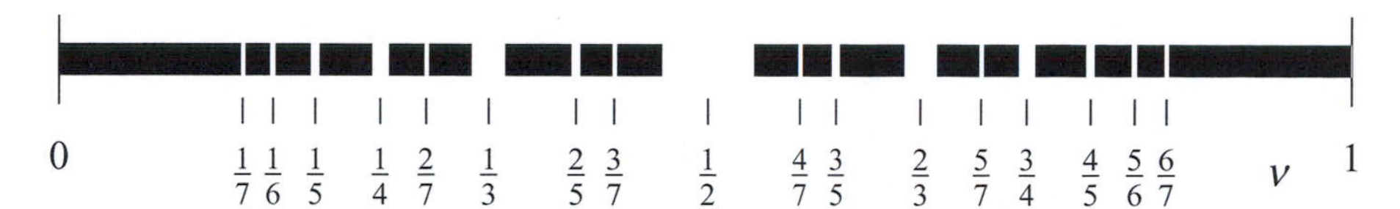

Fig. 7.23. KAM theorem: winding number intervals of surviving tori (black segments) represented on the interval $0 < \nu \leq 1$ of all possible winding numbers ($K = 0.2$). The removed intervals belonging to fractions with denominators greater than seven are so short that they cannot be seen in this resolution. For the sake of clarity, the removed interval around $\nu = 1$ is not marked.

The condition of the destruction of tori is formulated in precise form by the Kolmogorov–Arnold–Moser (KAM) theorem. This is valid for *weak* perturbations described by sufficiently smooth functions (the perturbation strength is measured by a dimensionless number, $\varepsilon \ll 1$). The KAM theorem states that tori are not destroyed by perturbations whose winding number, ν , satisfies

$$\boxed{\left|\nu - \frac{r}{s}\right| > \frac{K(\varepsilon)}{s^{5/2}}} \tag{7.41}$$

for *any* rational approximant, r/s, of the winding number ν. The constant, K, depends on the perturbation strength, ε, only, and tends to zero in the limit $\varepsilon \to 0$. Rational tori, and irrational ones whose winding number can be well approximated by rational numbers, typically do not survive. The preserved tori are the very irrational ones whose winding numbers can accurately be approximated by rational numbers with very large denominators only. The surviving tori are often called *KAM tori*.

According to condition (7.41), in the winding number interval $(0, 1]$ there exists a band of width $2K(\varepsilon)s^{-5/2}$ around every rational number, r/s (see Fig. 7.23), where perturbations generally destroy all the tori. (What exactly appears in place of the resonant tori will be discussed in Section 7.5.3.) Since there exist s different fractions with denominator s ($r = 1, 2, \ldots, s$), the length of the ν-intervals around rational winding numbers with denominator s that do not fulfil the KAM condition can be estimated as $2sK(\varepsilon)/s^{5/2} = 2K(\varepsilon)/s^{3/2}$. The total length of such intervals is given by

$$2\,K(\varepsilon)\sum_{s=1}^{\infty} s^{-3/2}. \tag{7.42}$$

The infinite series converges,[6] and since $K(\varepsilon)$ is small for small ε, the total length is always less than unity. In spite of the fact that an infinity

[6] The limiting value is 2.612.

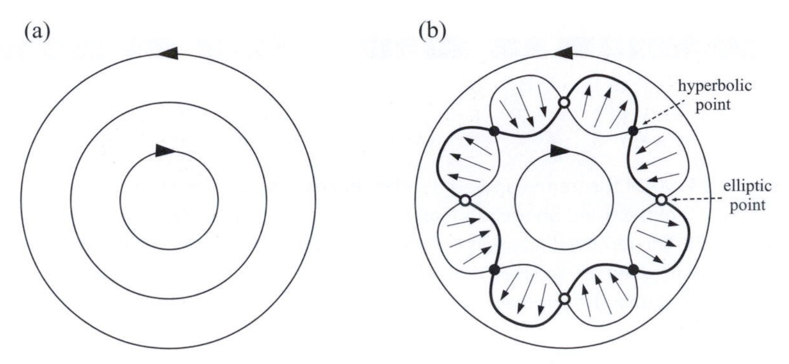

Fig. 7.24. The neighbourhood of a rational (resonant) torus. (a) Integrable system. The middle curve is a rational torus with winding number $\nu = r/s$; the inner and outer curves represent irrational tori. The arrows mark the displacement directions under the s-fold iterated map. (b) Perturbed system. The inner thin line represents a curve which moves under the s-fold iterated perturbed map in the radial direction only, as marked by the arrows. The bold line is the image of the thin line; the intersection points of the two curves are fixed points. The innermost and outermost curves represent surviving irrational tori, i.e. KAM tori (their deformation is, for simplicity, not indicated).

of tori and their neighbourhoods are destroyed, the majority of the tori survive for small ε.

The scenario found in the winding number interval faithfully reflects what happens in phase space: the surviving tori fill a *large* proportion of the entire phase space. Even though for small values of ε the phase space volume of the destroyed tori is small, it is not zero, which implies that chaos may appear due to an *arbitrarily weak* external perturbation on an originally non-chaotic system.[7] This is the mathematical background of the statement that chaos is not an exceptional, rather a *typical* temporal behaviour.

7.5.3 Remnants of resonant tori (microscopic chaos)

Consider a rational torus whose winding number is of the form $\nu = r/s$. Each point of the torus is a fixed point of the s-fold iterated unperturbed map. In general, the winding number changes continuously; therefore, on one side (say outside) of the rational torus there are tori with winding numbers greater than ν. We select a strongly irrational nearby torus on each side (see Fig. 7.24). Therefore, in the s-fold iterated map the points

[7] It may, for example, appear in the dynamics of a planet orbiting a not perfectly spherical celestial body, because the force is not exactly central (and the planet's angular momentum is not a conserved quantity).

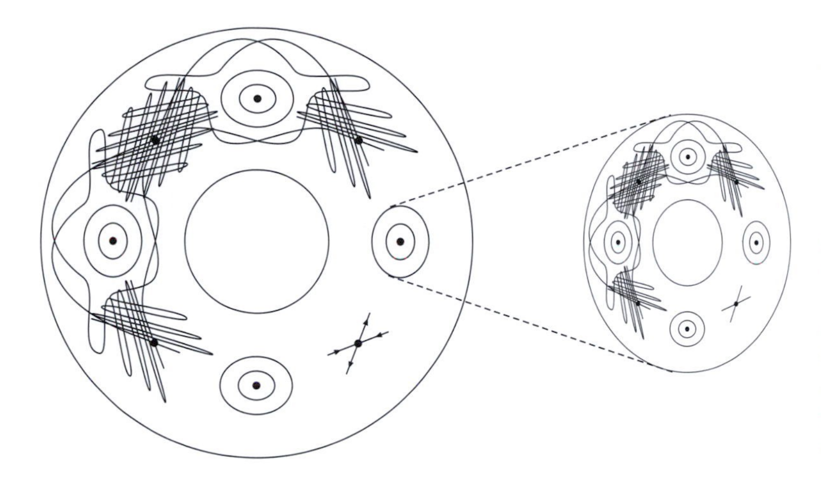

Fig. 7.25. Remnants of a destroyed resonant torus: a chain of elliptic and hyperbolic points (schematic diagram) around which quasi-periodic and chaotic motion dominates, respectively. Only a few of all the oscillating manifolds are shown. For better visibility, the extension of chaos (the amplitude of the oscillations) is enhanced.

on the outer (inner) irrational torus move anti-clockwise (clockwise), as indicated schematically in Fig. 7.24.

Under the effect of a weak perturbation, the irrational tori, according to the KAM theorem, remain as invariant curves (and become slightly deformed), but their winding direction does not change. The rational torus, however, is typically destroyed and ceases to exist as an invariant curve. The question is whether there remain any invariant points when this occurs. The original rational torus can be considered as a curve whose points do not rotate in the s-fold iterated map. Owing to continuity, in the vicinity of the original location of the torus there must exist points at which no rotation takes place, even in the presence of perturbation. These points can only move radially under the action of the s-fold iterated perturbed map. The union of all these points (a non-invariant curve) is marked by a thin solid line in Fig. 7.24(b). Due to area preservation, the image of this curve must be a curve that intersects the original one. The intersection points, whose number is even, are the *fixed points* of the s-fold iterated perturbed map. The number of such fixed points is, therefore, at least $2s$, but can also be an integer multiple of this. The stability of the fixed points can be read off from the displacement arrows around the intersection points and the direction of motion along the preserved tori. The remnant of a destroyed rational torus is a chain of elliptic and hyperbolic points alternately following each other. This is the *Poincaré–Birkhoff theorem*.

Hyperbolic points are the sources of chaos. As we have seen, generally speaking nothing keeps the stable and unstable manifolds from intersecting each other (Fig. 7.25). (In the original integrable case, the coincidence of the manifolds is due to a kind of inherent symmetry that ceases to exist when perturbation is switched on.)

The presence of homoclinic and heteroclinic points implies chaoticity (see Box 6.3). In the debris of the resonant torus thin chaotic bands therefore also appear along the manifolds of the hyperbolic points. The irrational tori act as barriers to chaos since they are invariant curves that cannot be crossed by trajectories in the plane of the map.

Smaller KAM tori are formed around the elliptic points of the debris. The chains of short segments in Fig. 7.7(b) represent, in fact, chains of small KAM tori whose vertical extension cannot be resolved. In these regions the motion is basically quasi-periodic, and the elements of the chains are called elliptic or regular islands. This is, however, not true everywhere, since our original argument applies here too: tori with rational winding numbers must be destroyed. Around them a chain of elliptic and hyperbolic points is formed again, but now on a smaller scale. If we magnify any of these small KAM tori, we see that the structure repeats itself.

7.5.4 Macroscopic chaos

The KAM theorem is valid for very weakly perturbed integrable systems. In a numerical simulation, such systems might appear to move everywhere regularly, and the value of the average Lyapunov exponent does not considerably differ from zero. Chaos – although present – is difficult to observe (microscopic chaos). For example, no extended chaos can yet be observed in Fig. 7.7(b).

When the perturbation parameter ε is increased, one generally finds that more and more KAM tori disappear,[8] and, consequently, the originally adjacent and separated chaotic bands merge. The proportion of the area of the elliptic islands inside them gradually decreases, and chaos becomes stronger.

A system is called *macroscopically chaotic* if chaotic bands fill a significant proportion of the entire phase space. We must not forget that, when this happens, the condition of the KAM theorem (weak perturbation) is generally no longer valid. For finite, but not too strong, perturbations, nevertheless, several elliptic islands are still present in the chaotic bands. Moreover, within each island, between two adjacent KAM tori, there are again chaotic bands and other elliptic islands, which are themselves similarly structured, and so on to arbitrarily small scales. The phase space structure then exhibits an easily observable self-similar character. This is clearly visible in Figs. 7.7(c), 7.10(d) and 7.15. Since chaotic and elliptic domains occupy a finite area larger than zero, the phase space structures constitute a *fat fractal* (see Section 2.2.3). The

[8] The surviving tori are called KAM tori, even for strong perturbations.

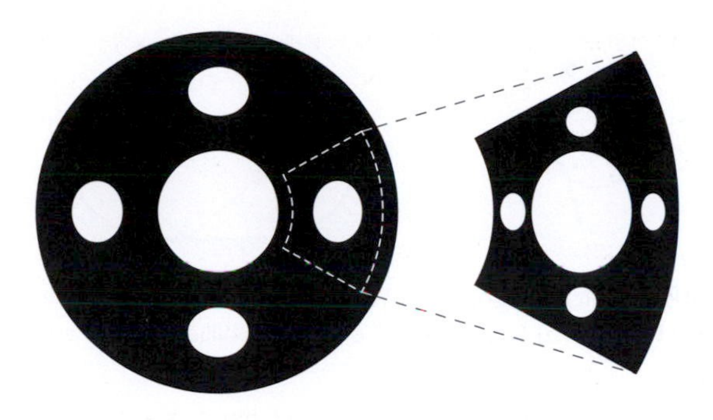

Fig. 7.26. Arrangement of chaotic bands (black) and elliptic islands (white) in macroscopic chaos (schematic diagram). The construction resembles that of Fig. 2.13, and indicates the fat fractal character of the phase space.

chaotic domain is pierced by elliptic regions on all scales (see Fig. 7.26). According to numerical investigations, the fat fractal exponent defined by equation (2.15) falls in the interval $\alpha = 0.3$ to 0.7.

7.6 Summary of the properties of conservative chaos

The properties of conservative chaos can be summarised as follows. Any chaotic band is, to a good approximation,

- the union of all the unstable (hyperbolic) periodic orbits lying in it, or
- the unstable or stable manifold of a single periodic orbit inside the band.

A more precise definition of a chaotic band is

- the union of all the hyperbolic periodic orbits and of all the homoclinic and heteroclinic points formed among the manifolds of these orbits within the band, or
- the union of the unstable *and* stable manifolds of all the unstable periodic orbits it contains.

The description makes the similarity between the structures of a chaotic band and a chaotic attractor quite clear. The difference is due to the fact that, in conservative systems, motion forwards and backwards in time is equivalent, and therefore both types of manifolds must be parts of chaotic bands. In dissipative systems it is the lack of time reversal invariance that leads to asymmetry: chaotic attractors contain only the unstable manifolds, and, due to phase space contraction, they are fractal objects of zero area.

The main properties of chaotic motion in conservative systems are again irregularity, unpredictability and organised but complex phase space structures. Characteristic numbers are again related to the individual properties as follows.

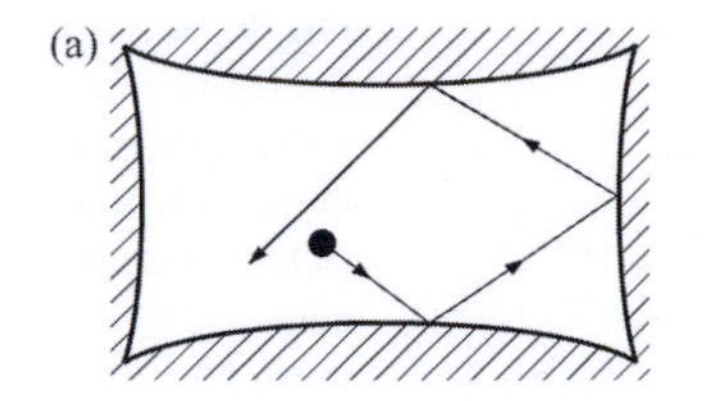

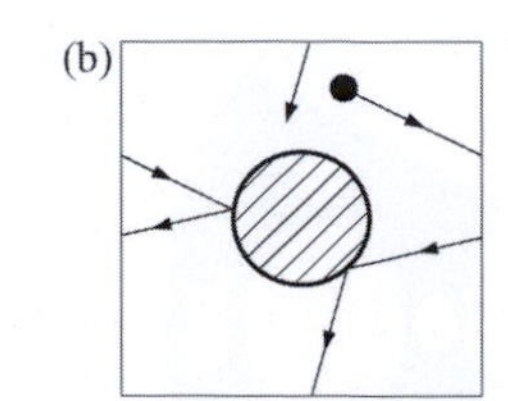

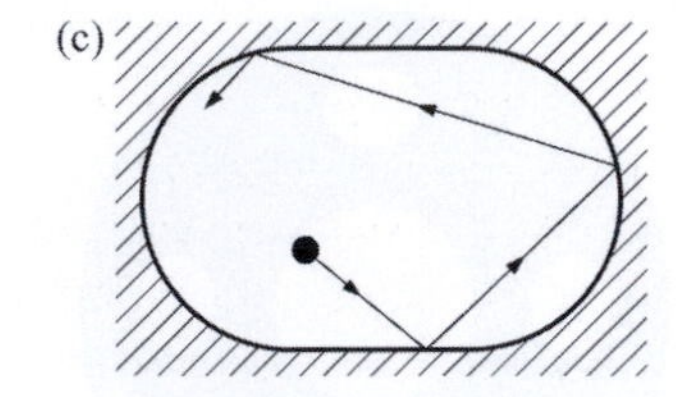

Fig. 7.27. Planar billiards in which chaotic behaviour extends over the entire phase space. (a) Concave billiard, (b) Sinai billiard, (c) stadium billiard.

- Irregularity: the topological entropy of a chaotic band is positive; $h > 0$.
- Unpredictability: the average Lyapunov exponent within a band is positive ($\bar{\lambda} > 0$), and the natural distribution in the chaotic bands is *uniform* (see Plate XX).[9]
- Structured phase space: the fractal dimension of chaotic bands is[10] $D_0 = D_1 = 2$, but they are fat fractals with a non-trivial fat fractal exponent α.

In a general conservative system the interwoven structure of chaotic and elliptic domains is an essential feature of complex dynamics, which has no counterpart in dissipative cases (this is exactly what the fat fractal property refers to).

7.7 Homogeneously chaotic systems

With an increasing perturbation strength, it may happen that even the last invariant phase space curve, the last KAM torus, breaks up. Since the KAM theorem does not apply, this is not necessarily a golden mean torus. Therefore, every single trajectory visits the entire phase space.

This behaviour can be observed in certain billiard problems, in the dynamics of a ball moving in a region of the horizontal plane bounded by vertical walls on which elastic bounces take place (Fig. 7.27). (The disc scatterers of Section 1.2.4, to be treated in detail in Section 8.2.3, define open billiard problems.)

The presence of a positive average Lyapunov exponent in some of the closed billiard problems is easy to understand by means of an optical analogy. The walls of a concave billiard problem (Fig. 7.27(a)) act as dispersing mirrors: incident parallel orbits diverge after the collision.

[9] This follows from equation (5.64): $P^* \equiv$ constant is always a solution for $J \equiv 1$.

[10] Since the negative average Lyapunov exponent is (-1) times the positive one, (7.4), $D_1 = 1 + \bar{\lambda}/|\bar{\lambda}'| = 2$ follows from the Kaplan–Yorke relation, (5.74).

The Sinai billiard problem (also known as Lorentz gas, Fig. 7.27(b)) is a model for a particle wandering among identical discs placed on a large square lattice. Assuming the dynamics to be translation-invariant, it is sufficient to study the cell presented in the Fig. 7.27(b), with periodic boundary conditions in both directions. The positive Lyapunov exponent is again the consequence of the disc's scattering effect. The stadium billiard problem (Fig. 7.27(c)) consists of two semi-circles with parallel straight line segments connecting them. The chaoticity of the motion is not so obvious here. Qualitatively speaking, it is partially due to the fact that, beyond their focal lengths, the focusing mirrors scatter the beams.

These billiard problems, together with the area preserving baker map, the cat map and the strongly non-linear standard map ($a \gg 1$), exhibit a special type of chaos, in which the natural distribution spreads uniformly over the entire available phase space,[11] similar to the motion of a gas molecule. The statistical properties of conservative systems with a few components resemble, in this case, those of systems consisting of 10^{23} particles (see Boxes 7.3–7.5), and are similar to the characteristics of noise.

Box 7.4 Ergodicity and mixing

A conservative system is called *ergodic* if randomly chosen trajectories visit the entire available phase space volume (whose size is only limited by a few external constraints, first of all by the conservation of energy), or, more precisely, if, sooner or later, it reaches an arbitrarily small neighbourhood of any point in phase space. This class of problems originates from the statistical physics of high-degree-of-freedom systems, where it is of central importance whether the time average of physical quantities equals the so-called ensemble average taken with respect to a stationary distribution. If it is true that the trajectories visit the entire phase space, the two averages are typically identical. It is also possible to speak of ergodicity over certain sub-sets of the phase space. In this sense, the motion is ergodic in *every single* chaotic band of conservative chaotic systems, even with a few variables.

Another important feature of many-particle systems is the *mixing property*. Qualitatively speaking, this means that an ensemble of particles initially bound to a certain region spreads as time passes in such a way that its shape ceases to be compact, i.e. it grows offshoots, and finally forms a 'uniform' net of very thin filaments over the entire phase space (Fig. 7.28).

The mixing property on an invariant set, D, of a two-dimensional area preserving map, M, can be formulated as follows. After a sufficiently large number of steps, the nth image, $M^n S$, of some domain, S, fills the same proportion of any domain, S', no matter how S' is chosen. This implies that the ratio

[11] In these exceptional situations, often arising as limiting cases of typical conservative chaos treated so far, the chaotic domain is not a fat fractal, since it is not interrupted by elliptic islands.

$\mathcal{A}(S' \cap M^n S)/\mathcal{A}(S')$ (where $\mathcal{A}$ denotes the area) is independent of S'. In particular, choosing S' to be the entire domain D implies $\mathcal{A}(D \cap M^n S) = \mathcal{A}(M^n S) = \mathcal{A}(S)$, so we obtain

$$\frac{\mathcal{A}(S' \cap M^n S)}{\mathcal{A}(S')} = \frac{\mathcal{A}(S)}{\mathcal{A}(D)}, \quad \text{for} \quad n \gg 1.$$

The amount of material initially in domain S is thus uniformly distributed over all sub-sets after a sufficiently long time, in the same proportion as if the amount of material in S were uniformly distributed over the entire invariant set, D (see also Fig. 7.3 and Box 7.1).

The mixing property implies ergodicity: each trajectory gets everywhere in mixing systems. On the other hand, chaotic motion always produces efficient mixing; in fact, in each chaotic band the dynamics is not only ergodic but also mixing.

Both ergodicity and mixing are properties valid on chaotic attractors (and saddles) of *dissipative* systems (in the equation above, the area of the phase space regions is to be replaced by their weight taken with respect to the *natural distribution*). Thus, as far as ergodicity and mixing are concerned, there is no difference between dissipative and conservative (or permanent and transient) chaos.

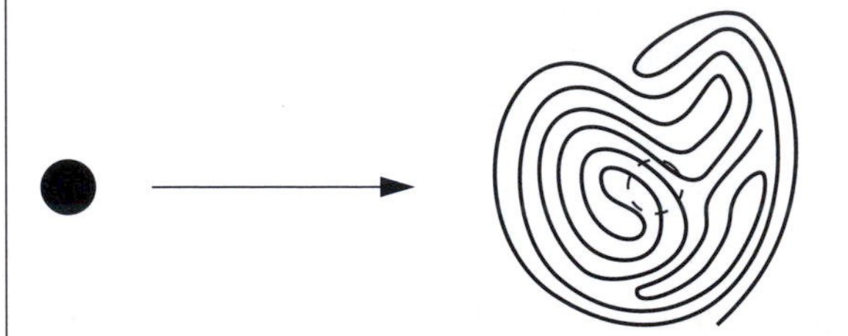

Fig. 7.28. Schematic representation of the mixing property. Any phase space domain spreads out in a filamentary manner. After a long time, the area ratio of the black regions to that of the white regions is equal over any sub-set.

Box 7.5 Conservative chaos and irreversibility

An important feature of many-particle systems is macroscopically observable irreversibility, i.e. the preferred direction of time. This is especially surprising in view of the fact that the microscopic equations of motion are invariant under time reversal.

The same behaviour is typical also for homogeneously chaotic conservative systems with a few degrees of freedom. In every single point of the phase space, motion starting forwards is equivalent to motion backwards in time. One might thus expect that if a phase space volume element is followed over n steps (until it becomes well mixed and its points are uniformly distributed in the entire phase space), and the iterations are then reversed and carried out backwards for the *same* number of steps, then finally the initial shape is recovered. Taking into account, however, that the final states are only known with a finite accuracy due to the *round-off errors* of the numerical simulation, it becomes obvious that, after a certain number of forward and backward iterations, n^*, the points do not trace out the initial shape, but remain dispersed (see Fig. 7.29). The reason for this is the sensitivity to the initial conditions. The critical iteration number, n^*, is simply the prediction time (see Section 5.4.2) related to the number representation error Δ, i.e. $n^* \approx \ln(1/\Delta)/|\bar{\lambda}|$. In general, the initial state of the system cannot be recovered due to round-off errors, or, more generally, due to observations of limited accuracy, even if the dynamics is time reversal invariant.

To use an everyday example, this is like stirring cocoa powder into semolina cream (molecular diffusion is negligible). If stirring is reversed, i.e. the original motion is performed backwards in time, the cocoa powder does not recover its original shape, but becomes even better mixed.

Thus, *irreversibility* is a joint consequence of *chaos and limited accuracy observations.* It is a manifestation of the lack of practical determinism and of the amplification of errors. The existence of chaotic motion also sheds new light on the understanding of the irreversible behaviour of high-degree-of-freedom systems.

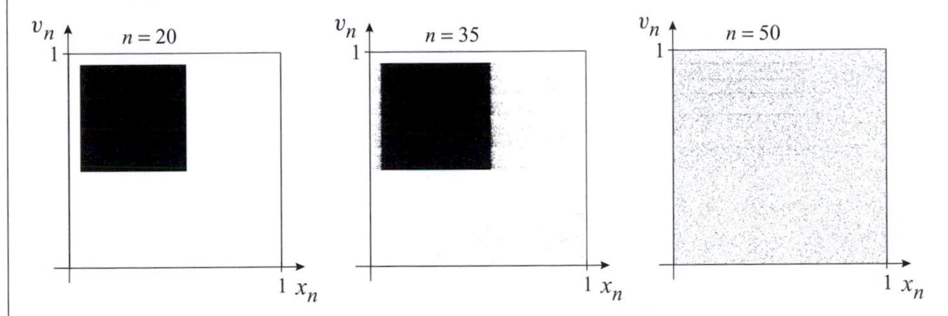

Fig. 7.29. Does it return? Image of a square-shaped phase space volume of the area preserving baker map ($b = 1/3$) after iterating it n steps forwards *and* backwards. For $n = 20$ the initial object is precisely recovered; for $n = 35$ the borders become somewhat blurred (a few points, however, remain far away); and for $n = 50$ the initial object is no longer recognisable. The numerical accuracy that has been used corresponds to an error $\Delta = 10^{-11}$. As $\bar{\lambda} = 0.636$ (see Problem 7.3), the value of n^* is found to be 40. With higher precision the blurring of the object occurs at larger values of n^*.

Chapter 8
Chaotic scattering

Scattering processes have played an important role in different sciences since the discovery of the atomic nucleus by directing a particle beam onto a thin layer of a solid and evaluating its deflection.[1] Scattering methods are now widely used in the investigation of material structures. Other phenomena, such as, for example, the motion of a comet, or the reflection of light on a set of mirrors, are also scattering processes (cf. Section 1.2.4). Perhaps the simplest example is provided by the motion of a particle under the effect of a force bounded to a finite region in space. In general, a scattering process is the dynamics of a conservative system that starts and ends with a very simple (usually uniform rectilinear) motion, typically far away from the region where interactions are strong (the scattering region). The well known classical examples of scattering all exhibit regular motion. The moral of Chapter 7 is, however, also valid in these cases: even the slightest perturbation makes the dynamics chaotic. Chaotic scattering is, therefore, typical.

Because of the simplicity of the initial and final states, chaotic behaviour can only extend to a finite domain of phase space, and it can only be transient. Chaotic scattering is therefore the manifestation of transient chaos in conservative systems. Consequently, it is related to the chaotic saddle (see Chapter 6) of a volume-preserving ($\sigma \equiv 0$ or $J \equiv 1$) dynamics. Since chaotic scattering is thus similar to dissipative transient chaos, this chapter is restricted to a brief overview of the specific features. For the sake of simplicity, only planar scattering will be considered.

[1] By E. Rutherford in 1911.

Scattering processes with regular and chaotic dynamics are fundamentally different. Through the example of the one-, two- and three-disc problems (open billiards), we demonstrate that the possibility of chaos suddenly arises with the appearance of a third disc. We also discuss the properties of chaotic scattering in (the conservative limit of) kicked systems and in motion developing under the effect of time-independent forces changing continuously in space. Finally, we summarise general relations characteristic of chaotic scattering.

8.1 The scattering function

A possible initial parameter of a scattering process is usually chosen to be the distance, b, at which the particle would pass by the centre of the scattering region (the origin) if there were no interactions present (see Fig. 8.1). This is called the *impact parameter*. A possible output parameter is the angle between the straight lines of the incident and outgoing segments of the orbit, the *angle of deflection*, θ. An important characteristic of the scattering process is the functional relationship between the angle of deflection and the impact parameter,

$$\boxed{\theta(b) = ?,} \tag{8.1}$$

the scattering function.

The chaotic nature of scattering manifests itself in a very complicated form of the scattering function. In such cases, a tiny change in b may result in a significant change in the angle of deflection. In addition, these

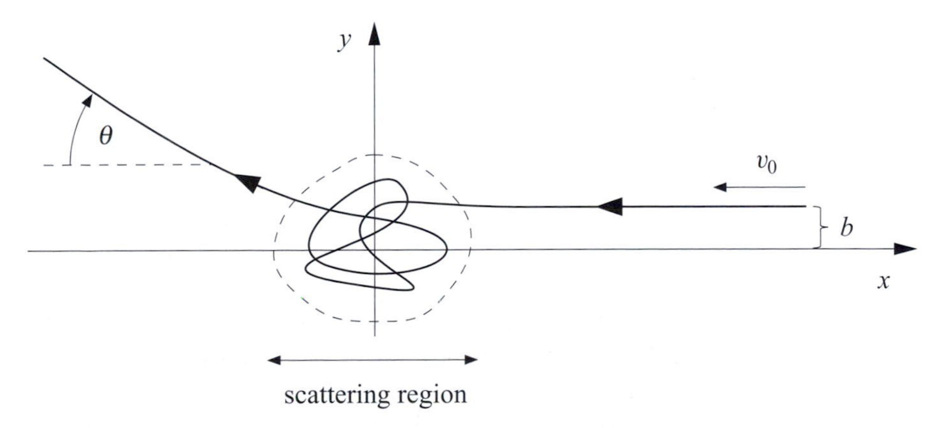

Fig. 8.1. Characteristics of a scattering process. The scattering region is the domain in which the interaction is significant. The particle is projected parallel to the x-axis. The incident segment of the orbit is a straight line at distance b from the x-axis; the deflected orbit tends asymptotically to a straight line under angle θ to the x-axis (θ increases clockwise and takes on values in $(-\pi, \pi)$).

singularities sit on a fractal set of the impact parameter. We study all this in detail via the example of scattering on discs.

8.2 Scattering on discs

8.2.1 One-disc problem

Scattering is caused by the elastic collision of a particle with a disc of radius R. The particle's orbit breaks on the disc in the same way as a light ray on a cylindrical mirror: the angles of incidence and reflection are identical, ϕ_0, (Fig. 8.2). Since there is no energy loss, the speed of the particle is constant and can be chosen to be unity: $|\mathbf{v}| = \mathbf{1}$. In the range $|b| \leq R$ (the only range in which collision occurs) the angle ϕ_0 is related to the impact parameter via $b = R \sin \phi_0$. For negative b values (impact below the x-axis), the angle of incidence/reflection is negative.

The angle of deflection is the supplementary angle of $2\phi_0$: $\theta = \pi - 2\phi_0$. The scattering function is therefore given by

$$\theta = \pi - 2 \arcsin \frac{b}{R}, \quad \text{for } |b| \leq R; \tag{8.2}$$

otherwise it is identically zero (Fig. 8.3).

8.2.2 Two-disc problem

The centres of two identical discs of radius R are chosen to be arrange symmetrically on the y-axis at distance $a > 2R$ from each other (Fig. 8.4). Let particles be incident parallel to the x axis. For impact parameter values larger than the distance of the disc centres from the

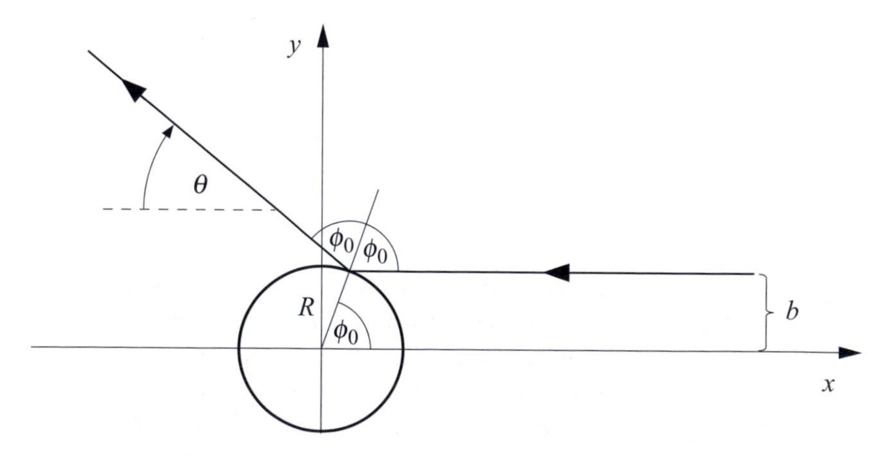

Fig. 8.2. Scattering on a single disc. The collision is elastic (no loss of energy): at the point of collision the angle, ϕ_0, between the incident straight line and the incidence norm is identical to the reflection angle.

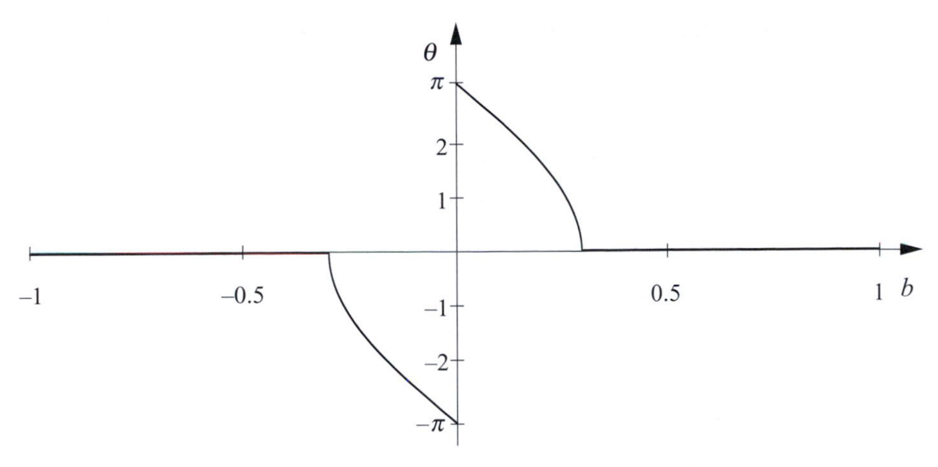

Fig. 8.3. The scattering function in the one-disc problem. Apart from the points $b = \pm R$, it is a smooth function without any singularity ($R = 0.3$).

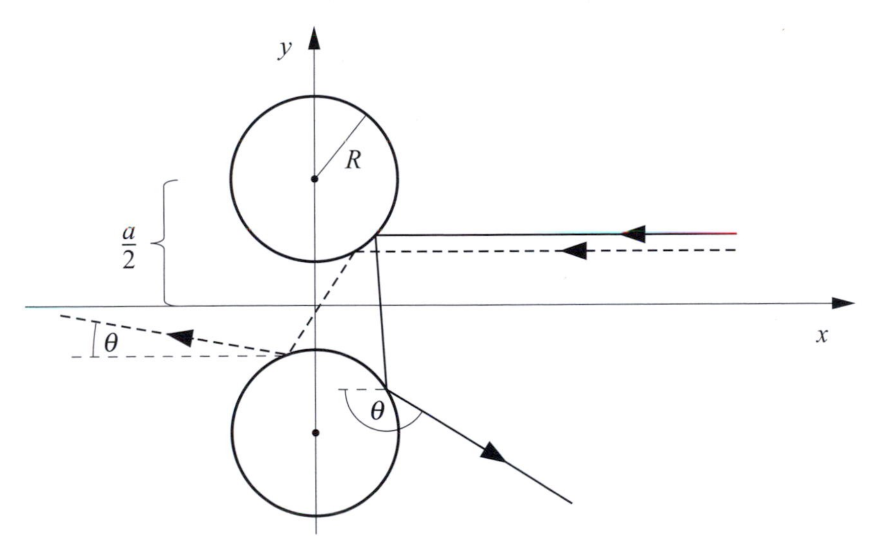

Fig. 8.4. Two-disc problem. One of the particles bounces back ($|\theta| > \pi/2$), and the other one (dashed orbit) is scattered forward ($|\theta| < \pi/2$).

origin, $|b| \geq a/2$, the scattering function is of the same form as in the one-disc problem, just the variable is shifted by $\pm a/2$. In the intermediate range $|b| < a/2$, a new possibility arises: the particle can bounce between the two discs. The scattering function, obtained in this range numerically (Fig. 8.5), displays the striking emergence of two impact parameter values ($\pm b_c$) where the deflection function is singular.

The explanation for this is that between the impact parameters of backward and forward scattering (the particle escapes, respectively, with

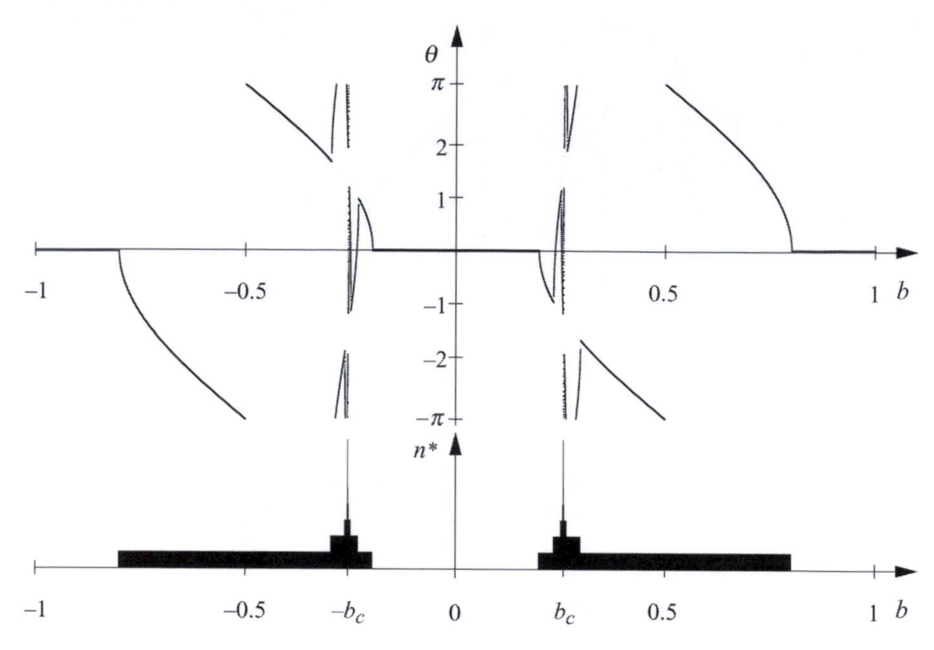

Fig. 8.5. Scattering in the two-disc problem ($R/a = 0.3$). The upper figure depicts the scattering function for particles injected along the x-axis. Between the disc centres there are two values, $\pm b_c$, where the deflection angle is singular. The lower figure shows the lifetime distribution: number, n^*, of collisions before escape (the height of the widest black band is unity).

a positive and a negative velocity component along the x-direction) shown in Fig. 8.4 there must exist – due to continuity – a single value that does not correspond to either type of scattering, rather to a motion trapped between the two discs. Accordingly, between the two discs there exists a periodic orbit bouncing to and fro along the y-axis. To use an optical analogy, the two discs correspond to two dispersing mirrors, the bouncing between them is therefore *unstable*. This orbit can thus be hit at parameters $\pm b_c$ only. With a b value close to the critical one, the particle orbit comes close to the periodic orbit, but ultimately it deviates, executing several bounces before escape. When approaching the critical value, the deflection angle therefore goes up and down more and more rapidly: the graph of the function $\theta(b)$ *accumulates* at $\pm b_c$. Exactly at the critical impact parameter, the lifetime of the scattered particle is infinitely long between the discs (Fig. 8.5, lower figure).

In order to understand the dynamics better, we derive a kind of Poincaré map relating the data of two consecutive collisions on the disc surfaces. Collision n is chosen to be characterised by the angle θ_n between the incident straight orbit and the x-axis (the same as the deflection

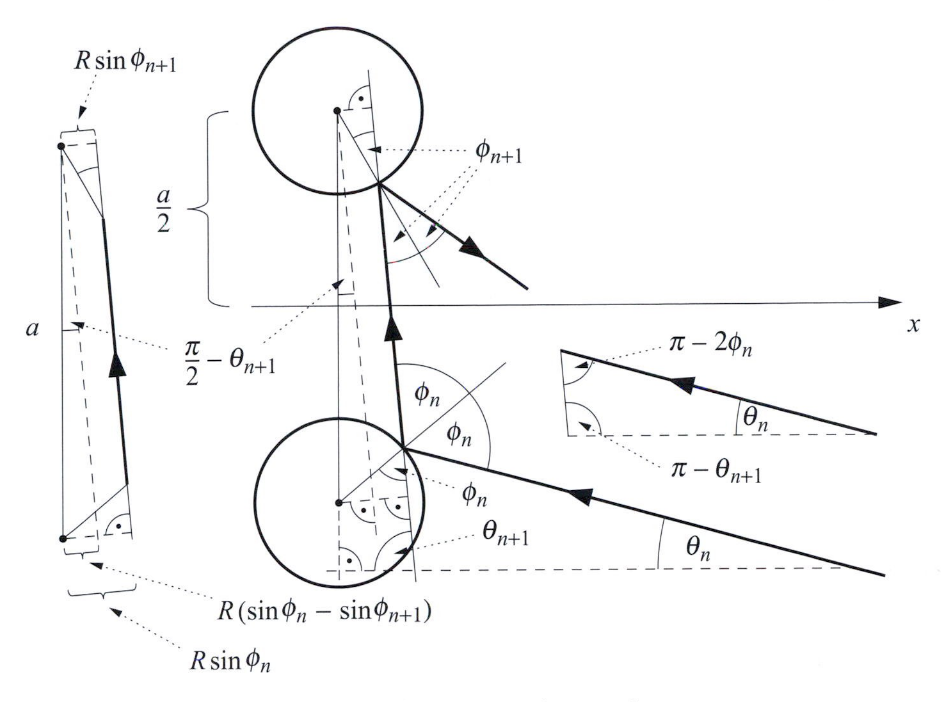

Fig. 8.6. Geometrical relations required to construct the scattering map.

angle after the previous collision) and the angle of incidence, ϕ_n (Fig. 8.6). Note that ϕ_n is positive if the incidence norm can be reached by an anti-clockwise rotation of the incident orbit by an angle of ϕ_n. Consider the triangle formed by the dashed horizontal line and the lines of the incident and reflected orbits in Fig. 8.6. The sum of the angles is π; therefore, $\theta_{n+1} = \theta_n - 2\phi_n + \pi$. Shifting the line of the reflected orbit parallel to itself so that it crosses the centre of one of the discs, we obtain a rectangular triangle of hypotenuse a and of shorter right-angle side $R(\sin\phi_n - \sin\phi_{n+1})$. Consequently (cf. Fig. 8.6), $a\sin(\pi/2 - \theta_{n+1}) = R(\sin\phi_n - \sin\phi_{n+1})$. For collisions occurring on the top disc downwards ($\theta_{n+1} < 0$), the left-hand side appears with opposite sign. The scattering map of the two-disc problem is therefore given by

$$\sin\phi_{n+1} = \sin\phi_n - \frac{a}{R}\operatorname{sgn}(\theta_{n+1})\cos\theta_{n+1}, \qquad \theta_{n+1} = \theta_n - 2\phi_n + \pi. \tag{8.3}$$

The phase portrait presented in Fig. 8.7 clearly indicates that points $(0, -\pi/2)$, $(0, \pi/2)$, corresponding to bounces between the discs along the y-axis, form a *hyperbolic* two-cycle. The initial co-ordinates, $x_0 =$ constant, $y_0 = b$, of the particle correspond to the collision co-ordinates, $\phi_0 = \arcsin[(b - (a/2)\operatorname{sign}(b))/R]$, $\theta_0 = 0$. The lifetime of

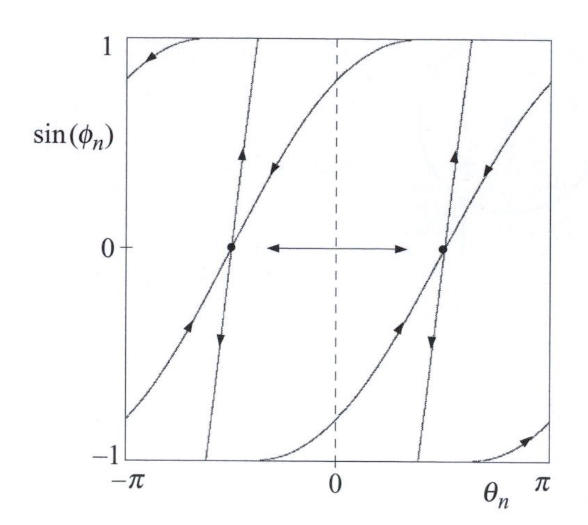

Fig. 8.7. Phase portrait of scattering map (8.3) on the $(\theta, \sin\phi)$-plane ($R/a = 0.3$) based on the manifolds of the two-cycle (black dots, connected by a double arrow). The line of initial condition, $\theta_0 = 0$, is marked by a dashed line.

the scattering process is the number, n^*, of collisions before the particle escaping the discs. The deflection angle, θ, is the angle of deflection after the last collision: $\theta = \theta_{n^*+1}$. The values b_c and $-b_c$ correspond to the intersections of the stable manifold of the two-cycle with the line $\theta_0 = 0$ of initial conditions.

Problem 8.1 Show that map (8.3) is area preserving in the variables θ_n, $\sin\phi_n$.

Problem 8.2 Determine the stability eigenvalues of the two-cycle of map (8.3) for one iteration.

8.2.3 The three-disc problem

Besides the discs on the y-axis, we now add a third disc of identical radius with its centre at the point $x = -\sqrt{3}/2$, $y = 0$. A fundamentally new scattering situation arises, since, after leaving the discs centred on the vertical axis, the particle can bounce back from the new disc and collide again with the previous ones. Accordingly, there also exist new types of trapped orbits: besides the two-cycles corresponding to pair-wise bouncing to and fro, the simplest type describes the bouncing around the discs along an orbit of the shape of a regular triangle (Fig. 8.8(b)). (There are two such triangular orbits, dependent on the winding direction.)

The Poincaré map relating the parameters of collision n and $n+1$ can be obtained from that of the two-disc problem. To this end, we have to be able to identify the disc on which collisions n and $n+1$ occur. By rotating the reference frame by 120° or −120°, so that the discs of collision n and $n+1$ fall on the new y-axis, (8.3) holds again. The motion can again be monitored on the plane θ_n, $\sin\phi_n$. The triangular orbits appear in this map as three-cycles and prove to be hyperbolic.

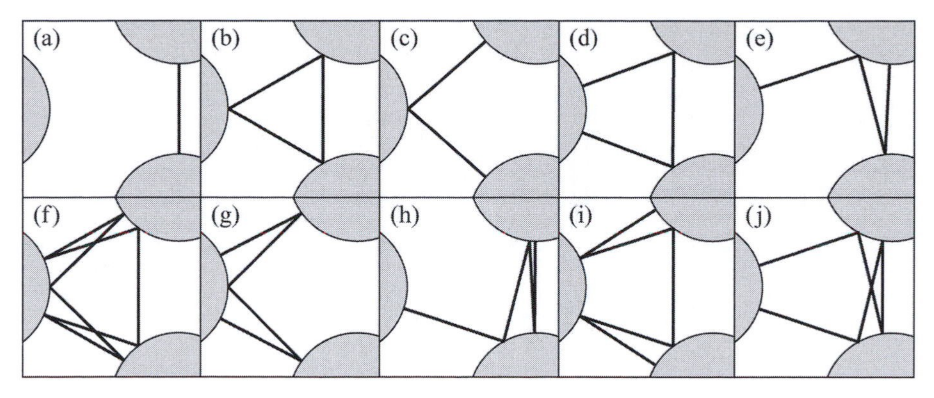

Fig. 8.8. A few simple periodic (all unstable) orbits of the three-disc problem ($R/a = 0.3$).

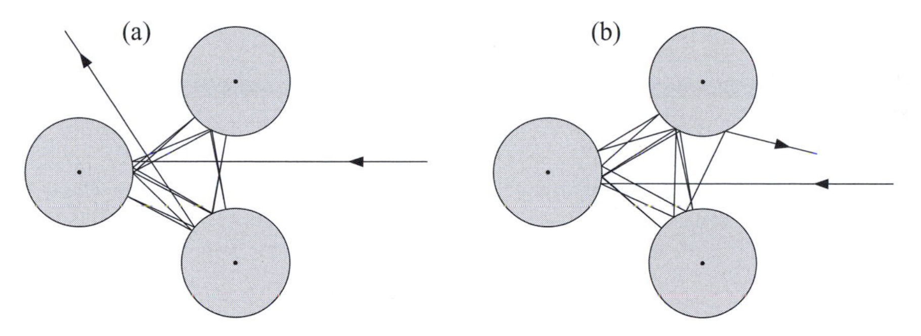

Fig. 8.9. Scattering orbits in the three-disc problem. In (a) and (b) the orbit approaches cycles (a), (c), (d), and (c), (d), (j) of Fig. 8.8, respectively. Both orbits escape after 14 collisions.

Problem 8.3 Determine the stability eigenvalues of the triangular orbit of Fig. 8.8(b) for one iteration. (Consider bounces along the triangular orbit and apply (8.3) with appropriate rotations).

In fact, an infinite number of trapped periodic orbits are present in the system, all of them hyperbolic. Some of these are shown in Fig. 8.8. A general scattering orbit can be considered as one wandering among the unstable periodic orbits before escape takes place, as illustrated by Fig. 8.9.

Owing to the presence of unstable cycles, the two critical points of the two-disc problem are replaced by an *infinity* of critical b values belonging to singularities. The scattering function is therefore very complicated in the range $|b| < a/2$ (see Fig. 8.10).

The infinite number of hyperbolic cycles form a chaotic saddle (Fig. 8.11) with fractal stable and unstable manifolds (Fig. 8.12). The intersection of its stable manifold with the line of the initial condition

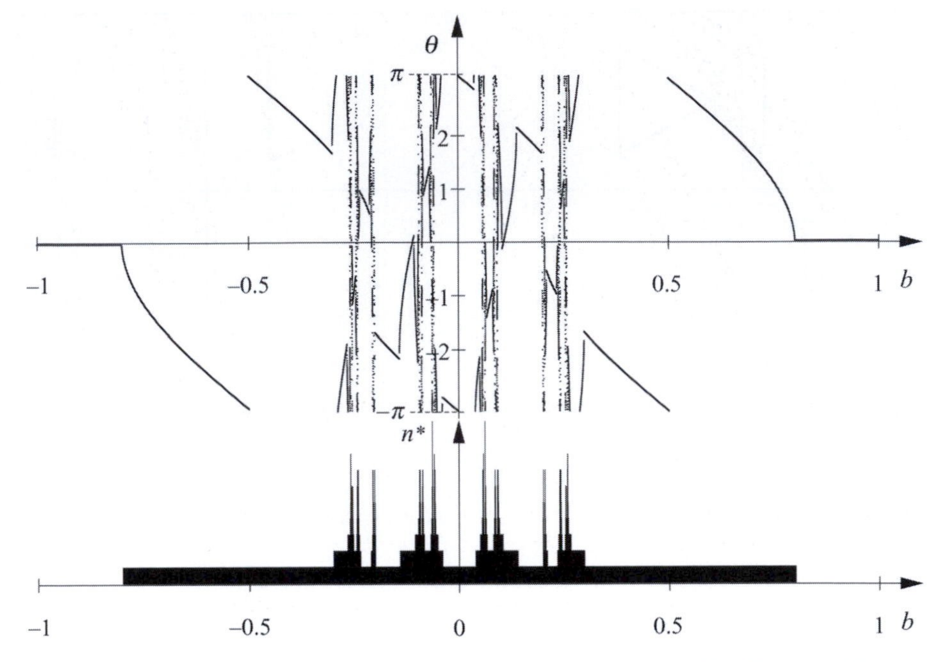

Fig. 8.10. Scattering in the three-disc problem ($R/a = 0.3$). The upper figure depicts the scattering function for particles injected along the x-axis. The lower figure is the lifetime distribution: number n^* of collisions before escape.

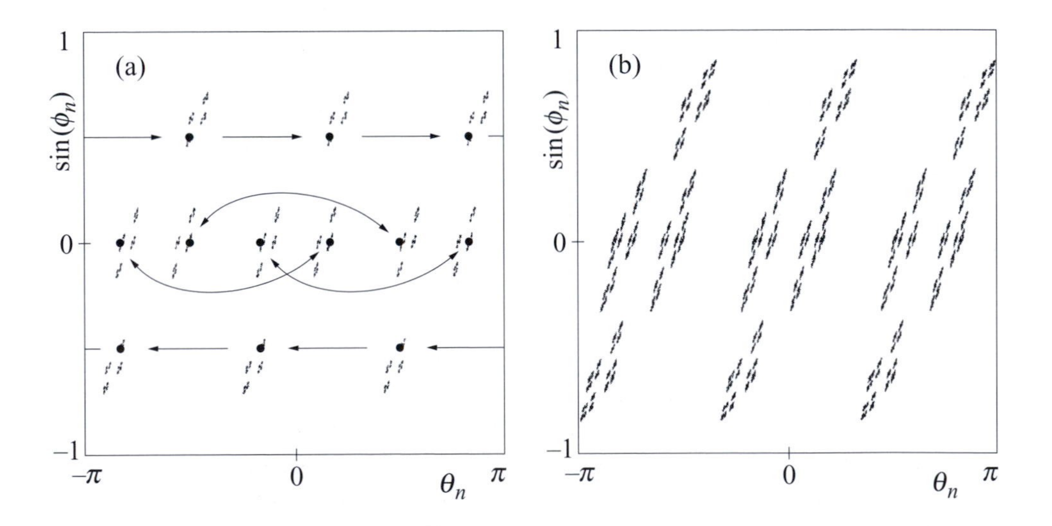

Fig. 8.11. Chaotic saddle of the three-disc problem in the phase plane (θ_n, $\sin\phi_n$): (a) $R/a = 0.3$; (b) $R/a = 0.4$. The mid-points of the trajectories not escaping up to $n_0 = 8$ and $n_0 = 12$ collisions have been plotted at collision number $n_0/2$, respectively (cf. Box 6.1). In (a) the two-cycles and the three-cycles are marked, along with their transformation under the scattering map.

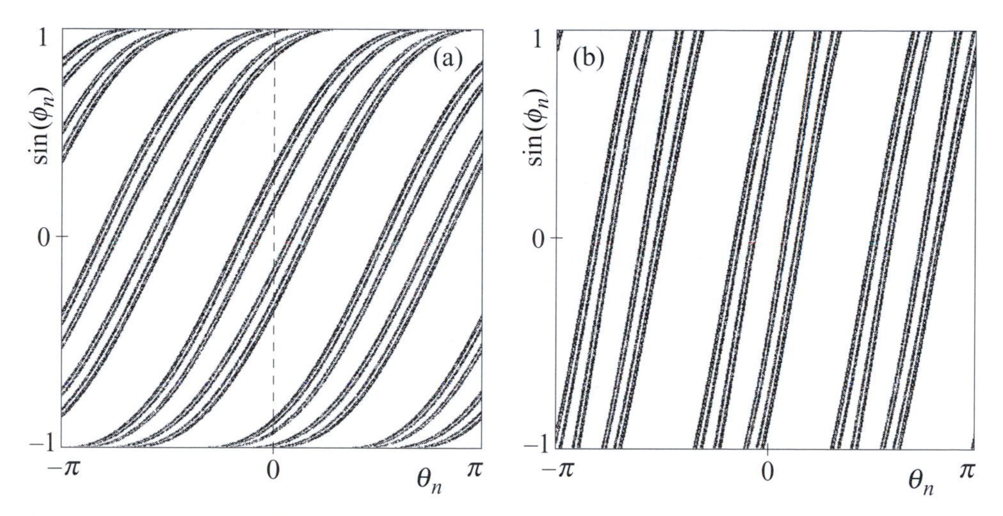

Fig. 8.12. Invariant manifolds of the three-disc problem ($R/a = 0.4$): (a) stable and (b) unstable manifold. The plots have been generated from trajectories not escaping before 12 collisions. Parts (a) and (b) are obtained at collision numbers $n = 0$ and $n = 12$, respectively (cf. Box 6.1). The line of initial condition, $\theta_0 = 0$, is marked by a dashed line in (a).

yields the b values where the scattering function is singular. Accordingly, the lifetime in these points is infinite (Fig. 8.10, lower figure).

Problem 8.4 In the limit $a \gg R$, the chaotic saddle tends to the chaotic saddle of an open area preserving baker map (for a dissipative analogy, see Problems 6.6 and 6.7). The fixed points, $H_{\pm}$, of the baker map correspond to the triangular and pair-wise bouncing between the discs.[2] Determine how the chaos parameters (the positive Lyapunov exponent, the partial fractal dimensions, the topological entropy) of the scattering process change with a/R in this limit.

Particles leave the scattering region across one of the three lines connecting the centres of the discs. Accordingly, we can speak about three different final escape routes of the scattering. If we assign a colour to each of these, we observe that the plane of the map splits into three *escape regions* (see Plate XXI). These are similar to the basins of attraction of dissipative systems, but they are not identical since there exist no attractors in this case. In chaotic scattering, the escape regions are separated by *fractal* boundaries. These boundaries contain the stable manifolds of the chaotic saddle underlying the scattering process. In fact, the stable manifold turns out to be on the boundary of all three colours simultaneously.

[2] Utilising the symmetries of the three-disc problem, the scattering map can be written in a reduced form in which both pair-wise and triangular bouncings appear as fixed points.

8.3 Scattering in other systems

Chaotic scattering can arise not only due to walls, ensuring elastic collisions, but also to a force, **F**, or potential, V, changing smoothly in space. In all cases where

$$\mathbf{F}(x, y) \quad \text{or} \quad V(x, y) \tag{8.4}$$

is not rotationally symmetric, and the force decays to zero for large distances (the potential tends to a constant), chaotic scattering may occur. A smooth analogue of the three-disc problem is provided by a potential forming three identical Gaussian hills of half-width R and height V_0, centred at the positions of the discs (Fig. 8.13):

$$V(x, y) = V_0 \left(e^{-[x^2+(y-a/2)^2]/(2R^2)} + e^{-[x^2+(y+a/2)^2]/(2R^2)} + e^{-[(x+\sqrt{3}a/2)^2+y^2]/(2R^2)} \right). \tag{8.5}$$

The motion now also depends on the total energy, $\mathcal{E} = v_0^2/2 < V_0$, of the particles. The instantaneous velocity of a particle starting far away with initial velocity $\mathbf{v}_0$ changes continuously with the position, and in the vicinity of the potential hills the orbit bends strongly. One can, however, still identify two cases that describe the basis of chaotic behaviour: (i) bouncing between pairs of hills and (ii) bouncing around among three hills. In a wide energy range below the hill-top, V_0, the character of the scattering is the same as in the three-disc problem: an infinite number of unstable periodic orbits exist between the hills, forming a chaotic saddle. It is a new feature that the properties of the saddle and of the

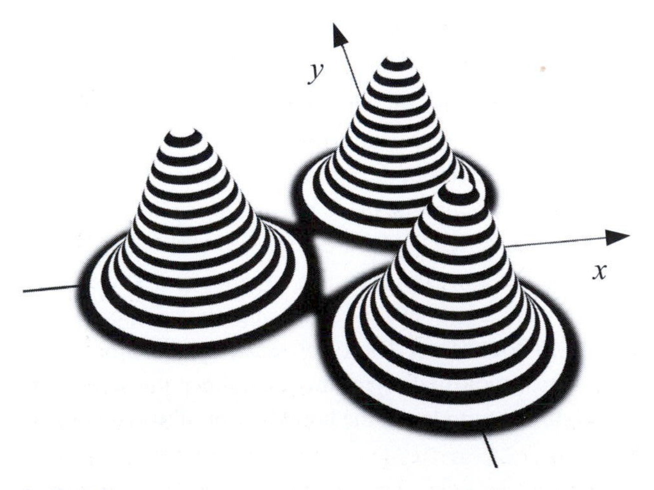

Fig. 8.13. Potential (8.5) corresponding to a 'softened' version of the three-disc problem ($a = 1$, $R = 1/6$, $V_0 = 6$). Black and white bands mark potential differences of size $\Delta V = 0.5$.

scattering function, such as, for example, the fractal dimension, *depend* on the total energy $\mathcal{E}$. Moreover, the chaoticity of the scattering *ceases* at energy $\mathcal{E}_c = V_0$ (in a kind of bifurcation). Bouncing is no longer possible for $\mathcal{E} > V_0$, and particles pass over the hill-tops in a regular motion.

Chaotic scattering can also develop in driven one-dimensional dynamics if, besides a time-independent attracting force, $F(x)$, a driving force, $F_d(x, t)$, is also present. In such cases, both forces have to decay at large distances. For charged particles, the driving can be due, for example, to illumination by laser light. Driving might force the particle to exhibit random 'wandering' for a long time in the region where both forces are significant. The incident and final states are, however, always represented by uniform motion along the x-axis. In one dimension, it is of course meaningless to speak of an angle of deflection, but the lifetime distribution and the phase space structures (in a stroboscopic map) remain well defined.

Instantaneous driving (for example, the repeated application of laser impulses) corresponds to kicking. The simplest models of this kind can be obtained from the kicked free motion discussed in Appendix A.1.5.

With a kicking amplitude function, $f(x)$, tending to zero for $|x| \to \infty$, the dimensionless scattering map is given by

$$x_{n+1} = x_n + v_n, \qquad v_{n+1} = v_n + f(x_{n+1}). \tag{8.6}$$

This is formally the same as (7.8) but x is no longer periodic. Chaotic scattering may occur for any non-linear kicking amplitude function.

Problem 8.5 Investigate the scattering map (8.6) for the amplitude function $f(x) = -7(x - x^2)e^{-x}$. Plot the manifolds of the chaotic saddle.

Problem 8.6 Investigate the scattering map (8.6) for the amplitude function $f(x) = -x(c - 1 - cx^2/2)e^{-x^2/2}$. Plot the chaotic saddle and its manifolds for $c = 10$.

Problem 8.7 Start a large number of particles along the line segment $2 < x_0 < 2.5$, $v_0 = -1$ of the scattering map (8.6) for the amplitude function $f(x) = -x(c - 1 - cx^2/2)e^{-x^2/2}$ with $c = 5$ (where a periodic island exists around the elliptic fixed point of the origin). Determine how the number of points not escaping the square $|x|, |v| < 2$ within n steps changes, with n, up to 300 steps. Plot the chaotic saddle and its unstable manifold.

Box 8.1 Chemical reactions as chaotic scattering

Chemical reactions can be considered as scattering processes. Imagine that a molecule, AB, travels along a straight line towards atom C, which is able to form a chemical bond with atom B (Fig. 8.14(a)). The molecule and the atom do not interact initially (they move uniformly), but when the distance between them becomes comparable to that between atoms A and B, all three pair-wise interactions may become of equal importance. In this state, we can no longer speak of molecule AB; the three participating atoms constitute a *transition complex* (Fig. 8.14(b)).

The interactions among the three bodies usually generate chaotic dynamics, which results, sooner or later, in one of the bodies escaping. Finally, an atom and a molecule remain. The two possible outcomes (Figs. 8.14(c) and (d)) are, however, fundamentally different: either a new molecule BC is formed and atom A is released (reaction takes place), or molecule AB reappears, along with atom C (AB 'bounces back' from C). Moreover, small differences in the initial conditions may lead to drastic differences in the final states: reactive and non-reactive processes exist that are strongly interwoven. The outcome of the scattering process has a fractal structure. Underlying the dynamics of complex ABC, there exist then an infinity of unstable periodic orbits, which provide the skeleton of a chaotic saddle.

The process described in the preceding paragraphs is a kind of *scattering*. The only difference between this scattering process and the case in Fig. 8.1 is that the scattering potential is not fixed in space; rather it moves together with the particles. We can therefore say that it is the theory of chaotic scattering that provides a suitable frame for the proper classical mechanical description of chemical reactions. Since chaos can only exist as long as all three atoms are close to each other, the lifetime of the transition complex is *identical* to the average lifetime, $\tau = 1/\kappa$, of the chaotic transients.

Similar phenomena occur in the collisions between atoms or ions. A particularly interesting case is that of a He^+ ion (He nucleus + one electron) and an electron. It has long been known that in the classical description they do *not* form a stable atom, due to the repulsion between the electrons. One of the important

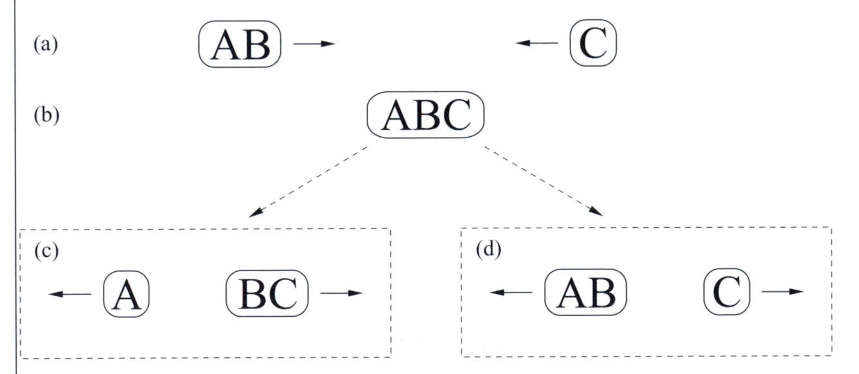

Fig. 8.14. Phases of the collision of molecule AB and atom C along a straight line (schematic diagram). (a) Initial state. (b) Collision state: transition complex ABC. (c) Reactive outcome. (d) Non-reactive outcome. The outcomes are sensitive to small differences in the initial conditions; reaction is one of the possible outcomes of a chaotic scattering process.

achievements of quantum mechanics is the explanation of the existence of a stable He atom. Recent classical studies show, however, that the system undergoes chaotic scattering and the ultimate state is always a He atom and an electron again. In analogy with chemical reactions (A=He^{++}, B=C=e^-), it is, however, also true that in the classical description a *metastable* He atom of finite lifetime is present. The discrepancy between the classical and the quantum world is, however, not so striking in the light of the discovery of the metastability of classical He atoms. It is appealing to see that the quantum mechanical states of the He atom are built upon the saddle of a transiently chaotic classical dynamics.

8.4 Summary of the properties of chaotic scattering

Chaotic scattering is the conservative limit of transient chaos. Accordingly, all the properties listed in Section 6.4 hold with the restriction of volume preservation ($\sigma = 0$, or $J = 1$). Consequently (cf. (7.4)), the average Lyapunov exponents are the opposites of each other: $\bar{\lambda}' = -\bar{\lambda}$. The partial information dimensions along the two manifolds are therefore obtained from (6.24) and (6.28) as follows:

$$\boxed{D_1^{(1)} = D_1^{(2)} = 1 - \frac{\kappa}{\bar{\lambda}}.} \tag{8.7}$$

The information dimensions of the scattering chaotic saddle and its manifolds are thus given by

$$\boxed{D_1 = 2\left(1 - \frac{\kappa}{\bar{\lambda}}\right)} \tag{8.8}$$

and

$$D_1^{(u)} = D_1^{(s)} = 2 - \frac{\kappa}{\bar{\lambda}}, \tag{8.9}$$

respectively. The fact that the partial dimensions and the dimensions of the stable and unstable manifolds are identical is a consequence of the time-reversal invariance of the conservative dynamics.

Besides these simplifying features, chaotic scattering also possesses a property specific to conservative systems. Often, alongside an infinite set of unstable periodic orbits, stable periodic orbits also exist, and the neighbouring dynamics is then bounded. Such orbits cannot be reached by particles incident from outside. They are elliptic, surrounded by KAM tori. Inside the scattering region one may then also find bands of permanent chaos, regular islands, and KAM tori (see Section 7.5.2). The set of all these bounded orbits is separated by an outermost KAM torus from the scattering trajectories. Some of the latter may come close to the outermost KAM torus and spend a long time there. Selecting these trajectories, we find that the number of non-escaped points decreases much more *slowly* than exponentially; it follows a power law (see Problem 8.6). In the number of all non-escaped points, the decay rule (6.1) of transient chaos therefore crosses over for long times into

$$\boxed{N(t) \sim t^{-\sigma}, \quad \text{or} \quad N(n) \sim n^{-\sigma},} \tag{8.10}$$

where σ is a positive exponent. This slow decay reflects the 'stickiness' of KAM tori, and is often observed in chaotic scattering. Since there is no exponential escape from the vicinity of the outermost tori (or, more formally, $\kappa = 0$ and consequently $D_0 = D_1 = 2$), the chaotic saddle is very dense in these regions. There is thus a smooth cross-over at the outermost KAM torus from a fractal chaotic saddle (transient chaos) to a fat fractal chaotic band (permanent chaos).

Chapter 9
Applications of chaos

In this chapter we briefly present how chaos appears in problems of a larger scale. We wish to illustrate by this (i) the ubiquity of chaos and (ii) that numerous research problems are still to be resolved. According to the introductory nature of this book, the selection is based on cases that are not too technically complicated. Solved problems are not provided in this chapter; we merely formulate questions that may encourage the reader to investigate the subject further. We emphasise that, for a given phenomenon, different aspects of chaos (permanent–transient, dissipative–conservative) may be present simultaneously.

We start our survey with two problems, one related to space research, the other to engineering practice, that have also played historically important roles: the gravitational three-body problem and the dynamics of a heavy asymmetric top. Next we turn to a simple model of the general atmospheric circulation, which nevertheless reflects important features of the weather. Finally, we overview the occurrence of chaotic behaviour related to fluid flows, and, in connection with this, we point out the relevance of chaotic mixing in environmental fluid flows. Further fields of application are discussed in the Boxes in this chapter.

9.1 Spacecraft and planets: the three-body problem

In the course of their motion, spacecraft are subject to the gravitational attraction of neighbouring celestial bodies. As gravitational interaction with the Earth decays slowly, the effect of at least two celestial bodies on the spacecraft have to be taken into account; i.e., that of the Earth–Moon, or (if the spacecraft moves further away) that of the Sun–Jupiter couple. Often, the interattraction of spacecraft with the other planets (for example Saturn) is not negligible either, but the three-body problem describing the motion of three gravitationally interacting bodies is a good model which shows the spacecraft dynamics to be, in general, chaotic. Consequently, the uncontrolled orbits of spacecraft may become complicated, and part of the engine's task would be to keep the motion exactly in the predefined direction. This same three-body problem provides insight into the issue of the stability of the Solar System, if the spacecraft is substituted by a smaller planet. The ubiquity of chaotic motion and the possibility of escape indicate instability. In fact, the birth of chaos science can be traced back to Poincaré's paper on the stability of the Solar System in the 1890s.

We consider the simplest version of the three-body problem, as Poincaré did, in which one of the bodies (the spacecraft or an asteroid, for example) is so small that its feedback on the other two is negligible. Furthermore, the large bodies are assumed to move on circular orbits in such a way that all three bodies remain permanently in the same plane.[1]

We work in a reference frame co-rotating with the large bodies around their centre of mass as the origin. The x-axis is chosen to contain the large bodies of mass m_1 and m_2 ($< m_1$). Their x co-ordinates are given by $x_1 = -r_0 m_2/(m_1 + m_2)$ and $x_2 = r_0 m_1/(m_1 + m_2)$, respectively, where r_0 is the distance between the bodies (Fig. 9.1). The gravitational force of magnitude $\gamma m_1 m_2/r_0^2$ (γ is the gravitational constant) acting between the bodies is counter-balanced by the centrifugal forces $m_1 x_1 \omega^2$ or $m_2 x_2 \omega^2$, where ω is the rotational angular velocity. From this, $\omega^2 = \gamma(m_1 + m_2)/r_0^3$. The gravitational forces acting on the small body of mass m at point (x, y) are of magnitudes $F_1 = \gamma m m_1/s_1^2$ and $F_2 = \gamma m m_2/s_2^2$, where the distances can be expressed as $s_i = \sqrt{(x - x_i)^2 + y^2}$, $i = 1, 2$ (see Fig. 9.1).

The x- and y-components of the gravitational forces are $-F_i(x - x_i)/s_i$ and $-F_i y/s_i$, respectively. In the rotating system, the small body is subjected to both the centrifugal force and the Coriolis force. For a positive angular velocity, the latter deviates the body to the right. In a motion of velocity $\dot{x}$ along the x-axis, for example, the y-component of this force is $-2m\omega\dot{x}$. Adding up all the forces, the equation of motion of the small body is given by

$$\ddot{x} = 2\omega\dot{y} + \omega^2 x - \frac{\gamma m_1(x - x_1)}{s_1^3} - \frac{\gamma m_2(x - x_2)}{s_2^3}, \tag{9.1}$$

$$\ddot{y} = -2\omega\dot{x} + \omega^2 y - \frac{\gamma m_1 y}{s_1^3} - \frac{\gamma m_2 y}{s_2^3}, \tag{9.2}$$

and we note that mass m no longer appears.

[1] This is the restricted, planar, circular three-body problem.

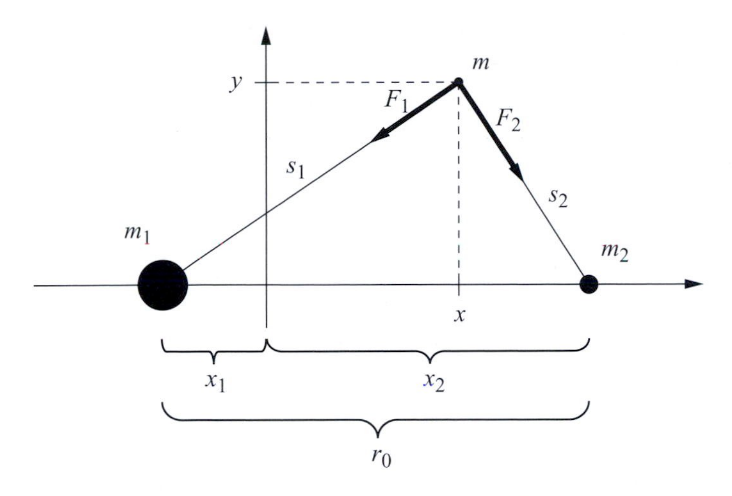

Fig. 9.1. Positions of the three bodies in a co-rotating frame.

The equations become simpler in a dimensionless form (see Appendix A.2). Measuring length and time in units of r_0 and $1/\omega$, respectively, we obtain

$$\ddot{x} = 2\dot{y} + x - \frac{\mu_1(x+\mu_2)}{s_1^3} - \frac{\mu_2(x-\mu_1)}{s_2^3}, \tag{9.3}$$

$$\ddot{y} = -2\dot{x} + y - \frac{\mu_1 y}{s_1^3} - \frac{\mu_2 y}{s_2^3}. \tag{9.4}$$

Here, $\mu_i \equiv m_i/(m_1+m_2)$, and the distances given in terms of the dimensionless co-ordinates are $s_1 = \sqrt{(x+\mu_2)^2+y^2}$ and $s_2 = \sqrt{(x-\mu_1)^2+y^2}$ (Fig. 9.2). The *only* remaining parameters are thus the reduced masses μ_i. Since $\mu_1 + \mu_2 = 1$, only one of these is independent. Usually, the ratio $\mu_2 = m_2/(m_1+m_2)$, i.e. the proportion of the lighter large body's mass to the total mass, is considered to be the parameter.

The position-dependent forces in the equations of motion (9.3) and (9.4) can be written as the gradient of a potential V:

$$\ddot{x} = 2\dot{y} - \frac{\partial V}{\partial x}, \qquad \ddot{y} = -2\dot{x} - \frac{\partial V}{\partial y}, \tag{9.5}$$

where

$$V(x,y) = -\frac{\mu_1}{s_1} - \frac{\mu_2}{s_2} - \frac{1}{2}(x^2+y^2) - \frac{1}{2}\mu_1\mu_2. \tag{9.6}$$

The first two terms represent the attracting gravitational potential; the third term is the repelling potential of the centrifugal force (similar in nature to a potential characterising an unstable state, see Fig. 3.5); and the last one is a constant. Together these generate an interesting potential landscape (Fig. 9.3). There exist five equilibrium states for the small body in the co-rotating system: the Lagrange points denoted by L_i. The first three of these, L_1 to L_3, are always unstable, and they correspond to saddle points of the potential function, $V(x,y)$. The last two, L_4 and L_5, are, for a sufficiently small mass ratio, μ_2, stable, and are at the bottoms of (rather shallow) local potential wells.

Fig. 9.2. Co-ordinates of the three bodies in dimensionless units. The points denoted by L_i $(i = 1, \ldots, 5)$ are the Lagrange points.

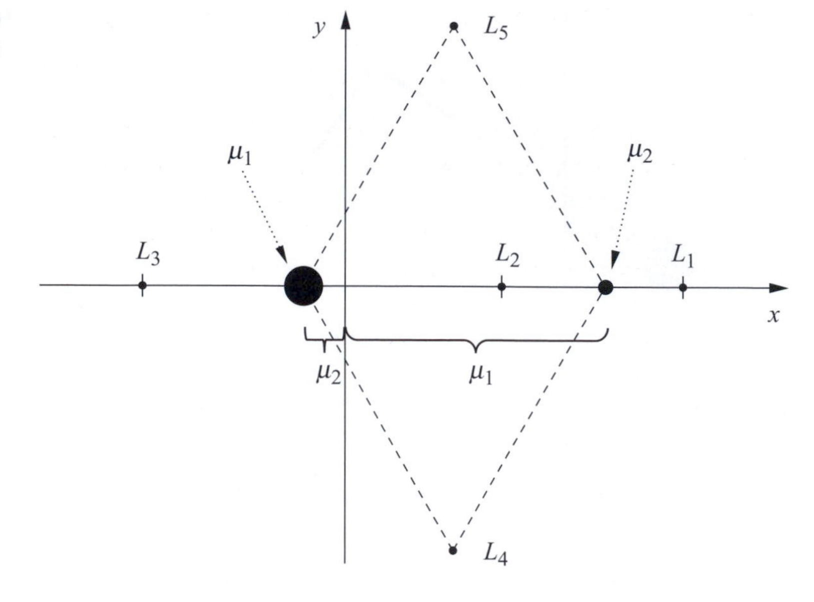

Fig. 9.3. Potential, $V(x, y)$, of the three-body problem ($\mu_2 = 10^{-1}$, $|x|, |y| \leq 1.5$ and the potential difference of the black and white bands is $\Delta V = 0.05$). The large celestial bodies are in the wells, which are of infinite depth. The saddle points of the landscape corresponding to the unstable Lagrange points L_1, L_2 and L_3 are marked by crosses. The Lagrange points L_4 and L_5 (black dots) are located in the ellipsoidal regions of the two hill-tops (they would become stable for $\mu_2 \leq 0.039$ only).

Since the Coriolis force does not do any work, the total energy,

$$\mathcal{E} = \frac{1}{2}(\dot{x}^2 + \dot{y}^2) + V(x, y), \tag{9.7}$$

of the small body in the rotating frame is constant.[2] The behaviour of the restricted three-body system is therefore a planar conservative motion in a potential $V(x, y)$, in the presence of a Coriolis force.

An energy range $\mathcal{E}_2 < \mathcal{E} < \mathcal{E}_1$ exists in which the motion is bounded and the small body can approach both large celestial bodies ($\mathcal{E}_i$ are the potential energies of the Lagrange points, L_i, $\mathcal{E}_3$ is always larger than $\mathcal{E}_1$, and $\mathcal{E}_4 = \mathcal{E}_5 = -1.5 > \mathcal{E}_3$). For energies higher than $\mathcal{E}_1$, the small body may escape and run out into infinity, while, for energies smaller than $\mathcal{E}_2$, it becomes bounded to one of the large celestial bodies. It is interesting to note that in the Sun–Earth–Moon system ($\mu_2 = 2.738 \times 10^{-6}$), the energy, $\mathcal{E} = -1.5006$, of the Moon is only slightly less than $\mathcal{E}_2 = -1.5004$; it is therefore only a matter of a tiny difference in energy that our Moon cannot wander into the proximity of the Sun.

In the following, we investigate the case of the reduced mass parameter $\mu_2 = 10^{-3}$ corresponding to a Sun–Jupiter–asteroid system. Here $\mathcal{E}_1 = -1.5198$, $\mathcal{E}_2 = -1.5205$, $\mathcal{E}_3 = -1.5010$, and in the energy range $-1.55 < \mathcal{E} < \mathcal{E}_1$ macroscopic chaos is present.

A Poincaré section is generated by recording the position, x, of the asteroid and its velocity, v, along the x-direction when it intersects the x-axis from below. In the map (x_n, v_n), it is clearly visible that, depending

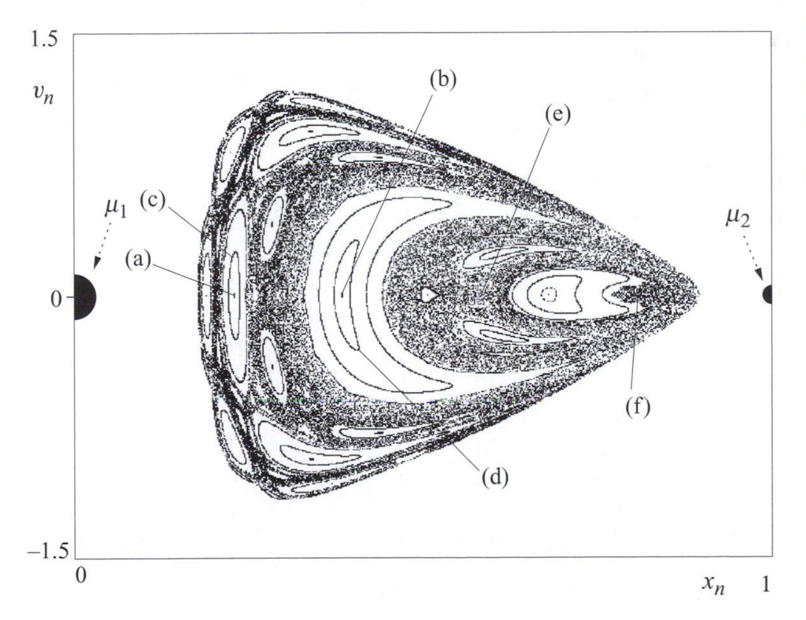

Fig. 9.4. Poincaré portrait of the three-body problem with total energy $\mathcal{E} = -1.525$ ($\mu_2 = 10^{-3}$). The orbits of trajectories (a)–(f) are given in fig 9.5.

[2] In astronomy, the quantity $C \equiv -2\mathcal{E}$ is called the Jacobi constant.

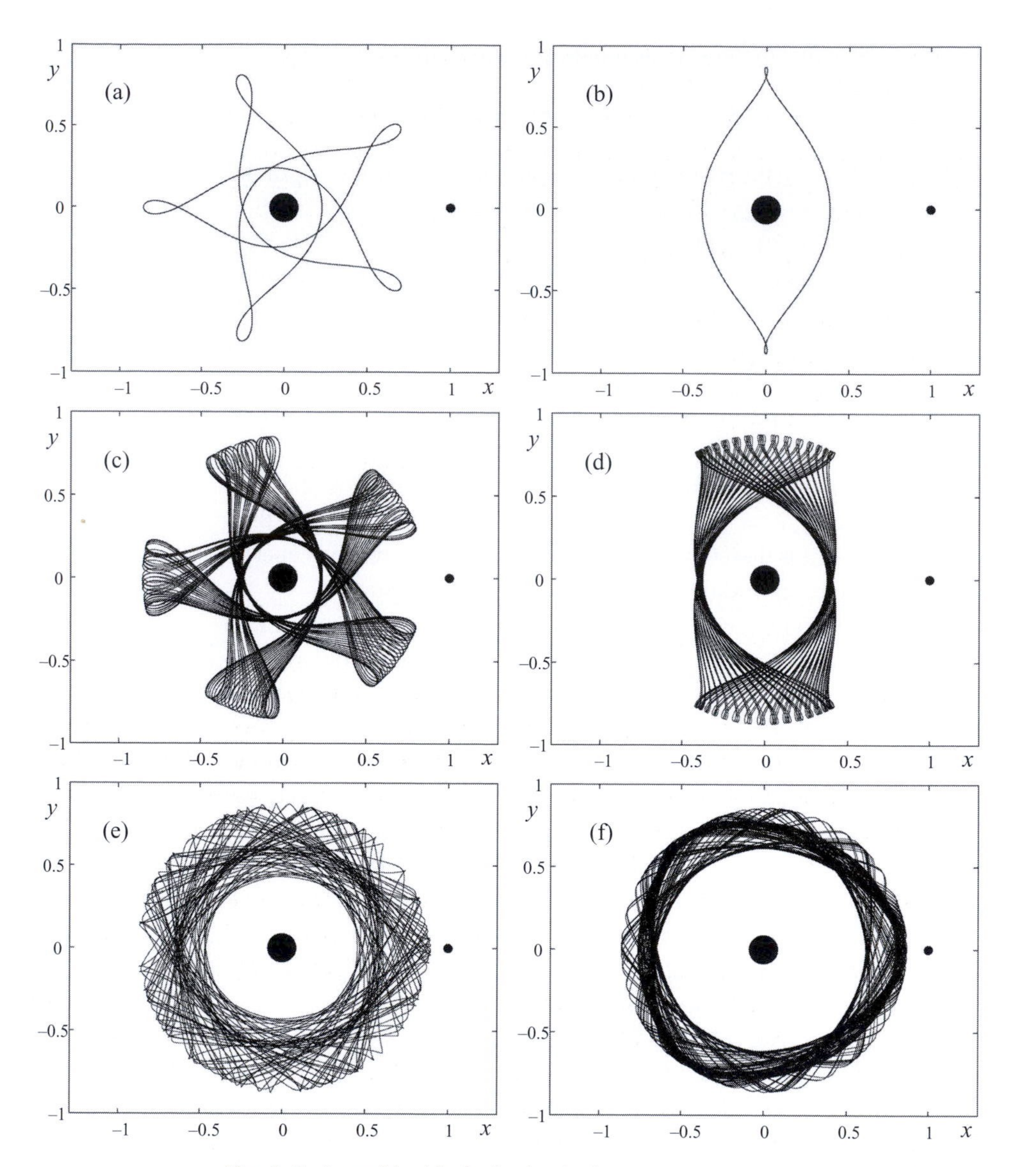

Fig. 9.5. Asteroid orbits in the (x, y)-plane corresponding to initial conditions (a)–(f) of Fig. 9.4. The quasi-periodic orbits, (c) and (d), can be obtained approximately by rotating the periodic orbits (a) and (b) to and fro around the heavy mass (the Sun) periodically. The amplitude of this rotation is proportional to the size of the quasi-periodic torus of the map which the orbit belongs to. In chaotic cases, (e) and (f), different basic patterns are mixed, depending on which elliptic island the corresponding trajectory has approached.

on the initial condition, periodic, quasi-periodic or chaotic motion occurs (Fig. 9.4). In Fig. 9.5 we also present a few orbits in real space chosen from different regimes of the map.

The complexity of the three-body problem, relevant from the points of view of both the history of science and current research, is of the same degree as that of any chaotic conservative system; for example, the spring pendulum or a ball moving in a vessel (see Section 7.4). In the same sense in which the limited occurrence of the quasi-periodic solutions of the three-body problem imply the instability of the Solar System, we can also say that the dynamics of conservative systems is usually unstable: any change in the energy or in the parameters may cause a fundamental change in the character of the motion.

Problems recommended for further studies

(i) Investigate the dynamics around the Lagrange point L_4 or L_5.

(ii) Study the chaotic scattering appearing in the three-body problem for $\mathcal{E} > \mathcal{E}_1$.

(iii) Investigate the effect of weak friction on the motion of an asteroid.

(iv) How is the motion of the small body modified under the occasional presence of some weak external force (the engines of the spacecraft are switched on)?

Box 9.1 Chaos in the Solar System

Recent space research observations and high performance computer simulations indicate that the Solar System hosts a plethora of chaotic phenomena. The dynamics of the outermost planet, Pluto, has been shown to be characterised by a positive average Lyapunov exponent of 1/(20 million years). The prediction time is thus about 20 million years, which may seem very long on human scales, but is surprisingly short on a time scale of several billion years, the lifetime of the Solar System. Due to Pluto's small mass, its feedback on the other planets is negligible. The entire system of the Sun and its nine planets has been found in various studies to be chaotic, and the results indicate that, on a scale of several million years, this manifests itself mostly in the motion of the inner planets: Mercury, Venus, Earth and Mars (the giant planets behave regularly), whose average Lyapunov exponent is found to be 1/(5 million years). This implies that an initial uncertainty of 1 km in the position of any of them increases to only 55 km in 20 million years, but it will be 5×10^8 km in 100 million years, which is larger than the Earth to Sun distance. The largest fluctuations are expected in the orbits of Mercury and Mars, and much smaller ones in the orbits of Venus and Earth. For Mercury, the fluctuation of the orbit's eccentricity can be as large as the average value. Nevertheless, the probability of the planets colliding with each other or escaping the Solar System is very small, even over a very long time scale.

The first chaotic phenomenon discovered by a satellite was the irregular rotation of Hyperion, a moon of Saturn. The moon is of elongated shape, and therefore a gravitational torque is acting on it while it is moving along its orbit. Such moons can also be considered as spinning-tops (see Section 9.2). Spacecraft *Voyager-1* observed that the intensity of the reflected light from Hyperion did not show any regularity. The rotation turned out to be chaotic, with a Lyapunov exponent of the order of 1/(10 days). A related phenomenon has been found by simulating the time dependence of the orientation of the rotation axis (tilt axis) of the planets. The inner planets seem to have gone through a period in which the dynamics of this axis was chaotic. Mars appears to be in this phase even today: the uncertainty in the position of its rotation axis can be $\pm 30^\circ$ over a few million years. These drastic changes led to the melting of the ice caps around the poles in a seemingly random sequence: the presence of dried-out water beds scattered over the surface of Mars can therefore be viewed as fingerprints of the chaoticity of Mars' dynamics. (Without the Moon's stabilising effect, even the Earth's axis would exhibit large chaotic fluctuations, which could have had catastrophic consequences for the development of life.)

The Solar System is full of small celestial bodies. Between Mars and Jupiter, several asteroids orbit the Sun. Their distribution is, however, not uniform. The density is much smaller in the so-called 'Kirkwood gaps' than outside them. The ratio of Jupiter's orbital period to that of the asteroids in the gaps is close to ratios of small integer numbers (resonances). The most significant Kirkwood gaps belong to the resonances $3/1$, $2/1$, $5/2$ and $7/3$ (the winding numbers are the reciprocals of these). The fact that very few asteroids are found around these resonances suggests that the tori corresponding to these winding numbers have been destroyed and that the respective asteroids moved away from the vicinity of Jupiter. This phenomenon appears to be in harmony with the KAM theorem (Section 7.5.2). Around other resonances (for example, $3/2$ and $1/1$), however, many asteroids are present. In fact, the KAM theorem does not apply to the Solar System since perturbations are too large due to the mutual gravitational interaction of all the constituents. Numerical simulations nevertheless indicate that after long times even the currently occupied resonances will become much less occupied.

Two groups of asteroids (the Trojans) are situated around the stable Lagrange points L_4 and L_5 of the Sun–Jupiter system. In spite of the stability, the dynamics can be chaotic quite close to these Lagrange points.

Close encounters of two asteroids orbiting a larger body lead to chaotic scattering: while far away, they hardly interact, but, upon approaching each other, a complicated motion may begin (due to an underlying chaotic saddle) and lead ultimately to a strong separation. This phenomenon may also play a role in the interaction of bodies constituting the rings of the giant planets.

Simulations of the observed asteroids that have a finite probability of a close encounter with the Earth lead to the conclusion that the dynamics of some of them is chaotic. In such cases only probabilistic predictions can be given concerning their future. A threat for our civilisation is not so much the chaoticity of the Solar System as a whole, but rather a potential collision with a large asteroid.

9.2 Rotating rigid bodies: the spinning top

Machines and technical equipment contain rotating rigid bodies, such as wheels and shafts. The rotation of these bodies can be regular only if they are exactly symmetric; otherwise a knocking motion develops that leads to the rapid abrasion and deterioration of the equipment. It is therefore important to inhibit chaos. How rare regular rotation is among asymmetric bodies is illustrated via the example of the heavy spinning top.

A rigid body with one fixed point is called a top. The mass distribution of a rigid body of arbitrary shape can be described by means of just three data, the principal moments of inertia (Θ_1, Θ_2, Θ_3). Figure 9.6(a) presents a general top of simple, rectangular, shape. In such a case, the principal moments of inertia are the moments taken with respect to axes

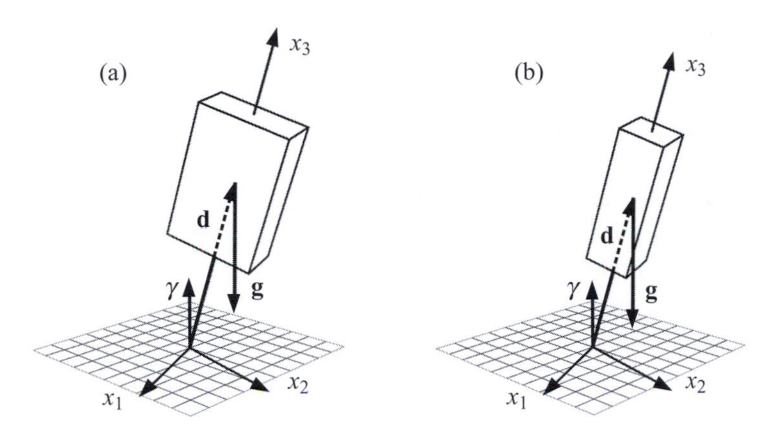

Fig. 9.6. Tops. (a) Asymmetric top. The principal moments of inertia with respect to the x_1-, x_2- and x_3-axes of the body frame are Θ_1, Θ_2 and Θ_3, respectively, and they are all different. (b) Symmetric top: $\Theta_1 = \Theta_2 \neq \Theta_3$. The grid represents a horizontal plane (not identical to the plane of x_1, x_2) and γ is a vertical unit vector. A massless rod is also part of the top: its end on the horizontal plane is the top's point of fixation.

parallel to the symmetry axes of the body. The co-ordinate system fixed to the body is oriented in such a way that the centre of gravity falls on the x_3-axis. The point of fixation is also on the x_3-axis, but in the case of a heavy top does not coincide with the centre of gravity. Gravitational force therefore exerts a torque about the point of fixation, chosen to be the origin.

A top is called symmetric if the moments of inertia with respect to the x_1- and x_2-axes are identical. These bodies are usually (but not necessarily) rotationally symmetric (everyday usage of the word 'top' implies that the top is symmetric, for example the humming-top). Figure 9.6(b) shows a pencil-shaped symmetric top.

We take a unit vector, γ, pointing vertically in a standing frame (see Fig. 9.6) and we define the positions of the x_1-, x_2- and x_3-axes of the body frame relative to this. The position of the centre of gravity is determined by vector **d**. For a complete description of the motion, the angular velocity vector, ω, of the body of total mass m must be known. In the absence of any dissipation, the equations of motion in a frame fixed to the body are given by:

$$\begin{pmatrix} \dot{\gamma}_1 \\ \dot{\gamma}_2 \\ \dot{\gamma}_3 \end{pmatrix} = \begin{pmatrix} \omega_3\gamma_2 - \omega_2\gamma_3 \\ \omega_1\gamma_3 - \omega_3\gamma_1 \\ \omega_2\gamma_1 - \omega_1\gamma_2 \end{pmatrix} \tag{9.8}$$

and

$$\begin{pmatrix} \Theta_1\dot{\omega}_1 \\ \Theta_2\dot{\omega}_2 \\ \Theta_3\dot{\omega}_3 \end{pmatrix} = \begin{pmatrix} (\Theta_2 - \Theta_3)\omega_2\omega_3 \\ (\Theta_3 - \Theta_1)\omega_3\omega_1 \\ (\Theta_1 - \Theta_2)\omega_1\omega_2 \end{pmatrix} + mg \begin{pmatrix} d_3\gamma_2 - d_2\gamma_3 \\ d_1\gamma_3 - d_3\gamma_1 \\ d_2\gamma_1 - d_1\gamma_2 \end{pmatrix}. \quad (9.9)$$

The first group of equations, (9.8), yields the velocity of vector $\boldsymbol{\gamma}$ viewed from the body frame, while the second group, (9.9), is the equation of rotation describing the change of the angular momentum due to the gravitational torque $\mathbf{M} = \mathbf{d} \times (-mg\,\boldsymbol{\gamma})$, in the same frame.[3] In the body frame, the components of the angular velocity vector, $\boldsymbol{\omega}$, of the fixed unit vector, $\boldsymbol{\gamma}$, and of $\mathbf{d}$ are $\omega_1, \omega_2, \omega_3, \gamma_1, \gamma_2, \gamma_3$ and $d_1 = 0, d_2 = 0, d_3 \equiv d$, respectively.

Equations (9.8) and (9.9) constitute a set of six first-order non-linear differential equations determining the dynamics of a top. The state of the system is characterised by variables $\gamma_1, \gamma_2, \gamma_3$ and $\omega_1, \omega_2, \omega_3$.

In the course of the top's motion there exist, in general, three conserved quantities: the energy,

$$\mathcal{E} = \frac{1}{2}(\Theta_1\omega_1^2 + \Theta_2\omega_2^2 + \Theta_3\omega_3^2) + mgd\gamma_3; \quad (9.10)$$

the vertical component of the angular momentum in the standing frame,

$$P = \gamma_1\Theta_1\omega_1 + \gamma_2\Theta_2\omega_2 + \gamma_3\Theta_3\omega_3; \quad (9.11)$$

and the length of vector γ,

$$|\gamma| = \sqrt{\gamma_1^2 + \gamma_2^2 + \gamma_3^2} \equiv 1. \quad (9.12)$$

In a general case, therefore, the phase space is three-dimensional; consequently, there is a possibility that the motion is chaotic. The aperiodic time dependence of $\omega_3(t)$ is illustrated in Fig. 9.7. The real-space motion of a top can be represented by the orbit traced out by the x_3-axis on the surface of an imagined sphere, as shown in Fig. 9.8.

We can again obtain a global overview of the flow by constructing a Poincaré map. We record ω_3 and γ_3 when $\gamma_1 = 0$ (the x_1-axis is exactly

[3] The derivation of these equations is part of standard courses and textbooks on rigid-body mechanics; therefore we do not give details. For the interested reader we briefly mention, however, the basic ideas. An arbitrary vector, $\mathbf{a}$, fixed in a frame that rotates at angular velocity $\boldsymbol{\omega}$ is seen from a standing frame to move at velocity $\dot{\mathbf{a}}' = \boldsymbol{\omega} \times \mathbf{a}$. The respective derivatives, $\dot{\mathbf{a}}'$ and $\dot{\mathbf{a}}$, of vector $\mathbf{a}$ in a standing frame and in a frame rotating with $\boldsymbol{\omega}$ are therefore related as $\dot{\mathbf{a}}' = \dot{\mathbf{a}} + \boldsymbol{\omega} \times \mathbf{a}$. Since $\dot{\gamma}' = 0$, $\dot{\gamma} = -\boldsymbol{\omega} \times \boldsymbol{\gamma}$ (see (9.8)), and since, in a standing frame, the time derivative of the angular momentum, $\mathbf{N} = (\Theta_1\omega_1, \Theta_2\omega_2, \Theta_3\omega_3)$, is the torque $\mathbf{M}$, thus $\dot{\mathbf{N}} = (\Theta_1\dot{\omega}_1, \Theta_2\dot{\omega}_2, \Theta_3\dot{\omega}_3) = \mathbf{M} - \boldsymbol{\omega} \times \mathbf{N}$, as in (9.9).

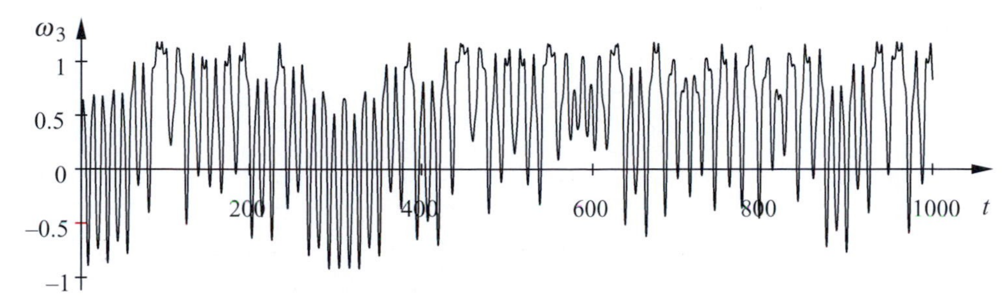

Fig. 9.7. Asymmetric heavy top: time dependence of the angular velocity component, ω_3, for moments of inertia $\Theta_1 = 2.5$, $\Theta_2 = 1.5$ and $\Theta_3 = 2$ ($\mathcal{E} = 1.7$, $P = 1.2$) and initial conditions $\gamma_1 = 0$, $\gamma_3 = -0.5$, $\omega_3 = 0$. The remaining variables, γ_2, ω_1 and ω_2, follow from the conservation laws. In all our examples, moments of inertia, energy and time are measured in units of md^2, mgd and $\sqrt{d/g}$, respectively.

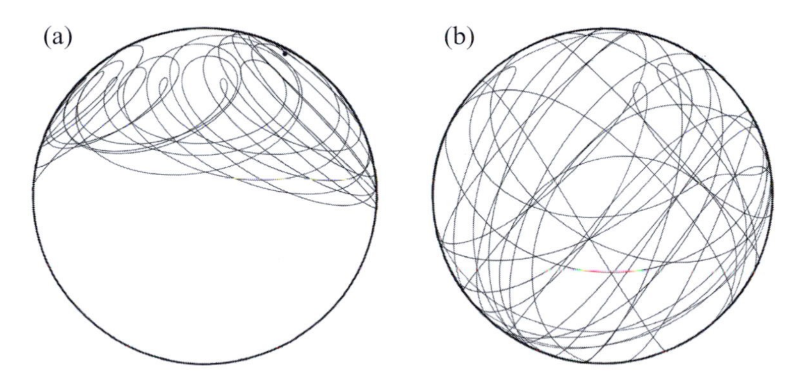

Fig. 9.8. Real-space motion of one point of the x_3-axis of the body frame traced out on the surface of a (non-transparent) sphere viewed from the side. The length of the simulations is $t = 200$ units. (a) Weak chaos: $\Theta_1 = 2.55$, $\Theta_2 = 2$, $\Theta_3 = 3$, $\mathcal{E} = 3$, $P = 2.5$, with initial conditions $\gamma_1 = 0$, $\gamma_3 = 0.5$, $\omega_3 = 1.1$. (b) Strong chaos with the same parameters and initial conditions as in Fig. 9.7. Note that in this latter case the centre of mass is often below the point of fixation.

horizontal) and when $\dot{\gamma}_1 > 0$. The phase portrait obtained in this manner (Fig. 9.9) exhibits the chaos of conservative systems.

Only two special cases are known with one more conserved quantity. The phase space dimension is then reduced to two, and the motion becomes regular.[4]

[4] For the sake of completeness, we also mention the special case of a free top, when the centre of gravity falls on the point of fixation. The resultant torque is then zero, and the angular momentum vector is constant. This implies two more conserved quantities, in total: the three components of the angular momentum, the energy and the length of vector $\boldsymbol{\gamma}$.

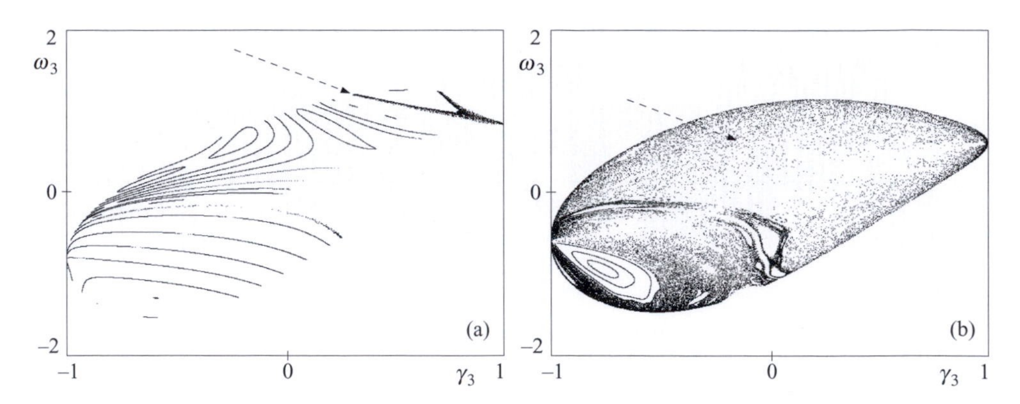

Fig. 9.9. Poincaré portraits γ_3, ω_3 of two chaotic tops. The moments of inertia, the energy and the angular momentum, P, in (a) and (b) are the same as in (a) and (b) of Fig. 9.8, respectively. The number of different initial conditions is (a) 18, (b) 10. The arrows point towards the chaotic bands from which the initial conditions of Fig. 9.8 are taken.

One of these cases is that of symmetric tops, where $\Theta_1 = \Theta_2 \neq \Theta_3$. The new conserved quantity is the projection of the angular momentum vector on the x_3-axis of the body frame, i.e.

$$P_3 = \Theta_3 \omega_3. \tag{9.13}$$

The symmetric top precesses in a quasi-periodic manner, during which the x_3-axis rotates at a constant angular velocity around the vertical axis γ, while it also oscillates in the vertical plane around some finite angle. The trajectories in the map (ω_3, γ_3) are correspondingly straight line segments (Fig. 9.10(a)). In the real-space representation (Fig. 9.11(a)), this corresponds to an orbit of wavy shape which does not close into itself. From our everyday experience, we take for granted that tops spin in a regular fashion. It is obvious from the consideration above that by fixing a weight onto the side of a symmetric humming top, its spinning may become chaotic.

The other case was discovered by Sonia Kovalevskaia in 1889. The principal moments of inertia are then $\Theta_2 = \Theta_3 = 2\Theta_1$. Although two moments are equal, this is not a symmetric case, since it is not the moments perpendicular to the axis of the centre of gravity that are identical. The dynamics appears to be more complicated than in the symmetric case, but is, nevertheless, quasi-periodic (see Figs. 9.10(b) and 9.11(b)). The new conserved quantity is called the Kovalevskaia constant (K):

$$K = (\Theta_1(\omega_3^2 - \omega_2^2) - mgd\gamma_3)^2 + (\Theta_2\omega_2\omega_3 - mgd\gamma_2)^2, \tag{9.14}$$

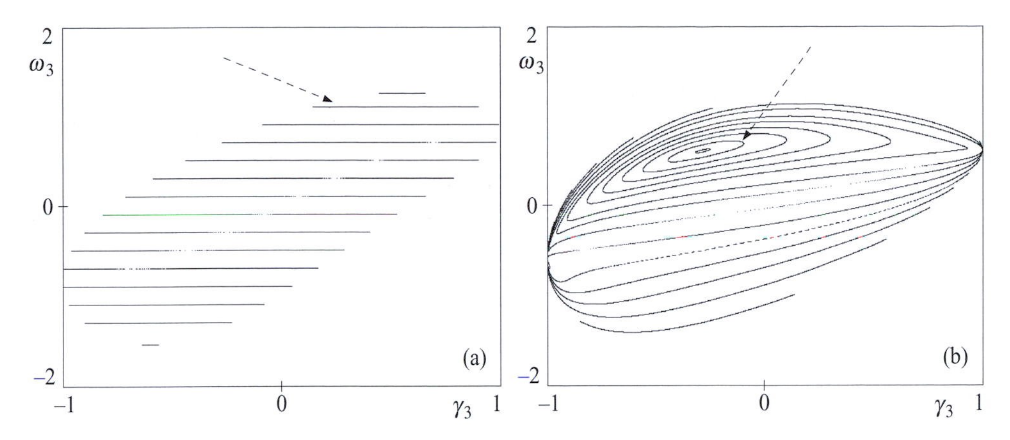

Fig. 9.10. Poincaré portraits γ_3, ω_3 of two integrable cases. (a) Symmetric top: $\Theta_1 = \Theta_2 = 2$, $\Theta_3 = 3$, $\mathcal{E} = 3$, $P = 2.5$, 15 different initial conditions. (b) The Kovalevskaia top: $\Theta_1 = 1$, $\Theta_2 = 2$, $\Theta_3 = 2$, $\mathcal{E} = 1.7$, $P = 1.2$, 15 different initial conditions. For an explanation of the arrows, see Fig. 9.11.

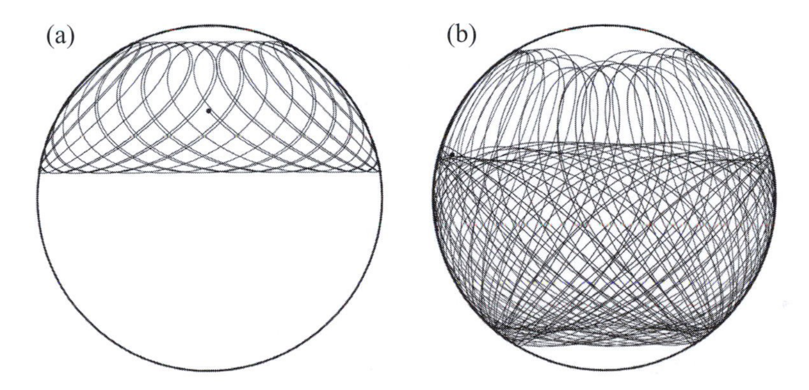

Fig. 9.11. Real-space motion of the (a) symmetric and (b) Kovalevskaia tops of Fig. 9.10. (Side views as in Fig. 9.8.) (a) $\gamma_1 = 0$, $\gamma_3 = 0.5$, $\omega_3 = 1.1$ ($P_3 = 3.3$), $t = 150$; (b) $\gamma_1 = 0$, $\gamma_3 = -0.1$, $\omega_3 = 0.7$ (the dimensionless Kovaleskaia constant $K/(mgd)^2 = 0.021$), $t = 500$. The arrows shown in Fig. 9.10 point towards the tori from which these initial conditions are taken.

whose constancy can be verified via substitution. This quantity has no direct physical meaning, in contrast to energy or the components of the angular momentum, and this is exactly what makes the Kovalevskaia's discovery so peculiar. It is remarkable that we are aware of no further moment of inertia ratios Θ_1/Θ_3, Θ_1/Θ_2 for which an asymmetric top's dynamics would be regular.

Problems recommended for further studies

(i) Investigate the dynamics of heavy tops that differ slightly from the symmetric case or the Kovalevskaia case.

(ii) Study the effect of air drag on asymmetric heavy tops.

Box 9.2 Chaos in engineering practice

Machines and equipment are designed (with a 'linear' mindset) to operate as linearly as possible. Outside the desired operational regime, where non-linear effects play an essential role, chaos is, however, quite common. Chaotic oscillations appear in a realm of motion ranging from polishing machines, through metal cutters and centrifuges, to moored vessels driven by steady waves.

An interesting phenomenon of vehicle dynamics, having attracted attention for decades, is that of shimmying wheels. (The original meaning of 'shimmy' is a dance dating back to the 1930s.) Shimmy is a lateral vibration of towed wheels which leads to unstable rolling. When pushing a shopping trolley we sometimes find that the front wheel starts dancing and the cart becomes difficult to manoeuvre. The same phenomenon can also be observed on the front wheels of bicycles and motorcycles, and on aircraft nose-wheels, and has led to serious accidents. Vehicle manufacturers therefore take precautions to avoid shimmy by utilising special equipment. Theoretical investigations show that the dynamics of shimmying wheels is chaotic. The chaos is of transient type, but the lifetime can be rather long. Vehicles designed to be immune against shimmy nevertheless remain quite sensitive: for example, fixing a small bag to the rear seat may make a motorcycle susceptible to shimmying.

A related problem is that of trailers. Even a single trailer can exhibit irregular lateral motion. Perturbations arising due to sudden severe side winds or sudden steering constraints (resulting, for example, from the appearance of a dog in the road) may result in lateral oscillations of truck trailers, which resembles shimmying (see Fig. 9.12). Such oscillations may develop to the point of overturning: accidents like this are commonplace with private caravans. In most countries the use of a second trailer is prohibited because it may drift into the neighbouring traffic lane due to non-linear oscillations.

The dynamics preceding the capsize of a vessel subjected to periodic sea waves or, more generally, escaping from metastable states (potential wells) is often (transiently) chaotic.

A wide range of control methods is available to keep machines within the desired operational modes. Small-scale chaos, however, often proves inevitable in control devices, which assure large-scale periodicity. One example is the balancing of our own bodies.

The buckling of long elastic rods or cables has long been a central problem of engineering. Although this is basically a static problem, an interesting analogy relates it to dynamics. The shape of such systems is described by ordinary differential equations, in which derivatives are taken with respect to the arc length

Fig. 9.12. A truck trailer combination on the stability limit, by Knorr Bremse.

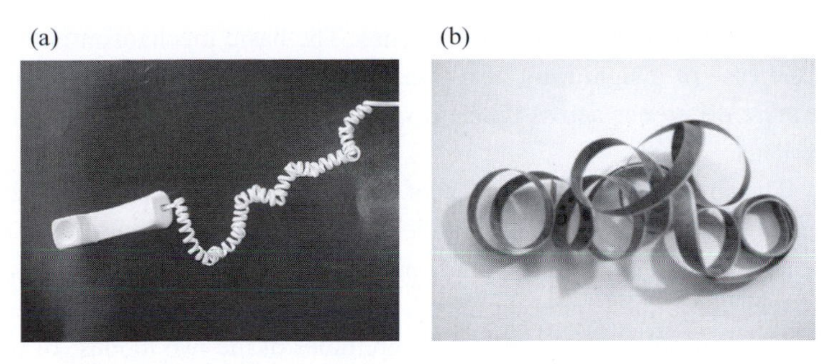

Fig. 9.13. Coiled elastic structures are spatial analogues of chaotic trajectories. (a) Telephone cord (photograph by G. Károlyi). (b) Unfolded serpentine.

along the rod. If we identify arc length with time, the static problem can be mapped onto a dynamical one. In particular, the equation for a stretched and twisted anisotropic rod turns out to be equivalent to that of an asymmetric spinning top. The case of a rod with a periodically varying cross-section corresponds, for example, to that of a torque-free asymmetric top with one of the principal moments of inertia changing periodically in time. The possible spatial shapes of such a rod can thus be qualitatively as complicated as the trajectories of a chaotic top. This is quite understandable by recalling a telephone cord or a vine tendril, not to mention a stretched electric wire that has been stored in a coil. These studies are relevant in fields ranging from macromolecular structures, such as, for example, DNA coils, to marine pipelines and communication cables, and they provide the opportunity to define purely spatial chaos in the context of buckling rods (Fig. 9.13).

9.3 Climate variability and climatic change: Lorenz's model of global atmospheric circulation

In order to describe accurately the instantaneous state of the Earth's entire atmosphere, a very large number (nearly 10^7) of variables has to be used. Predicting the future of the system (i.e. forecasting the weather) is based on the integration of the governing hydrodynamical equations using modern numerical codes. Special techniques have been developed to facilitate the handling and interpretation of so many variables. A conceptual understanding of the essential behaviour of the atmosphere is, however, often supported by models with *few* variables. Naturally, these are only suitable for reflecting a few general features, and, contrary to the problems studied so far, they are of heuristic nature. One of the simplest models, which nevertheless provides considerable insight into the subject, is Edward Lorenz's model of general circulation. It mimics the behaviour of the Earth's atmosphere averaged over mid latitudes on one (for example the northern) hemisphere, with just three variables. The model formulates the balance between the energy of the prevailing west wind and the energy transported polewards by large-scale atmospheric vortices: the cyclones and anti-cyclones. The basic mechanism is that, in the case of a significant heat transfer to the north, the average temperature difference across the hemisphere decreases and the west wind weakens.

The dimensionless variable, x, of the model is the velocity of the west wind averaged over the mid latitudes. Its magnitude is proportional to the north–south temperature gradient, i.e. to the temperature difference between the Equator and the North Pole. The two other, also dimensionless, variables are y and z, the amplitudes of the two modes[5] of the poleward heat-transport, which takes place via the large vortices, and the energies transported by them are taken to be $y^2/2$ and $z^2/2$, respectively. The time evolution of these variables is determined by the differential equations[6]

$$\dot{x} = -y^2 - z^2 \qquad - ax + aF, \tag{9.15a}$$

$$\dot{y} = xy - bxz - y + G, \tag{9.15b}$$

$$\dot{z} = xz + bxy - z + G', \tag{9.15c}$$

where all the parameters are positive and $a < 1, b > 1$.

The terms in the first 'column' on the right-hand side of equations (9.15) express that the energy, $(y^2 + z^2)/2$, transported polewards decreases the kinetic energy, $x^2/2$, of the west wind[7] (and, consequently, the temperature difference). If the model consisted of these terms only, the total energy, $\mathcal{E} = (x^2 + y^2 + z^2)/2$, would be constant. This energy is not modified by the terms in the second 'column' either; these characterise the energy exchange between the modes y and z. Parameter values b greater than unity express that this exchange is faster than that between the vortices and the west wind.

The terms in the third 'column' describe damping due to atmospheric dissipation. In (9.15a), the choice $a < 1$ reflects that the damping of the west wind is weaker than that of the vortices. The coefficient of y and z is -1, indicating that time is also dimensionless. The time unit is the damping time of the vortices, which can be considered to be five days in the atmosphere. The terms in the final 'column' correspond to constant driving due to the temperature differences originating from the uneven warming of the Earth by the Sun. In a stationary case, without non-linear

[5] Modes are basically different spatial forms of energy transport; y and z are similar to variables A and B in the water-wheel dynamics, the respective coefficients of the sine and cosine terms of the angle dependence in equation (5.94).

[6] In their mathematical structure, these are similar to the equations for the water-wheel (Section 5.7) or of the Lorenz model of Box 5.6, although the phenomenon is completely different.

[7] To see this, multiply the first equation by x.

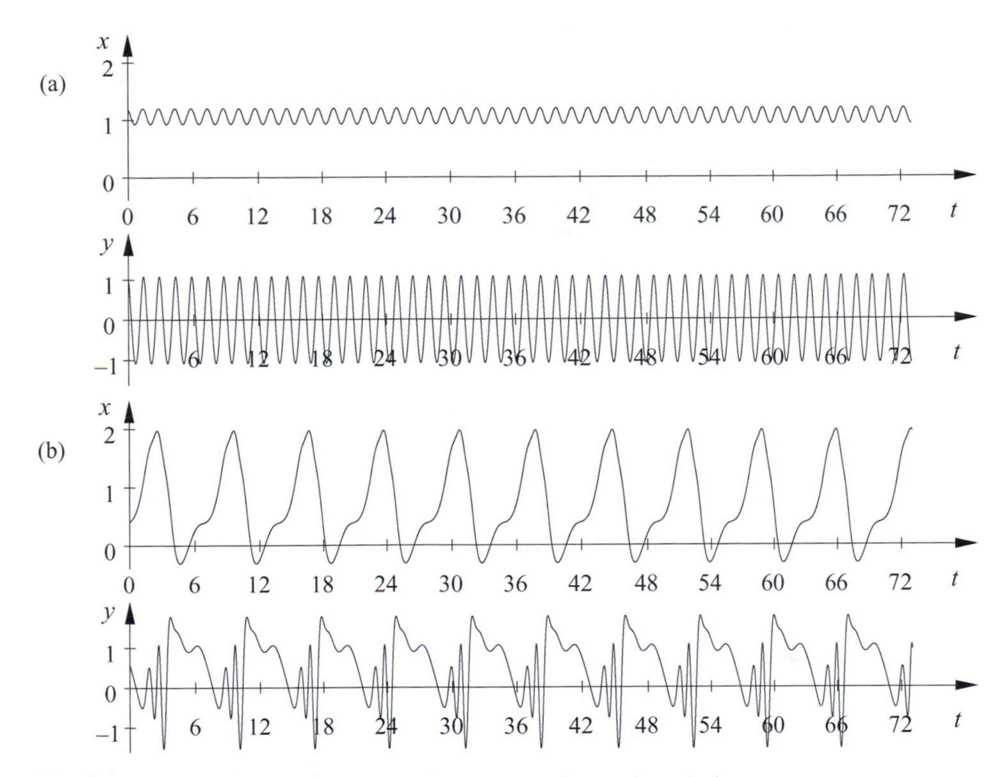

Fig. 9.14. The two limit cycle attractors (component z is not shown) of permanent summer, $F = 6$. (a) Inactive summer (small amplitude, nearly sinusoidal limit cycle). (b) Active summer (large amplitude, non-trivial limit cycle). The signals are shown over 73 time units corresponding to one year (6 units $\approx$ 1 month).

interactions, variables x, y and z would take the values F, G and G', respectively. Therefore, F can be considered as the average value of the west wind's velocity or of the temperature difference between the Equator and the North Pole. Similarly, G and G' are the driving forces of the poleward transport. Since the temperature difference between the Equator and the Pole is larger in winter than in summer, larger values of F can be assigned to winter situations. The standard parameters of the model are $F = 6$ (summer), $F = 8$ (winter), $a = 0.25, b = 4, G = 1$ and $G' = 0$. In the full model, equations (9.15), energy $\mathcal{E}$ is not constant. For small values it grows due to the driving, while for large ones it decreases due to damping. There may therefore exist non-trivial stationary states, i.e. attractors.[8]

[8] The phase space contraction rate is $\sigma = a + 2 - 2x$, which is negative for $x > 1 + a/2$. For $x > 1 + a/2$, repellors may thus also exist.

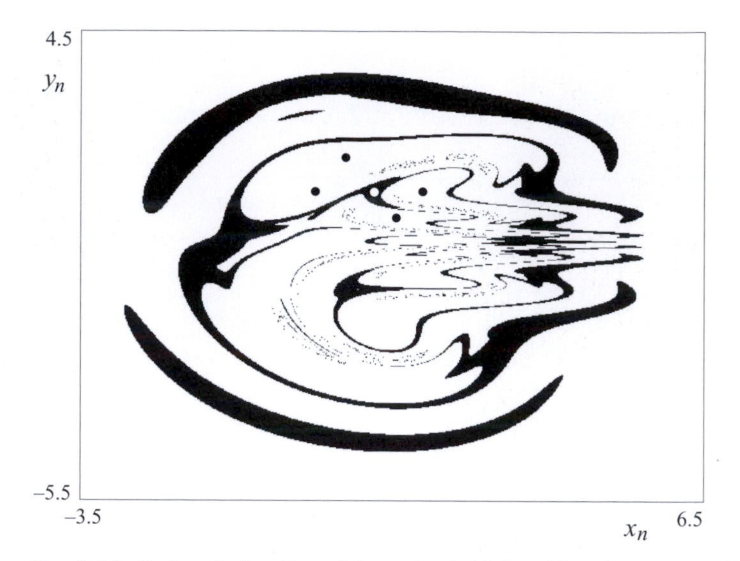

Fig. 9.15. Basins of attraction of the active (white) and inactive summer (black) in the plane of initial conditions with $z = 0$ ($F = 6$). The active summer attractor is represented on this plane by four points marked by black dots (cf. Fig. 9.14(b), y vs. t). The inactive summer attractor appears as a white dot. Note that the inactive basin is much smaller than the active one and that it does not extend towards large values of the variables.

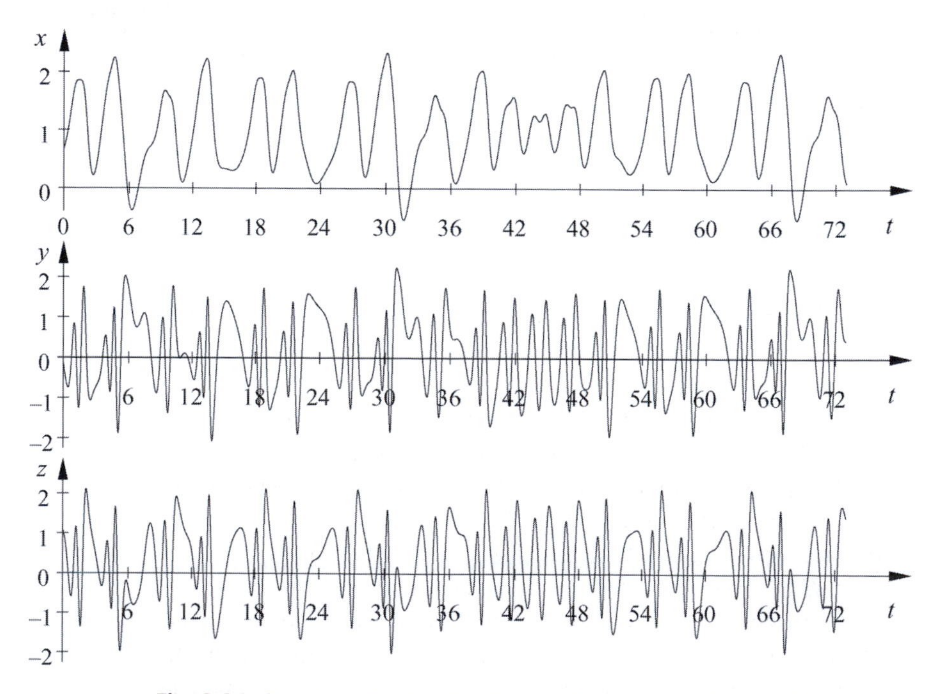

Fig. 9.16. Signals on the chaotic attractor of permanent winter at $F = 8$.

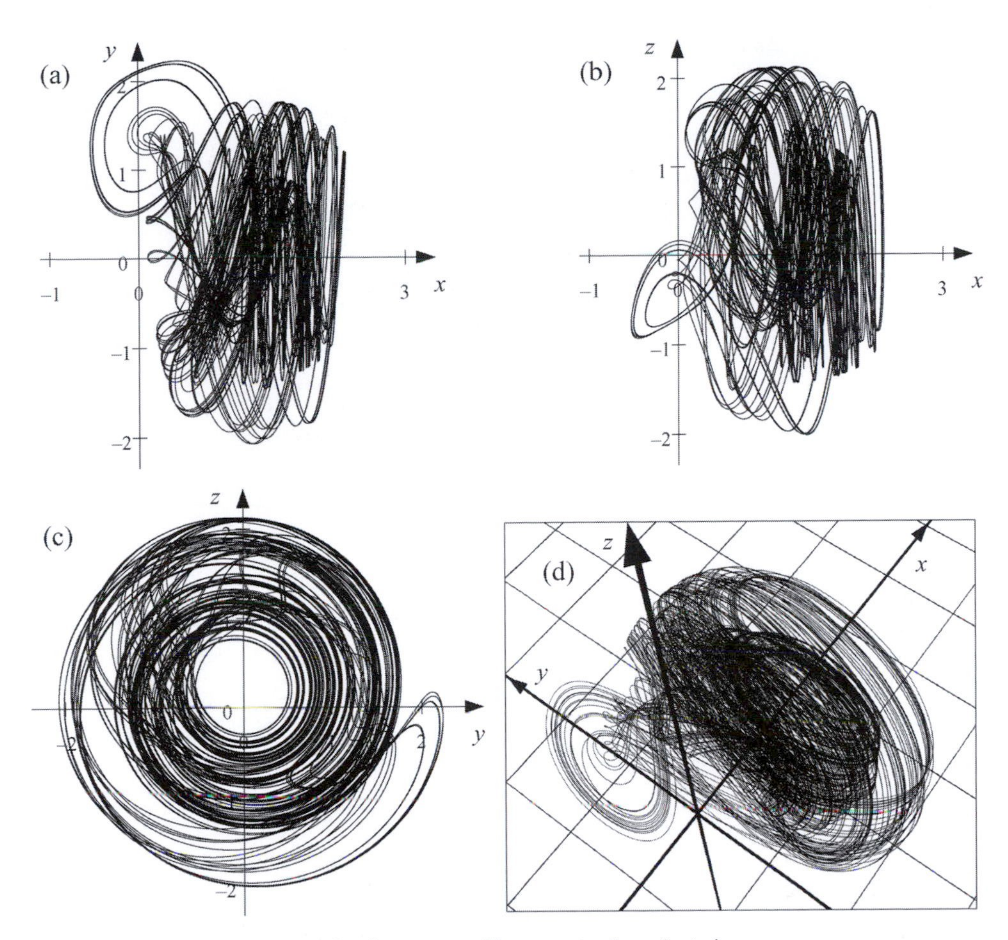

Fig. 9.17. The winter attractor in phase space. The attractor is projected on planes (a) (x, y), (b) (x, z), (c) (y, z) and (d) in a stereoscopic view.

For $F = 6$ (permanent summer) two simple attractors (limit cycles) co-exist (see Fig. 9.14), the system is therefore bistable. One of the limit cycles has a smaller amplitude and is nearly sinusoidal, while the other is less regular. They can be considered to characterise two different types of summer: inactive and active summers, respectively.

Although permanent chaos cannot be found, the convergence towards the simple attractors is preceded by transient chaos. Accordingly, the basin boundary between the two attractors is a fractal (Fig. 9.15).

For $F = 8$ (permanent winter), a single attractor exists, and it is chaotic: the climate exhibits strong internal variability. Figures 9.16 and 9.17 present the chaotic time series of the three variables and the chaotic attractor in different views, respectively. The small loop on the attractor

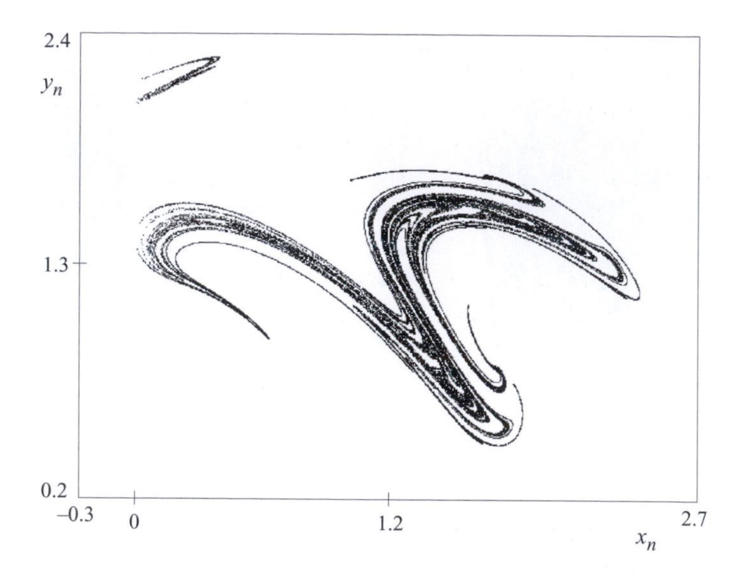

Fig. 9.18. The chaotic attractor ($F = 8$) of permanent winter on the Poincaré map (x_n, y_n).

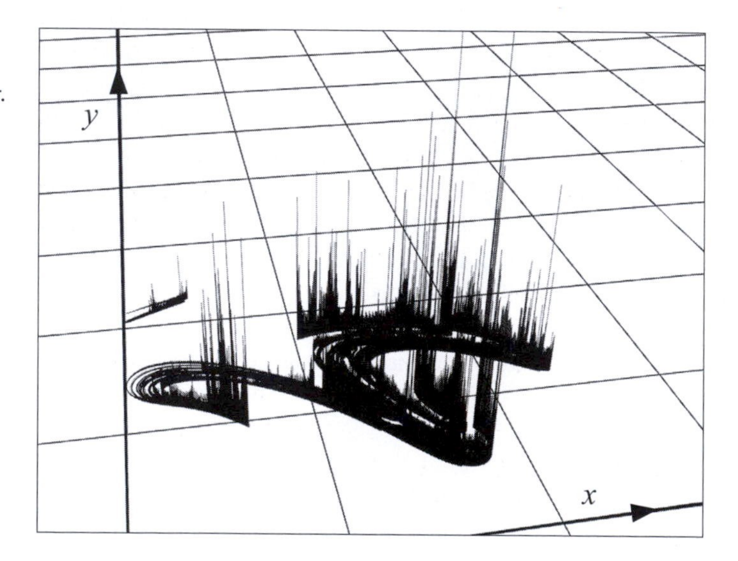

Fig. 9.19. Natural distribution on the chaotic attractor of permanent winter. A single trajectory of 10^7 intersection points is generated and the points are plotted on a grid of size $\varepsilon = 1/500$.

is of special interest since it is in a region with x-values falling around the origin. This corresponds to a situation in which the prevailing wind might change direction.

We also represent the chaotic attractor on a Poincaré section, which can be chosen to be the $z = 0$ plane ($\dot{z} > 0$) as shown in Fig. 9.18. The small loop of Fig. 9.17 appears here as a small isolated block around $x_n = 0$, $y_n \approx 2$. From a single long run, the natural distribution can also be determined (Fig. 9.19 and Plate XXII). It is clearly visible that the distribution is rather irregular. Huge spikes appear nearly everywhere indicating that the probability of two slightly different weather conditions can differ drastically. Furthermore, extreme events, such as, for example, very strong winds or east winds (large positive or very small negative x_n-values, respectively), carry very low probabilities only, but being parts of the attractor, they definitely occur in a sufficiently long observation period.

The cases treated so far indicate that the variability of the global circulation may be chaotic *without* any change of the parameters (the system is autonomous). It is, however, natural to take into account the annual periodicity of the heating.[9] This is, to a good approximation, a sinusoidal driving that can be written as

$$F(t) = F_0 + A\cos(\omega t). \tag{9.16}$$

Here $\omega = 2\pi/73$ is the annual frequency according to the five-day time unit and A is the amplitude of the temperature gradient's oscillation. With the values $F_0 = 7$, $A = 2$, the averages of the three winter and summer months are just around $F = 8$ and $F = 6$, respectively. In this continuously driven system, no periodic behaviour can be observed at all. Active and inactive summers follow each other, separated by winters, in an irregular sequence (Fig. 9.20). The variability of the climate becomes stronger. This can be interpreted as a consequence of the wandering on the attractor of permanent winter during the winter months. When summer sets in, the system is left at random in the basin of one of the permanent summer's limit cycles. Naturally, due to the time dependence of F, none of these attractors is present in an exact sense. In fact, the chaotic attractor of the driven system lives in a four-dimensional phase space (three-dimensional map) and contains, with certain probabilities, behaviour similar to that represented by the constant-F attractors. The model reflects the general feature that periodic (seasonal) variation of temperature gradients makes long-term weather periodicity impossible.

Problems recommended for further studies

(i) What would the general circulation be like for weak solar radiation? Does a steady west wind always exist?

(ii) Is there a climatic change if – by modelling anthropogenic effects – parameters F_0 and G become shifted, or more extreme temperature fluctuations appear (A increases)?

[9] Daily fluctuations are not taken into account by the model since the time unit is longer than a day.

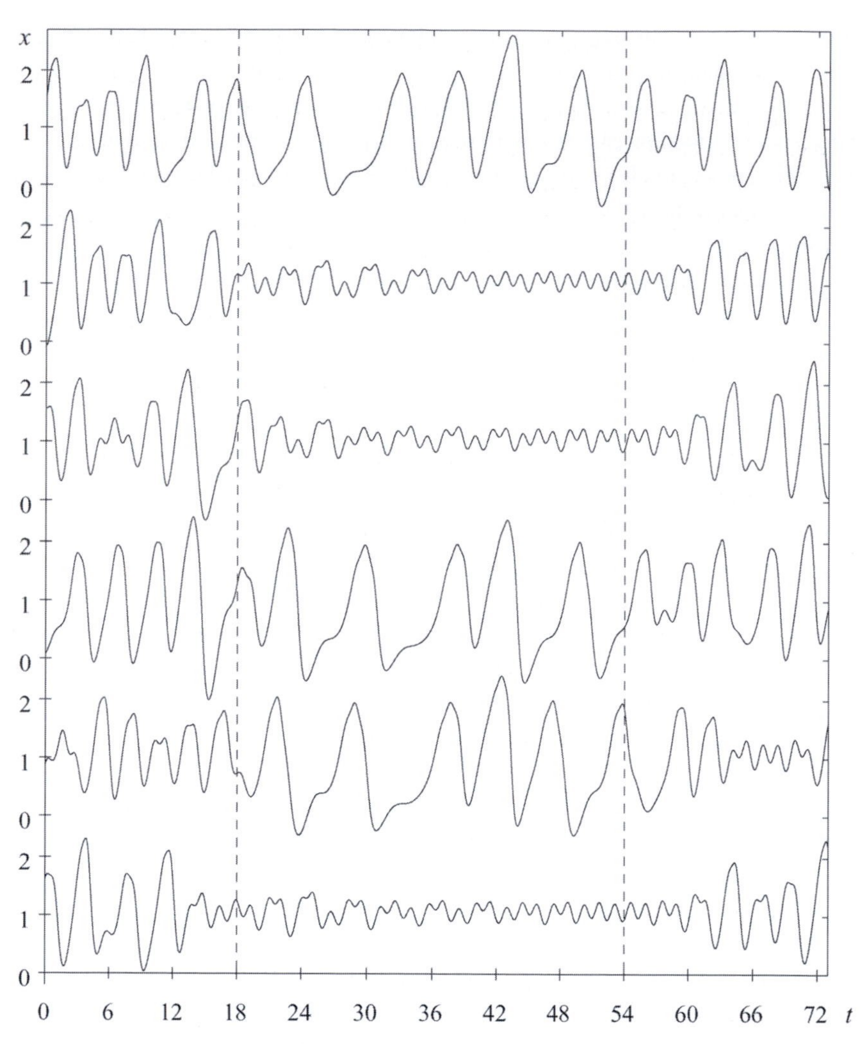

Fig. 9.20. Chaotic attractor of the seasonally driven model: active and inactive summers appear in a random sequence ($F_0 = 7, A = 2$). The signal is shown over a period of six years. The dashed lines indicate the winter/summer boundaries.

Box 9.3 Chaos in different sciences

Meteorology: The detailed analysis of meteorological data does not indicate that the atmosphere is chaotic in the sense used throughout the book; i.e., the atmosphere does *not* possess a low-dimensional chaotic attractor. The dynamics of the atmosphere is, nevertheless, unpredictable. This is utilised by the so-called 'ensemble forecasts', operational at all leading weather services. In this framework, numerical runs of the prognostic codes are carried out for the same time interval of a few days with about 50 equally possible different initial atmospheric conditions (in very much the same way as we follow the evolution of droplets in phase space (see, for example, Fig. 6.18). The divergence of different trajectories can be used to indicate situations of better or worse predictability. In the latter case, probabilities can, and should, be associated with the different possible atmospheric events (rainfall or sunshine, for example). The maximum average prediction time proves to be of the order of a week. According to a recent observation, at certain geographical locations and at certain times the atmosphere may, however, behave as if it were a system composed of only a few elements. The identification of the locations of low-dimensional behaviour and of the local positive Lyapunov exponents can lead to improved forecasting. Another method is based on data obtained from different observation platforms (for example aeroplanes) sent to these geographical locations of pure predictability ('targeted' observations) with the aim of providing precise initial conditions for a further forecast. In this way, the prediction of certain storms, for example, has considerably improved.

Geophysics: Chaotic processes also occur in geophysical phenomena, typically on the same long time scale as those found in astronomy (see Box 9.1). A notable example is the Earth's magnetic field, which is known to have repeatedly changed polarity over geophysical times. The dynamics resembles that of the water-wheel or the Lorenz model. The motion of the Earth's crust (lithospheric plates) driven by the convection in the Earth's mantle can be considered to be chaotic, which explains in part why earthquakes break out so unpredictably.

Plasmas and lasers: The dynamics of charged particles in electric and magnetic fields is typically chaotic, as can be observed in particle accelerators, in fusion equipment or in interstellar plasma. Even the amplitude of magnetohydrodynamic waves may change chaotically due to non-linear interactions among different wave components. A somewhat similar phenomenon is the chaotic light emission of lasers.

Electronic circuits: Any electronic circuit containing non-linear elements may produce a voltage of chaotic time dependence. Some of these circuits are surprisingly simple. More generally, the auto-excitation of electronic circuits often corresponds to the signal becoming chaotic. Although auto-excited systems were once considered to be noisy, in view of chaos science, this behaviour is now believed to be of deterministic origin.

Nanotechnology: The technological progress made since the mid 1990s has made possible the manufacture of mesoscopic semiconductor samples of a few microns in size. To these samples, within which electrons can move freely, channels are connected. Electrons coming in through one channel and leaving through another experience chaotic scattering. The whole problem is thus analogous to that of open billiards. On this micro-scale, the laws of quantum mechanics are valid. Nevertheless, in certain quantum features, parameters of classical dynamics do show up; for example, the width of the conductance correlation function is determined by the average lifetime of transient chaos. Another recent development is the need for miniaturistion of flow devices used in biotechnology, which led to the fabrication of channels of a few hundred microns in cross-section. In these micromixers efficient mixing of materials is only possible with chaotic particle dynamics (see Section 9.4).

Acoustics: A basic law of room acoustics states that sound intensity decays exponentially in time. In view of the phenomenon of transient chaos and chaotic scattering, this can be interpreted as the consequence of an escaping process: some of the chaotically bouncing sound rays escape the room by entering small holes in the wall (where their energy is attenuated), in very much the same way as if a window or door were left open. A concert hall is, from this point of view, an open billiard. Most musical instruments are intended to produce clear sounds containing the fundamental notes and their harmonics. At strong sound intensities, however, non-linear effects generate unpleasant sounds: playing fortissimo on wind instruments, for example, is therefore a continuous fight against chaos. Some other instruments, such as gongs, always generate a broadband sound spectrum, a sign of chaos. Many acoustic phenomena, among other vocal instabilities, can be analysed by means of dynamical system techniques (bifurcation diagrams, Poincaré maps, etc.).

Chemical reactions: (see also Box. 8.1): A class of macroscopical chemical and biochemical reactions exhibits time periodic behaviour, for example in the form of a periodic change of colour. With different parameters, these have been shown experimentally to exhibit chaotic oscillations (changes of colour) occurring irregularly in time. To obtain permanent chaos, a permanent flux of fresh material is needed. Without this, reactions go towards a state of thermal equilibrium, which is only compatible with a time-independent chemical state (no change of colour). It has been found, however, that, with initial conditions very different from those of the asymptotic state, non-trivial temporal behaviour can set in, which can even be chaotic. Thus, transient chaos can be a precursor of thermal equilibrium .

Medicine, physiology: Chemical chaos might be the basis of the fact that some physiological phenomena are also related to chaotic behaviour. One of the best known examples is the electric activity of the heart. The ECG signal measuring this appears to be periodic for a healthy person, but its detailed analysis reveals chaotic deviations from exact periodicity. It is generally true that healthy physiological functions exhibit small chaotic fluctuations about the normal values (similar to chaos in engineering control; see also Box. 9.2), and chaotic components disappear in the case of illness only. The evolution of certain epidemics in large cities is also a chaotic process. For example, New York measles epidemics have been shown to be a transiently chaotic phenomenon.

Biology, ecology: The change in the number of living organisms in a given area may also be chaotic. A thoroughly investigated example is that of a kind of flour beetle whose population appears in either a larval, pupal or adult form. The interaction between the populations of these life stages is non-linear due to cannibalism (eggs are eaten by larvae and adults and pupae are eaten by adults), and can be described by a three-variable map (one iteration step corresponding to two weeks, the average duration of all the life stages). Laboratory cultures of these beetles can be utilised, and some parameters, such as, for example, the rate of mortality and cannibalism can be set precisely. The experimental observations have been shown to be in perfect agreement with the model predictions. This biological system produces several phenomena of non-linear dynamics, for example chaotic attractors, co-existence of several attractors, fractal basin boundaries and chaotic transients. In population dynamics, the number of individuals is an integer. When this number is small (the population is close to extinction), and environmental or demographic noise is weak, integer-valued maps appear to be more adequate tools to describe the dynamics than traditional maps or differential equations. Chaotic phenomena may accompany numerous other biological processes, such as the dynamics of food chains, animal behaviour and competition between species.

Internet: The competition between different internet applications (for example e-mail) being transmitted over the same network may lead to a chaotic time dependence of the transfer protocol. This might be an explanation for the occasionally very long waiting times.

Economy: Economy as a whole is a very complex system. In the spirit of the book, it cannot be considered therefore to be chaotic: it is more complicated than that. Nevertheless, certain elementary economical processes can be described by simple, non-linear mathematical models, and these can exhibit chaotic behaviour.

Box 9.4 Controlling chaos

The chaotic behaviour of equipment is, in certain cases, undesirable from the point of view of the user, and is thus to be avoided. It is therefore worth knowing that chaos *is controllable*; i.e., chaotic dynamics can be converted into regular, periodic motion. One of the most often used control method, the OGY algorithm, developed in 1990 by Ott, Grebogi and Yorke, utilises exactly the chaotic character of the dynamics to be controlled. The two essential facts on which the algorithm is based are as follows:

- that chaotic sets (attractors, bands or saddles) contain an infinite number of hyperbolic periodic orbits,
- and that, due to the existence of a natural distribution, the state of the system visits arbitrarily small neighbourhoods of any periodic orbit sooner or later (ergodicity, cf. Box 7.4).

An important feature of the algorithm is that it only applies *small* controlling effects. Such a control is economical, since tiny external perturbations are sufficient to allow the system to reach the desired behaviour.

The first step of the OGY algorithm is to choose a desired periodic orbit on the chaotic set, with the aim of directing the system's dynamics towards the dynamics of this orbit. Next, a small neighbourhood of the orbit, for example a circle around one of the cycle points, is taken, and control is applied only if the state point enters this neighbourhood. Within this region, the linearised dynamics must be known, in particular the eigenvalues and the stable and unstable manifolds of the hyperbolic cycle should be determined. The control itself is maintained by changing one of the system parameters *proportionally* to the distance of the state point from the closest point of the pre-selected cycle. Instead of the original map $\mathbf{r}_{n+1} = M(\mathbf{r}_n, \mu)$, a modified map, $\mathbf{r}_{n+1} = M(\mathbf{r}_n, \mu + \delta\mu_n) \equiv M_n(\mathbf{r}_n, \mu)$, is applied with a parameter perturbation $\delta\mu_n$ proportional to the distance $\mathbf{r}_n$ from the cycle point. Note that the originally autonomous system is thus converted into a non-autonomous one,

$$M \to M_n,$$

whose form thus also depends on the time instant. A *feedback* has been created between the instantaneous state of the system and its parameter. The parameter perturbation is chosen so that the image under the action of M_n falls on the *stable manifold* of the pre-selected cycle in the linearised dynamics (see Fig. 9.21). Knowing the map M and its linearised version, the required $\delta\mu_n$ can be explicitly determined. In a piece-wise linear system, control would be maintained by such a single control step since the iterated point then approaches the cycle along the straight line of the stable manifold. In general, however, the algorithm should be kept active in later steps also, since the linear approximation is not exact and therefore the image point falls somewhat off the exact manifold. The controlled dynamics converges towards the pre-selected cycle quite quickly, and remains in its vicinity over a long time. By means of a parameter perturbation, we can thus convert an *unstable cycle* into an *attracting limit cycle*.

The OGY algorithm has been applied widely. In several cases, successful experimental chaos controls have been implemented in, among others, electronic circuits, chemical reactions, lasers and biological systems (like the heart).

Other ways of manipulating chaos makes it possible to direct the dynamics to certain regions of the chaotic set, to synchronise the performance of chaotic systems and to carry out secret communication with coded chaotic signals. All these extend the possible applications of chaos.

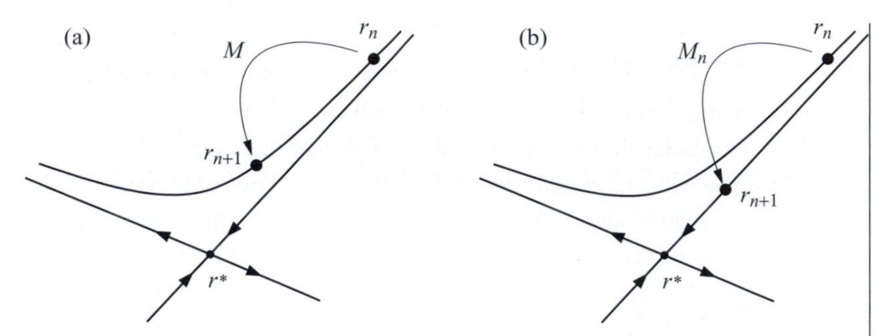

Fig. 9.21. The OGY control (schematic diagram). The pre-selected periodic orbit is a hyperbolic fixed point, $\mathbf{r}^*$, of the two-dimensional map, M. Point $\mathbf{r}_n$ falls within the control region. (a) The map without applying control. The curve is a hyperbola; the straight lines represent the manifolds in a linear approximation. (b) The map in the presence of control: the image point falls on the straight-line approximant of the stable manifold. The applied parameter perturbation, $\delta\mu_n$, proves to be proportional to the deviation of $\mathbf{r}_n$ from the straight line of the stable manifold.

9.4 Vortices, advection and pollution: chaos in fluid flows

The description of a fluid[10] flow implies knowledge of the velocity field, i.e. knowledge of the velocity, $\mathbf{v}$, in any point, $\mathbf{r}$, of space and at any instant, t, of time. The velocity field, $\mathbf{v}(\mathbf{r}, t)$, can be obtained via experimental observations and/or by solving numerically the basic equations of hydrodynamics. In the following we assume that velocity changes smoothly both in time and in space, i.e. the flow is not turbulent (cf. Box. 10.1).

One possibility for the hydrodynamical appearance of chaos is that the velocity at any fixed position changes chaotically in time. We emphasise that this only implies that at the specified position the velocity is not periodic or quasi-periodic: the velocity field does not repeat itself.

The other possibility is more surprising. It is related to the phenomenon of advection, i.e. to the issue of the motion of a dye particle or a pollution grain in the flow. Before the appearance of chaos science, it was believed that in simple, for example in time-periodic flows, advection was also simple. The fact that this is usually not true can be read off from the equation of motion.

Let $\mathbf{r}(t)$ denote the position of the particle at time instant t. Our aim is to determine the complete function, $\mathbf{r}(t)$, i.e. the orbit of the particle in a

[10] Gases are also considered to be fluids.

given velocity field, $\mathbf{v}(\mathbf{r}, t)$. In a wide range of the practical applications, the advecting particle can be assumed to be small and light; consequently, it adopts the velocity of the surrounding medium *instantaneously*. The velocity, $\dot{\mathbf{r}}$, must therefore equal the velocity, $\mathbf{v}$, of the fluid at position $\mathbf{r}(t)$ and time instant t. The equation of motion[11] of advection is therefore

$$\dot{\mathbf{r}} = \mathbf{v}(\mathbf{r}, t). \tag{9.17}$$

Since $\mathbf{v}(\mathbf{r}, t)$ is given, solving the advection problem implies solving an ordinary differential equation that is usually non-linear. The condition for the appearance of chaos is that the phase space be at least three-dimensional. Consequently, the motion of advected particles is usually chaotic both in *three-dimensional* (even in stationary fluid flows) and in *two-dimensional time-dependent* fluid flows. In the latter case, chaos already appears for the simplest, periodic time dependence.

A special feature of the advection problem (9.17) is that its variables are the space co-ordinates themselves, thus (see Section 3.5.1) the phase space *coincides* with the configurational space (which is to be augmented with the phase axis for time-periodic flows). Therefore, fractal structures usually appearing only in phase space can be observed with the naked eye or on photographs in the context of advection!

The phase space contraction rate, according to (3.55), is $\sigma = \mathrm{div}\,\mathbf{v}$. It is known from hydrodynamics that the divergence of the velocity field is proportional to the density change. In compressible media, advection may lead to the appearance of attractors, i.e. to the accumulation of the particles at certain positions. The typical velocity of fluids in advection problems is, however, much lower than the speed of sound in the medium; the fluid can therefore be considered to be *incompressible*: $\mathrm{div}\,\mathbf{v} = 0$. Accordingly, advection dynamics exhibits the chaos of *conservative* systems (open or closed, depending on the nature of the flow).

In connection with the mixing of dyes and the spreading of pollutants, it is usually not the motion of a single particle that is of interest, rather it is that of an ensemble of particles, of a dye droplet. We call this the *droplet dynamics*. We have seen that a droplet slowly flowing out of a given domain always traces out the unstable manifold of a hyperbolic set, which, for chaotic dynamics, possesses a fractal structure. An important consequence of this is that droplets or pollutant stains spread, in general, by forming *fractal-like patterns*. In the advection problem, the abstract concept of unstable manifold materialises in a *visible* object. The following sections illustrate different facets of chaotic advection dynamics.

[11] Note that this is not a Newtonian equation, since force and acceleration do not appear in it.

9.4.1 Tank with two outlets (advection in open flows)

First, let us examine the effect of a single outlet in a large, flat container, located at the origin. The velocity of the flow towards the outlet at distance r is $v_r = -Q/r$, where $Q > 0$ is a constant, the strength of the outflow. From everyday experience, it is known that a rotational flow accompanies drainage. Consequently, a flow of velocity, v_φ, perpendicular to the radius is also present, and it is assumed to have the form $v_\varphi = K/r$, where $K > 0$ is the strength of the circulation. In the velocity field given by the above two relations, the equation of advection, (9.17), implies that the respective radial and peripheral velocities, $\dot{r}$ and $r\dot{\varphi}$, of the particle equal v_r and v_φ:

$$\dot{r} = v_r = -\frac{Q}{r}, \qquad \dot{\varphi} = \frac{v_\varphi}{r} = \frac{K}{r^2}. \tag{9.18}$$

The solution with initial condition (r_0, φ_0) at $t = 0$ is given by

$$r(t) = (r_0^2 - 2Qt)^{1/2}, \qquad \varphi(t) = \varphi_0 - \frac{K}{Q}\ln\frac{r(t)}{r_0}. \tag{9.19}$$

A particle starting from distance r_0 reaches the outlet in time $r_0^2/(2Q)$, along a spiralling orbit.

At this point, it is useful to turn to a representation in terms of complex numbers: $z = r\exp(i\varphi)$. The complex form of solution (9.19) (with initial condition $z_0 = r_0 \exp(i\varphi_0)$) is given by

$$z(t) = z_0\left(1 - \frac{2Qt}{|z_0|^2}\right)^{(1-iK/Q)/2}. \tag{9.20}$$

In the tank with two outlets (see Fig. 1.23 and Section 1.2.5) the outlets are located at points $-a$, and a on the x-axis, and they are alternately open for a period of $T/2$. At time instant $t = 0$, the left outlet is opened. The flow is thus periodic with period T.

The position of the particle after time $T/2$ is obtained from (9.20) by placing the outlet at point $-a$:

$$z(T/2) = (z_0 + a)\left(1 - \frac{QT}{|z_0 + a|^2}\right)^{(1-iKQ)/2} - a. \tag{9.21}$$

After the next half-period (the origin corresponds now to point $+a$), the position is given by

$$z(T) = (z(T/2) - a)\left(1 - \frac{QT}{|z(T/2) - a|^2}\right)^{(1-iK/Q)/2} + a. \tag{9.22}$$

Thus, we have obtained the map relating the initial point to the position taken after time T.

It is useful to choose the length unit to be a (the outlets are then at points $z = \pm 1$). Clearly, the advection dynamics is determined by two dimensionless parameters, given by

$$\eta = QT/a^2 \quad \text{and} \quad \xi = K/Q, \tag{9.23}$$

i.e. the dimensionless strengths of outflow and circulation, respectively.

Since the flow is periodic, the positions at instants $t = nT$ and $t = (n+1)T$ are related by the same rule. With the notation $z_n \equiv z(nT)$, the stroboscopic map of an advected particle is given by

$$\left.\begin{aligned} z_{n+1} &= (z'_n - 1)\left(1 - \frac{\eta}{|z'_n - 1|^2}\right)^{1/2 - i\xi/2} + 1, \\ z'_n &= (z_n + 1)\left(1 - \frac{\eta}{|z_n + 1|^2}\right)^{1/2 - i\xi/2} - 1, \end{aligned}\right\} \tag{9.24}$$

where the auxiliary variable, z'_n, denotes the position of the particle at instant $t = (n + 1/2)T$.

The advection map, (9.24), yields the position of the particle at the instants when the left outlet is opened. Particles within a circle of radius $R = \sqrt{\eta}$ around the outlet leave the system in half a period, and an identical sink core is formed around the other outlet in the next half period. The advection map possesses therefore two extended, but non-chaotic, attractors (although in the continuous-time dynamics, only the centres of the outlets are attractors). Outside the outlets, velocity is divergence-free; consequently, the map is area preserving outside of the sink cores. The flow is *open*, i.e. fluid flows across the observational domain, and never returns after leaving it (through one of the outlets). The advection dynamics therefore exhibits the properties of the transient chaos of open conservative systems (chaotic scattering).

Figure 9.22 presents the chaotic saddle of the map and its manifolds for parameters $\eta = 0.5, \xi = 10$ (the ones also used in Figs. 1.24 and 1.25, and in Plate XI). These figures clearly illustrate our previous statement that a spreading dye droplet traces out the unstable manifold: the last colour plate, taken after four periods, is practically identical to the image of Fig. 1.25 taken one period later (even though the initial droplets are not identical), and in the given resolution both agree with the unstable manifold seen in Fig. 9.22(c).

The stable manifold of Fig. 9.22(b) can, in principle, also be traced out by dye. The map is invertible outside the sink core, and the inverted map describes advection due to a fluid being injected alternately every half period at the positions of the initial outlets and flowing circularly. An impurity droplet placed into this flow spreads along the stable manifold of the original problem.

Because attractors are present, the basin boundaries can be defined. Particles coming close to the chaotic saddle or its stable manifold are trapped: for a long time they cannot decide which outlet to take. The

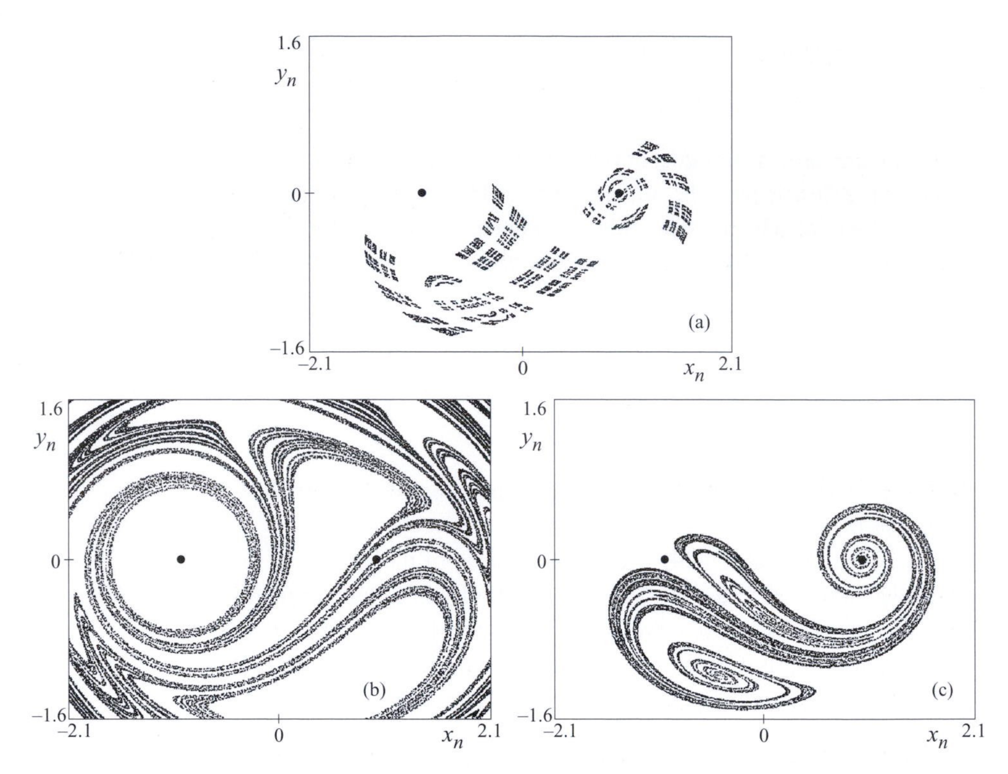

Fig. 9.22. Invariant sets of the advection dynamics in a tank with two outlets ($\eta = 0.5$, $\xi = 10$). (a) Chaotic saddle, (b) stable manifold, (c) unstable manifold obtained by plotting the points $n = 8$, $n = 0$ and $n = 15$ of trajectories not escaping the region shown in 15 periods (see Box 6.1). The outlets are marked by black dots.

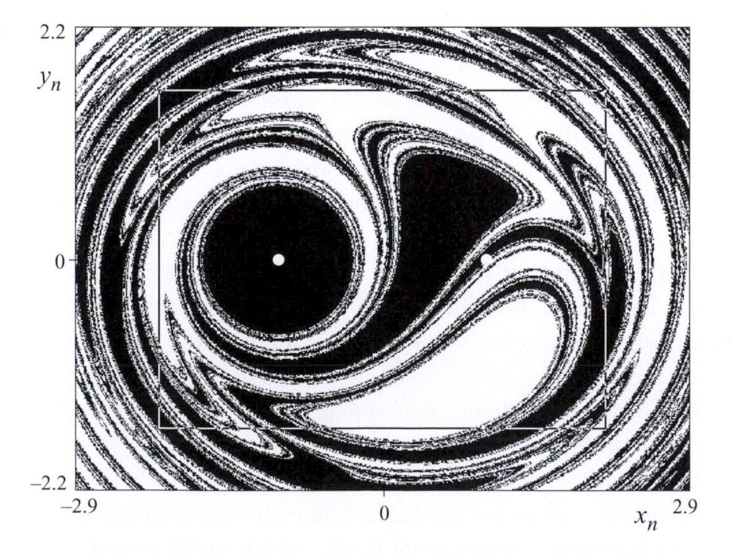

Fig. 9.23. Basins of attraction of the two outlets (initial positions corresponding to flowing out through the left outlet are marked in black; $\eta = 0.5$, $\xi = 10$). The frame within the figure indicates the size of Fig. 9.22.

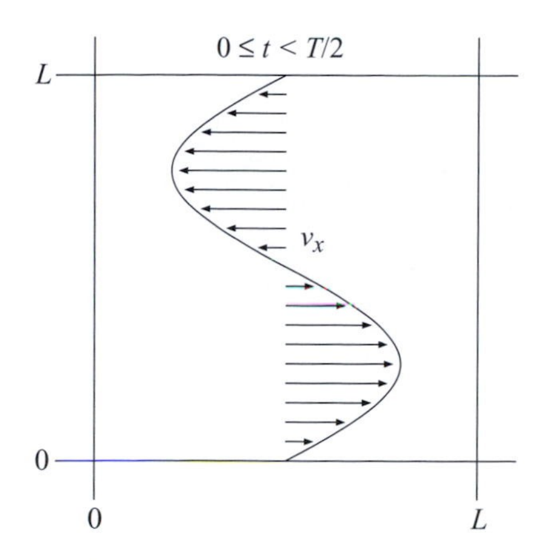

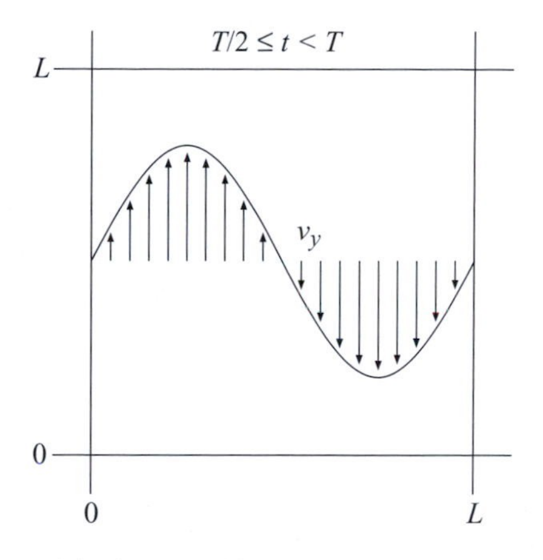

Fig. 9.24. The shear-flow model (9.25), (9.26): sinusoidal velocity profiles alternate each half period.

fractal boundary therefore coincides with the stable manifold of the saddle (Fig. 9.23).

9.4.2 Advection in closed flows

In order to model a spatially extended flow, we consider a square region of linear size L and repeat it periodically in the plane. Within each square, the flow is assumed to be shear-like: the velocities are opposite in opposite halves of the square. Let the flow be periodic in time in such a way that, in the first and second half periods of length $T/2$, the velocities are parallel to the x- and y-directions, respectively. Possibly the simplest shear form is given in terms of a sine function of the co-ordinates. The chosen velocity field is therefore

$$v_x = U\sin(2\pi y/L), \quad v_y = 0, \quad \text{if} \quad 0 \le t < T/2, \tag{9.25}$$

$$v_x = 0, \quad v_y = U\sin(2\pi x/L), \quad \text{if} \quad T/2 \le t < T, \tag{9.26}$$

where U is the maximum velocity (Fig. 9.24).

The displacement of a particle situated at position (x_n, y_n) after n periods occurs in the x-direction during the first half period and is of magnitude $(UT/2)\sin(2\pi y_n/L)$. After reaching the end-point with co-ordinate x_{n+1}, the y-displacement is $(UT/2)\sin(2\pi x_{n+1}/L)$. Measuring length in units of L, the advection map is of the following form:

$$x_{n+1} = x_n + a\sin(2\pi y_n), \quad y_{n+1} = y_n + a\sin(2\pi x_{n+1}), \tag{9.27}$$

where

$$a = \frac{UT}{2L} \tag{9.28}$$

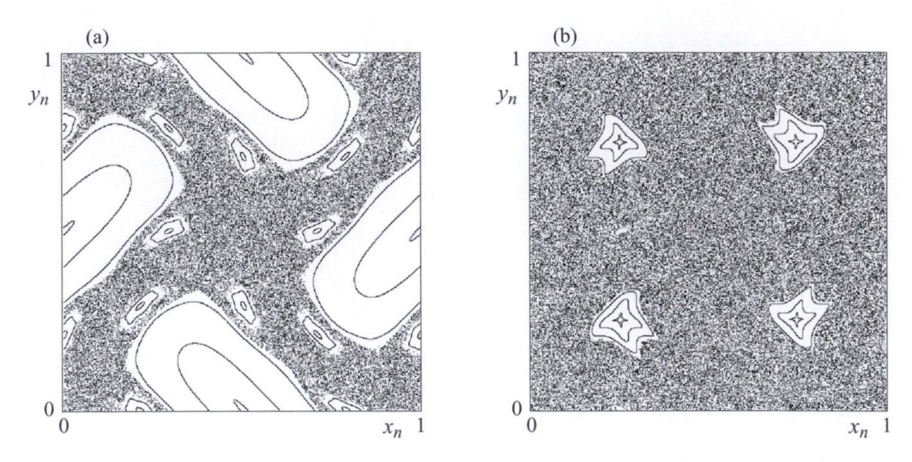

Fig. 9.25. Phase portrait of map (9.27) for parameters (a) $a = 0.3$ and (b) $a = 0.5$. (The number of initial conditions is 11 and 7, respectively.) The transport barriers (KAM tori) are much more extended in (a) than in (b).

is the only dimensionless parameter. Since the velocity field is periodic with a period of unity in both x and y, neighbouring squares are equivalent. The position of a particle can thus also be represented within the unit square, i.e. map (9.27) can be applied with periodic boundary conditions at x and at $y = 1$. The flow is then *closed* and the exiting fluid returns to the observed domain. (When a selected particle leaves the unit square, it is considered to be advected back immediately at the opposite side.) Map (9.27) is a prototype of the advection dynamics in general closed flows.

The advection map is area preserving because of the incompressibility of the fluid. The phase portrait of advection dynamics is therefore similar to that of closed conservative systems. Nested in the chaotic domain, KAM tori appear (Fig. 9.25). In the context of advection, these regular islands correspond to the appearance of what are called *transport barriers*. These are domains that no particle coming from outside can enter and no particle within can leave. In real flows, these are often formed by long-lived vortices.

In the course of spreading, fractal-like objects seem to appear for some time as the drop starts to follow the unstable manifold. Since, however, the latter is two-dimensional is an area preserving map, the particles constituting the drop become uniformly distributed after a long time. They cannot cover the entire fluid surface, however, since KAM tori are impermeable (see Figs. 9.26 and 9.27). In a closed fluid, therefore, the smaller the total area of regular islands, the more effective the stirring. Consequently, when designing a good mixer, the aim is to find parameters for which chaos is most extended (in our case, for example, for $a = 0.8$).

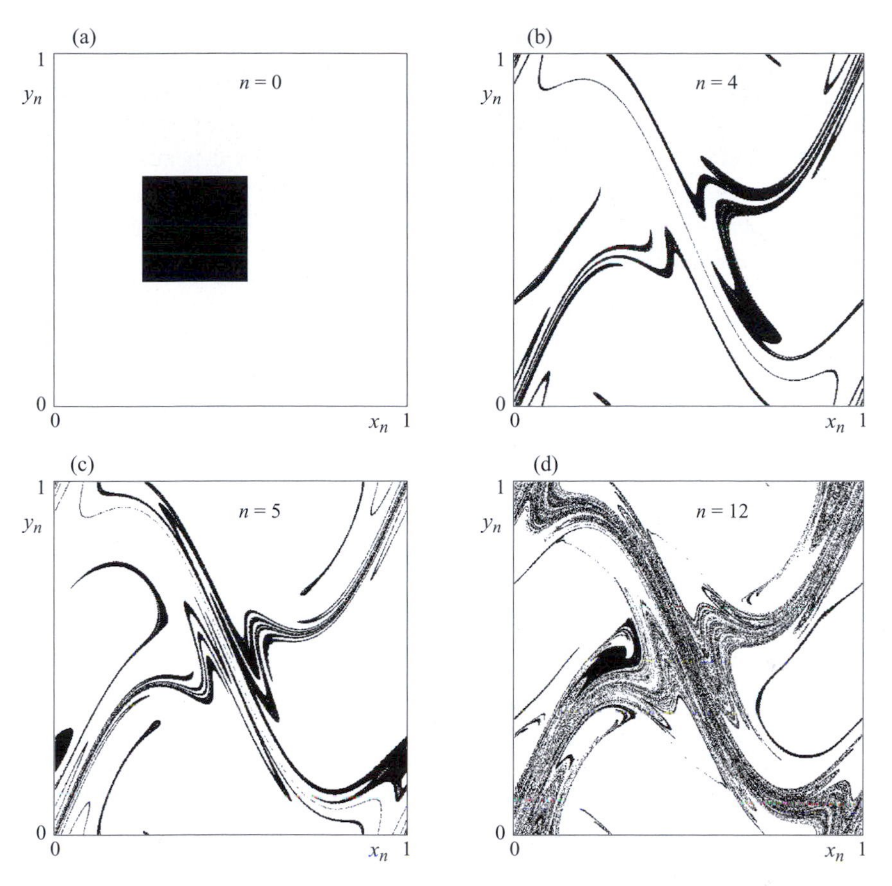

Fig. 9.26. Droplet dynamics in the shear flow model ($a = 0.3$). (a) Initial droplet of 90 000 points. The shape of the droplet is shown after (b) 4, (c) 5 and (d) 12 time units.

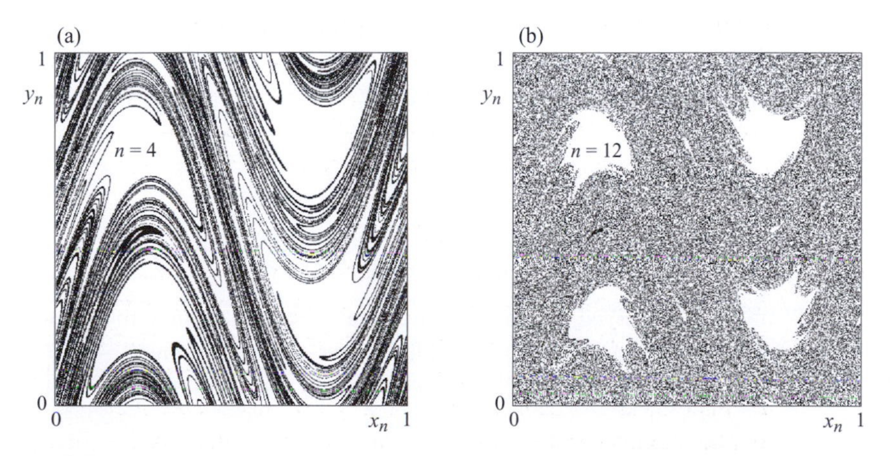

Fig. 9.27. Droplet dynamics in the case of more efficient stirring ($a = 0.5$). The shape of the same droplet as in Fig. 9.26 is shown after (a) 4 and (b) 12 time units.

9.4.3 Vortex dynamics

Long-lived vortices often develop in fluids. The flow around them is circular, and the velocity increases towards the vortex centre. In inviscid shallow fluids,[12] the absolute value of the velocity depends on the distance, r, from the centre as K/r (as in the case of the outlets, except that there is no radial flow). The velocity components around a single vortex centred at the origin are given by

$$v_x = -K\frac{y}{r^2}, \quad v_y = K\frac{x}{r^2}, \tag{9.29}$$

where $r = \sqrt{x^2 + y^2}$. If there is more than one vortex, none of them remains in place, since each vortex is advected by the flow generated by the others. The motion of the vortex centres can only be determined from the solution of a differential equation.

In the presence of N vortices, let (x_i, y_i) and K_i denote the positions of the vortex centres and the vortex strengths, respectively. In order to derive the differential equation, note that, at the position of a vortex centre, the x-component of the fluid's velocity is the sum of the contributions, v_x, of all the other vortices. For example, for vortex 1 at position (x_1, y_1), this resultant velocity is given by (cf. (9.29)) $\sum_{j=2}^{N} v_{x,i} = -\sum_{j=2}^{N} K_j(y_1 - y_j)/r_{1,j}^2$, where $r_{i,j} = \sqrt{(x_i - x_j)^2 + (y_i - y_j)^2}$ is the distance between vortices i and j. The vortex centres take on the fluid velocity instantaneously, just like the advected particles considered so far. The equation of motion of the N-vortex problem is therefore given by

$$\dot{x}_i = -\sum_{j\neq i}^{N} K_j\frac{y_i - y_j}{r_{i,j}^2}, \quad \dot{y}_i = \sum_{j\neq i}^{N} K_j\frac{x_i - x_j}{r_{i,j}^2}. \tag{9.30}$$

The phase space of the N-vortex problem is $2N$-dimensional. There exist, however, conserved quantities. One of these is the vortex interaction energy

$$\mathcal{E} = -\sum_{j\neq i}^{N} K_i K_j \ln r_{i,j}, \tag{9.31}$$

but the quantities

$$\sum_i^N K_i x_i, \quad \sum_i^N K_i y_i, \quad \sum_i^N K_i(x_i^2 + y_i^2), \quad \sum_i^N K_i(x_i\dot{y}_i - y_i\dot{x}_i) \tag{9.32}$$

are also conserved, as can be verified via direct substitution. In the four-vortex problem, the number of independent equations is thus $8 - 5 = 3$; consequently, *four* or more vortices usually move *chaotically* in each

[12] Or in three-dimensional flows where velocity does not depend on the third co-ordinate.

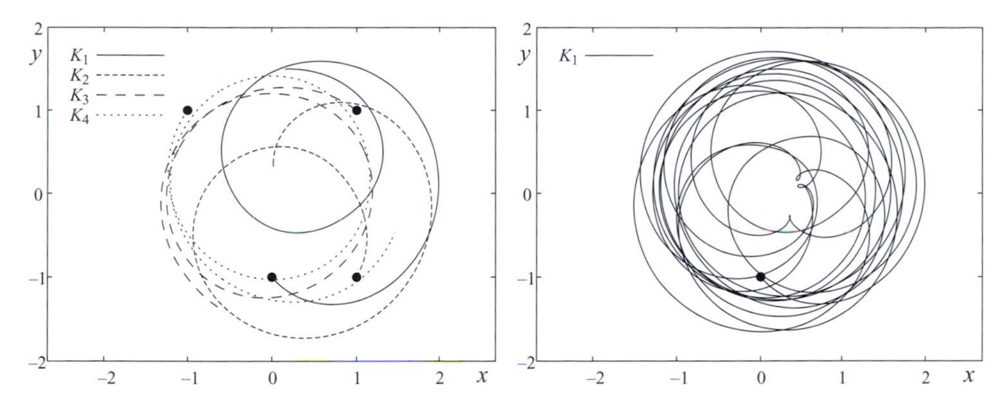

Fig. 9.28. Chaotic dynamics of four vortices: orbits of the vortex centres are obtained by solving (9.30) in a dimensionless form ($K_1 = K_2 = K_3 = K_4 = 1$). The initial positions are marked by black dots. (a) Orbits of all four vortices over 10 time units. (b) Orbits of vortex 1 over 100 time units.

other's flow field (Fig. 9.28). The dynamics of the two- and three-vortex problem is always regular, i.e. periodic and quasi-periodic

Knowing the orbits $(x_i(t), y_i(t))$, the velocity field of the fluid is obtained from (9.29) at an arbitrary point (x, y) (outside of the vortex centres) as follows:

$$v_x(x, y, t) = -\sum_{i=1}^{N} K_i \frac{y - y_i(t)}{r_i^2}, \quad v_y(x, y, t) = \sum_{i=1}^{N} K_i \frac{x - x_i(t)}{r_i^2}, \tag{9.33}$$

where $r_i = \sqrt{(x - x_i)^2 + (y - y_i)^2}$ is the distance from vortex i. The time dependence of the velocity field is therefore of the same character as the dynamics of the vortex centres. Thus, in general, the velocity field generated by four vortices changes chaotically at any point of the fluid.

The equation of an advected particle in the velocity field of N vortices is, according to (9.17), given by

$$\dot{x} = -\sum_{i=1}^{N} K_i \frac{y - y_i(t)}{r_i^2}, \quad \dot{y} = \sum_{i=1}^{N} K_i \frac{x - x_i(t)}{r_i^2}. \tag{9.34}$$

This is a two-dimensional motion driven by the dynamics of the vortex centres.

An advected particle has no feedback on the vortices. It moves as a vorticity-free vortex centre. If one of the vorticities, K_i, of an N-vortex problem is zero, then we speak of a restricted N-vortex problem (in terminology analogous to that of the gravitational N-body problem; see Section 9.1). Particle advection in the field of $N - 1$ vortices is therefore a special case of the N-vortex problem. Chaoticity is not affected by

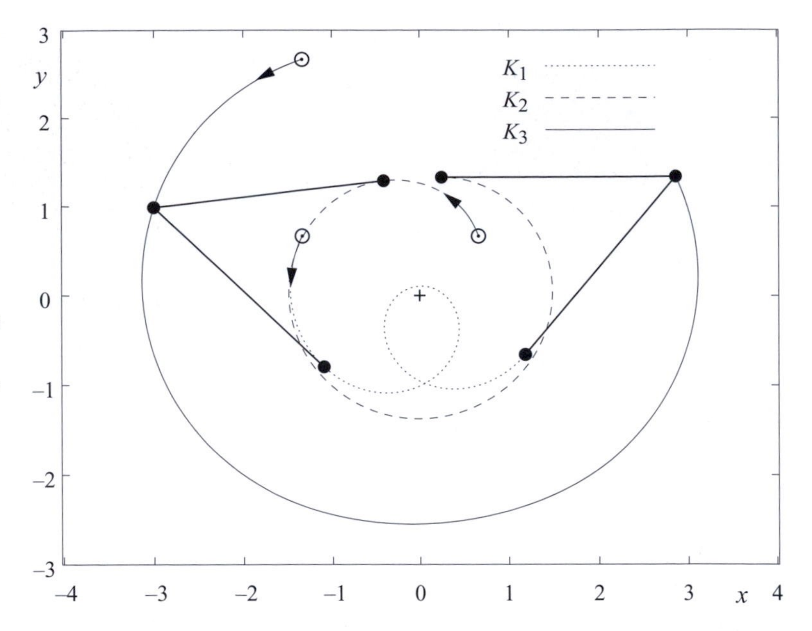

Fig. 9.29. Dynamics of three vortices ($K_1 = K_2 = 1, K_3 = -0.5$). From the initial positions (shown by open circles), an isosceles-triangle configuration is reached after 3.4 time units. The same shape (but of opposite orientation) appears after an additional 13.2 time units. The period is thus $T = 26.4$. The origin is the centre of vorticity, $\mathbf{T}_0$ (marked with a cross).

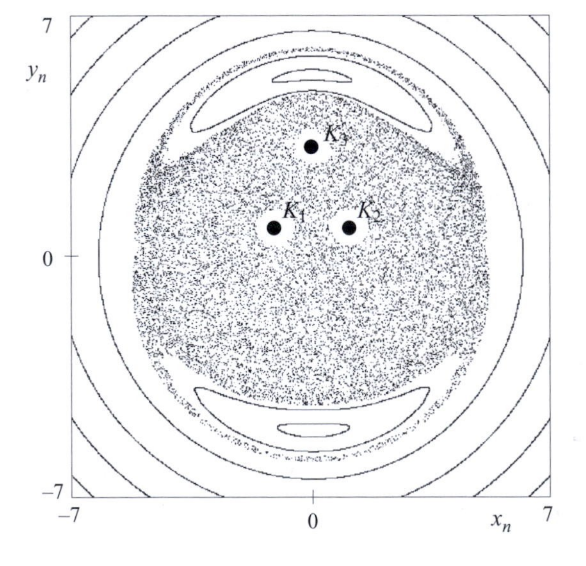

Fig. 9.30. Advection dynamics in the field of three vortices on a stroboscopic map corresponding to the isosceles-triangle configurations of the vortices ($K_1 = K_2 = 1, K_3 = -0.5$); 13 trajectories are plotted.

this restriction; consequently, *advection* is typically *chaotic* in the field of *three vortices*.

In order to characterise advection properly, we have to know how the three vortices move. Starting from any initial condition (provided $\sum_{i=1}^{3} K_i \neq 0$), the vortices sooner or later reach an isosceles-triangle-shaped configuration. This same configuration is then repeated with some

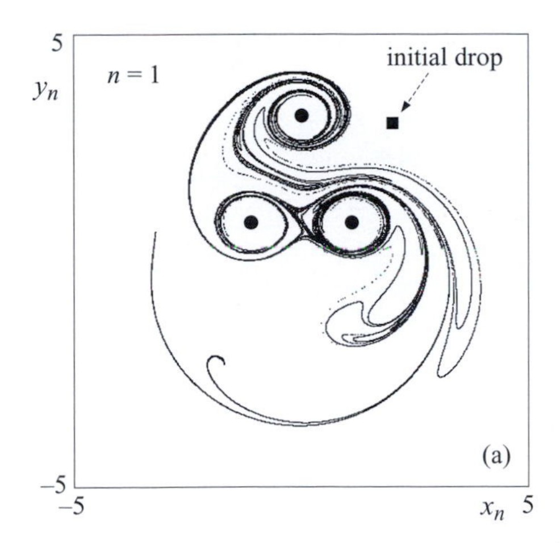

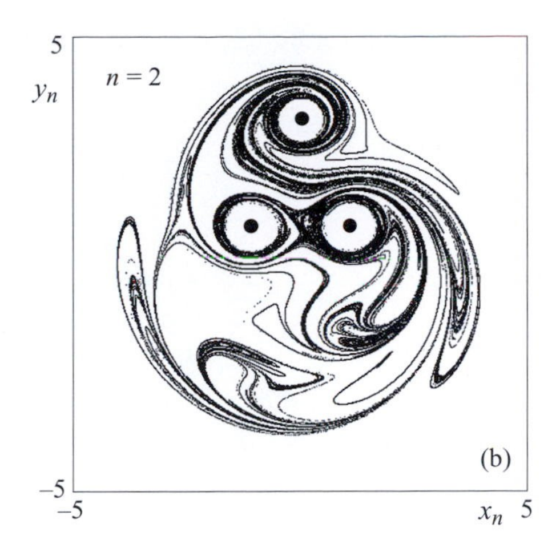

Fig. 9.31. Droplet dynamics. The shape of a 10^6 point droplet shown as a small black square in (a) after (a) 1 and (b) 2 periods, T. Stroboscopic map and parameters are as in Fig. 9.30.

period T, but not at the same location (Fig. 9.29). In a reference frame whose origin is fixed to the centre of vorticity $\mathbf{r}_0 = \sum_{i=1}^{3} K_i \mathbf{r}_i / \sum_{i=1}^{3} K_i$, and whose y-axis goes through vortex 3, the vortex dynamics appears to be periodic with period T.

It is this frame of reference in which the advection dynamics is easiest to follow in the form of a stroboscopic map. Chaos is present among the vortices, but not in the immediate vicinity of the vortex centres, where rotation is very strong (Fig. 9.30). The droplet dynamics can also conveniently be investigated in this frame (Fig. 9.31).

Problems recommended for further studies

(i) Study the chaotic scattering of vortex pairs ($K_2 = -K_1$, $K_4 = -K_3, \ldots$).
(ii) Investigate chaotic advection in the velocity field of four arbitrary vortices.
(iii) Model advection dynamics in chaotically time-dependent flows by replacing one of the fixed parameters, μ, of advection maps (for example (9.24) and (9.27)) with parameters changing randomly from iteration to iteration, $\mu \to \mu_n \equiv \mu + \delta\mu_n$, where $\delta\mu_n$ is a random number over a finite short interval.

Box 9.5 Environmental significance of chaotic advection

When material is mixed in fluids, the flow of the medium plays a much more important role than diffusion. In a medium perfectly at rest, it would take *days* for it to spread over a few metres by means of pure diffusion. (Typical orders of magnitude of diffusion coefficients in water and air are 10^{-9} and 10^{-5} m^2/s,

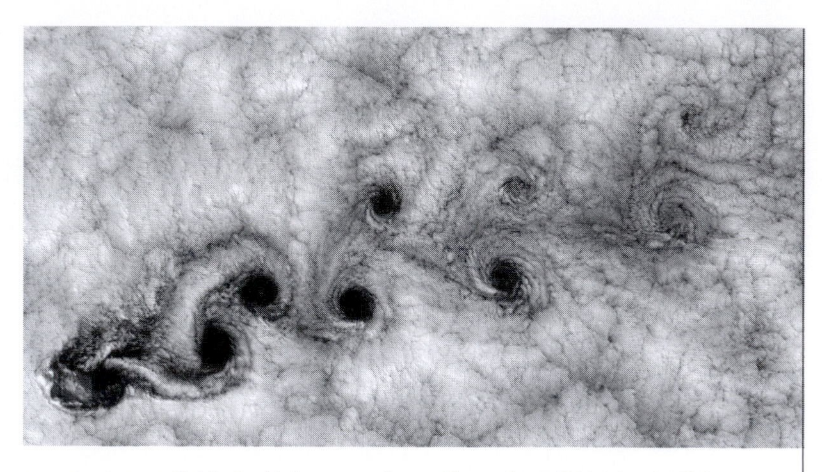

Fig. 9.32. Mountain acting as a skyscraper. Behind a high mountain on Alexander Selkirk Island (Chile) in the Southern Pacific Ocean, the wind generates vortices on September 15, 1999 (NASA archive, http://visibleearth.nasa.gov/cgi-bin/viewrecord?617). In this nearly periodic open flow, advection is chaotic and a saddle develops in the wake. In the cloud layer, the unstable manifold of this saddle is traced out, which surrounds large isolated vortices, similar to the dye in Fig. 9.31.

respectively, and, since the average distance between two originally adjacent particles grows in time as $\Delta x(t) = \sqrt{2Dt}$, this yields, for the two respective cases, centimetres and metres over a day.) Thus, if upon entering a home we immediately smell what is being prepared in the kitchen, this is due *not* to diffusion, but to the *advection* generated by the slight, ever-present, air currents. Advection is, however, usually chaotic. It occurs in everyday phenomena (for example, the stirring of milk into coffee, the rising smoke of cigarettes and the mixing of dyes) in much the same way as on larger scales.

Advected materials spread in our environment via chaotic advection. Consequently, in time-dependent smooth flows, all materials, including pollutants, spread along *filamentary, fractal-like patterns*. As we have seen, these are related to unstable manifolds; therefore, we can say that numerous environmental phenomena exhibit the marks of unstable manifolds. Figure 9.32 presents the pattern developing in the wake of an obstacle, in the so-called 'von Kármán vortex street'. In this practically time-periodic open flow, advection is chaotic and a saddle exists in the wake. Particles may be trapped and may spend quite long times behind obstacles. This chaotic scattering process explains that pollutants may *accumulate* in the wake of bridge pillars, piers or islands.

Plate XXIII exhibits fractal-like patterns in a less regular flow, traced out by drifting ice-plates on the sea surface.

In many environmental problems, advected particles are not passive but rather undergo some reactions. As a consequence of chaotic advection, chemical or biological reactions taking place in fluid flows are basically different in character from those occurring in non-moving media. Plates XXIV and XXV–XXVI illustrate by examples taken from the ocean and the atmosphere, respectively, that the distribution of the product (plankton and ozone) is fractal-like in such cases, and is expected to be bound to some kind of unstable manifold.

The naive idea that pollutants have a compact distribution is therefore not realistic. The typical Lyapunov exponent of large scale advection in the atmosphere is of the order 1/(a few days) (in the ocean it is $1/(4 \text{ weeks})$), which implies that a pollutant with an initial radius of 1 km may wind around the Earth along a given latitude in about one month ($e^{10} = 22\,026$). Moreover, its distribution is filamentary; therefore, an accurate prediction of whether a certain geographical site will be affected is very difficult.

In summary, chaotic advection sheds new light on many environmental phenomena and the problem of pollution, and can be considered as one of the most appealing fields of application of chaos science.

Chapter 10
Epilogue: outlook

In this concluding chapter, we present a brief overview of some phenomena and concepts, the detailed investigation of which is beyond the scope of this introductory book, but whose inclusion may provide (along with the bibliography) further understanding.

First and foremost, we emphasise that chaotic behaviour can be observed in *laboratory experiments*. The validity of the physical laws determining the motion of macroscopic systems is beyond doubt; consequently, the phenomena found in numerical simulations are also present in the real world. The chaotic feature of many of our examples (magnetic pendulum, ball bouncing on a double slope or on a vibrating plate, or the mixing of dyes) can be demonstrated by relatively simple equipment. In the cases of the periodically driven pendulum, the spring pendulum, the driven bistable system or chaotic advection, the chaos characteristics have been determined by precise laboratory measurements, and the transitions towards chaos have also been investigated. In other branches of science, numerous processes are also known whose chaoticity is supported by observational or experimental evidence (see Box 9.3).

In this book we have presented the simplest forms of chaos and interpreted them as the consequence of hyperbolic periodic orbits. In general, however, *non-hyperbolic* effects also play a role due to the existence of orbits whose local Lyapunov exponents are zero. One example of this is the algebraic (non-exponential) decay of the lifetime distribution in chaotic scattering due to the existence of KAM tori (see equation (8.10)). Another manifestation of non-hyperbolic behaviour is *intermittency* in dissipative systems: the interruption of long-lasting, apparently periodic, motions by chaotic 'bursts' of large amplitudes. This also provides a possible route towards chaos.

An information-theoretic approach to chaos and the related concept of symbolic dynamics and *dynamical entropies* (generalisations of the topological entropy) make possible a very detailed characterisation of chaos and deepen the link between temporal disorder and fractal order in phase space.

We have restricted ourselves to chaos in three-dimensional flows (two-dimensional maps). In systems with *higher-dimensional* phase spaces (three or more dimensional stroboscopic or Poincaré maps) chaos has further novelties in store. The unstable manifolds of the hyperbolic orbits can then be surfaces, implying the positivity of at least two Lyapunov exponents. In high-dimensional systems quasi-periodic motion can also be attracting, and *torus attractors* might appear as new types of regular attractors. The transition to chaos may also occur via the destruction of torus attractors. In conservative cases, with more than three-dimensional phase spaces, the tori do not act as barriers of transport. Therefore, a motion spreading over large distances of phase space, the so-called *Arnold diffusion*, might be present, even for very weak perturbations.

Deterministic chaos is the temporal behaviour of simple systems made up of few components. The temporal behaviour of many-component systems, systems of *high degrees of freedom*, is necessarily *more* complicated than chaos. The origin of the random behaviour lies in the huge number of components, which would be impossible to examine individually. Random motion of this origin, perceived by macroscopic observers, is called *noise*, and the underlying microscopic dynamics is sometimes referred to (using the terminology of the kinetic theory of gases) as *molecular chaos*. Molecular and deterministic chaos are, however, different in several aspects. The concepts of deterministic chaos can, nevertheless, also be useful elements in the description of systems of high degrees of freedom (and of molecular chaos). Thus, for example, positive average Lyapunov exponents might exist. Their number often increases with the number of the components, together with the dimension of characteristic phase space structures. Even though the dynamics of both types of chaos is unpredictable, *low-dimensional fractal* phase space structures, so characteristic of deterministic chaos, are typically *not present* in high degree of freedom systems.

In investigating deterministic chaos, we study the simplest form of irregular temporal behaviour originating only from the non-linearity of the internal dynamics. In systems with high degrees of freedom, this is supplemented by the complexity arising from the large number of variables. The complicated behaviour of systems consisting of many components therefore inevitably possesses features that cannot be understood on the basis of deterministic chaos alone. As far as the behaviour of such systems is concerned, the knowledge of deterministic chaos is only a first step.

Understanding deterministic chaos requires us to make a fundamental change in our entire scientific view of Nature. We have to accept that motion described by simple non-linear equations is usually complicated, and also that the solution can only be obtained using computers. The science of mechanics, which appeared a short while ago to be closed, has to be extended by the concepts discussed in the book. The same holds for all branches of science whose basic phenomena can be described

Box 10.1 Turbulence and spatio-temporal chaos

Turbulence of fluids is the most complicated form of fluid motion, irregular not only in time but also in *space*. At the time when 'chaos theory' was beginning to take shape (in the 1970s and 1980s), it was widely accepted that the understanding of chaos would lead to the understanding of turbulence as well. This expectation has proved to be an exaggeration, and, even though the most important features of chaos are understood, turbulence still requires further exploration.

The temporal and spatial behaviour of fluid motion is described by partial differential equations. By approximating the space co-ordinates with a finite grid, these can be replaced by a set of ordinary differential equations, but their number is still extremely large. Fluid dynamics is therefore a problem of high degrees of freedom.

At a not too strong energy input to fluid flows, or, more generally, to any spatio-temporal process described by partial differential equations (for example, chemical reactions or excitation waves in living tissues) it might occur that only certain groups of the degrees of freedom are active. In such special cases the motion is effectively low-dimensional. For the observer this might appear as more or less regular *spatial patterns* (waves) passing through the system, whose repetition is not exactly periodic in time. Such *spatio-temporal chaos* thus manifests itself in the random-like occurrence of certain local spatial structures, and might possess an underlying low-dimensional chaotic attractor. These structures are related to novel phenomena, such as non-linear waves, spikes, fronts, boundary effects, coherent behaviour. As for deterministic chaos, the presence of spatio-temporal chaos cannot be detected by just looking at the equations and parameters. Instead, its presence can only be shown through measurement and/or simulation. Some phenomena that appeared in first approximation to be purely temporally chaotic (for example the behaviour of certain populations or epidemics) have turned out, on closer inspection, to exhibit spatio-temporal chaos.

In the general case of strong energy input, many degrees of freedom are active both in fluids and in other spatio-temporal processes. In particular, in the extreme situation of turbulence, it is evident that nearly all the degrees of freedom take part in the entire dynamics. Several other phenomena (certain dysfunction of the heart, electric activity of the brain, price fluctuations of markets and stock markets, for example) that were once suspected to be chaotic have turned out (in view of accurate recent investigations) to be high degree of freedom problems with noisy or turbulence-like dynamics.

by ordinary differential equations. In all of these, the problem of unpredictability, attached so far to meteorology only, appears, and the use of probabilistic concepts becomes unavoidable. Chaos requires a novel way of thinking, one imposed upon us by the phenomenon itself. Thus, in understanding chaos, we act – eventually, without being conscious of it – in the spirit of Aristotle:[1]

> For it is the mark of an educated man to look for precision in each class of things just so far as the nature of subject admits.

[1] Barnes, J. (ed.) *The Complete Works of Aristotle (Nicomachean Ethics)*: *The Revised Oxford Translation*. Bollingen Series vol. LXXI, no. 2. Princeton: Princeton University Press, 1991, p. 1730.

Appendix

A.1 Deriving stroboscopic maps

A.1.1 Harmonically driven motion around a stable state

Continuous-time dynamics

A particular solution of the linear, inhomogeneous equation (4.9) can be checked to be given by

$$x_P(t) = A\cos(\Omega t + \varphi_0 - \delta); \tag{A.1}$$

the amplitude, A, and the phase, δ, are given by

$$A = \frac{f_0}{\sqrt{(\omega_0^2 - \Omega^2)^2 + \alpha^2\Omega^2}} \tag{A.2}$$

and

$$\delta = \arctan\left(\frac{\alpha\Omega}{\omega_0^2 - \Omega^2}\right), \tag{A.3}$$

respectively. The sharp maximum of the amplitude around $\omega_0 \approx \Omega$ represents the well known phenomenon of resonance. Note that the particular solution does not depend on the initial conditions.

The complete solution also contains the general solution of the homogeneous part, so

$$x(t) = A\cos(\Omega t + \varphi_0 - \delta) + A_1 e^{-(\alpha/2)t}\cos(\omega_\alpha t) + A_2 e^{-(\alpha/2)t}\sin(\omega_\alpha t). \tag{A.4}$$

The last two terms coincide with the solution for the damped harmonic oscillator (see equation (3.34)) and describe a decaying component; therefore the first term yields the long-term behaviour. Taking into account that damping is very weak ($\alpha/2 \ll \omega_0$), the velocity is found to be

$$\begin{aligned} v(t) = &-A\,\Omega\sin(\Omega t + \varphi_0 - \delta) - A_1\omega_0 e^{-(\alpha/2)t}\sin(\omega_0 t) \\ &+ A_2\omega_0 e^{-(\alpha/2)t}\cos(\omega_0 t). \end{aligned} \tag{A.5}$$

Terms proportional to α have been neglected, and ω_α in (3.33) has been substituted by the natural frequency, ω_0, in the arguments of the trigonometric functions. The asymptotic motion,

$$x^*(t) = A\cos(\Omega t + \varphi_0 - \delta), \quad v^*(t) = -A\Omega\sin(\Omega t + \varphi_0 - \delta), \tag{A.6}$$

represents a limit cycle attractor.

According to (A.4) and (A.5), the arbitrarily chosen initial conditions (x_0, v_0) uniquely determine the amplitudes A_1 and A_2 :

$$x_0 = A\cos\delta + A_1, \quad v_0 = A\Omega\sin\delta + A_2\omega_0. \tag{A.7}$$

The stroboscopic map

In a map taken at instants $t = nT$ ($\Omega t = 2n\pi$) with $\varphi_0 = 0$, the limit cycle appears with co-ordinates

$$x^* = A\cos\delta, \quad v^* = A\Omega\sin\delta, \tag{A.8}$$

since the trigonometric functions in (A.6) take on the same values, independent of n, at these times. The amplitudes A_1, A_2 expressed in terms of the limit cycle co-ordinates are thus given by

$$A_1 = x_0 - x^*, \quad A_2 = \frac{v_0 - v^*}{\omega_0}. \tag{A.9}$$

The position $x_1 = x(T)$ and velocity $v_1 = v(T)$ of the state arising after one period are given by (A.4) and (A.5) at $t = T$. Along with (A.9), which brings in the initial conditions, equations (A.4) and (A.5) yield the stroboscopic map relating the instants $n = 0$ and $n = 1$:

$$x_1 = x^* + E\left(C(x_0 - x^*) + \frac{S}{\omega_0}(v_0 - v^*)\right), \tag{A.10}$$

$$v_1 = v^* + E\left(-S\omega_0(x_0 - x^*) + C(v_0 - v^*)\right), \tag{A.11}$$

where the shorthand notations given in (4.12) have been used. Since x_0 and v_0 are arbitrary, they can also represent the co-ordinates at the nth stroboscopic level. The above rule then yields the co-ordinates at the $(n+1)$st snapshot: with substitutions $0 \to n$ and $1 \to n+1$, we obtain equations (4.10) and (4.11).

A.1.2 Harmonically driven motion around an unstable state

Continuous-time dynamics

The equation of motion is now given by

$$\ddot{x} = s_0^2 x - \alpha\dot{x} + f_0\cos(\Omega t + \varphi_0), \tag{A.12}$$

which is valid for small displacements, x, around the original equilibrium state ($x = 0$), where s_0 is the repulsion parameter, f_0 is the driving amplitude assumed to be constant and, for very weak damping, $\alpha/2 \ll s_0$. Thus the solution of the problem can be obtained from that presented in Section A.1.1 via the substitution

$\omega_0 \to is_0$. The particular solution of equation (4.9) is still a limit cycle of the form of (A.6) but now

$$A = -\frac{f_0}{\sqrt{(s_0^2 + \Omega^2)^2 + \alpha^2\Omega^2}}, \qquad \delta = \arctan\left(\frac{\alpha\Omega}{s_0^2 + \Omega^2}\right). \tag{A.13}$$

Since the denominator contains the sum of the squares of the repulsion parameter and the driving frequency, resonance cannot be present. In the general solution, trigonometric functions are replaced by hyperbolic functions, which describe an exponential increase of the deviation: the limit cycle is therefore unstable.

The stroboscopic map

For the stroboscopic map taken at instants $t = nT$ ($\varphi_0 = 0$) we obtain from (4.10) and (4.11), by means of the aforementioned substitution,

$$x_{n+1} \equiv M_1(x_n, v_n) = x^* + EC'(x_n - x^*) + \frac{ES'}{s_0}(v_n - v^*), \tag{A.14}$$

$$v_{n+1} \equiv M_2(x_n, v_n) = v^* + ES's_0(x_n - x^*) + EC'(v_n - v^*). \tag{A.15}$$

The co-ordinates of the fixed point corresponding to the limit cycle are still given by (A.8), with A and δ given by (A.13), and the shorthand notation given in (4.22) has been used.

A.1.3 Kicked harmonic oscillator

The map can be decomposed into two steps. In the first step we monitor the motion from the values (x_n, v_n) taken right after the nth kick to the instant just before the next kick at time $(n+1)T$, when the co-ordinates are $(x_{n+1}, \tilde{v}_{n+1})$. The notation already expresses the fact that the kick itself does not influence the position co-ordinate, so x_{n+1} is the position immediately after the $(n+1)$st kick. The same argument does not hold for the velocity; therefore the second step is to find out the velocity, v_{n+1}, after the kick in terms of the velocity, $\tilde{v}_{n+1}$, before it.

In order to determine the relation $(x_n, v_n) \to (x_{n+1}, \tilde{v}_{n+1})$, we note that no external driving force acts on the body in the open interval between two kicks. The time evolution within this interval of length T can be obtained from the non-driven oscillator result, (3.34), and its derivative. Alternatively, we can take formulae (4.10) and (4.11) of the sinusoidally driven oscillator in the limit of vanishing driving amplitude, $x^* = v^* = 0$. In any case, we obtain

$$x_{n+1} = ECx_n + \frac{ES}{\omega_0}v_n, \qquad \tilde{v}_{n+1} = -ES\omega_0 x_n + ECv_n, \tag{A.16}$$

where the notation given in (4.12) has been used. The co-ordinates (x_n, v_n) after the nth kick and the parameters of the oscillator thus uniquely determine the parameters before the next kick. Naturally, (A.16) is a linear relation, since it does not yet include the kick, which is the source of non-linearity in the problem.

The effect of the kick is that it simply shifts the velocity by $uI(x_{n+1})$ instantaneously. Thus, the velocity after the kick is given by

$$v_{n+1} = \tilde{v}_{n+1} + uI(x_{n+1}). \tag{A.17}$$

By using (A.16), we thus arrive at the kicked oscillator map, (4.28).

A.1.4 Kicked frictionless harmonic oscillator

For negligible friction, $\alpha \to 0$, $E \to 1$ and the stroboscopic map is given by

$$x_{n+1} = Cx_n + \frac{S}{\omega_0} v_n, \quad v_{n+1} = -S\omega_0 x_n + Cv_n + uI(x_{n+1}). \tag{A.18}$$

A.1.5 Kicked free motion

The stroboscopic map for kicked free motion is obtained from (A.18) by taking the limit $\omega_0 \to 0$ ($C \to 1$, $S/\omega_0 \to T$), and it appears in the following form:

$$x_{n+1} = x_n + v_n T, \quad v_{n+1} = v_n + uI(x_{n+1}) \tag{A.19}$$

(which is easily derivable directly from the equation $\ddot{x} = 0$, which is valid between kicks).

A.2 Writing equations in dimensionless forms

Equations describing physical phenomena usually contain dimensional quantities. Writing such equations in *dimensionless forms*, in which neither the variables nor the parameters carry units (i.e. they are dimensionless) is important for two reasons.

- The dimensionless forms clearly indicate the *independent* parameters of the problem. The number of independent parameters is usually smaller than those of the original dimensional equation. A full solution always contains the exploration of the parameter space, which is practically impossible in the dimensional forms.
- Computers work with dimensionless numbers. The *numerical solution* of equations is therefore only possible in dimensionless forms. One could of course choose some set of the parameters and omit the units of all the dimensional quantities. This is, however, conceptually unclear, and excludes the numerical exploration of the parameter space.

For the sake of simplicity, we present the method for a second-order differential equation. Consider the problem

$$\ddot{x} \equiv \frac{d^2x}{dt^2} = a(x, \dot{x}, t; \mu). \tag{A.20}$$

The right-hand side shows us that acceleration, a, is known as a function of position, velocity and time. Symbol μ represents the parameter set. Eliminating the dimensions means that instead of position x and time t, dimensionless variables x' and t' are introduced via

$$x = Lx', \quad t = Tt', \tag{A.21}$$

where L and T are some (as yet unknown) length and time parameters of the system, respectively. The length and time units are thus chosen to be L and T, respectively, and the primed quantities represent unitless values. Accordingly, the dimensional velocity can be written as follows:

$$\dot{x} \equiv \frac{dx}{dt} = \frac{L}{T}\frac{dx'}{dt'} = \frac{L}{T}\dot{x}', \tag{A.22}$$

where $\dot{x}'$ denotes the dimensionless velocity. The original differential equation then takes the following form:

$$\frac{d^2x'}{dt'^2} \equiv \ddot{x}' = \frac{T^2}{L} a\left(Lx', \frac{L}{T}\dot{x}', Tt'; \mu\right). \tag{A.23}$$

We now choose parameters L and T in such a way that the parameter dependence is *simplest*; i.e. wherever possible, the pre-factor is unity. Since there are two quantities to choose, the number of the parameters can be decreased by two. Thus, a new parameter set, μ', is obtained based on a natural length and time scale (L and T) of the system.

The right-hand side of (A.23) can then be written as $a'(x', \dot{x}', t'; \mu')$, i.e.

$$\ddot{x}' = a'(x', \dot{x}', t'; \mu'). \tag{A.24}$$

This is the required *dimensionless* equation. All the systems with different μ but identical dimensionless μ' can be considered to be *equivalent*, since their motion only appears different because of the difference between their natural length and time units. These types of motion are called dynamically similar.

Note that it is sufficient to solve the dimensionless problem, since this automatically provides the solution to the original problem. Let the solution of (A.24) with the dimensionless initial condition x_0', v_0' be $x'(t') = f(x_0', v_0', t'; \mu')$. The dimensionless velocity is then $v'(t') = dx'(t')/dt' \equiv g(x_0', v_0', t'; \mu')$. From (A.21), we obtain, for the original problem,

$$x(t) = Lf\left(\frac{x_0}{L}, \frac{v_0 T}{L}, \frac{t}{T}; \mu'\right), \quad v(t) = \frac{L}{T} g\left(\frac{x_0}{L}, \frac{v_0 T}{L}, \frac{t}{T}; \mu'\right), \tag{A.25}$$

where $x_0 = Lx_0'$ and $v_0 = v_0' L/T$. The dimensional solution can be expressed in terms of functions f and g, and the basic parameter dependence is thus via the dimensionless parameter μ'. We make the following remarks.

- In systems with more variables, *all* position co-ordinates have to be re-scaled by the same factor, $x_i = Lx_i'$. If different masses are involved, they also have to be written into a dimensionless form as $m_j = Mm_j'$, where M is some characteristic mass of the system. In this case the number of the parameters can be decreased by *three*. (It is impossible to have more than three because there are only three independent fundamental dimensions: length, time and mass.)
- The dimensionless forms are *not* unique in systems where more than one characteristic length, time or mass parameter is present. If, for example, there are two characteristic times, T_1 and T_2, both choices $T = T_1$ and $T = T_2$ are acceptable, and T_1/T_2 then appears among the dimensionless parameters. Fortunately, it does not matter which choice is made as the dimensionless forms are basically equivalent.
- The method is conceptually similar to that used in the theory of hydrodynamical similarity of fluid flows, in which often a stationary, but spatially extended, problem is considered and therefore the boundary conditions of the corresponding partial differential equations would also have to be taken into account.

We illustrate the general procedure with examples.

A.2.1 Example 1: simple pendulum

The equation of motion for the angle of deflection φ, $\ddot{\varphi} = -(g/l)\sin\varphi$, contains one parameter (g/l). The angle variable is dimensionless; it is therefore meaningless to re-scale it. With the dimensionless time, $t' = t/T$,

$$\ddot{\varphi} \equiv \frac{d^2\varphi}{dt'^2} = -\frac{T^2 g}{l}\sin\varphi. \tag{A.26}$$

Making the choice $T = \sqrt{l/g}$, this becomes

$$\ddot{\varphi} = -\sin\varphi. \tag{A.27}$$

Thus, the problem does not possess any dimensionless parameter. It is useless, for example, to solve the original problem for different lengths l, since l only plays a role in the re-scaling of time. The motion of all pendulums is similar. The natural time unit, T, is proportional to the period, $2\pi\sqrt{l/g}$, of the small-amplitude swings of the pendulum.

A.2.2 Example 2: spring pendulum

Consider a pendulum whose thread is a spring of rest length l_0 and natural frequency ω_0. Let the point of suspension be the origin. The instantaneous thread length, l, and the angular deflection, φ, fulfil the dimensional equations derived in Section 7.4.2:

$$\ddot{l} = l\dot{\varphi}^2 - \omega_0^2(l - l_0) + g\cos\varphi, \quad l\ddot{\varphi} = -2\dot{l}\dot{\varphi} - g\sin\varphi, \tag{A.28}$$

which include three parameters, g, ω_0 and l_0. With the dimensionless variables $l = Ll', t = Tt'$,

$$\ddot{l}' = l'\dot{\varphi}^2 - \omega_0^2 T^2\left(l' - \frac{l_0}{L}\right) + \frac{gT^2}{L}\cos\varphi, \quad l'\ddot{\varphi} = -2\dot{l}'\dot{\varphi} - \frac{gT^2}{L}\sin\varphi. \tag{A.29}$$

A simple form is obtained by choosing

$$L = l_0, \quad T = \frac{1}{\omega_0}. \tag{A.30}$$

The length is then measured in units of the rest length, l_0, and the time unit is proportional to the natural period of the spring. Thus,

$$\ddot{l}' = l'\dot{\varphi}^2 - (l' - 1) + q\cos\varphi, \quad l'\ddot{\varphi} = -2\dot{l}'\dot{\varphi} - q\sin\varphi, \tag{A.31}$$

where

$$q = \frac{g}{\omega_0^2 l_0}. \tag{A.32}$$

The problem thus has a *single* dimensionless parameter, q, which is the ratio of the squares of the two characteristic times, $T_1 = 1/\omega_0$ and $T_2 = \sqrt{l_0/g}$. This faithfully represents the fact that our system contains the dynamics of both a harmonic oscillator and a simple pendulum (being nevertheless much richer than any of these, since it displays chaotic motion). All systems with the same q are equivalent from the point of view of dynamics.

Since two characteristic times exist, we can also choose $T = T_2 = \sqrt{l_0/g}$. In this case, we obtain the dimensionless form

$$\ddot{l}' = l'\dot{\varphi}^2 - \frac{1}{q}(l' - 1) + \cos\varphi, \quad l'\ddot{\varphi} = -2\dot{l}'\dot{\varphi} - \sin\varphi. \tag{A.33}$$

These are formally different from, but equivalent to, (A.31), and the dimensionless parameter is again given by q.

A.2.3 Example 3: driven non-linear oscillator

Consider a non-linear spring characterised by the force law $F(x) = -\omega_0^2 x - \varepsilon_0 x^3$ ($\varepsilon_0 > 0$). The equation of motion of a body fixed to the spring, driven harmonically with amplitude, f_0, and angular frequency, Ω, in the presence of a linear friction, is given by

$$\ddot{x} = -\alpha\dot{x} - \omega_0^2 x - \varepsilon_0 x^3 + f_0 \cos(\Omega t). \tag{A.34}$$

In this equation for an anharmonic oscillation there are five parameters. Using (A.21), the dimensionless equation is given by

$$\ddot{x}' = -\alpha T\dot{x}' - \omega_0^2 T^2 x' - \varepsilon_0 L^2 T^2 x'^3 + \frac{f_0 T^2}{L}\cos(\Omega T t'). \tag{A.35}$$

The force law itself defines a characteristic length by the displacement beyond which the non-linear term dominates. Let $\tilde{x}$ denote the displacement at which the linear contribution equals the cubic one, i.e. $\omega_0^2\tilde{x} = \varepsilon_0\tilde{x}^3$. This yields $\tilde{x} = \omega_0/\sqrt{\varepsilon_0}$.

Choosing this as the length unit ($L_1 = \tilde{x}$), and taking the period $T = T_0 = 2\pi/\Omega$ of the driving as the time unit, we obtain

$$\ddot{x}' = -B\dot{x}' - C^2(x' + x'^3) + I\cos(2\pi t'), \tag{A.36}$$

with

$$B = \alpha T_0, \quad C = \omega_0 T_0, \quad I = \frac{f_0 T_0^2\sqrt{\varepsilon_0}}{\omega_0.}. \tag{A.37}$$

The system is characterised by three independent parameters: the dimensionless friction coefficient, B, the dimensionless natural frequency, C, and the dimensionless driving amplitude, I. Our choice of length scale implies that the dimensionless force contains the linear and the cubic terms with equal weight.

Several different dimensionless forms exist that are equivalent to the above. By measuring time in units of $1/\omega_0$, we obtain, for example,

$$\ddot{x}' = -\frac{B}{C}\dot{x}' - (x' + x'^3) + \frac{I}{C^2}\cos\left(\frac{2\pi}{C}t'\right). \tag{A.38}$$

Another length scale, L_2, is determined by choosing $f_0 T^2/L = 1$ in (A.35). With $T = T_0 = 2\pi/\Omega$, $L_2 = (2\pi)^2 f_0/\Omega^2$,

$$\ddot{x}' = -B\dot{x}' - C^2(x' + I^2 x'^3) + \cos 2\pi t'. \tag{A.39}$$

The independent dimensionless parameters are unchanged.

A.3 Numerical solution of ordinary differential equations

Most ordinary differential equations possess unique solutions for given initial conditions. Numerical methods make use of this property: starting from the initial condition and proceeding with finite, but small, *time steps*, the solution is rolled up, i.e. becomes approximated by an iterated sequence. Due to the finiteness of the time step, *numerical errors* are generated in each step. A reliable estimation of these errors is an important part of the numerical procedure. When choosing the algorithm, one seeks some kind of optimum to reduce both the run-time and the error. In the following, we briefly summarise the principle of the numerical solution and present the most important algorithms.

A.3.1 First-order equations

The solution, $x(t)$, of the *dimensionless* equation,

$$\dot{x} \equiv \frac{dx}{dt} = f(x, t), \tag{A.40}$$

is to be determined for initial condition $x(t_0) = x_0$ at the time instant $t = t_0$ and with f as a given bivariate function. The numerical solution provides the function $x(t)$ at some discrete instants t_n. In the simplest case, these are equi-distant with a dimensionless time step h: $t_n = t_0 + nh$. Our task is to determine $x_{n+1} \equiv x(t_n + h)$ in terms of $x_n \equiv x(t_n)$ of the previous instant and the function f. The Taylor expansion of $x(t_n + h)$ yields

$$x_{n+1} = x_n + \sum_{j=1}^{\infty} \frac{1}{j!} \frac{d^j x(t_n)}{dt^j} h^j. \tag{A.41}$$

Knowing the differential equation (A.40), the time derivatives, $x(t)$, can be expressed in terms of function f. The first two derivatives are, for example,

$$\frac{dx}{dt}(t_n) = f(x_n, t_n), \quad \frac{d^2x}{dt^2}(t_n) = f_t(x_n, t_n) + f_x(x_n, t_n)\frac{dx}{dt}(t_n), \tag{A.42}$$

where the subscripts indicate partial derivatives of f with respect to x or t.

In any numerical procedure, one has to truncate the Taylor expansion. Keeping the first N terms, we can write, with an error of order h^{N+1}, that

$$x_{n+1} = x_n + \sum_{j=1}^{N} \frac{1}{j!} \frac{d^j x(t_n)}{dt^j} h^j. \tag{A.43}$$

We then say that an *algorithm of order* N is applied. Since higher-order partial derivatives of f are complicated to determine in analytic forms, we express the right-hand side of (A.43) in terms of function f itself. In certain well chosen points near (x_n, t_n), the f-values are evaluated and sum (A.43) then appears as a suitably weighted average. Depending on the choice of the points, different methods exist.

Euler method (a first-order method, $N = 1$)

$$x_{n+1} = x_n + hf(x_n, t_n). \tag{A.44}$$

The error in one step is of the order of h^2.

Second-order Runge–Kutta method

The algorithm is given by

$$\left.\begin{aligned} x_{n+1} &= x_n + \frac{1}{2}(k_1 + k_2), \\ k_1 \equiv hf(x_n, t_n), &\quad k_2 \equiv hf(x_n + k_1, t_n + h). \end{aligned}\right\} \tag{A.45}$$

Using (A.42), the series expansion, (A.43), of x_{n+1} up to second order in h ($N = 2$) can be checked to be consistent with (A.45). The error in one step is proportional to h^3.

Third-order Runge–Kutta method

The algorithm is given by

$$\left.\begin{aligned} x_{n+1} &= x_n + \frac{1}{6}(k_1 + 4k_2 + k_3), \\ k_1 \equiv hf(x_n, t_n), &\quad k_2 \equiv hf\left(x_n + \frac{k_1}{2}, t_n + \frac{h}{2}\right), \\ k_3 &\equiv hf\left(x_n + 2k_2 - k_1, t_n + h\right). \end{aligned}\right\} \tag{A.46}$$

Expanding this up to order h^3, form (A.43) with $N = 3$ is recovered. The error in one step is thus of the order of h^4.

Fourth-order Runge–Kutta method

The algorithm is given by

$$\left.\begin{aligned} x_{n+1} &= x_n + \frac{1}{6}(k_1 + 2k_2 + 2k_3 + k_4), \\ k_1 \equiv hf(x_n, t_n), &\quad k_2 \equiv hf\left(x_n + \frac{k_1}{2}, t_n + \frac{h}{2}\right), \\ k_3 \equiv hf\left(x_n + \frac{k_2}{2}, t_n + \frac{h}{2}\right), &\quad k_4 \equiv hf(x_n + k_3, t_n + h). \end{aligned}\right\} \tag{A.47}$$

A direct verification is again possible, although somewhat lengthy. The error per time step is of the order of h^5.

In practice, the fourth-order Runge–Kutta method is a good compromise: the algorithm is not too long, and the accuracy is typically good enough to allow us to choose (dimensionless) time steps $h \gtrsim 10^{-2}$. The method is suitable for the simulation of most chaotic problems.

A.3.2 Sets of first-order equations

By using the vector notation $\mathbf{x} = (x_1, x_2, \ldots, x_r)$, the solution, $\mathbf{x}(t)$, of the equation

$$\dot{\mathbf{x}} = \mathbf{f}(\mathbf{x}, t) \tag{A.48}$$

is to be found with initial conditions $\mathbf{x}(t_0) = \mathbf{x}_0$. The basic idea of the numerical solution is the same as for a single equation. Here, we only present the formulae of the fourth-order Runge–Kutta method:

$$\left.\begin{array}{ll} \mathbf{x}_{n+1} = \mathbf{x}_n + \frac{1}{6}(\mathbf{k}_1 + 2\mathbf{k}_2 + 2\mathbf{k}_3 + \mathbf{k}_4), & \\ \mathbf{k}_1 \equiv h\mathbf{f}(\mathbf{x}_n, t_n), & \mathbf{k}_2 \equiv h\mathbf{f}\left(\mathbf{x}_n + \frac{\mathbf{k}_1}{2}, t_n + \frac{h}{2}\right), \\ \mathbf{k}_3 \equiv h\mathbf{f}\left(\mathbf{x}_n + \frac{\mathbf{k}_2}{2}, t_n + \frac{h}{2}\right), & \mathbf{k}_4 \equiv h\mathbf{f}(\mathbf{x}_n + \mathbf{k}_3, t_n + h). \end{array}\right\} \tag{A.49}$$

The error of one step is again proportional to h^5.

A.3.3 Accumulation of errors

So far, we have dealt with the error made in one step. Since dimensionless forms have been used, this corresponds to a kind of relative error in the original problem. In an algorithm of order N, this error is of order h^{N+1}. The errors accumulate in time, but this accumulation is not necessarily linear. The estimate h^{N+1} yields the absolute value of the error, but errors occur with different signs. Their number therefore increases in a similar way to the average displacement in a random walk. If the consecutive signs are completely independent, the total error increases with the square root of the number, n, of the steps. If the signs are somewhat correlated, the increase can be more rapid. In a simple estimate, we therefore assume that the error grows with some power, σ, of the number, n, of steps ($\sigma \leq 1$). Let the program run until the accumulated error reaches a threshold value, Δ (for example 10^{-2} (1 %)). The number, n_0, of steps required for this can be estimated from $n_0^\sigma h^{N+1} = \Delta$, which yields $n_0 = \left(h^{-(N+1)}\Delta\right)^{1/\sigma}$. Since one time step is of dimensionless length, h, the run-time is approximately

$$t_0 = \Delta^{1/\sigma} h^{-(N+1)/\sigma+1} \tag{A.50}$$

dimensionless physical time units. With a 1% total accumulated error and a time step of one-hundredth ($\Delta = h = 10^{-2}$), a fourth-order method ($N = 4$) can run up to 10^6 or 10^{14} time units with a linear ($\sigma = 1$) or a square root ($\sigma = 1/2$) error growth, respectively. Since chaotic sets (attractors, bands or saddles) are typically well traced out within 10^4–10^5 time units, numerical methods can determine the chaotic sets to a very good accuracy.

Besides the error originating from the truncation of the Taylor series, another source of error is the round-off error. Its total contribution increases with the number of steps. It is therefore not advisable to choose the time step, h, to be too small. On the other hand, it cannot be too large since the algorithm may become unstable (see Box 7.2). In an intermediate range, numerical solutions are effective even for long runs. In autonomous conservative systems, energy conservation can, for example, be used to estimate the accuracy of the solution: if the numerically determined total energy remains around the initial energy value over the run, the method is reliable. In chaotic systems, the aforementioned errors become exponentially magnified along *the unstable direction*, in spite of the reliability of the method. Consequently, numerical solutions cannot provide accurate individual trajectories on chaotic sets. They are, however, well suited for determining the shape of the chaotic set and its natural distribution (see Box 5.4).

A.4 Sample programs

The following two sample programs are to help readers with elementary programming skills and having yet no experience with numerical simulations of dynamical systems. The programs are written in Pascal since both the structure of the codes and the handling of the graphical display is particularly simple there. Having understood the logic, rewriting to other programming languages or other environments is easy. These examples also illustrate that the numerical investigation of deterministic chaos is not at all a demanding task from the point of view of programming.

A.4.1 The Pascal program map

```
program map;

{Simulation of the dimensionless map}
{x'1=M1(x1,x2), x'2=M2(x1,x2) with parameters a and b.}
{The points of the map are displayed on a}
{screen of 640x480  pixels, the origin of the}
{co-ordinate system is the centre of the screen.}

{Turbo Pascal program to run under  DOS}
{or  in the DOS prompt of Windows.}


uses Graph,crt;              {For the use of graphics.}

{Variables and their meaning:}

const

  x10=0.1;      {Choice of the initial point.}
  x20=0.1;

  Niter=20000;  {Setting the number of iterations.}

  xmax=5; {Determining the area to be plotted on the
           screen.}
  ymax=4;

var
  a,b            : real;    {Parameters of the map.}
  m              : longint; {Counter of the iterations.}
```

```
  x1,x2,xx1,xx2   : real;     {Phase plane co-ordinates.}

  px,py           : integer; {Pixel co-ordinates on the screen.}

  ox,oy           : integer; {Pixel co-ordinates of the origin}
                             {on the screen.}
  sx,sy           : real;     {Scale factors for the plot.}

  grDriver,grMode: integer; {Variables needed to set}
                             {the graphics.}
  label finish;

procedure wait; {Waiting for a key to be hit.}

  var ch:char;
  begin
    repeat until keypressed;
    while keypressed do ch:=readkey;
  end;


{Program body:}
begin
  clrscr;
  {Parameter input:}
  write('a? ');
  readln(a);
  write('b? ');
  readln(b);

  {Opening the graphics window:}
  grDriver := Detect;
  InitGraph(grDriver,grMode,'c:\tp\bgi'); {At }
         {'c:\tp\bgi', the path to folder bgi}
         {of Turbo Pascal (here c:\tp) should be given.}

  {Determining the scale factors to plot the desired}
  {area on screen:}
  ox:=320;
  oy:=240;
  sx:=ox/xmax;
  sy:=oy/ymax;

  {Initial values:}
  x1:=x10;
  x2:=x20;
```

```
  {Plotting the initial point on the screen:}
  px:=ox+round(x1*sx);
  py:=oy-round(x2*sy);
  PutPixel(px,py,11); {The value 11 of the last argument}
                      {corresponds to a light blue colour.}

  {Iteration of the map up to step number Niter:}
  for m:=1 to Niter
    do
    begin
      xx1:=x1;
      xx2:=x2;
      x1:=M1(xx1,xx2);  {Here, in place of M1 and M2 the forms of the mapping}
                          {functions should be specified.}
      x2:=M2(xx1,xx2);
      if (abs(x1)>xmax) or (abs(x2)>ymax) then goto finish;
      {Plotting  the point if it has not left the screen:}
      px:=ox+Round(x1*sx);
      py:=oy-Round(x2*sy);
      PutPixel(px,py,14); {The value 14 of the last argument}
                                      {corresponds to a yellow colour.}
    end;

  finish:
  {Waiting until a key is hit (meanwhile the plot}
  {remains visible  on the screen):}
  if not(m<Niter) then wait;
  CloseGraph; {Closing the graphich window.}
  if m<Niter then  begin
     writeln('The iterated point left the screen after ',m,' steps!');
     wait; end;
end.
```

A.4.2 The Pascal program diffeq

```
program diffeq;

{Solution of the dimensionless set of equations}
{dx1/dt=F1(x1,...,x4), dx2/dt=F2(x1,...,x4),}
{dx3/dt=F3(x1,...,x4), dx4/dt=F4(x1,...,x4)}
{with parameters a and b}
{by means of the fourth order Runge-Kutta method.}
{The flow is displayed  on the (x1,x3) plane,}
{on a screen consisting of 640x480 pixels. The origin of}
```

```
{the co-ordinate system is the centre of the screen.}
{Turbo Pascal program to run under  DOS}
{or in the DOS prompt of Windows.}

uses Graph,crt;              {For the use of graphics.}

{Variables and their meaning:}
const
  nmax=4;    {Dimension of the phase space.}

  a=1;       {Setting the parameters.}
  b=2;

  x10=0.1;     {Initial conditions: xi0 (i=1,...4).}
  x20=0.1;
  x30=0.1;
  x40=0.1;

  h=0.01;                {Dimensionless time step of the numerical algorithm.}
  cntmax=10000;          {Maximum number of time steps.}
  xmax=5;                {Size of the area to be plotted on the screen.}
  ymax=4;

type vector=array[1..nmax] of double;

var
  x             : vector;
  t             : real;
  cnt           : longint;
  ox,oy         : integer; {Pixel co-ordinates of the origin}
                                      {on the screen.}
  sx,sy         : real;      {Scale factors for the plot.}
  grDriver,grMode: integer; {Variables needed to set the graphics.}

procedure wait; {Waiting for a key to be hit.}

  var ch:char;
  begin
    repeat until keypressed;
    while keypressed do ch:=readkey;
  end;

procedure derivative(x:vector; var xp: vector);
  {Output xp yields the time derivative dx/dt in the phase space point x}
  {(a vector of dimension nmax=4).}
```

```
  begin
    xp[1]:=F1(x[1],x[2],x[3],x[4]); {Here, in place of F1,...,F4,}
    xp[2]:=F2(x[1],x[2],x[3],x[4]); {the forms of the right-hand}
    xp[3]:=F3(x[1],x[2],x[3],x[4]); {sides should be specified.}
    xp[4]:=F4(x[1],x[2],x[3],x[4]);
  end;

procedure RungeKutta4(var x:vector; h:real);
  {Vector x is evolved by a time step h}
  {according to the fourth order Runge--Kutta method.}
  {Procedure derivative is used to evaluate the increments of  x.}
  var
    k1,k2,k3,k4,y: vector;
    i           : integer;
  begin
    derivative(x,k1);
    for i:=1 to nmax do y[i]:=x[i]+h*k1[i]/2;
    derivative(y,k2);
    for i:=1 to nmax do y[i]:=x[i]+h*k2[i]/2;
    derivative(y,k3);
    for i:=1 to nmax do y[i]:=x[i]+h*k3[i];
    derivative(y,k4);
    for i:=1 to nmax do x[i]:=x[i]+h*(k1[i]+2*(k2[i]+k3[i])+k4[i])/6;
  end;


{Program body:}
begin
  {Opening the graphics window:}
  grDriver := Detect;
  InitGraph(grDriver,grMode,'c:\tp\bgi'); {At }
         {'c:\tp\bgi', the path to folder bgi}
         {of Turbo Pascal (here c:\tp) should be given.}

  {Determining  the scale factors to plot the desired area}
  {on the screen:}
  ox:=320;
  oy:=240;
  sx:=ox/xmax;
  sy:=oy/ymax;

  {Initial values:}
  x[1]:=x10;
  x[2]:=x20;
  x[3]:=x30;
  x[4]:=x40;
```

```
  {Plotting  the initial point on the (x1,x3) plane:}
  PutPixel(ox+round(x[1]*sx),oy-round(x[3]*sy),11);
  {Value 11 corresponds to a light blue colour.}

  {Carrying out a  Runge-Kutta step and}
  {plotting the new state on the (x1,x3) plane:}
  cnt:=1;  {Setting the time counter.}
  repeat
    t:=cnt*h; {If time t is needed for some purpose.}
    RungeKutta4(x,h);
    PutPixel(ox+round(x[1]*sx),oy-round(x[3]*sy),14);
    {Value 14 corresponds to a yellow colour.}
  cnt:=cnt+1;
  until cnt>cntmax;

  wait; {Waiting for a key to be hit.}

  CloseGraph; {Closing the graphics window.}
end.
```

A.5 Numerical determination of chaos parameters

We present simple methods for determining the chaos parameters, the application of which does not require any special numerical expertise. As particular examples, we choose the roof attractor ($E = 0.7, a = 1.77$, cf. Fig. 5.28) and the roof saddle ($E = 0.7, a = 2.3$, cf. Fig. 6.18), but the algorithms can successfully be applied to any two-dimensional maps.

A.5.1 Topological entropy

For numerical measurements, monitoring a line segment in phase space is the most suitable (see (5.52) and (6.15)). Since the line becomes stretched and multiply folded, its length, $L(n)$, can best be determined by prescribing a small threshold and inserting additional points whenever the distance between neighbouring points grows beyond this value in order to keep the distance of neighbours under the threshold.

In the case of the roof attractor, the stretching is followed over 12 iterates (Fig. A.1). The initial straight line segment connects points $(2, -1)$ and $(0, 1)$ and consists of 20 000 points. (The choice of the segment is arbitrary; it does not affect the outcome of the measurement.) The threshold is taken to be twice the initial distance,[1] $\Delta r = 0.14 \times 10^{-5}$.

The topological entropy of a saddle can be determined in a similar way, except that the length of the line segment has to be determined in a fixed region containing

[1] The piecewise linearity of the map also makes an exact determination of the instantaneous length possible. The accuracy of the numerical method can thus be estimated: a deviation shows up in the fourth digit of the topological entropy only.

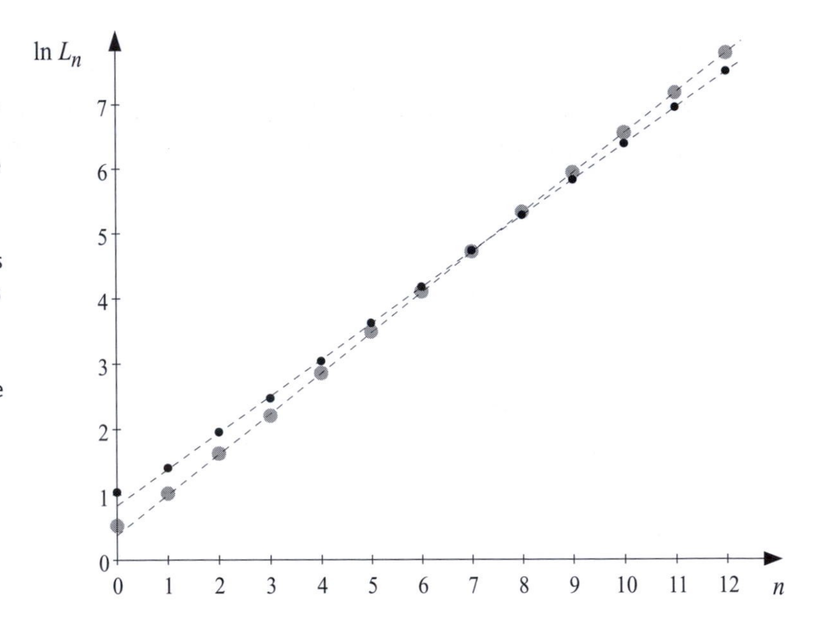

Fig. A.1. Measuring topological entropy. Logarithm of the length, L_n, of a line segment vs. number, n, of iterates on the roof attractor (black dots) and around the roof saddle (grey dots). The dashed lines represent fitted straight lines of slope h with $h = 0.556$ and $h = 0.692$, respectively, for the roof attractor and the roof saddle.

the saddle. We choose the region above the basic branch and below the topmost branch of H_-'s stable manifold (see Fig. 6.24). The saddle proves to be more chaotic than the attractor (its h is found to be larger; see also Table A.1).

A.5.2 Average Lyapunov exponents

The positive local and average Lyapunov exponents can be determined from the growth rate of the distance, $\Delta r_n = \Delta r_0 \exp(\lambda(x, y)n)$, between point pairs chosen arbitrarily on a chaotic set. Since the initial distance, Δr_0, is finite in practice, two points quickly become separated, due to exponential divergence, to distances comparable to the size of the chaotic attractor or saddle. An exponential growth can thus be observed over finite times only (in our particular example, up to about 60 and 40 iterates on the attractor and saddle, respectively; see Fig. A.2). The local Lyapunov exponent characterising the point of the initial condition can be read off from the slope of the curves $\ln \Delta r(n)$ vs. n over the first iterates (Fig. A.2). Later, the curves start oscillating around a straight line (dashed), the slope of which corresponds to the average Lyapunov exponent. In order to determine the average Lyapunov exponent accurately, several point pairs have to be investigated.

On attractors, this can be achieved by using a long reference trajectory (a simulated trajectory $x_n, v_n, n = 0, 1, \ldots$); each time the distance between a test and a reference trajectory exceeds a certain threshold, say Δr_{th}, the point of the test trajectory is pushed back to the corresponding reference trajectory point so that its new co-ordinates are $x_n + \Delta x_0$ and $v_n + \Delta v_0$. Let $\Delta r_n(i)$ denote the distance of the two points at iteration n after the ith push-back. Index n grows until $\Delta r_n(i)$ reaches the threshold value. After M push-backs, let M_n denote how many point pairs are present below the threshold ($\Delta r_n(i) < \Delta r_{\text{th}}$) at the nth iterate. Consequently,

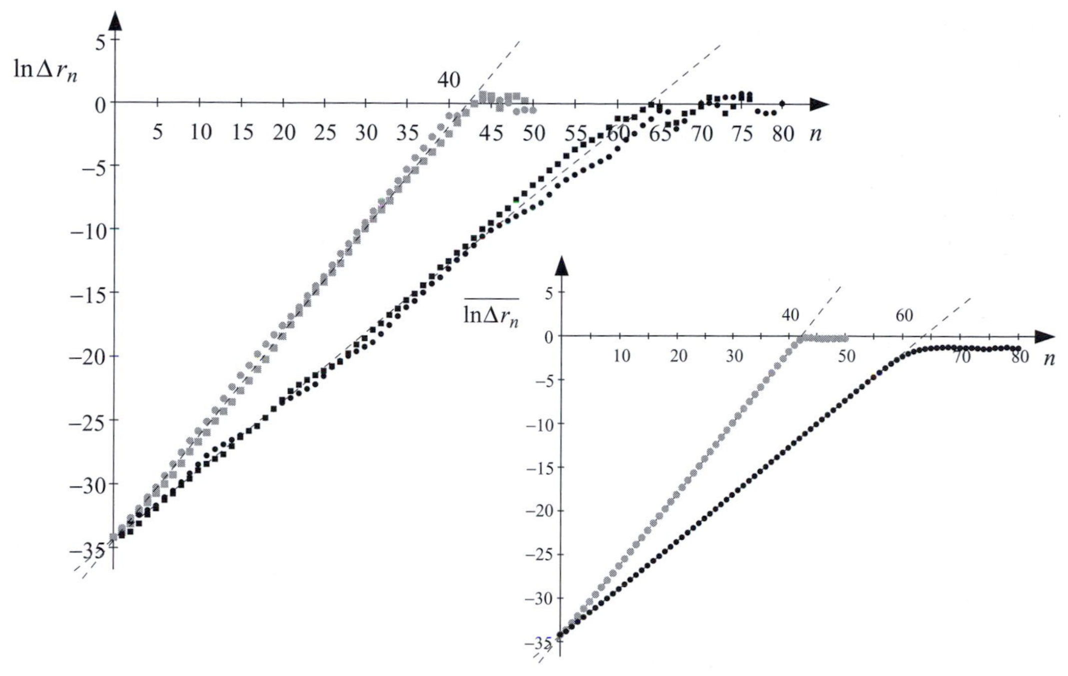

Fig. A.2. Measuring the positive average Lyapunov exponents. Black (grey) symbols represent the distances of point pairs on the attractor (saddle). Circles and squares mark the distances of trajectory pairs starting around (0, 0) and (−1, −1), respectively. The initial differences are $\Delta x_0 = \Delta v_0 = 10^{-15}$ ($\Delta r_0 = \sqrt{2} \times 10^{-15}$). The slopes of the dashed lines are the average Lyapunov exponents. The inset shows results obtained by the reference trajectory method on the attractor and by using 10^4 trajectory pairs around the saddle. The threshold is $\Delta r_{\text{th}} = 0.5$ and the number of push-backs is $M = 10^4$ on the attractor. Around the saddle, trajectories with lifetimes of at least 50 have been selected. In both cases, $\Delta x_0 = \Delta v_0 = 10^{-15}$. The slopes of the straight lines fitted to $\overline{\ln \Delta r_n}$ (dashed lines) are $\bar{\lambda} = 0.540$ and $\bar{\lambda} = 0.819$ for the attractor and saddle, respectively.

$M_n \leq M$. The average of the logarithm of the distances is given by $\overline{\ln \Delta r_n} \equiv (1/M_n) \sum_{i=1}^{M_n} \ln \Delta r_n(i)$. For M, $M_n \gg 1$ this quantity grows to a good approximation linearly in n, with a slope corresponding to the average Lyapunov exponent, $\bar{\lambda}$ (see the inset of Fig. A.2).[2]

For transient chaos, it is impossible to find a single long reference trajectory, and one therefore chooses $M \gg 1$ initial conditions with a lifetime of at least 50 around the saddle, and does not apply any push-backs. Trajectories are started from these points and from points at a distance of $\Delta x_0 = \Delta v_0 = 10^{-15}$ away. An average is taken over trajectory pairs which do not escape for a sufficiently long time (see the inset of Fig. A.2).

[2] In this same measurement, the topological entropy can also be obtained by taking the logarithm of the *average distance*: the quantity $\ln((1/M_n) \sum_{i=1}^{M_n} \Delta r_n(i))$ grows linearly with n, with slope h.

Table A.1. *Chaos parameters of the roof attractor and roof saddle.*

The escape rate is measured as indicated in Fig. 6.2 (see also Fig. 6.17). Note that all the data are consistent with the general relations (5.68), (6.19), (5.74) and (6.29).

	Roof attractor[a]	Roof saddle[b]
h	0.556	0.692
$\bar{\lambda}$	0.540	0.819
$\bar{\lambda}'$	−1.253	−1.532
D_1	1.429	1.293
κ	0	0.132

[a] $E = 0.7, a = 1.77$.
[b] $E = 0.7, a = 2.3$.

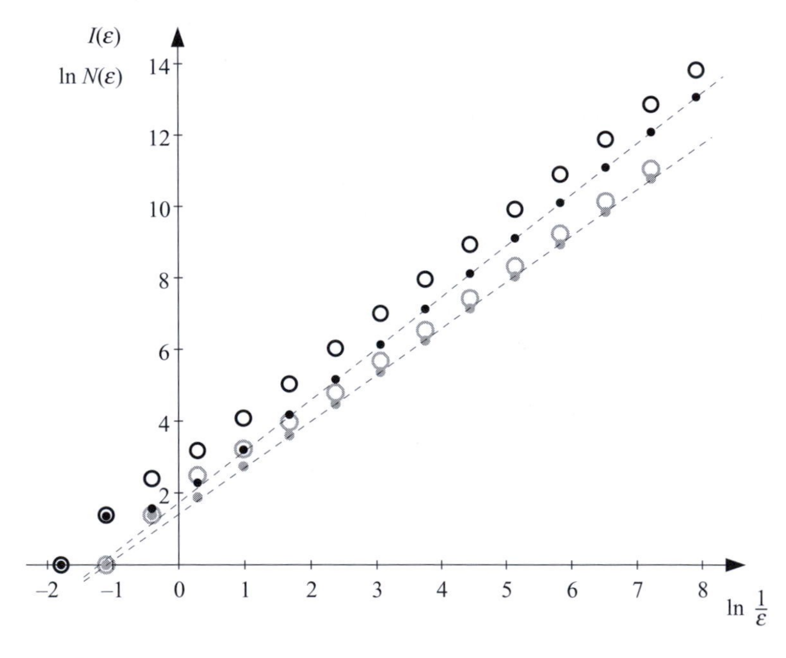

Fig. A.3. Measuring dimensions. The information content, $I(\varepsilon)$, of the natural distributions on the attractor (black dots) and on the saddle (grey dots) vs. resolution ε. The slopes of the straight lines fitted to the dots are the information dimensions $D_1 = 1.429$ and $D_1 = 1.293$, respectively. Empty circles mark the values of $\ln N(\varepsilon)$, where $N(\varepsilon)$ is the number of non-empty boxes in a coverage. The slope of a straight line fitted to these data for small ε yields the fractal dimension; in our case, $D_0 \approx D_1$.

The negative average Lyapunov exponent can be determined in a similar way from the contraction rates (the growth rate of the distances in the time-reversed dynamics). In the case of constant Jacobians, however, no extra simulation is needed, since $\bar{\lambda}' = \ln J - \bar{\lambda}$ (see (5.58)), and, in the roof map, $J = E^2$ (see Table A.1).

A.5.3 Fractal properties

Fractal and information dimensions can be measured in similar ways. A chaotic set containing, in a numerical simulation, N_{tot} points, distributed according to the natural distribution, is covered by grids whose resolution is refined. For the fractal dimension, the number, $N(\varepsilon)$, of boxes containing at least one point is to be recorded at each resolution, ε. The slope of $\ln N(\varepsilon)$ vs. $\ln(1/\varepsilon)$ is, for small ε, the fractal dimension, D_0 (see Section 2.1.2). For the information dimension, the box probabilities, $P_i(\varepsilon)$ are required, which can be given approximately as n_i/N_{tot}, where n_i is the number of points falling in box i. The information dimension follows from the scaling of the information content, $I(\varepsilon) = -\sum_i P_i(\varepsilon) \ln P_i(\varepsilon) \approx -\sum_i (n_i/N_{\text{tot}}) \ln(n_i/N_{\text{tot}})$, with ε (see 2.18).

Under identical conditions (number of points, resolution), the information dimension is easier to measure. This is because the natural distribution is, in general, strongly inhomogeneous: the probabilities are very small around many points of the chaotic set. In the course of a simulation, only a few points fall in these regions, which are therefore difficult to plot satisfactorily. Fortunately, in the determination of the information dimension, these regions have a negligible weight.

For the roof attractor, the natural distribution is determined from $N_{\text{tot}} = 10^7$ points. The sides of a 6×6 square centred at the origin are divided into 2^k ($k = 0, 1, \ldots, 14$) parts; the ε-values are thus 6×2^{-k} (see Fig. A.3). The empty circles represent the quantity $\ln N(\varepsilon)$. They form a slightly bent curve, the asymptotic ($\varepsilon \ll 1$) slope of which, however, is close to that of $I(\varepsilon)$. We therefore conclude that the fractal dimension is approximately equal to the information dimension.

For the saddle, the natural distribution is determined from points started at random in a 3×3 square centred at the origin, and the co-ordinates of those points that do not leave the square within 20 iterates are marked at the tenth iterate (see Box 6.1). The resolution is refined up to the level $k = 12$ ($\varepsilon = 3 \times 2^{-k}$) (see Table A.1).

Solutions to the problems[1]

Solution 2.1 The perimeter of a regular n-sided polygon of side length $\varepsilon = 2\sin(\pi/n)$ is $P = n\varepsilon$. In terms of the resolution, ε, $P = \pi\varepsilon/\arcsin(\varepsilon/2)$. For fine resolutions (small ε), the inverse sine function can be expanded as $\arcsin x \approx x + x^3/6$, and $P = 2\pi(1 - \varepsilon^2/24)$. Thus, the perimeter converges to the well known value 2π with the square of the resolution. (For less regular curves or less regular approximations, the correction is typically proportional to ε.)

Solution 2.3 In a coverage of the unit interval with segments of length ε, the number of segments is, in general, $[1/\varepsilon] + 1$, where $[x]$ denotes the integer part of x. For a small resolution, this is proportional to $1/\varepsilon$, the dimension given by equation (2.3) of a straight line segment is therefore indeed $D_0 = 1$.

In the case of the right-angled isosceles triangle, divide, for the sake of simplicity, one of the legs of unit length into n equal parts: $\varepsilon = 1/n$. Cover the triangle with boxes of width ε. There are n boxes in the column exactly at the other leg, $n - 1$ boxes in the next one, etc. The total number of boxes in the coverage is $\sum_{i=0}^{n}(n - i) = n(n + 1)/2$. Expressing this in terms of the resolution, $N(\varepsilon) = \varepsilon^{-2}(1 + \varepsilon)/2 \sim \varepsilon^{-2}$, consequently $D_0 = 2$.

Solution 2.6 With resolution $\varepsilon = r^n$ corresponding to the nth step of the construction, $P = 4^{n+1} \times r^n$, $A = 4^n \times r^{2n}$. The ratio P/A is always proportional to $1/\varepsilon$; it therefore always increases with refining resolution. The perimeter itself diverges, however, for $r > 1/4$ only, i.e. for $D_0 > 1$. For $r = 1/4$, the dimension of the Cantor cloud is exactly 1! The set is, of course, not a smooth curve in this case either, but it possesses one property in common with usual smooth curves: its observed perimeter does not depend on the resolution.

Solution 2.10 $\beta = (D_0 - 1)/(2 - D_0)$. An essential new property of fractals appears here: the relation $P \sim A^{1/2}$ valid for traditional objects is not recovered for any value of D_0 (not even the signs of the exponents agree!).

Solution 2.11 Remove the centre of the unit cube in such a way that eight identical cubes of size $1/3$ remain at the corners. Repeat this for each small cube. The volume observed at resolution $\varepsilon = 3^{-n}$ is $(8/27)^n$. In the Menger sponge, on the contrary, 20 cubes remain, and the observed volume is therefore $(20/27)^n$, which decreases much more slowly than $(8/27)^n$.

[1] Password-protected solutions which do not appear in the book are available online at www.cambridge.org/9780521839129.

Solution 2.12 In the nth step, the total length (volume) is $V_n = \Pi_{j=1}^{n}(1-\lambda_j)$. There are $N(\varepsilon) = 2^n$ intervals of length $\varepsilon = V_n/2^n$. Based on (2.4), the fractal dimension is the limit of

$$D_0 = \ln 2 \,/\, (\ln 2 - (\ln V_n)/n)$$

for large n. The dimension is unity if $(\ln V_n)/n$ converges to zero. This is always the case with fat fractals whose V_n tends to a finite value.

Solution 2.13 In general, in the nth step, the length (volume) of the preserved intervals is $V_n = \Pi_{j=1}^{n}(1-3^{-j})$. The total length, V, is obtained in the limit $n \to \infty$ as $V = \Pi_{j=1}^{\infty}(1-3^{-j})$. Since the factors tend rapidly to unity, the product is finite. Its numerical value is $V = 0.560$.

Consider now the difference $V_n - V$ and take the logarithm $\ln(V_n - V) = \ln V_n + \ln(1 - \Pi_{j=n+1}^{\infty}(1-3^{-j}))$. The first term is finite since V_n possesses a limit. In the second term, for sufficiently large n, $\Pi_{j=n+1}^{\infty}(1-3^{-j}) \approx 1 - \sum_{j=n+1}^{\infty} 3^{-j}$, which is, according to the summation rule for geometric series, $1 - 3^{-n}/2$. The smallest distance occurring in the nth step is the size of the holes: $\varepsilon = 3^{-n}2^{-n+1}\Pi_{j=1}^{n-1}(1-3^{-j})$ in between the intervals of length $2^{-n}\Pi_{j=1}^{n}(1-3^{-j})$. We therefore cover the set with intervals of this size. For large n, the infinite product tends to the total length V, and it is true, to a good approximation, that $\ln\varepsilon \approx -n\ln 6$. The exponent of the fat fractal is therefore given by

$$\alpha = \frac{\ln(V_n - V)}{\ln\varepsilon} = \frac{\ln V_n - n\ln 3 - \ln 2}{n \ln 1/6} = \frac{\ln 3}{\ln 6} = 0.613.$$

Solution 2.15 The total probability of the boxes of content p_m is, in the nth step, $N_m p_m$. According to Stirling's formula, the logarithm of this quantity is given by

$$\ln(N_m p_m) = n\ln n - m\ln m - (n-m)\ln(n-m) + m\ln p_1 + (n-m)\ln p_2.$$

The extremum is located at a value m^*, where the derivative of $\ln(N_m p_m)$ with respect to m vanishes. The condition for this is $\ln(n/m^* - 1) = \ln(p_2/p_1)$, which yields $m^* = p_1 n$. The number of boxes belonging to m^* is $N^* \equiv N_{m^*}$, for which $\ln N^* = -n(p_1\ln p_1 + p_2\ln p_2)$. Simultaneously, $\ln p_{m^*} = n(p_1\ln p_1 + p_2\ln p_2)$, and thus $\ln(N^* p_{m^*}) = 0$. At the accuracy of Stirling's formula, the total probability of the typical boxes is unity! As the resolution at level n is $\varepsilon = 3^{-n}$, from equation (2.16) $D_1 = \ln N^*/(n\ln 3)$, and thus

$$D_1 = -\frac{p_1\ln p_1 + p_2\ln p_2}{\ln 3}.$$

For $p_1 \neq 1/2$, the information $-(p_1\ln p_1 + p_2\ln p_2)$ is always smaller than $\ln 2$, the information content of the homogeneous case. Consequently, D_1 is always less than the fractal dimension, $D_0 = \ln 2/\ln 3$, of the support, and it is smaller, the smaller p_1 is.

Solution 3.2 From (3.15), the velocity is $v = \lambda_+ c_+ \exp(\lambda_+ t) + \lambda_- c_- \exp(\lambda_- t)$. The combination $v - \lambda_\mp x$ is therefore proportional to $\exp(\lambda_\pm t)$, and consequently both $(v - \lambda_- x)^{\lambda_-}$ and $(v - \lambda_+ x)^{\lambda_+}$ depend on time as $\exp(\lambda_-\lambda_+ t)$, from which (3.17) follows.

Solution 3.5 The velocity from (3.35) is given by

$$\frac{dx}{dt} = v(t) = -\frac{\alpha}{2}x(t) + \omega_\alpha A e^{-(\alpha/2)t} \cos(\omega_\alpha t + \delta).$$

In the polar co-ordinates $\varrho = \sqrt{x^2 + [v + (\alpha/2)x]^2/\omega_\alpha^2}$, $\phi = \omega_\alpha t$, this leads to the equation,

$$\varrho = A e^{-\alpha/(2\omega_\alpha)\phi},$$

of a logarithmic spiral.

Solution 3.8 Going up to second-order terms in the Taylor expansion:[2]

$$F(x) \approx F'(x^*)(x - x^*) + \frac{F''(x^*)}{2}(x - x^*)^2. \tag{P.1}$$

The force resulting from the second term is negligible compared with the first if

$$|x - x^*| \ll \left|\frac{2F'(x^*)}{F''(x^*)}\right| = \left|\frac{2s_0^2}{F''(x^*)}\right|. \tag{P.2}$$

The displacement must therefore be much less than $|2F'(x^*)/F''(x^*)|$ in order that equation (3.39) be a good approximation.

If (P.2) holds for the initial position, and the initial speed is not too high, this condition is usually fulfilled around stable fixed points during the entire motion. This is not the case around unstable fixed points. Along the unstable direction, where the distance $\Delta x(t) \equiv x(t) - x^*$ from the fixed point increases according to $\Delta x(t) = \Delta x_0 \exp(\lambda_+ t)$, the range of validity of the linear approximation is, from (P.2), given by

$$t \ll \frac{1}{\lambda_+} \ln \left|\frac{2s_0^2}{\Delta x_0 F''(x^*)}\right|. \tag{P.3}$$

The length of this time interval can be increased by choosing the initial position closer to the origin, but this increase is logarithmically slow. The dominant dependence is the inverse proportionality to the instability exponent, λ_+; for identical initial distances, the particle remains in the vicinity of the less unstable points for a longer period of time.

Solution 3.10 The fixed points $(x^*, 0)$, $(-x^*, 0)$ are elliptic, while the origin is hyperbolic. The equation of trajectories follows from the law of energy conservation, $v^2/2 + V(x) \equiv \mathcal{E} =$ constant. This yields

$$v(x) = \pm\sqrt{2(\mathcal{E} + bx^2 - dx^4)}.$$

For negative energies the trajectories are closed curves bound to the right or left well, which become, of course, ellipses upon approaching the well bottoms (Fig. P.1). At zero energy, the trajectory is $v(x) = \pm x\sqrt{s_0^2 - 2dx^2}$, where s_0 is the repulsion parameter given in (3.43). In the vicinity of the origin, this equation takes

[2] If the force law is centrally symmetric, i.e. if the derivative, $F''(x^*)$, of the force also vanishes (the potential curve is axially symmetric in the fixed point), the Taylor expansion has to be extended up to third-order.

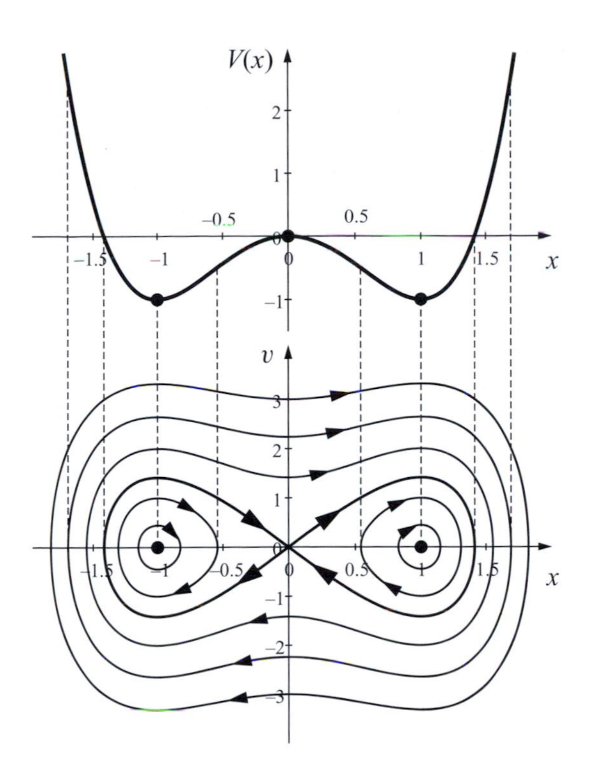

Fig. P.1. Double-well potential and trajectories of the frictionless motion ($a = 4$, $x^* = 1$; $b = 2$, $d = 1$). On the separatrix (thick line) the energy is zero. This curve separates the motion of negative energy bound to the wells from that of positive energy passing over the hill.

the linearised form $v(x) = \pm s_0 x$, in accordance with the general behaviour around hyperbolic points. The curve defined by the non-linear equation is also valid for points far from the origin and yields there the continuation of the local stable and unstable curves, the stable and unstable manifolds. The two manifolds (contrary to the frictional case) coincide and form a separatrix.

Solution 3.12 In the frictionless case, the point $(x_-^* = c, 0)$ is elliptic and the origin is hyperbolic. The equation of the trajectories is given by

$$v(x) = \pm\sqrt{2(\mathcal{E} - acx^2/2 + ax^3/3)}.$$

The curve $\mathcal{E} = 0$ is a separatrix, which separates three types of motion: periodic behaviour in the well, approaching the potential hill from the right and, for positive energies, winding around the well and running out into infinity (Fig. P.2).

In the presence of friction, the left branches of the stable and unstable manifolds of the origin no longer coincide: the unstable manifold leads into the fixed point attractor, while the stable manifold extends to minus infinity. The two stable manifold branches surround the basin of attraction of the fixed point. Trajectories starting outside of this domain all run out into infinity and the ship capsizes. The state $(x^* = \infty,\ v^* = \infty)$ can be considered as a second (simple) attractor.

Solution 3.13 As long as h is greater than the unstretched length of each spring, $h > l_0$, the position $x = 0$ obviously represents a stable equilibrium state. If the

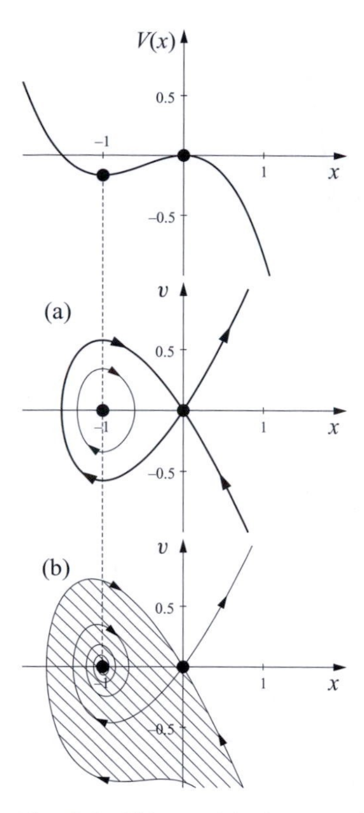

Fig. P.2. 'Ship capsizing': potential and phase portrait in the (a) frictionless and (b) frictional cases ($a = 1$, $c = -1$, $\alpha = 0.2$).

springs' end-points are moved closer to the x-axis than the unstretched length, then the equilibrium state of the body will be at $\pm\sqrt{l_0^2 - h^2}$, corresponding to unstretched springs. For a general displacement x, the length of the springs is $\sqrt{h^2 + x^2}$, and the absolute value of the forces is therefore $k(\sqrt{h^2 + x^2} - l_0)$, and their x-component (Fig. 3.14(a)) is $-k(\sqrt{h^2 + x^2} - l_0)x/\sqrt{h^2 + x^2}$. Consequently, the resultant force is horizontal and of magnitude

$$F(x) = -2kx\left(1 - \frac{l_0}{\sqrt{h^2 + x^2}}\right). \tag{P.4}$$

For equilibrium points x^*, $F(x^*) = 0$. For $h < l_0$, there exist three solutions, $x^* = \sqrt{l_0^2 - h^2}$, 0 and $-x^*$, and the origin is obviously unstable. At the value $h_c = l_0$, a pitchfork bifurcation occurs (Fig. P.3). The general parameter μ can be chosen, for example, as the difference $\mu = l_0 - h$.

In the neighbourhood of the bifurcation point, h hardly differs from l_0; therefore, the possible x-values are small and h can be replaced by l_0, except for expressions containing $(h - l_0)$. Using the approximate relation

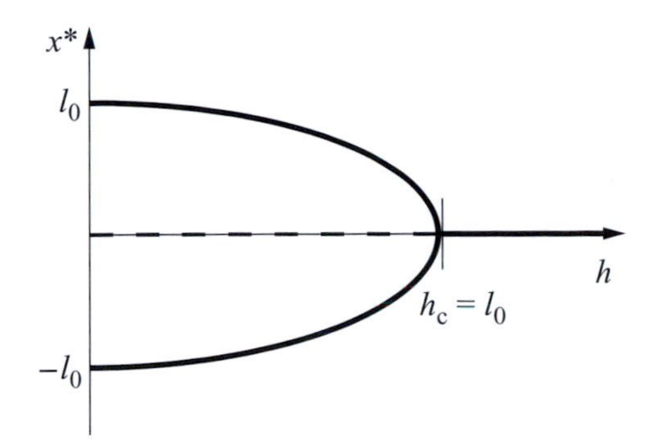

Fig. P.3. Pitchfork bifurcation of the system presented in Fig. 3.14(a) in terms of the parameter h.

$1/\sqrt{1+(x/h)^2} \approx 1-(1/2)(x/h)^2$, the force given in (P.4) becomes

$$F(x) \approx -2k\left(1-\frac{l_0}{h}\right)x - \frac{k}{l_0^2}x^3 \approx -\frac{k}{l_0^2}x\left[x^2 - 2l_0(l_0-h)\right]$$
$$\approx -\frac{k}{l_0^2}x(x-x^*)(x+x^*). \qquad \text{(P.5)}$$

This corresponds to force (3.41) with the fixed point $x^* \approx \sqrt{2l_0(l_0-h)}$.

Solution 3.17 With a horizontal rod of length $d \ll l$, the centrifugal acceleration is $(d + l\sin\varphi)\Omega^2$. The resultant force perpendicular to the rod is given by

$$F(\varphi) = \Omega^2(d + l\sin\varphi)\cos\varphi - g\sin\varphi.$$

Bifurcation occurs at small angular deviations; we therefore expand the force in powers of φ:

$$F(\varphi) \approx \Omega^2 d\left(1-\frac{\varphi^2}{2}\right) + \left[\varphi(\Omega^2 l - g) - \frac{\Omega^2 l}{2}\varphi^3\right]\left(1-\frac{\varphi^2}{6}\right).$$

The terms in the round brackets proportional to φ^2 are negligible compared to unity,[3] and we obtain

$$F(\varphi) \approx \Omega^2 l\left(\frac{d}{l} + \varphi\left(1-\frac{g}{\Omega^2 l}\right) - \frac{\varphi^3}{2}\right). \qquad \text{(P.6)}$$

For small angular velocities, there is a single equilibrium state. This can only occur if the graph of $F(\varphi)$ does not intersect the φ-axis again; i.e., if the minimum of $F(\varphi)$ is beyond the φ-axis (Fig. P.4). An extremum corresponds to a value φ_0, for which $F'(\varphi_0) = 0$. Consequently, the minimum is taken at

$$\varphi_0 = -\sqrt{\frac{2}{3}\left(1-\frac{g}{\Omega^2 l}\right)}.$$

[3] The final result indicates that φ is of the order of $(d/l)^{1/3}$.

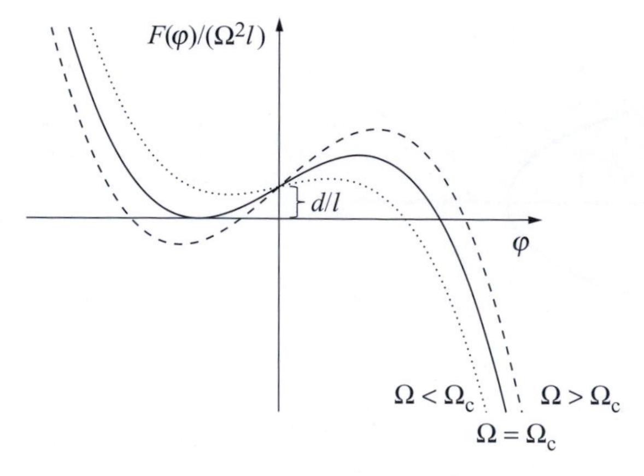

Fig. P.4. Graph of the force law around the critical angular velocity, Ω_c. The dotted, continuous and dashed lines correspond to the cases $\Omega < \Omega_c$, $\Omega = \Omega_c$ and $\Omega > \Omega_c$, respectively.

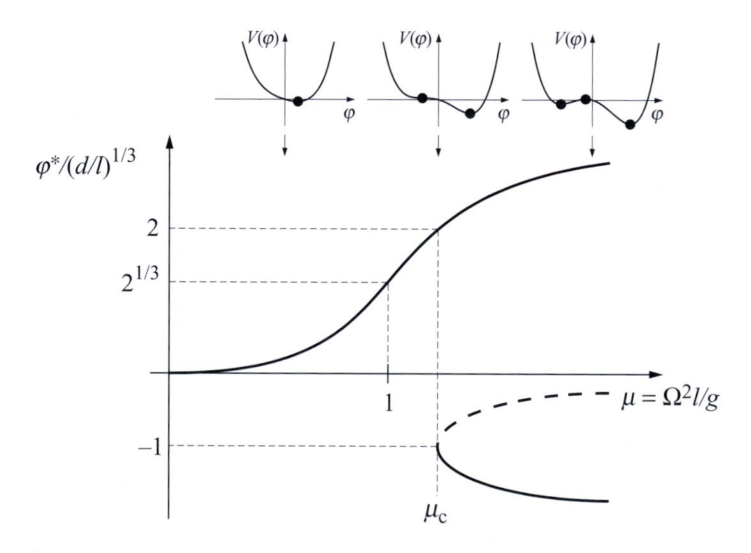

Fig. P.5. Distorted pitchfork bifurcation of the merry-go-round model. The potential corresponding to the force law is also plotted at three different angular velocities.

The corresponding force is $F(\varphi_0) = \Omega^2 l(\varphi_0^3 + d/l)$. At a certain critical angular velocity, Ω_c, this value is exactly zero, which yields

$$\Omega_c = \sqrt{\frac{g}{l}} \frac{1}{\sqrt{1-(3/2)(d/l)^{2/3}}} \approx \sqrt{\frac{g}{l}} \left(1 + \frac{3}{4}\left(\frac{d}{l}\right)^{2/3}\right).$$

At the critical angular velocity a new equilibrium state appears at the angular deviation $\varphi^* = -(d/l)^{1/3}$ (when the positive equilibrium state is at $2(d/l)^{1/3}$). The complete bifurcation diagram is sketched in Fig. P.5.

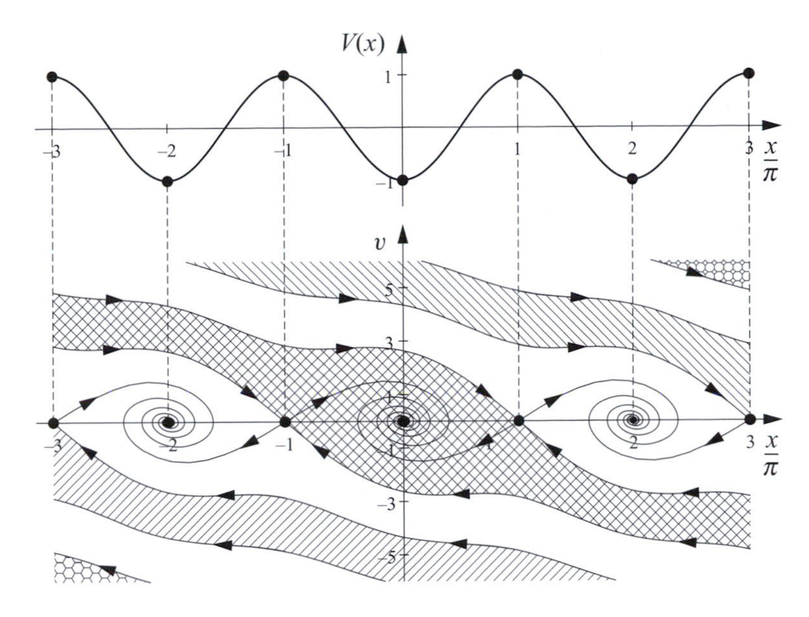

Fig. P.6. Potential and phase portrait of the weakly damped motion on a bumpy road. The basins of every second attractor are shaded. The phase portrait repeats itself periodically. The parameters are $A = 1$, $\alpha = 0.3$.

Solution 3.19 For weak friction, the points $x_+^* = 2\pi n$ (n is an integer), corresponding to the bottoms of the holes, are spiral attractors. According to the relation $F'(x_+^*) = \omega_0^2$, the natural frequency around the stable equilibrium points is $\omega_0 = \sqrt{A}$. On the other hand, the points $x_-^* = \pi + 2\pi n$, the local maxima of the potential, are hyperbolic with repulsion parameter $s_0 = \sqrt{A}$. A body starting from the top of a bump with a negligible initial speed is not able to get over the next bump; it therefore comes to a standstill at the bottom of one of the neighbouring holes after a few oscillations. Accordingly, the unstable manifold branches of the hyperbolic points go into the neighbouring spiral attractors, while the stable manifolds separate the basins of attraction (Fig. P.6).

Solution 3.22 As friction increases, the stationary motion slows down and the limit cycle comes closer to the x-axis. The basin of attraction of the fixed points broadens. At a critical friction coefficient, the limit cycle reaches the hyperbolic points from above (Fig. P.7). This situation arises when a body starting from one of the bumps with zero initial velocity can just reach the top of the next bump to its right. At this critical value the top branches of the stable and unstable manifolds coincide (Fig. P.7). For stronger friction no sliding down is possible: all bodies come to a halt in one of the holes, depending on the initial velocity (the snow is too soft for skiing). In this case both branches of the unstable manifold lead into a fixed point attractor (Fig. P.8).

Solution 3.26 From (3.64) with $\mathrm{Tr}\, A \equiv -\sigma$,

$$\lambda_\pm = \frac{-\sigma \pm \sqrt{\sigma^2 - 4 \det A}}{2}.$$

The hyperbolic points fall in the domain of negative determinants (Fig. P.9). The stability of the system is thus determined by the sign of $\det A$. For $\det A > 0$, both eigenvalues are real for $\sigma > 2\sqrt{\det A}$. Consequently, above (below) the parabola

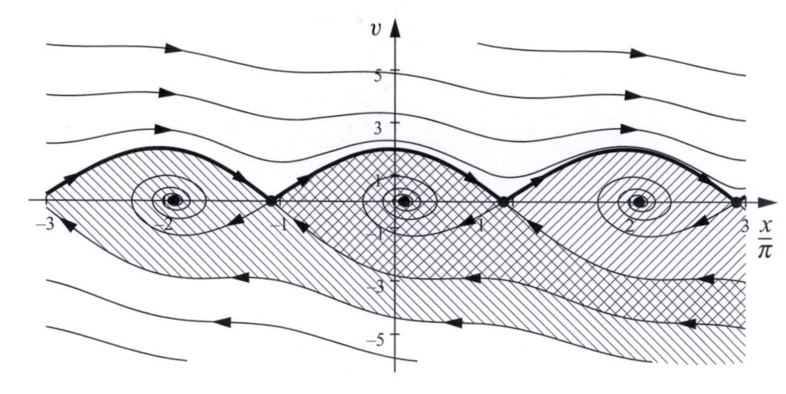

Fig. P.7. Phase portrait of the motion on a bumpy slope at the critical friction coefficient. The bottom branches of the stable manifolds again provide the basin boundaries of the spiral attractors ($A = 1$, $F_0 = 0.25, \alpha = \alpha_c = 0.198$).

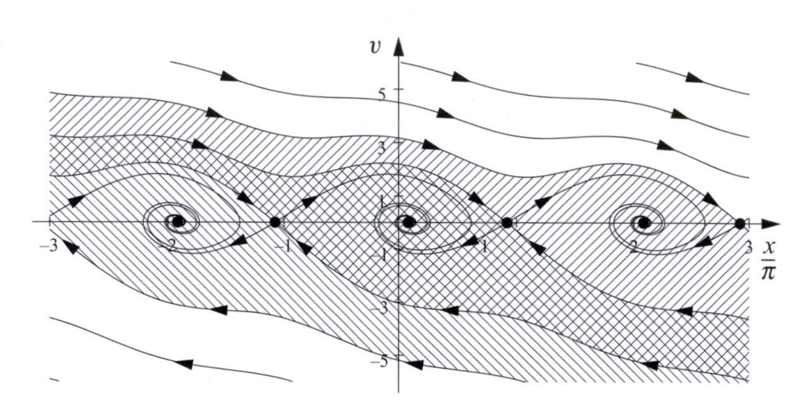

Fig. P.8. Phase portrait of the motion on a bumpy slope with strong friction. The limit cycle has disappeared, and all initial conditions lead to motion coming to rest in the holes ($A = 1$, $F_0 = 0.25$, $\alpha = 0.3$).

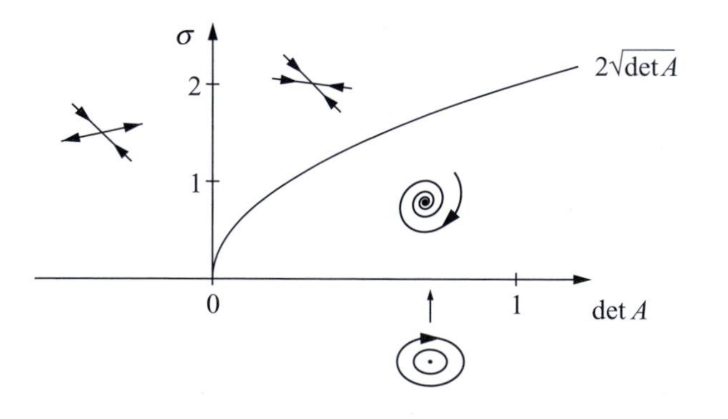

Fig. P.9. Character of the fixed points of a two-dimensional flow in different regions of the parameter plane $\sigma \geq 0$, det A.

$\sigma = 2\sqrt{\det A}$ a node (spiral) attractor exists (see Fig. P.9). (In an expanding phase space, $\sigma < 0$, the time-reversed version of the above behaviour appears.)

Solution 3.27 The character of the phase portraits depends on parameter α (see Fig. P.10). The origin is always a hyperbolic fixed point. The phase space contraction rate is $\sigma = \alpha - x_1$; point $(1, 0)$ is therefore a repellor for $\alpha < 1$ and an attractor otherwise. For $\alpha < 0.865$, the repellor is so strong that it spins the trajectories out into infinity. For $0.865 < \alpha < 1$, the repellor is surrounded by a

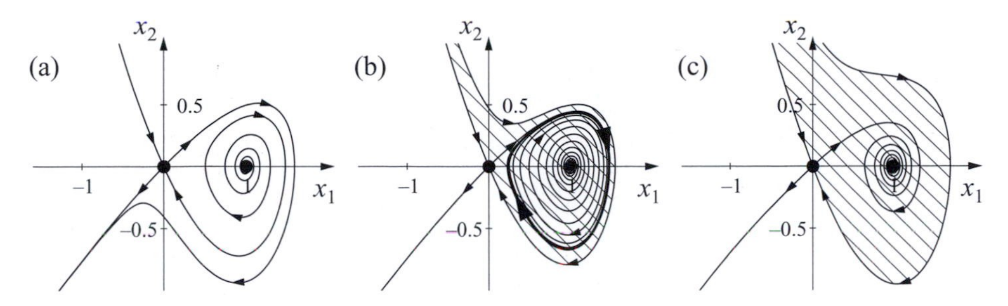

Fig. P.10. Three characteristic phase portraits obtained numerically: (a) $\alpha = 0.8$, (b) $\alpha = 0.9$, (c) $\alpha = 1.1$. In (b) the limit cycle is plotted with a thick line.

limit cycle attractor, whose basin of attraction is bounded by the stable manifolds of the origin. In the range $\alpha > 1$, the fixed point $(1, 0)$ is the only attractor.

Solution 4.2 From (4.20), the polar co-ordinate $\varrho_n = \sqrt{(\Delta x_n^2 + \Delta v_n^2/\omega_0^2)}$ and the polar angle $\phi_n = \omega_0 nT$ are found to be related by

$$\varrho_n = Ae^{-\alpha/(2\omega_0)\phi_n}, \tag{P.7}$$

where $A = \varrho_0$. The iterations jump on a logarithmic spiral.

Solution 4.3 The map is linear; its matrix is given by

$$L = \begin{pmatrix} E\left(C + S\dfrac{\alpha}{2\omega_\alpha}\right) & \dfrac{ES}{\omega_\alpha} \\ -ES\omega_\alpha\left(1 + \left(\dfrac{\alpha}{2\omega_\alpha}\right)^2\right) & E\left(C - S\dfrac{\alpha}{2\omega_\alpha}\right) \end{pmatrix}.$$

The abbreviations S and C now contain $\omega_\alpha = \sqrt{[\omega_0^2 - (\alpha/2)^2]}$ replacing ω_0. The eigenvalues are $\Lambda_\pm = E(C \pm iS) = e^{(-\alpha/2 \pm i\omega_\alpha)T}$.

Solution 4.7 The stroboscopic map of the kicked oscillator, for an arbitrary driving period T and friction coefficient α, is given by

$$\begin{aligned} x_{n+1} &= E\left(C + \frac{\alpha}{2\omega_\alpha}S\right)x_n + \frac{ES}{\omega_\alpha}v_n, \\ v_{n+1} &= -ES\left(\omega_\alpha + \frac{\alpha^2}{4\omega_\alpha}\right)x_n + E\left(C - \frac{\alpha}{2\omega_\alpha}S\right)v_n + uI(x_{n+1}), \end{aligned} \tag{P.8}$$

(cf. the solution to Problem 4.3), where abbreviations (4.12) are used with $\omega_\alpha = \sqrt{\omega_0^2 - (\alpha/2)^2}$ replacing ω_0. The inverse of the map is given by

$$\begin{aligned} x_{n+1} &= \frac{1}{E}\left(C - \frac{\alpha}{2\omega_\alpha}S\right)x_n - \frac{S}{\omega_\alpha E}(v_n - uI(x_n)), \\ v_{n+1} &= \frac{S}{E}\left(\omega_\alpha + \frac{\alpha^2}{4\omega_\alpha}\right)x_n + \frac{1}{E}\left(C + \frac{\alpha}{2\omega_\alpha}S\right)(v_n - uI(x_n)). \end{aligned} \tag{P.9}$$

Solution 4.11 The eigenvalues of a general two-dimensional linear map, L, are given by

$$\Lambda_\pm = \frac{\text{Tr } L \pm \sqrt{(\text{Tr } L)^2 - 4J}}{2}.$$

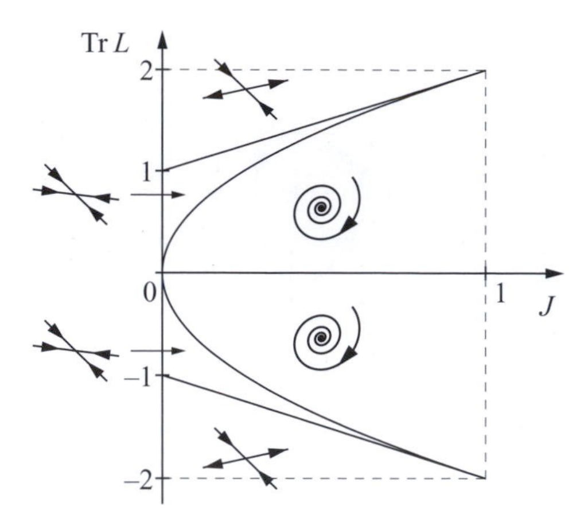

Fig. P.11. Character of the fixed points of a two-dimensional map on the parameter plane $(J, \mathrm{Tr}L)$ $(0 < J \leq 1)$.

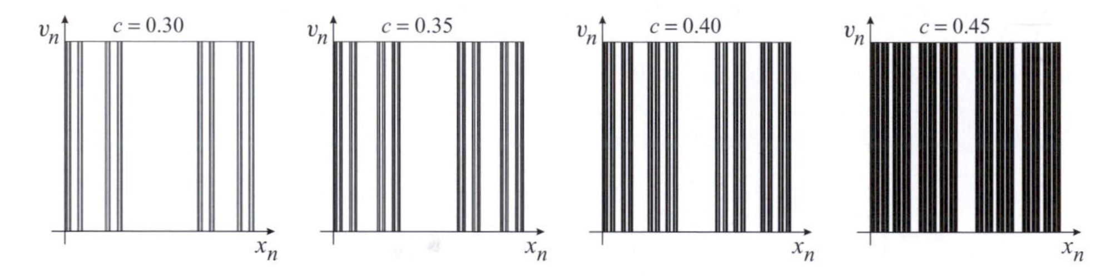

Fig. P.12. Baker attractors for different contraction parameters c (0.30, 0.35, 0.4, and 0.45). As the parameter increases, dissipation decreases and the attractor appears to be more and more extended, although its area is always zero. The respective fractal dimensions are $D_0 = 1.576$, 1.660, 1.756 and 1.868.

This is real for $|\mathrm{Tr}\ L| > 2\sqrt{J}$. Within the parabola arc, $\pm 2\sqrt{J}$, there exist spiral attractors. Outside of the parabola, the eigenvalues are real. Above the line $1 + J$ or below $-1 - J$ the fixed points are hyperbolic; within these lines, outside the parabola, the fixed points are nodes (see Fig. P.11).

Solution 5.1 See Fig. P.12

Solution 5.3 Lay a vertical segment of unit length, the basic branch, onto each two-cycle point and apply the twice iterated map $B^2(x_n, v_n)$ to them. Since the two-cycle is on the attractor, the unstable manifold emanating from the cycle elements is also part of the attractor (Fig. P.13). By means of further iterations, we would obtain the entire unstable manifold of the two-cycle more and more accurately, and the same curve would trace out the chaotic attractor with increasing accuracy.

Solution 5.5 In the m-fold iterated map, the velocity axis can be divided into 2^m identical intervals, with different functional dependences determining the new

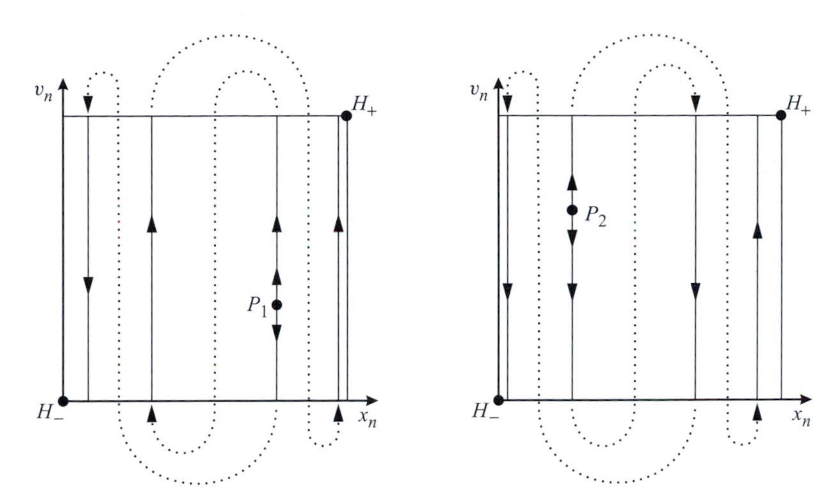

Fig. P.13. Unstable manifolds of the two-cycle: first step of the construction. Map $B^2(x_n, v_n)$ has been applied once to the vertical segment of unit length crossing points P_1, P_2 of the two-cycle. The dotted lines mark the connections ($c = 1/3$).

Table P.1. *Stability characteristics of the kicked oscillators for different values of the derivative f'^* of the amplitude function at the fixed point.*

$\lvert f'^* \rvert < 2E$	$2E < \lvert f'^* \rvert < 1 + E^2$	$1 + E^2 < \lvert f'^* \rvert$
Spiral attractor	node attractor	hyperbolic point

velocity values. The map expresses the fact that each velocity interval of length 2^{-m} is stretched by a factor of 2^m and mapped onto the entire length (0, 1). Accordingly, in the first interval $v_n \leq 2^{-m}$, $v_{n+m} = 2^m v_n$, in the next, $v_{n+m} = 2^m(v_n - 2^{-m})$, and in the kth interval $v_{n+m} = 2^m(v_n - k2^{-m})$, where $k = 1, 2, \ldots, 2^m$. The velocity co-ordinate of the fixed point of the m-fold iterated map is therefore the solution of the equation $v^* = 2^m v^* - k$. This leads to $v^* = k/(2^m - 1)$. (For the twice iterated map, see Problem 5.2.) Since k goes up to 2^m, there exist 2^m different fixed points altogether.

Solution 5.6 The construction rule of the asymmetric baker map of Fig. 5.16 projected on the x-axis corresponds to the construction of the two-scale Cantor set of Fig. 2.7 with $r_1 = c_1$, $r_2 = c_2$. The entire chaotic attractor is a set of Cantor filaments with one of the partial dimensions equal to unity (see Section 2.2.2). The other partial dimension is therefore $D_0 - 1$, which satisfies Eq. (2.9). The full fractal dimension, D_0, is therefore the solution to the equation $c_1^{D_0-1} + c_2^{D_0-1} = 1$.

Solution 5.9 The condition for a node attractor is $2E < |f'^*| < 1 + E^2$. If the eigenvalues given in (5.29) are complex ($|f'^*| < 2E$), spiral attractors are obtained. The classification of the kicked oscillator's fixed points is presented in Table P.1.

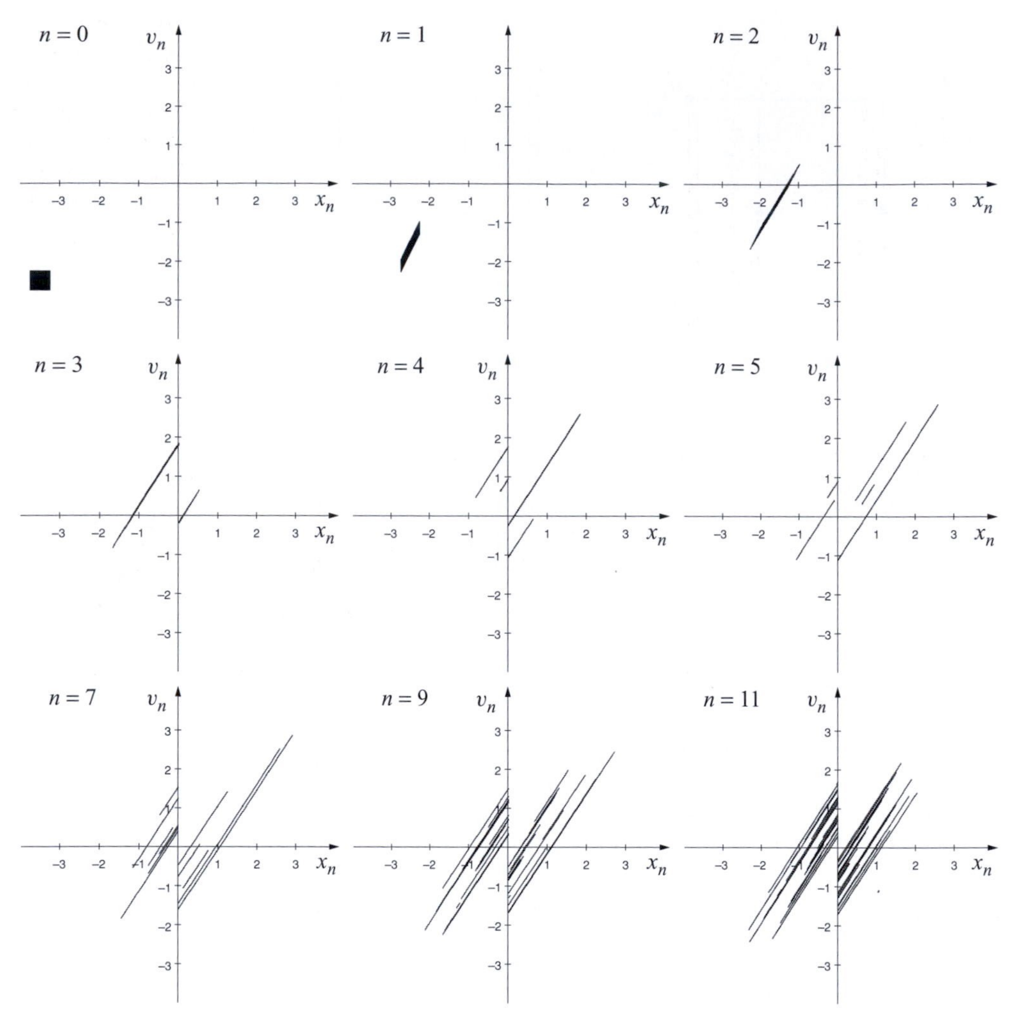

Fig. P.14. Contraction of the phase space volume onto the sawtooth attractor. The chosen phase space domain is a square of size 0.5 concentrated around the point $(-3.5, -2.5)$ $(a = 1.95,\ E = 0.8)$.

Solution 5.14 See Fig. P.14.

Solution 5.15 See Fig. P.15.

Solution 5.17 The equation for the unstable basic branch of point P_1 is given by

$$v_n = v_{1+}(x_n) = \Lambda_+(x_n - x_1^*) - x_1^*, \tag{P.10}$$

where Λ_+ is given by (5.29) with $f'^* = a$. The equation of the basic branch of P_2 is the same, except that x_1^* is replaced by x_1^*. The intersection of this basic branch with the v_n-axis is the point $(0, x_1^*(\Lambda_+ + 1))$, which is mapped into $(x_1^*(\Lambda_+ + 1),$ $ax_1^*(\Lambda_+ + 1) - 1)$. Substituting Λ_+ and x_1^*, we can verify that the image point

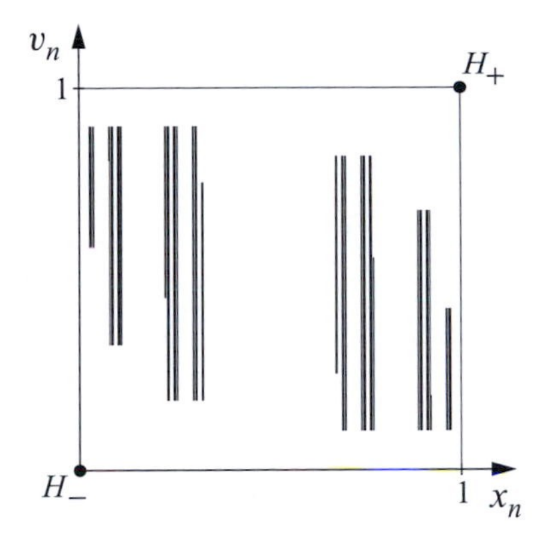

Fig. P.15. Baker attractor with expansion factor $a = 1.8$ ($c = 1/3$). Note the similarity with the sawtooth attractor of Fig. 5.22, but in this case all the straight segments are vertical.

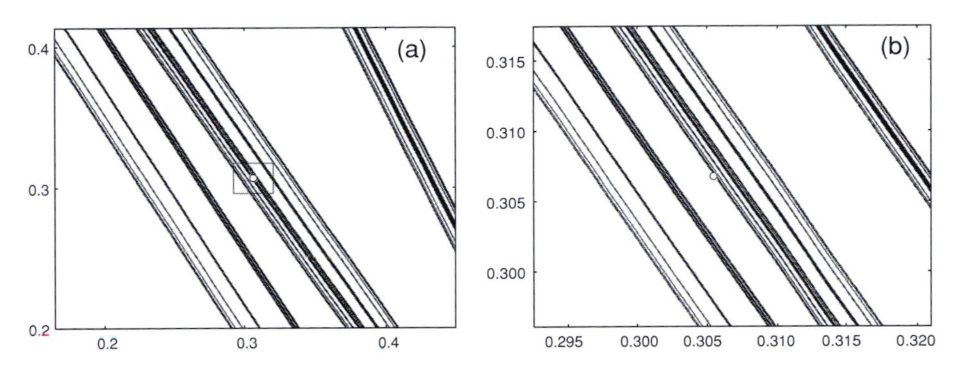

Fig. P.16. The roof attractor around H_+. (a) Rectangle in Fig. 5.28 magnified $\times 30$. (b) Rectangle in part (a) magnified by a further factor of 10.

satisfies equation (P.10). The end points of the basic branch of P_1 are therefore given by

$$\left(0, -x_1^*(1 + \Lambda_+)\right), \quad \left(x_1^*(1 + \Lambda_+), \, ax_1^*(1 + \Lambda_+) - 1\right),$$

and the end-points of the basic branches of P_2 are mirror images of these points with respect to the origin.

Solution 5.19 See Fig. P.16.

Solution 5.21 See Fig. P.17.

Solution 5.23 See Fig. P.18.

Solution 5.24 See Fig. P.19.

Solution 5.25 See Fig. P.20.

Solution 5.26 See Fig. P.21.

Solution 5.28 The values of P_0 fall in the interval $(1 - \gamma/2, 1 + \gamma/2)$. In one step, the x-intervals are reduced by a factor of c and the values of the functions are increased by a factor $1/(2c)$ in both phase space columns of width c: P_1 takes on

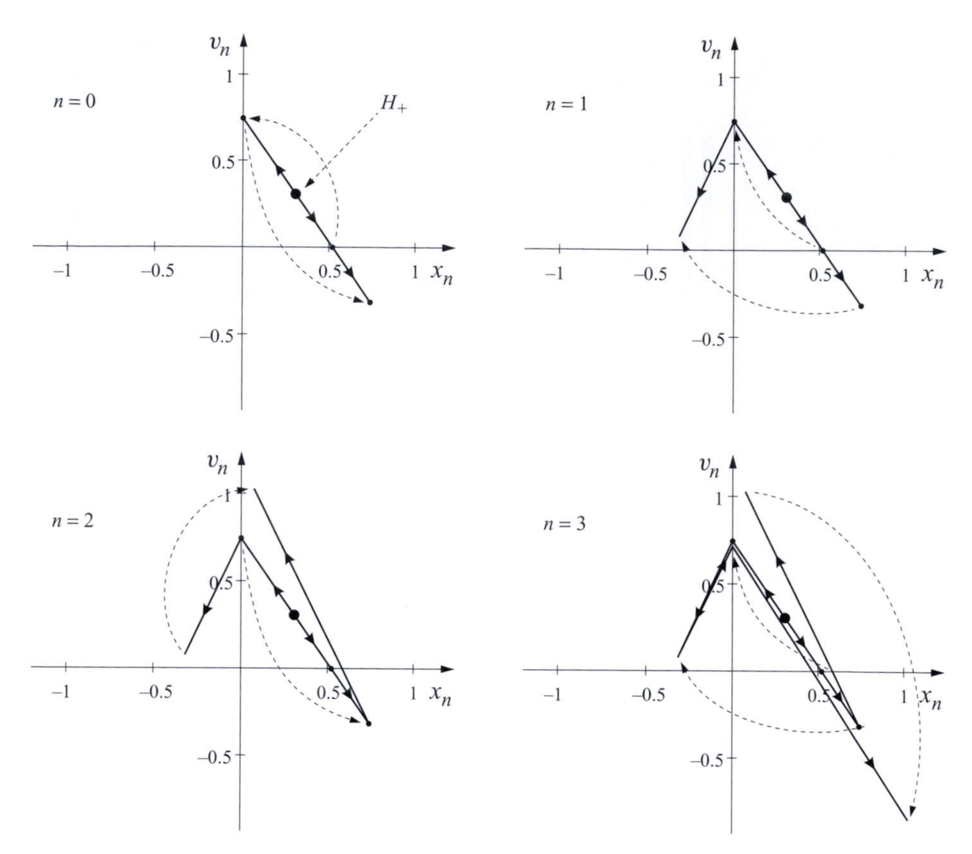

Fig. P.17. Images of the unstable basic branch of H_+ over three successive iterates. The mappings of a few characteristic points are marked with dashed lines.

values between $(1 - \gamma/2)/(2c)$ and $(1 + \gamma/2)/(2c)$ (it changes linearly in between). On the columns of width c^n obtained in n steps, function P_n takes on values between $(1 - \gamma/2)/(2c)^n$ and $(1 + \gamma/2)/(2c)^n$. For a very large n, the columns are so narrow that the distribution within one column cannot be resolved. Then only the total weight of each column, the integral of the distribution within one column counts. This is the product of the mean value $1/(2c)^n$ and the width, 2^{-n}, just like when starting from a homogeneous initial distribution.

Solution 5.31 Since the mapping function, $f(x)$, possesses two branches, the same image point can be reached from two points, i.e. each point can have two pre-images (this is why the map cannot be inverted). Let the two pre-images of a point x', chosen in the $(n+1)$th step, be x_1 and x_2 (Fig. P.22). The number $P_{n+1}(x')\,dx'$ of particles on a short interval of length dx' around the image point must be identical to the total number $P_n(x_1)\,dx_1 + P_n(x_2)\,dx_2$ of the particles on the intervals of lengths dx_1 and dx_2 around x_1 and x_2. Here, dx_1 and dx_2 are the lengths of the two pre-images of the interval of length dx'; dx'/dx_1 and dx'/dx_2 are therefore the magnitudes of the derivatives of the mapping function at positions

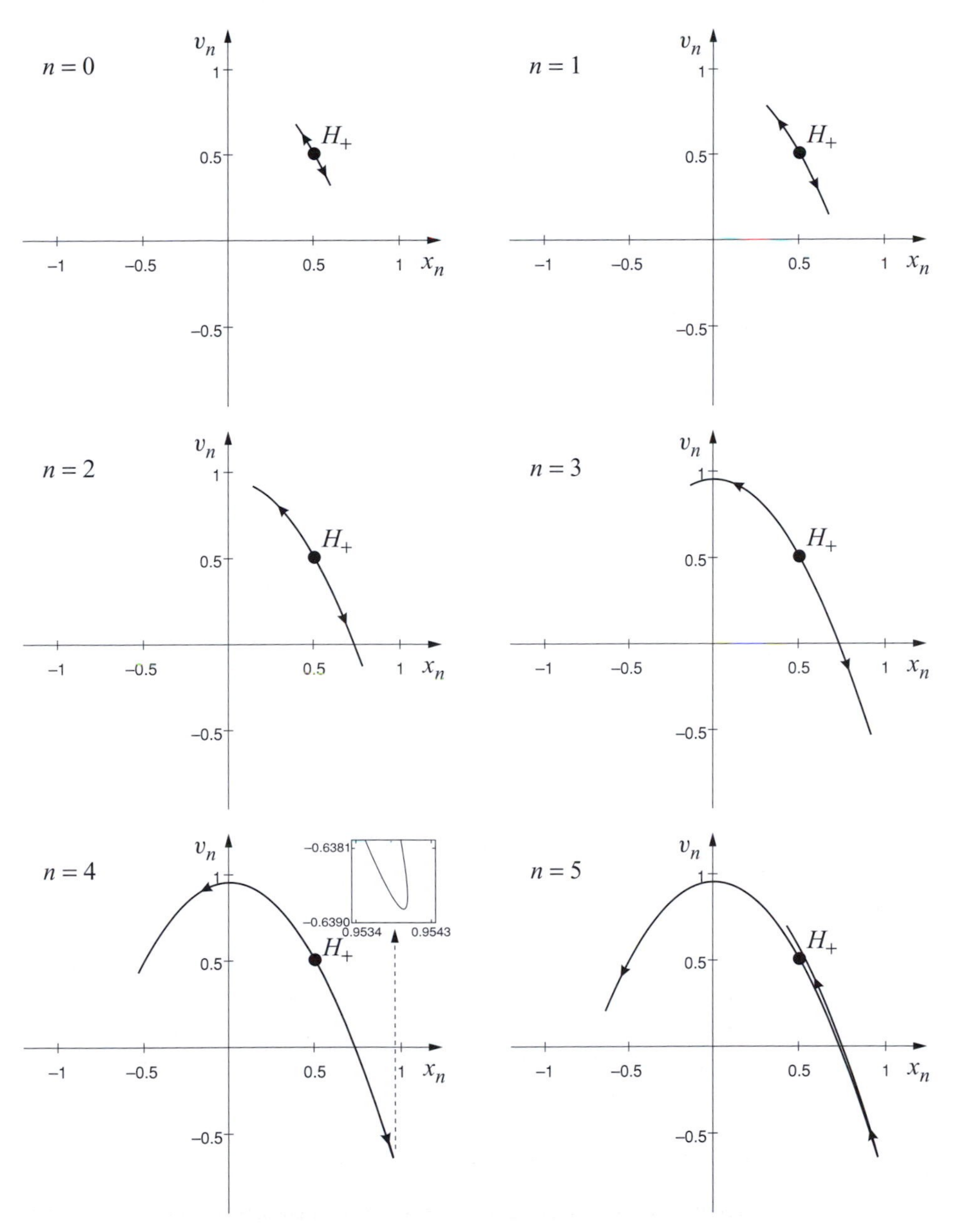

Fig. P.18. Stretching of a short piece of the unstable manifold emanating from point H_+ in five successive iterates. The unstable manifold is continuous and does not contain straight line segments. An enlargement of the point appearing to be a break-point in $n = 4$ shows that the unstable manifold *does not break* ($a = 1.8$, $E = 0.25$.)

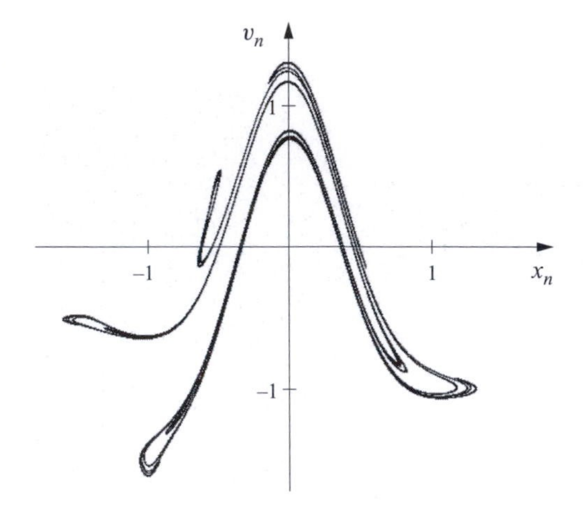

Fig. P.19. Chaotic attractor of an oscillator kicked with the bell amplitude function $f(x) = a(e^{-4x^2} - 1) + 1$ $(a = 2,\ E = 0.7)$.

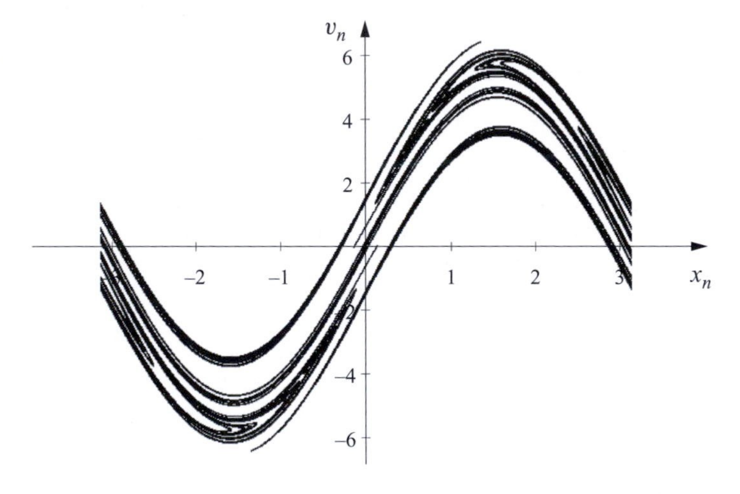

Fig. P.20. Chaotic attractor of an oscillator kicked with the sine amplitude function $f(x) = a \sin x$ $(a = 5,\ E = 0.7)$.

x_1 and x_2, respectively. Thus, $P_n(x_1)\,dx'/\,|f'(x_1)| + P_n(x_2)\,dx'/\,|f'(x_2)| = P_{n+1}(x')\,dx'$, from which the probability distribution after the $(n+1)$th step is given by

$$P_{n+1}(x') = \sum_{x=f^{-1}(x')} \frac{P_n(x)}{|f'(x)|}.$$

The summation is carried out over all the values x that are pre-images of x'. In this form, the equation is valid for mapping functions with any number of branches. For the natural distribution, $P_n = P_{n+1} \equiv P^*$. In the sawtooth map, (a), and roof map, (b), of constant slope, the density of the natural distribution is $P^* =$ constant; the distribution is uniform. In the parabola map, (c), assuming the symmetry of the distribution, we obtain the relation

$$P^*(x') = 2\frac{P^*\left(\sqrt{(1-x')/2}\right)}{4\sqrt{(1-x')/2}}.$$

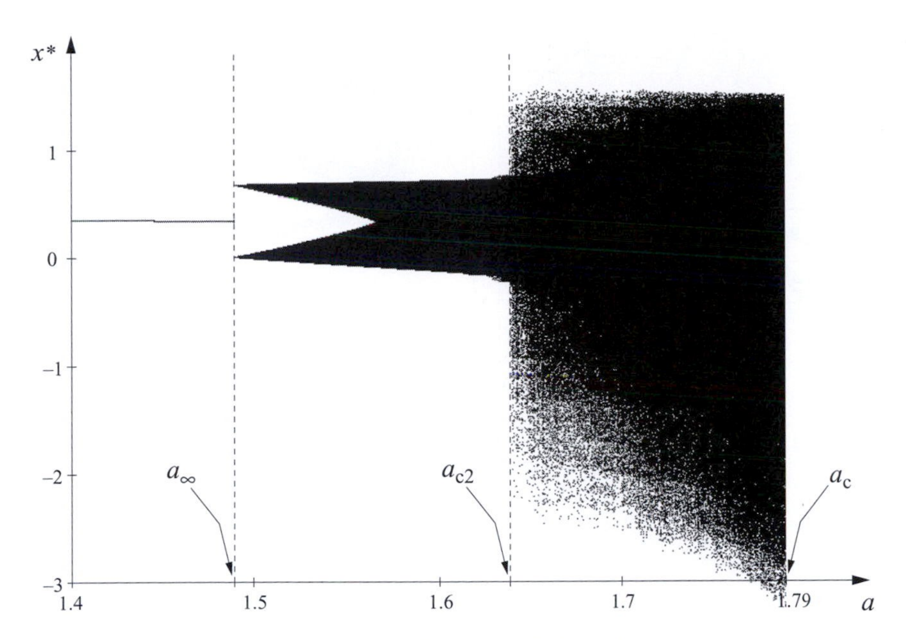

Fig. P.21. Bifurcation diagram of the roof map ($E = 0.7$). At the value $a_\infty = 1.490$ chaos appears abruptly (as if a period-doubling cascade were compressed into a single point). The chaotic domain is not interrupted by periodic windows. At $a = a_{c2} = 1.632$ the attractor suddenly widens; for values $a > a_c = 1.7898$ there exist no finite attractors.

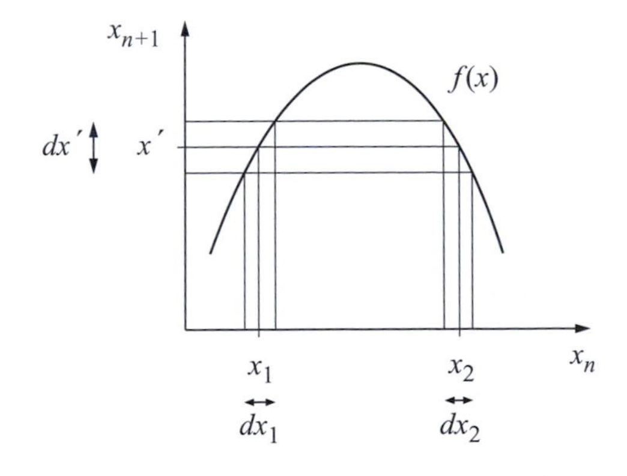

Fig. P.22. A one-dimensional map, $f(x)$, in the neighbourhood of point x' and of its two pre-images.

One can check via substitution that the solution of this equation is $P^*(x') \sim 1/\sqrt{1-x'^2}$. The natural distribution is position-dependent in this case. One-dimensional maps model the behaviour developing along the *unstable* manifolds of the chaotic attractors of two-dimensional invertible maps. The examples illustrate that the natural distribution is continuous along the unstable direction.

Solution 5.32 See Fig. P.23.

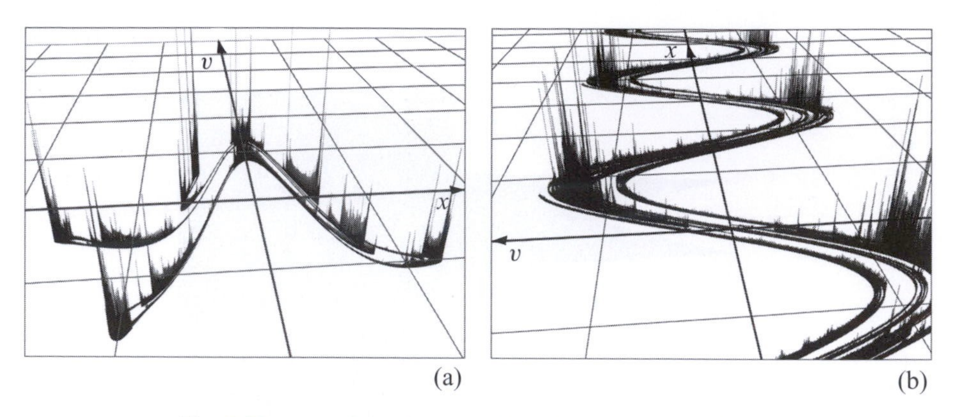

Fig. P.23. Natural distributions of the chaotic attractor of the oscillator kicked with (a) a bell and (b) a sine amplitude. The respective parameters are the same as in Problems 5.24 and 5.25. These distributions can be seen from different viewpoints in Plates XV and XVI.

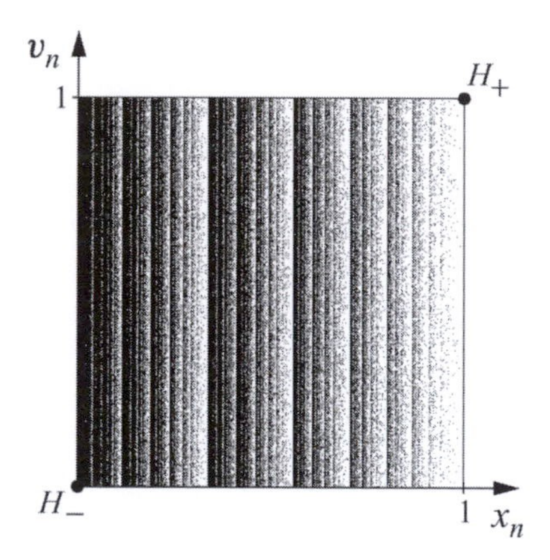

Fig. P.24. Asymmetric baker attractor with parameters $c_1 = 1 - b$, $c_2 = b$, $b = 2/3$. The fractal dimension is $D_0 = 2$, but the natural distribution is highly inhomogeneous. The filaments are visited with different probabilities. Observing the dynamics for even longer times, the white regions would also become black.

Solution 5.38 With the notation $b = 1/2 - \delta$, we obtain, for $|\delta| \ll 1$,

$$-\overline{\ln J} = -b\ln((1-b)/b) - (1-b)\ln(b/(1-b)) \approx 8\delta^2,$$

$$D_1^{(2)} = \frac{b\ln b + (1-b)\ln(1-b)}{b\ln(1-b) + (1-b)\ln b} \approx 1 - \frac{8\delta^2}{\ln 2}.$$

Irrespective of the sign of the asymmetry parameter, δ, on average the phase space volume decreases. The system is therefore dissipative, and the partial information dimension is always less than unity (the full information dimension is less than 2), even though the attractor fills the entire phase space: $D_0 = 2$ (see the solution to Problem 5.6, with $c_1 = 1 - b$, $c_2 = b$). See Fig. P.24

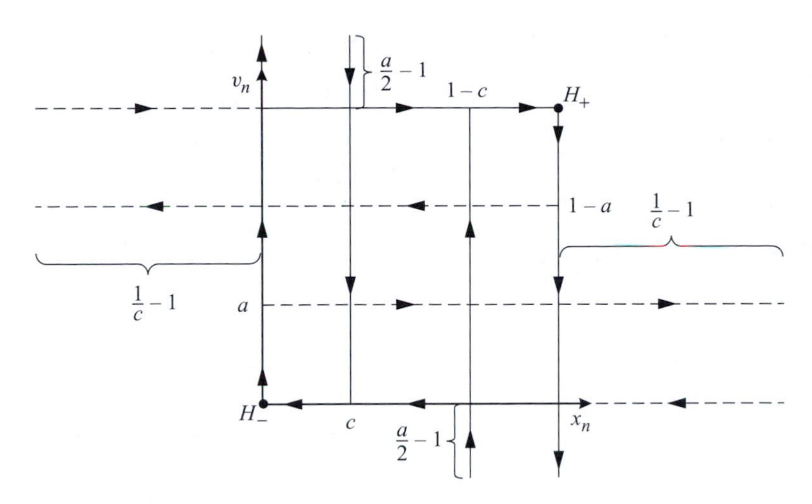

Fig. P.25. The first segments of the unstable (continuous line) and stable (dashed line) manifolds of the fixed points. Both manifolds extend beyond the unit square.

Solution 5.40 The equations linearised around the origin are given by

$$\Delta \dot{x} = \sigma(\Delta y - \Delta x), \quad \Delta \dot{y} = r\Delta x - \Delta y, \quad \Delta \dot{z} = -\Delta z.$$

The stability matrix is therefore given by

$$A = \begin{pmatrix} -\sigma & \sigma & 0 \\ r & -1 & 0 \\ 0 & 0 & -1 \end{pmatrix}.$$

The eigenvalue equation is $(\lambda + 1)(\lambda^2 + \lambda(\sigma + 1) + \sigma(1 - r)) = 0$, with solutions $\lambda_3 = -1$ and

$$\lambda_{1,2} = \frac{-1 - \sigma \pm \sqrt{(\sigma + 1)^2 + 4\sigma(r - 1)}}{2}.$$

All the eigenvalues are negative for $r < 1$. Since for $r > 1$ one of the eigenvalues is positive, the origin is hyperbolic in this range.

Solution 6.3 See Fig. P.25.

Solution 6.4 Under iteration of the map, a two-scale Cantor set with parameters c_1 and c_2 is formed along the x_n-axis (see Section 2.2.1). Its dimension, the partial dimension $D_0^{(2)}$ of the saddle, is the solution of (cf. equation (2.9))

$$c_1^{D_0^{(2)}} + c_2^{D_0^{(2)}} = 1.$$

Points that will never escape lie on horizontal line segments (the stable manifold) which constitute, along the v_n-axis, a two-scale Cantor set with parameters $1/a_1$ and $1/a_2$. The partial dimension, $D_0^{(1)}$, of the unstable direction therefore fulfils the following equation:

$$a_1^{-D_0^{(1)}} + a_2^{-D_0^{(1)}} = 1.$$

Solution 6.5 The image of the upper rectangle containing 'A' in Fig. P.26 is not only stretched and compressed in this map but is also rotated by 180°. The fixed points are $H_- = (0, 0)$, $H_+ = (1/(1 + c_2), a_2/(1 + a_2))$; the two-cycle is

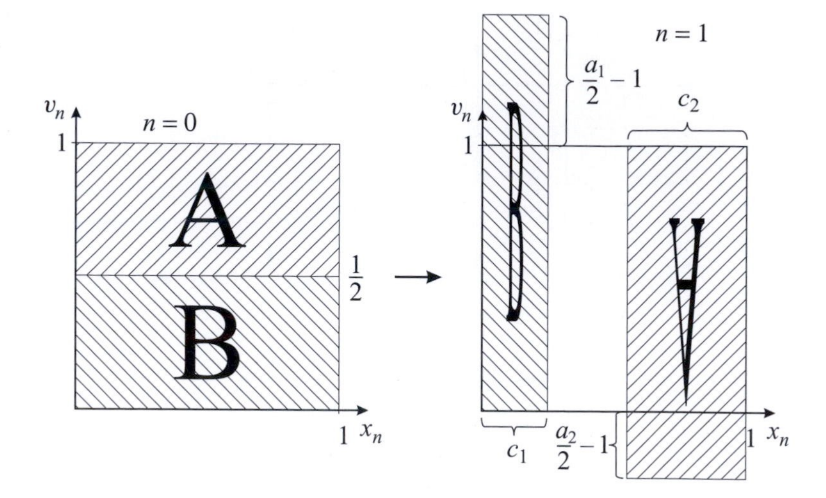

Fig. P.26. Action of the asymmetric baker map, (6.14), on the unit square ($a_1 = 3$, $a_2 = 2.5$, $c_1 = 0.25$, $c_2 = 0.45$). Capital letters A and B illustrate the different orientations of the two rectangles after one iteration.

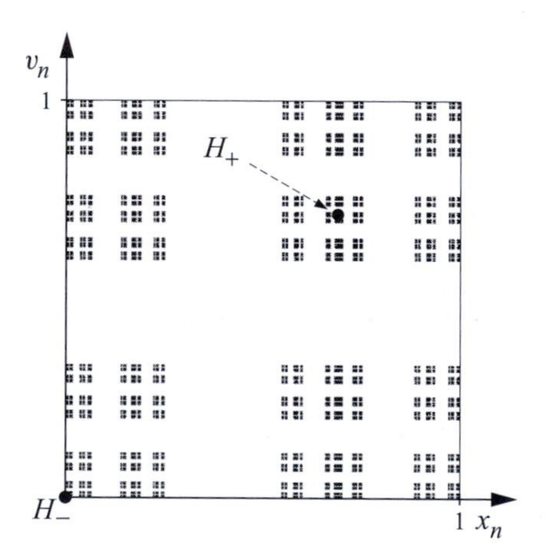

Fig. P.27. Saddle of the asymmetric baker map (6.14) ($a_1 = 3$, $a_2 = 2.5$, $c_1 = 0.25$, $c_2 = 0.45$).

$P_1 = (1/(1 + c_1c_2), a_2/(1 + a_1a_2))$, $P_2 = (c_1/(1 + c_1c_2), a_1a_2/(1 + a_1a_2))$. Owing to the rotation by 180°, the fixed point H_+ is now not at the upper right corner of the saddle (see Fig. P.27).

Even though the positions of the cycle points have changed, their stability eigenvalues are the same as in (6.13); consequently, *all* the chaos parameters, including the partial fractal dimensions, are the same as in map (6.13). The saddles of the two maps shown in Fig. 6.9 and Fig. P.27 are, however, not identical.

Solution 6.6 Introducing the new variables $x' = (a^2 - (1 + E^2)^2)/(2a)(x - x_-^*)$, $v' = (a^2 - (1 + E^2)^2)/(2a)(v - x_-^*)$, where $x_-^* = v_-^* = 1/(1 + E^2 - a)$, the map becomes $x'_{n+1} = v'_n$, and

$$v'_{n+1} = -E^2 x'_n + a v'_n, \quad \text{for} \quad v'_n < (1 + E^2 + a)/(2a),$$

$$v'_{n+1} - 1 = -E^2(x'_n - 1) - a(v'_n - 1), \quad \text{otherwise.}$$

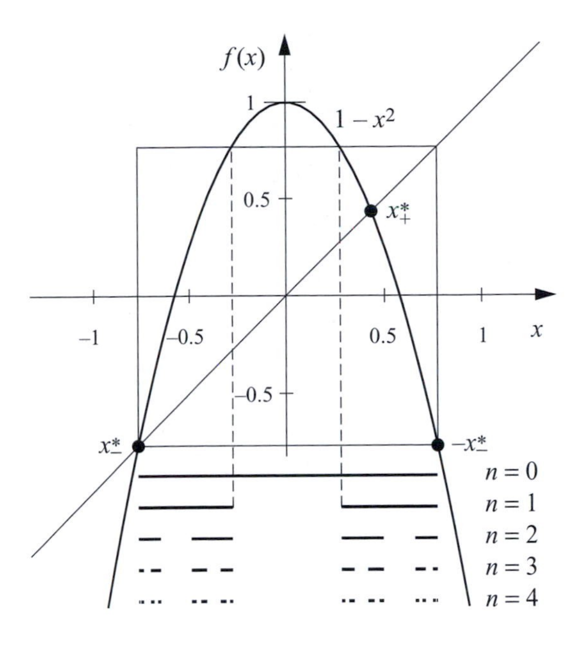

Fig. P.28. Condition for transient chaos in one-dimensional maps: function $f(x)$ stretches beyond the square of side length $2|x_-^*|$ centred at the origin. Below the square, intervals are plotted with lifetimes of at least n. For $n \to \infty$, these intervals approach the chaotic set, a repellor. Here, $f(x) = 1 - 3x^2$.

The eigenvalues in the lower and upper half-planes are $(a \pm \sqrt{a^2 - 4E^2})/2$ and $(-a \pm \sqrt{a^2 - 4E^2})/2$, respectively

In the limit $a \gg 1$, the switching between the two formulae happens precisely along the line $v_n' = 1/2$. The eigenvalue with absolute value larger (smaller) than unity is a (E^2/a) in both half-planes. Since these are the respective slopes of the unstable and stable manifolds (see Fig. 5.21), these manifolds consist in practice of vertical and horizontal lines, as for map (6.3). Accordingly, the non-linearity parameter, a, plays the role of the stretching rate, a, of the baker map, and E^2/a corresponds to c. According to equations (6.5)–(6.9), the escape rate and the Lyapunov exponent are, in leading order, proportional to $\ln a$, and the fractal dimension is proportional to the reciprocal of $\ln a$.

Solution 6.8 The fixed points of a one-dimensional map, $x_{n+1} = f(x_n)$, are the solutions of the equation $x^* = f(x^*)$. Transient chaos emerges if both fixed points are unstable and the maximum of the function is $f(0) > |x_-^*|$. In the first iteration step every point escapes from the interval I in which $f(x) > |x_-^*|$. Points escaping after n steps are in the nth pre-image of I. The sequence of the pre-images shows that on an ever-growing portion of interval $(-x_-^*, x_-^*)$, is removed similar to the construction of a Cantor set. Points that never escape perform chaotic dynamics, and form a chaotic repellor. See Fig. P.28.

Solution 6.13 The equation of the unstable manifold emanating from H_- is $v_+(x_n) = x_-^* + \Lambda_+(x_n - x_-^*)$, where $\Lambda_\pm = (a \pm \sqrt{a^2 - 4E^2})/2$, and the fixed point co-ordinates are given by (5.35). The upper end-point of this branch is located on the v_n-axis (see Fig. 5.22), namely, it is $(0, x_-^*(1 - \Lambda_+))$. This falls on the stable

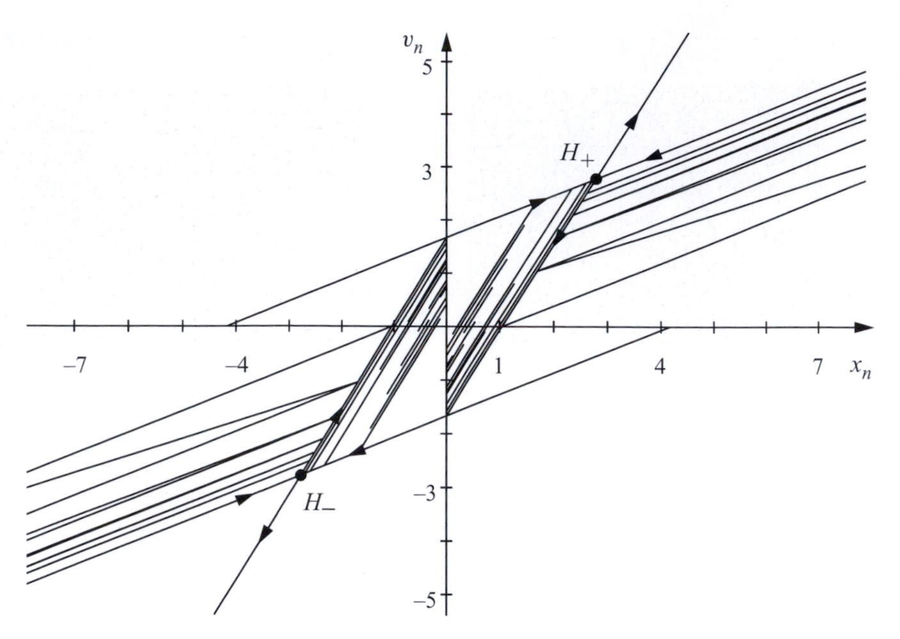

Fig. P.29. Sawtooth attractor in crisis ($a_c = 2$, $E = 0.8$).

manifold of H_+, on $v_-(x_n) = x_+^* + \Lambda_-(x_n - x_+^*)$, if $x_-^*(1 - \Lambda_+) = x_+^*(1 - \Lambda_-)$. This yields $\Lambda_- + \Lambda_+ = 2$, from which $a_c = 2$ (irrespective of the value of E).

Solution 6.14 See Fig. P.29.

Solution 6.19 See Fig. P.30.

Solution 6.20 See Fig. P.31.

Solution 6.21 See Fig. P.32.

Solution 7.3 Below the line $v_n = b$, the vertical (or horizontal) distance of point pairs increases (or decreases) in one step by a factor of $1/b$ (or b). For point pairs above the line, this factor is $1/(1 - b)$ (or $(1 - b)$). Due to the uniform natural distribution, deformation of the first and second type occurs in proportions b and $(1 - b)$ of all the cases, respectively. Thus, the average of the logarithm of the stretching factors, i.e. the average of the Lyapunov exponents, is given by $\bar{\lambda} = -b \ln b - (1 - b) \ln (1 - b)$. Similarly, the negative average exponent is given by $\bar{\lambda}' = b \ln b + (1 - b) \ln (1 - b) = -\bar{\lambda}$.

The inverse map is given by

$$(x_{n+1}, v_{n+1}) = \left(\frac{x_n}{b}, bv_n\right), \qquad \text{for} \quad x_n \leq b,$$

$$(x_{n+1}, v_{n+1}) = \left(1 + \frac{x_n - 1}{1 - b}, 1 + (1 - b)(v_n - 1)\right), \qquad \text{for} \quad x_n > b.$$

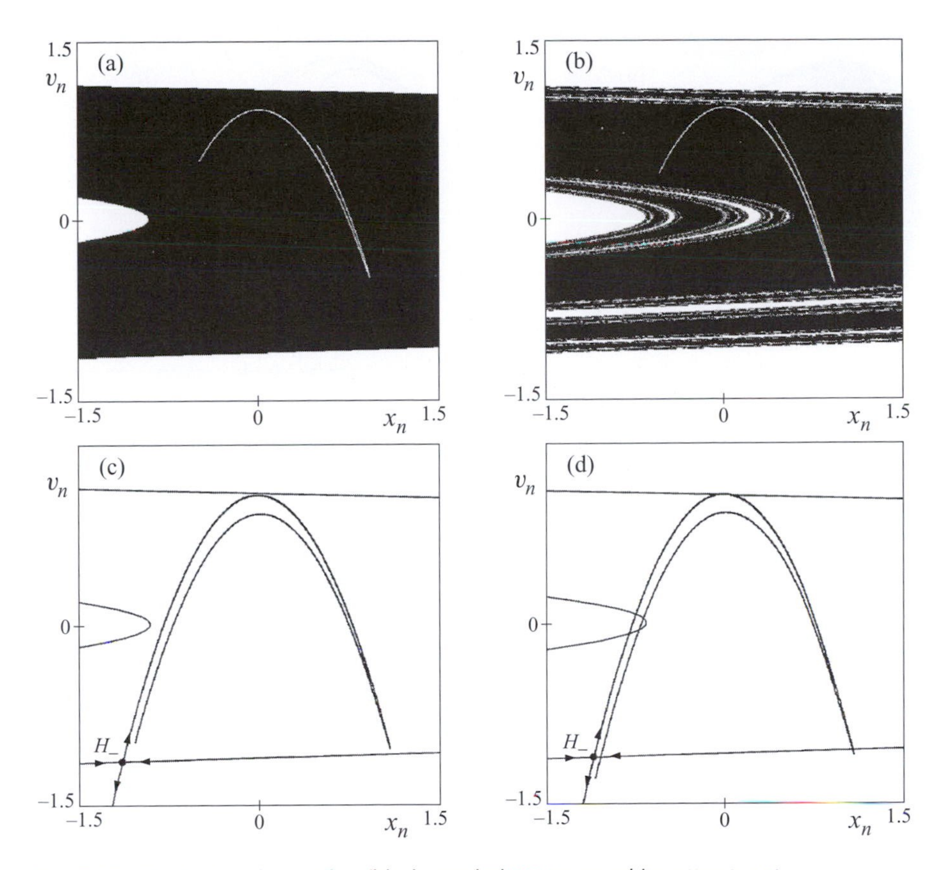

Fig. P.30. The basins of attraction (black: parabola attractor, white: attractor at infinity) at (a) $a = 1.74$ and (b) $a = 1.80$ ($E = 0.32$). Manifolds of the fixed point, H_-, on the boundary for (c) $a = 1.74$ and (d) $a = 1.80$ indicate that, in the first case, there are no homoclinic points (no horseshoe) formed.

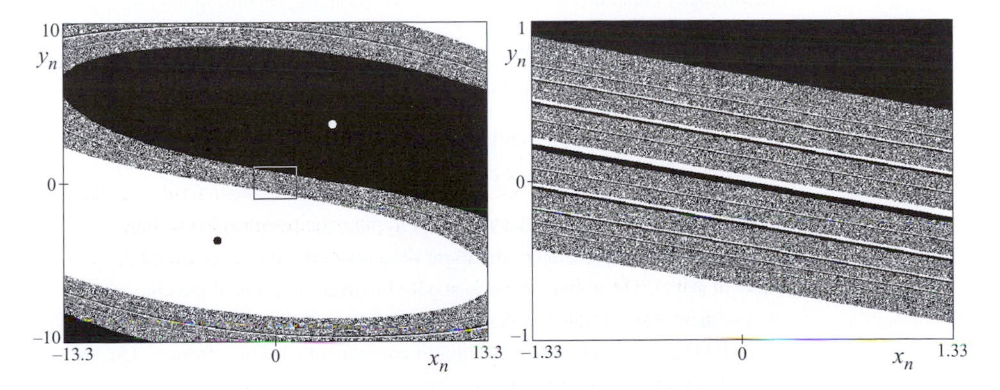

Fig. P.31. Basins of attraction (black and white) of the two fixed point attractors (white and black dots) for $r = 14$, $\sigma = 10$ ($z = r - 1$). The fractal basin boundary is the stable manifold of the saddle in Fig. 6.27. The right-hand panel is a magnification of the small rectangle in the left-hand panel.

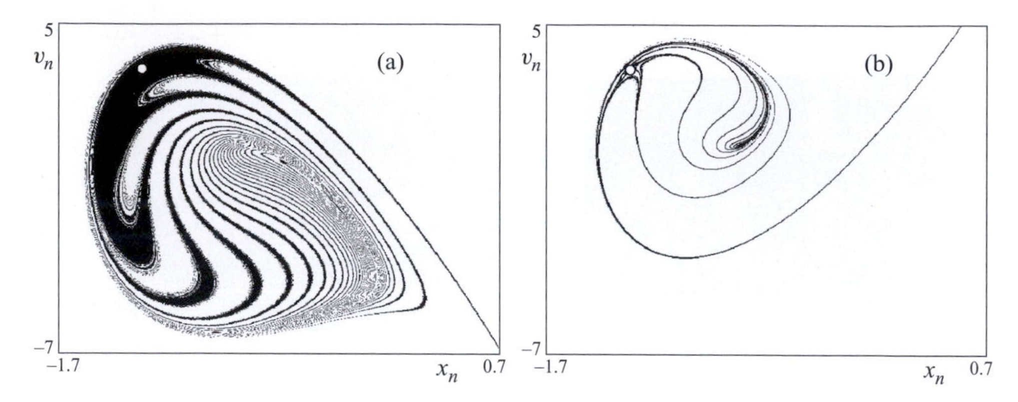

Fig. P.32. Basin of attraction of (a) the stable cycle (white dot) in black and (b) the unstable manifold on the stroboscopic map at $\alpha = 0.6$, $f_0 = 3.8$, $A = 55$.

By exchanging x and v, this map becomes equivalent to the original map, (7.5) and (7.6), and, as a consequence, the average Lyapunov exponents of the two maps are identical.

Solution 7.10 The velocity components, (u_n, w_n), can be expressed in terms of the horizontal and vertical components as follows (see Fig. 7.11):

$$u_n = v_{xn} \cos\alpha + v_{yn} \sin\alpha, \quad w_n = -v_{xn} \sin\alpha + v_{yn} \cos\alpha, \tag{P.11}$$

and vice versa as

$$v_{xn} = u_n \cos\alpha - w_n \sin\alpha, \quad v_{yn} = u_n \sin\alpha + w_n \cos\alpha. \tag{P.12}$$

The flight time, (7.16), is thus

$$gt_n = 2\frac{w_n}{\cos\alpha}. \tag{P.13}$$

The velocity components before the next collision are, according to (7.15), given by

$$\left.\begin{aligned} \tilde{u}_{n+1} &= u_n - gt_n \sin\alpha = u_n - 2w_n \text{tg}\,\alpha, \\ \tilde{w}_{n+1} &= w_n - gt_n \cos\alpha = -w_n. \end{aligned}\right\} \tag{P.14}$$

Since w changes sign at the collision, whereas u does not, the velocity immediately after the collision is given by equations (7.17).

Solution 7.11 With the initial velocities u_n, w_n, the combination $v_{xn}\tan\alpha + v_{yn}$ occurring in the expression of the flight time is $(u_n \sin(2\alpha) + w_n \cos(2\alpha))/\cos\alpha$ in view of (P.11). The impact velocity should be given in terms of the components perpendicular and parallel to the slope (which corresponds to taking $-\alpha$ instead of α in (P.11) and then changing the sign of component u). Thus, from (7.15), the velocity before the collision is given by

$$\left.\begin{aligned} \tilde{u}_{n+1} &= -v_{xn} \cos\alpha + (v_{yn} - gt_n) \sin\alpha, \\ \tilde{w}_{n+1} &= v_{xn} \sin\alpha + (v_{yn} - gt_n) \cos\alpha. \end{aligned}\right\} \tag{P.15}$$

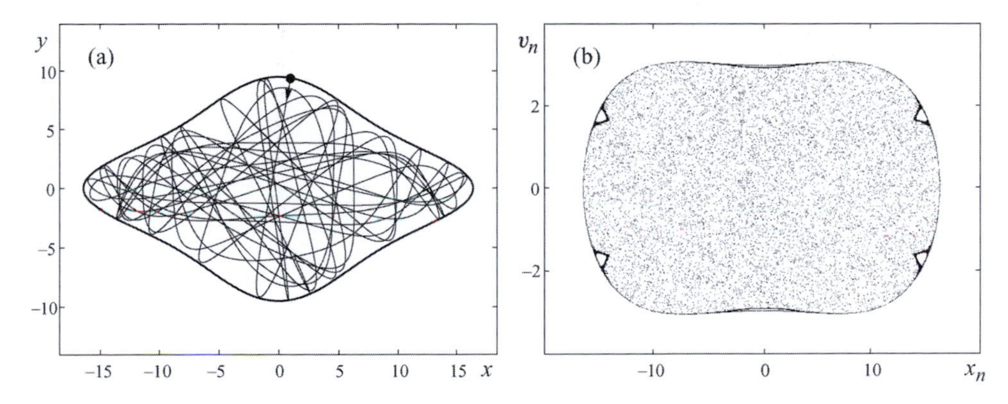

Fig. P.33. (a) Particle orbit starting from $x_0 = 1$, $\mathcal{E} = 5$. (b) Poincaré portrait defined by $y = 0$, $\dot{y} > 0$, $\mathcal{E} = 5$ generated from three different initial conditions. The majority of the phase space belongs to a single chaotic band.

Expressing v_{xn} and v_{yn} in terms of the decomposition (P.12), corresponding to the original slope, and substituting the flight time yields

$$\left.\begin{aligned} \tilde{u}_{n+1} &= -u_n + w_n \operatorname{tg}\alpha - \sqrt{D_n}\sin\alpha, \\ \tilde{w}_{n+1} &= -\sqrt{D_n}\cos\alpha. \end{aligned}\right\} \tag{P.16}$$

After the collision, $u_{n+1} = \tilde{u}_{n+1}$, $w_{n+1} = -\tilde{w}_{n+1}$, and (7.22) is recovered.

Solution 7.12 All this follows from the energy conservation, (7.21). Taking into account that the potential energy can only be positive, in the dimensionless form $u_n^2 + w_n^2 \equiv u_n^2 + z_n \leq 1$.

Solution 7.15 See Fig. P.33.

Solution 8.1 In terms of the variable $x \equiv \sin\phi$, map (8.3) is of the form

$$x_{n+1} = x_n - \frac{a}{R}\operatorname{sgn}(\theta_{n+1})\cos\theta_{n+1}, \qquad \theta_{n+1} = \theta_n - 2\arcsin x_n + \pi.$$

Its derivative matrix is given by

$$\begin{pmatrix} 1 - \operatorname{sgn}(\theta_{n+1})\sin\theta_{n+1}\frac{2a}{R\sqrt{1-x_n^2}} & \frac{a}{R}\operatorname{sgn}(\theta_{n+1})\sin\theta_{n+1} \\ -\frac{2}{\sqrt{1-x_n^2}} & 1 \end{pmatrix}, \tag{P.17}$$

whose Jacobian is unity.

Solution 8.2 The two-cycle of map (8.3) is given by the points $(x_n, \theta_{n+1}) = (0, \pi/2)$ and $(0, -\pi/2)$. The derivative matrix (P.17) is in both points

$$\begin{pmatrix} 1 - 2\frac{a}{R} & \frac{a}{R} \\ -2 & 1 \end{pmatrix},$$

with eigenvalues

$$\Lambda_{\pm} = 1 - \frac{a}{R} \mp \frac{a}{R}\sqrt{1 - 2\frac{R}{a}}.$$

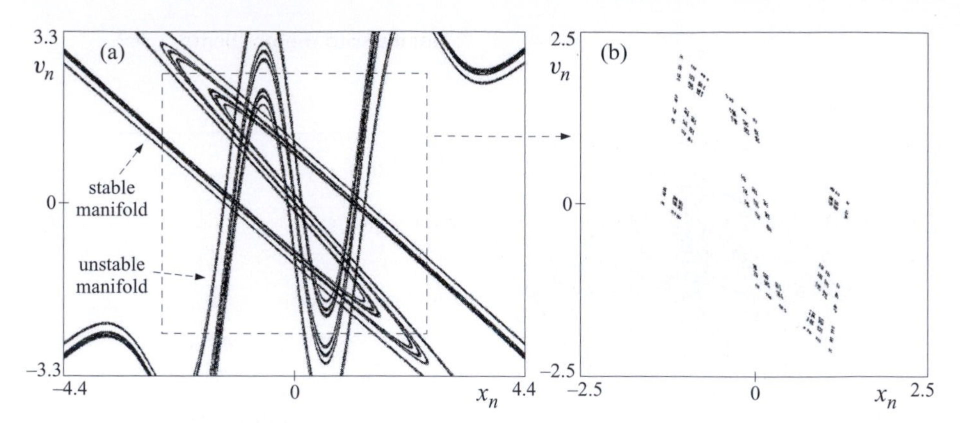

Fig. P.34. (a) Manifolds and (b) the chaotic saddle of map (8.6) with $f(x) = -x(c - 1 - cx^2/2)e^{-x^2/2}$ for $c = 10$. Points of trajectories of lifetime at least $n = 10$ in the depicted region are plotted at $n = 0$ and $n = 10$ in (a). The saddle is traced out by points $n = 5$.

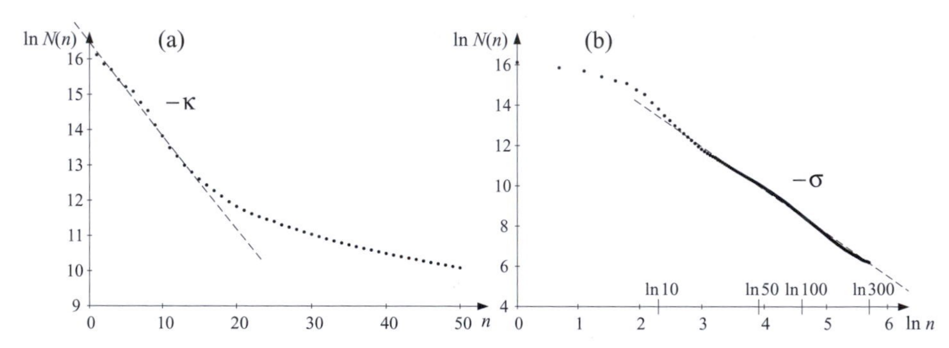

Fig. P.35. Distribution of the number of survivors, $N(n)$ vs n. (a) The short time decay is exponential with $\kappa = 0.263$. (b) The long time behaviour ($n > 30$) is a power law decay: $N(n) \sim n^{-\sigma}$ with $\sigma = 2.131$.

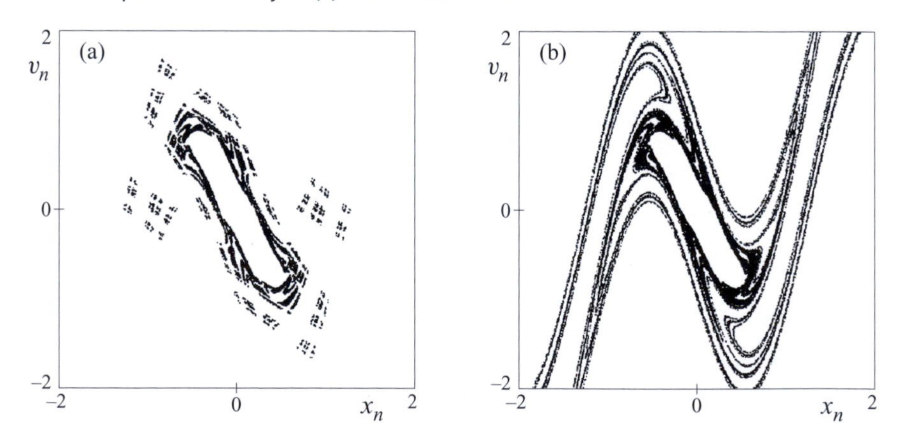

Fig. P.36. (a) Points of the saddle (obtained as the $n = 20$ points of trajectories not escaping up to $n = 30$ steps) accumulate around the outermost KAM torus surrounding the elliptic origin, and it is this dense accumulation which leads to a deviation from the exponential decay to a much slower power law decay shown in Fig. P.35(b). The part of the saddle which appears to be a Cantor cloud is responsible for the short time exponential behaviour of Fig. P.35(a). (b) The unstable manifold (obtained as the $n = 30$ points) is accordingly a Cantor filament away from the KAM torus, but is space filling around the torus.

The eigenvalues of the two-fold iterated map are $\Lambda_\pm^2$ in the points of the two-cycle. Since $a > 2R$, $|\Lambda_+| > 1$ and $|\Lambda_-| < 1$; the two-cycle is therefore hyperbolic.

Solution 8.3 The bounce of the (counter-clockwise running) triangular orbit with the bottom disk is characterized as a point $(x_n, \theta_{n+1}) = (-1/2, \pi/2)$ of map (8.3). The derivative matrix (P.17) is in this point

$$\begin{pmatrix} 1 - \frac{4a}{\sqrt{3}R} & \frac{a}{R} \\ -\frac{4}{\sqrt{3}} & 1 \end{pmatrix}. \tag{P.18}$$

After rotating the reference frame by $\pm 120°$, the same derivative matrices are found at the points of the other two bounces. The eigenvalues

$$\Lambda_\pm = 1 - \frac{2a}{\sqrt{3}R} \mp \frac{2a}{\sqrt{3}R}\sqrt{1 - \frac{\sqrt{3}R}{a}}$$

of (P.18) are thus the eigenvalues of the three-cycle for one iteration. Note that $|\Lambda_+| > 1$, $|\Lambda_-| < 1$; the triangular orbit is therefore hyperbolic.

Solution 8.6 See Fig. P.34.

Solution 8.7 See Figs. P.35 and P.36.

Bibliography

General bibliography

Textbooks on chaos

Ott, E. *Chaos in Dynamical Systems*, 2nd edn. Cambridge: Cambridge University Press, 2002.

Schuster, H.-G. *Deterministic Chaos: An Introduction*, 3rd (extended) edn. (Weinheim: VCH, 1995); 4th (revised and enlarged) edn. with W. Just (Weinheim: VCH. 2005)

Strogatz, S. H. *Nonlinear Dynamics and Chaos – With Applications to Physics, Biology, Chemistry, and Engineering.* Reading, MA: Addison Wesley, 1994.

Korsch, H. J. and Jodl, H.-J. *Chaos – A Program Collection for the PC*, 2nd edn. Berlin: Springer, 1998.

Alligood, K. T., Sauer, T. D. and Yorke, J. A. *Chaos: An Introduction to Dynamical Systems.* Berlin: Springer, 1998.

Baker, G. L. and Gollub, J. *Chaotic Dynamics: An Introduction*, 2nd edn. Cambridge: Cambridge University Press, 1996.

Nusse, H. E. and Yorke, J. A. *Dynamics: Numerical Explorations.* Berlin: Springer, 1994.

Jackson, E. A. *Perspectives of Nonlinear Dynamics*, Vols. I–II. Cambridge: Cambridge University Press, 1990.

Hilborn, R. C. *Chaos and Nonlinear Dynamics: An Introduction for Scientists and Engineers*, 2nd edn. Oxford: Oxford University Press, 2000.

Szemplinska-Stupnicka, W. *Chaos, Bifurcations and Fractals Around Us.* Singapore: World Scientific, 2003.

Textbooks on mechanics with chapters on chaos

Hand, L. N. and Finch, J. D. *Analytical Mechanics.* Cambridge: Cambridge University Press, 1998.

Josè, J. V. and Saletan, E. J. *Classical Dynamics: A Contemporary Approach.* Cambridge: Cambridge University Press, 1998.

Goldstein, H., Poole, C. P. and Safko, J. L. *Classical Mechanics*, 3rd edn. Englewood Cliffs, NJ: Prentice Hall, 2002.

Advanced texts on chaos

Guckenheimer, J. and Holmes, F. *Nonlinear Oscillations, Dynamical Systems, and Bifurcations of Vector Fields.* New York: Springer, 1983.

Lichtenberg, A. J. and Lieberman, M. A. *Regular and Stochastic Motion.* New York: Springer, 1983.

Wiggins, S. *Chaotic Transport in Dynamical Systems.* Berlin: Springer, 1992.

Beck, C. and Schlögl, F. *Thermodynamics of Chaotic Systems.* Cambridge: Cambridge University Press, 1993.

Zaslavsky, G. *Physics of Chaos in Hamiltonian Systems.* Singapore: World Scientific, 1998.

Hao, B.-L. and Zheng, W.-M. *Applied Symbolic Dynamics and Chaos*. Singapore: World Scientific, 1998.

Gutzwiller, M. C. *Chaos in Classical and Quantum Mechanics*. Berlin: Springer, 1990.

Cvitanović, P. *et al*. *Chaos: Classical and Quantum*, http://www.nbi.dk/ChaosBook/. Copenhagen: Niels Bohr Institute, 1999.

Stöckmann, H.-J. *Quantum Chaos*: *An Introduction*. Cambridge: Cambridge University Press, 1999.

Thompson, J. M. T. and Stewart, H. B. *Nonlinear Dynamics and Chaos*, 2nd edn. Chichester: Wiley, 2002.

Mori, H. and Kuramoto, Y. *Dissipative Structures and Chaos*. New York: Wiley, 2002.

Gilmore, R. and Lefranc, M. *The Topology of Chaos: Alice in Stretch and Squeezeland*. New York: Wiley, 2002.

Kantz, H. and Schreiber, T. *Nonlinear Time Series Analysis*, 2nd edn. Cambridge: Cambridge University Press, 2003.

Popular books on chaos

Gleick, J. *Chaos: Making a New Science*. New York: Penguin Books, 1987.

Ruelle, D. *Chance and Chaos*. Princeton: Princeton University Press, 1991.

Lorenz, E. N. *The Essence of Chaos*. Seattle: University of Washington Press, 1993.

Peterson, I. *Newton's Clock: Chaos in the Solar System*. New York: Freeman, 1993.

Diacu, F. and Holmes, P. *Celestial Encounters: the Origins of Chaos and Stability*. Princeton: Princeton University Press, 1996.

Stewart, I. *Does God Play Dice: The New Mathematics of Chaos*, 2nd edn. Oxford: Blackwell, 2002.

Differential equations

Borelli, R. L. and Coleman, C. S. *Differential Equations, a Modeling Perspective* New York: Wiley, 2004.

Arnold, V. I. *Mathematical Methods of Classical Mechanics*. Berlin: Springer, 1978.

Bibliography to selected sections

1.2.1 Irregular oscillations, driven pendulum – the chaotic attractor

Ueda, Y. 'Explosion of strange attractors exhibited by Duffing's equation', *Ann. N. Y. Acad. Sci.* **357**, 422 (1980).

Dooren, R. van, 'Chaos in pendulum with forced horizontal support motion', *Chaos, Solitons, and Fractals* **7**, 77 (1996).

1.2.2 Magnetic and driven pendulums, fractal basin boundary – transient chaos

Grebogi, C., Ott, E. and Yorke, J. A. 'Chaos, strange attractors, and fractal basin boundaries in nonlinear dynamics', *Science* **238**, 632 (1987).

Peitgen, H. O., Jürgens, H. and Saupe, D. *Chaos and Fractals: New Frontiers of Science*. Berlin: Springer, 1992, Sect. 12.8.

1.2.3 Body swinging on a pulley, ball bouncing on slopes – chaotic bands

Lehtihet, H. E. and Miller, B. N. 'Numerical study of a billiard in a gravitational field', *Physica* **D21**, 93 (1986).

Adams, C. S., Sigel, M. and Mlynek, J. 'Atom optics', *Phys. Rep.* **240**, 143 (1994).

1.2.4 Ball bouncing between discs, mirroring Christmas-tree ornaments – chaotic scattering

Berry, M. V. 'Reflections on a Christmas-tree bauble', *Phys. Edu.* **7**, 1 (1972).

Walker, J. 'The distorted images seen in Christmas-tree ornaments and other reflecting balls', *Sci. Am.* **259** (6), 84 (1984).

Korsch, H. J. and Wagner, A. 'Fractal mirror images and chaotic scattering', *Computers in Phys.* Sept./Oct., 497 (1991).

Sweet, D., Ott, E. and Yorke, J. A. 'Complex topology in chaotic scattering: a laboratory experiment', *Nature* **399**, 315 (1999).

1.2.5 Spreading of pollutants – an application of chaos

Aref, H., Jones, S. W., Mofina, S. and Zawadski, I. 'Vortices, kinematics and chaos', *Physica* **D37**, 423 (1989).

Károlyi, G. and Tél, T. 'Chaotic tracer scattering and fractal basin boundaries in a blinking vortex-sink system', *Phys. Rep.* **290**, 125 (1997).

2 Fractal objects

Mandelbrot, B. B. *The Fractal Geometry of Nature*. New York: Freeman, 1983.

Tél, T. 'Fractals, multifractals, and thermodynamics', *Z. Naturforsch.* **43a**, 154 (1988).

Vicsek, T. *Fractal Growth Phenomena*. Singapore: World Scientific, 1989.

Falconer, K. *Fractal Geometry*, 2nd edn. New York: Wiley, 2003.

Peitgen, H.-O. *et al. Fractals for the Classroom*, *Strategic Activities* I., II. Berlin: Springer, 1991, 1992.

Korvin, G. *Fractal Models in the Earth Sciences*. Amsterdam: Elsevier, 1992.

2.2.3 Thin and fat fractals

Farmer, J. D. 'Sensitive dependence on parameters in nonlinear dynamics', *Phys. Rev. Lett.* **55**, 351 (1985).

Umberger, D. K. and Farmer, J. D. 'Fat fractals on the energy surface', *Phys. Rev. Lett.* **55**, 661 (1985).

Eykholt, R. and Umberger, D. K. 'Fat fractals in nonlinear dynamical systems', *Physica* **30D**, 43 (1988).

2.3 Fractal distributions

Farmer, J. D., Ott, E. and Yorke, J. A. 'The dimension of chaotic attractors', *Physica* **7D**, 153 (1983).

5.3 Parameter dependence: the period doubling cascade

Feigenbaum, M. J. 'Quantitative universality for a class of nonlinear transformations', *J. Stat. Phys.* **19**, 25 (1978).

'The universal metric properties of nonlinear transformations', *J. Stat. Phys.* **21**, 669 (1979).

5.4.1 The measure of complexity: topological entropy

Newhouse, S. and Pignataro, T. 'On the estimation of topological entropy', *J. Stat. Phys.* **72**, 1331 (1991).

5.4.6 Link between dynamics and geometry: the Kaplan–Yorke relation

Kaplan, J. L. and Yorke, J. A. *Chaotic Behavior of Multidimensional Diference Equations*. Lecture Notes in Mathematics **730**. Berlin: Springer, 1979, p. 204.

5.7 The water-wheel

Strogatz, S. H. *Nonlinear Dynamics and Chaos – With Applications to Physics, Biology, Chemistry, and Engineering*. Reading, MA: Addison Wesley, 1994, Chap. 9.

6.1 The open baker map

Szépfalusy, P. and Tél, T. 'New approach to the problem of chaotic repellers', *Phys. Rev.* **A34**, 2520 (1986).

Tél, T. 'Transient chaos', in Hao Bai-lin, ed. *Directions in Chaos*, Vol. 3. Singapore: World Scientific, 1990, pp. 149–211.

6.3.3 Link between dynamics and geomerty: the Kantz–Grassberger relation

Kantz, H. and Grassberger, P. 'Repellers, semi-attractors, and long-lived chaotic transients', *Physica* **D17**, 75 (1985).

Hsu, G.-H., Ott, E. and Grebogi, C. 'Strange saddles and the dimension of their invariant manifolds', *Phys. Lett.* **A127**, 199 (1988).

6.5 Parameter dependence: crisis

Grebogi, C., Ott, E. and Yorke, J. A. 'Chaotic attractors in crisis', *Phys. Rev. Lett.* **48**, 1507 (1982).

'Critical exponents of chaotic transients in nonlinear dynamical systems', *Phys. Rev. Lett.* **57**, 1284 (1986).

6.7 Other types of crises, periodic windows

Grebogi, C., Ott, E. and Yorke, J. A. 'Crises, sudden changes in chaotic attractors and chaotic transients', *Physica* **D7**, 181 (1983).

6.8 Fractal basin boundaries

McDonald, S. W., Grebogi, C., Ott, E. and Yorke, J. A. 'Fractal basin boundaries', *Physica* **D17**, 125 (1985).

Grebogi, C., Ott, E. and Yorke, J. A. 'Metamorphoses of basin boundaries in nonlinear dynamical systems', *Phys. Rev. Lett.* **56**, 1011 (1986).

Kennedy, K. and Yorke, J. A. 'Basins of Wada', *Physica* **D51**, 213 (1991).

7.5.2 The KAM theorem

Kolmogorov, A. N. 'General theory of dynamical systems in classical mechanics', in *Proceedings of the 1954 International Congress of Mathematics* Amsterdam: North Holland, 1957.

Arnold, V. I. 'Small denominators II: Proof of a theorem by A. N. Kolmogorov on the preservation of conditionally-periodic motion under a small perturbation of the Hamiltonian', *Russ. Math. Surveys* **18**, 9 (1963). [*Uspekhi Mat. Nauk* **18**, 13 (1963).]

Moser, J. 'On invariant curves of area preserving mappings of an annulus', *Nachr. Akad. Wiss. Gttingen Math. Phys.* **1**, 1 (1962).

7.7 Homogeneously chaotic systems

Bunimovich, L. A. 'On the ergodic properties of nowhere dispersing billiards', *Commun. Math. Phys.* **65**, 295 (1979).

Bunimovich, L. A. and Sinai, Ya. G. 'Statistical properties of the Lorentz gas with periodic configuration of scatterers', *Commun. Math. Phys.* **78**, 479 (1981).

Szász, D. *Hard Ball Systems and the Lorentz Gas*, Encyclopaedia of Mathematical Sciences, Vol. 101. Berlin: Springer, 2000.

8.1 The scattering function

Eckhardt, B. 'Irregular scattering', *Physica* **D 33**, 89 (1988).
Ott, E. and Tél, T. 'Chaotic scattering: an introduction', *Chaos* **3**, 417 (1993).

8.2 Scattering on discs

Eckhardt, B. 'Fractal properties of scattering singularities', *J. Phys.* **A20**, 5971 (1987).
Gaspard, P. and Rice, S. A. 'Scattering from a classically chaotic repellor', *J. Chem. Phys.* **90**, 2225 (1989).
Poon, L., Campos, S., Ott, E. and Grebogi, C. 'Wada basin boundaries in chaotic scattering', *J. Bifurcation Chaos* **6**, 251 (1996).

8.3 Scattering in other systems

Jung, C. and Scholz, H. J. 'Cantor set structure in the singularities of classical potential scattering', *J. Phys.* **A 20**, 3607 (1987).
Bleher, S., Grebogi, C. and Ott, E. 'Bifurcation to chaotic scattering', *Physica* **D45**, 87 (1990).
Burghardt, I. and Gaspard, P. 'The molecular transition state: from regular to chaotic dynamics', *J. Phys. Chem.* **99**, 2732 (1995).
Gaspard, P. *Chaos, Scattering, and Statistical Mechanics.* Cambridge: Cambridge University Press, 1998.
Lau, Y.-T., Finn, J. M. and Ott, E. 'Fractal dimension in nonhyperbolic chaotic scattering', *Phys. Rev. Lett.* **66**, 978 (1991).
Lai, Y.-C., Grebogi, C., Blümel, R. and Kan, I. 'Crisis in chaotic scattering', *Phys. Rev. Lett.* **71**, 2212 (1993).

9.1 Spacecraft and planets: the three-body problem

Winter, O. C. and Murray, C. D. 'Atlas of the planar, circular, restricted three-body problem. I. Internal orbits', QMC Math Notes 16. London: Queen Mary College, 1994.
Richter, P. H. 'Chaos in cosmos', *Rev. Mod. Astron.* **14**, 53 (2001).
Szebehelyi, V. *Theory of Orbits.* New York: Academic Press, 1967.

9.2 Rotating rigid bodies: the spinning top

Dullin, H. R., Juhnke, M. and Richter, P. H. 'Action integrals and the energy surfaces of the Kovalevskaya top', *Int. J. Bifurcation Chaos* **4**, 1535 (1994).

9.3 Climate variability and climatic change

Lorenz, E. N. 'Irregularity: a fundamental property of the atmosphere', *Tellus* **36A**, 98 (1984).

'Can chaos and intransivity lead to interannual variability?' *Tellus* **42A**, 378, (1990).

Provenzale, A. and Balmforth, N. J. 'Chaos and structures in geophysics and astrophysics'. Lecture notes, http://gfd.whoi.edu/proceedings/1998/PDFvol1998.html; 1999.

9.4 Vortices, advection and pollution: chaos in fluid flows

Aref, H. 'Stirring by chaotic advection', *J. Fluid. Mech.* **143**, 1 (1984).

Ottino, J. M. *The Kinematics of Mixing: Stretching, Chaos, and Transport.* Cambridge: Cambridge University Press, 1989.

9.4.2 Advection in closed flows

Pierrehumbert, R. T. 'Tracer microstructure in the large-eddy dominated regime', *Chaos Solitons Fractals* **4**, 1091 (1994).

Alvarez, M. M., Muzzio, F. J., Cerbelli, S., Adrover, A. and Giona, M. 'Self-similar spatiotemporal structure of intermaterial boundaries in chaotic flows', *Phys. Rev. Lett.* **81**, 3395 (1998).

Ottino, J. M. and Wiggins, S. 'Designing optimal micro-mixers', *Science* **305**, 485 (2004).

9.4.3 Vortex dynamics

Newton, P. *The N Vortex Problem.* Berlin: Springer, 2001.

10 Epilogue: outlook

Simple experiments in mechanics

Moon, F. C. *Chaotic Vibrations.* New York: Wiley, 1987.

Moon, F. C. and Holmes, P. J. 'A magnetoelastic strange attractor', *J. Sound. Vib.* **69**, 339 (1979).

Meissner, H. and Schmidt, G. 'A simple experiment for studying the transition from order to chaos', *Am. J. Phys.* **54**, 800 (1986).

Briggs, K. 'Simple experiments in chaotic dynamics', *Am. J. Phys.* **55**, 1083 (1987).

Mello, T. M. and Tuffilaro, N. M. 'Strange attractors of a bouncing ball', *Am. J. Phys.* **55**, 316 (1987).

Tuffilaro, N. M. 'Nonlinear and chaotic string vibrations', *Am. J. Phys.* **57**, 408 (1989).

Zimmerman, R. L. 'The electronic bouncing ball', *Am. J. Phys.* **60**, 378 (1992).

Shinbrot, T., Grebogi, C., Wisdom, J. and Yorke, J. A. 'Chaos in a double pendulum', *Am. J. Phys.* **60**, 491 (1992).

Blackburn, J. A. and Baker, G. L. 'A comparison of commercial chaotic pendulum', *Am. J. Phys.* **66**, 821 (1998).

Berdahl, J. P. and Lugt, K. V. 'Magnetically driven chaotic pendulum', *Am. J. Phys.* **69**, 821 (2001).

Ditto, W. L. (ed.) *Proceedings of the 5th Experimental Chaos Conference.* Singapore: World Scientific, 2001.

Bocaletti, S. *et al.* (eds.) *Experimental Chaos. 6th Experimental Chaos Conference.* Melville, NY: AIP Conference Proceedings, 2002.

In, V. *et al.* (eds.) *Experimental Chaos. 7th Experimental Chaos Conference.* Melville, NY: AIP Conference Proceedings, 2003.

Bocaletti, S. *et al.* (eds.) *Experimental Chaos. 8th Experimental Chaos Conference.* Melville, NY: AIP Conference Proceedings, 2004.

Chaos experiments with fluids

Gollub, J. P. and Swinney, H. L. 'Onset of turbulence in a rotating fluid', *Phys. Rev. Lett.* **35**, 927 (1975).

Swinney, H. L. and Gollub, J. P. 'Characterization of hydrodynamic strange attractors', *Physica* **18D**, 448 (1986).

Libchaber, A. 'From chaos to turbulence in Bénard convection', *Proc. Roy. Soc. London* **A413**, 633 (1987).

Sommeria, J., Meyers, S. D. and Swinney, H. L. 'A laboratory simulation of Jupiter's Great Red Spot', *Nature* **331**, 689 (1988).

Solomon, T. H. and Gollub, J. P. 'Chaotic particle transport in time-dependent Rayleigh-Bénard convection', *Phys. Rev.* **A 38**, 6280 (1988).

Sommerer, J. C. and Ott, E. 'Particles floating on a moving fluid: a dynamically comprehensive physical fractal', *Science* **259**, 335 (1993).

Sommerer, J. C., Ku, H. C. and Gilreath, H. E. 'Experimental evidence for chaotic scattering in a fluid wake', *Phys. Rev. Lett.* **77**, 5055 (1996).

Rothstein, D., Henry, E. and Gollub, J. P. 'Persistent patterns in transient chaotic fluid mixing', *Nature* **401**, 770 (1999).

Voth, G. A., Haller, G. and Gollub, J. P. 'Experimental measurements of stretching fields in fluid mixing', *Phys. Rev. Lett.* **88**, 254 501 (2002).

Solomon, T. H. and Mezić, I. 'Uniform resonant chaotic mixing in fluid flows', *Nature* **425**, 376 (2003).

Nuqent, C. R., Quarles, W. M. and Solomon, T. H. 'Experimental studies of pattern formation in a reaction-advection-diffusion system', *Phys. Rev. Lett.* **93**, 218 301, 2004.

11.3 Numerical solution of ordinary differential equations

Press, W. H., Flannery, B. P., Teukolsky, S. A. and Vetterling, W. V. *Numerical Recipes in C^{++}.* Cambridge: Cambridge University Press, 2002.

Bibliography to selected boxes

Box 1.1 Brief history of chaos

Poincaré, H. *Les méthodes nouvelles de la mécanique céleste I–III.* Paris: Gauthier-Villars, 1892, 1893, 1899.

Kovalevskaia, S. 'Sur le problème de la rotation d'un corps solide d'un point fixe', *Acta Math.* **12**, 177 (1889).

Li, T. Y. and Yorke, J. A. 'Period three implies chaos', *Am. Math. Monthly* **82**, 983 (1975).

Ueda, Y. *The Road to Chaos.* Santa Cruz: Aeral Press, 1993.

Yorke, J. A. and Grebogi, C. (eds.) *The Impact of Chaos on Science and Society.* New York: United Nations University Press, 1996.

Abraham, R. and Ueda, Y. (eds.) *The Chaos Avant-Garde. Memories of the Early Days of Chaos Theory.* Singapore: World Scientific, 2001.

Box 2.1 Brief history of fractals

Mandelbrot, B. B. *The Fractal Geometry of Nature.* New York: Freeman, 1983.

Grassberger, P. and Procaccia, I. 'Measuring the strangeness of strange attractors', *Physica* **9D**, 189 (1983).

Grebogi, C., McDonald, S. W., Ott, E. and Yorke, J. A. 'Final state sensitivity: an obstruction to predictability', *Phys. Lett. A* **99**, 415 (1983).

'Exterior dimension of fat fractals', *Phys. Lett. A* **110**, 1 (1985).

Farmer, J. D. 'Sensitive dependence on parameters in nonlinear dynamics', *Phys. Rev. Lett.* **55**, 351 (1985).

The Selected Papers of A. Rènyi. Budapest: Akadémiai Kiadó, 1976.

Barnsley, M. F. *Fractals Everywhere*. Orlando: Academic Press, 1988.

Box 4.1 The world of non-invertible maps

Peitgen, H.-O. and Richter, P. H. *The Beauty of Fractals.* Berlin: Springer, 1986.

Csordás, A., Györgyi, G., Szépfalusy, P. and Tél, T. 'Statistical properties of chaos demonstrated in a class of one-dimensional maps', *Chaos* **3**, 31 (1993).

Mira, C., Gardini, L., Barugola, A. and Cathala, J. C. *Chaotic Dynamics in Two-Dimensional Noninvertible Maps.* Singapore: World Scientific, 1996.

Box 5.1 Hénon-type maps

Hénon, M. 'A two-dimensional mapping with a strange attractor', *Commun. Math. Phys.* **50**, 69 (1976).

Lozi, R. 'Un attracteur étrange(?) du type attracteur de Hénon', *J. Physique* **39** (C5), 9 (1978).

Tél, T. 'Invariant curves, attractors and phase diagram of a piecewise linear map with chaos', *J. Stat. Phys.* **33**, 195 (1983).

'Fractal dimension of the strange attractor in a piecewise linear two-dimensional map', *Phys. Lett.* **119 A**, 65 (1983).

Box 5.2

Gleick, J. *Chaos: Making a New Science.* New York: Penguin Books, 1987.

Lorenz, E. N. *The Essence of Chaos.* Seattle: University of Washington Press, 1993.

Box 5.6 The Lorenz model

Lorenz, E. N. 'Deterministic nonperiodic flow', *J. Atmos. Sci.* **20**, 130 (1963).

Sparrow, C. *The Lorenz Equations: Bifurcations, Chaos, and Strange Attractors.* New York: Springer, 1982.

Box 6.1 How do we determine the saddle and its manifold?

Lai, Y.-C., Tél, T. and Grebogi, C. 'Stabilizing chaotic scattering trajectories using control', *Phys. Rev.* **E48**, 709 (1993).

Box 6.3 The horseshoe map

Smale, S. 'Differentiable dynamical systems', *Bull. Am. Math. Soc.* **73**, 747 (1967).

Box 6.4 Other aspects of chaotic transeints

Gaspard, P. and Nicolis, G. 'Transport properties, Lyapunov exponents, and entropy per unit time', *Phys. Rev. Lett.* **65**, 1693 (1990).

Gaspard, P. and Dorfman, J. R. 'Chaotic scattering theory, thermodynamical formalism and transport coefficients', *Phys. Rev.* **E52**, 3525 (1995).

Bleher, S., Grebogi, C., Ott, E. and Brown, R. 'Fractal boundaries for exit in Hamiltonian dynamics', *Phys. Rev.* **A38**, 930 (1988).

Sommerer, J. C., Ditto, W., Grebogi, C., Ott, E. and Spano, M. L. 'Experimental confirmation of the scaling theory for noise induced crises', *Phys. Rev. Lett.* **66**, 1947 (1991).

Xu, B., Lai, Y.-C., Zhu, L. and Do, Y. 'Experimental characterization of transition to chaos in the presence of noise', *Phys. Rev. Lett.* **90**, 164 101 (2003).

Box 7.1 The origin of the baker map

Hopf, E. 'On causality, statistics and probability', *J. Math. Phys.* **13**, 51 (1934). *Ergodentheorie*. Berlin: Springer, 1937.

Farmer, J. D., Ott, E. and Yorke, J. A. 'The dimension of chaotic attractors', *Physica* **7D**, 153 (1983).

Box 7.4 Ergodicity and mixing

Simányi, N. and Szász, D. 'Hard ball systems are completely hyperbolic', *Ann. Math.* **149**, 35 (1999).

Szász, D. *Hard Ball Systems and the Lorentz Gas*, Encyclopaedia of Mathematical Sciences, Vol. 101. Berlin: Springer, 2000.

Box 7.5 Conservative chaos and irreversibility

Shepelyansky, D. L. 'Some statistical properties of simple classically stochastic quantum systems', *Physica* **8D**, 208 (1983).

Gaspard, P. *Chaos, Scattering, and Statistical Mechanics.* Cambridge: Cambridge University Press, 1998.

Dorfman, J. R. *An Introduction to Chaos in Nonequilibrium Statistical Mechanics.* Cambridge: Cambridge University Press, 1999.

Vollmer, J. 'Chaos, spatial extension, transport, and non-equilibrium thermodynamics', *Phys. Rep.* **372**, 131 (2002).

Chaos and Irreversibility. Focus issue, *Chaos* **8**, 309 (1998).

Klages, R. *Microscopic Chaos, Fractals and Transport in Nonequilibrium Statistical Mechanics*. Singapore: World Scientific, 2006.

Box 8.1 Chemical reactions as chaotic scattering

Noid, D. W., Gray S. and Rice, S. A. 'Fractal behavior in classical collision energy transfer', *J. Chem. Phys.* **84**, 2649 (1986).

Kovács, Z. and Wiesenfeld, L. 'Chaotic scattering in reactive collisions – a classical analysis', *Phys. Rev.* **E51**, 5476 (1995).

Yamamoto, T. and Kaneko, K. 'Helium atom as a classical three-body problem', *Phys. Rev. Lett.* **70**, 1928 (1993).

Bümel, R. and Reinhardt, W. P. *Chaos in Atomic Physics.* Cambridge: Cambridge University Press, 1997.

Box 9.1 Chaos in the Solar System

Wisdom, J. 'The Urey Prize Lecture: Chaotic Dynamics in the Solar System', *Icarus* **72**, 241 (1987).

Sussman, G. J. and Wisdom, J. 'Numerical evidence that the motion of Pluto is chaotic', *Science* **241**, 433 (1988).

'Chaotic evolution of the Solar System', *Science* **257**, 56 (1992).

Laskar, J. 'A numerical experiment on the chaotic behavior of the Solar System', *Nature* **338**, 237 (1989).

'Large scale chaos and marginal stability in the Solar System', *Celest. Mech. Dyn. Astron.* **64**, 115 (1996).

Laskar, J. and Robutel, P. 'The chaotic obliquity of the planets', *Nature* **361**, 608 (1993).

Sándor, Z., Érdi, B. and Efthymiopoulos, C. 'The phase space structure around L_4 in the restricted three-body problem', *Celest. Mech. Dyn. Astron.* **78**, 113 (2000).

Sándor, Z., Balla, R., Téger, F. and Érdi, B. 'Short time Lyapunov indicators in the restricted three-body problem', *Celest. Mech. Dyn. Astron.* **79**, 29 (2001).

Petit, J.-M. and Hénon, M. 'Satellite encounters', *Icarus* **66**, 536 (1986).

Moons, M. 'Review of the dynamics in the Kirkwood gaps', *Celest. Mech. Dyn. Astron.* **65**, 175 (1997).

Malhotra, R. 'Chaotic planet formation', *Nature* **402**, 599 (1999).

Murray, N. and Holman, M. 'The origin of chaos in the outer Solar System', *Science* **283**, 1877 (1999).

Contopoulos, G. I. *Order and Chaos in Dynamical Astronomy.* Berlin: Springer, 2002.

Box 9.2 Chaos in engineering practice

Moon, F. C. *Chaos and Fractal Dynamics: An Introduction for Applied Scientists and Engineers.* New York: Wiley, 1992.

Kapitaniak, T. *Chaos for Engineers: Theory, Applications, and Control.* Berlin: Springer, 2000.

Thompson, J. M. T. and Stewart, H. B. *Nonlinear Dynamics and Chaos*, 2nd edn. Chichester: Wiley, 2002.

Szalai, R., Stépán, G. and Hogan, S. J. 'Global dynamics of low immersion high-speed milling', *Chaos* **14**, 1069 (2004).

Thompson, J. M. T. 'Complex dynamics of compliant off-shore structures', *Proc. Roy. Soc. Lond.* **A 387**, 407 (1983).

Stépán, G. 'Chaotic motion of wheels', *Vehicle Syst. Dyn.* **20**, 341 (1991).

Appell-Gibbs equations for classical wheel shimmy – an energy view, *J. Comp. Appl. Mech.* **3**, 85 (2002).

Spyrou, K. J. and Thompson, J. M. T. (eds.) 'The nonlinear dynamics of ships', special issue, *Phil. Trans. Roy. Soc. Lond.* **A 358**, 1733 (2000).

Haller, G. and Stépán, G. 'Micro-chaos in digital control', *J. Nonlin. Sci.* **6**, 415 (1996).

Enikov, E. and Stépán, G. Micro-chaotic motion of digitally controlled machines', *J. Vibr. Control*, **4**, 427 (1998).

Stépán, G. and Kollár, L. 'Balancing with reflex delay', *Math. Comp. Mod.* **31**, 199 (2000).

Stirling, J. R. and Zakynthinaki, M. S. 'Stability and maintenance of balance following a perturbation from quiet stance', *Chaos* **14**, 96 (2004).

Domokos, G. and Holmes, P. 'Euler's problem, Euler's method, and the standard map; or the discrete charm of buckling', *J. Nonlin. Sci.* **3**, 109 (1993).

Davies, M. A. and Moon, F. C. '3-D spatial chaos in the elastic and the spinning top: Kirchhoff analogy', *Chaos* **3**, 93 (1993).

Heiden, G. H. M. van der, Champneys, A. R. and Thompson, J. M. T. 'The spatial complexity of localised buckling in rods with non-circular cross-section', *SIAM J. Appl. Math.* **59**, 198 (1998).

Károlyi, G. and Domokos, G. 'Symbolic dynamics of infinite depth: finding invariants for BVPs', *Physica* **D 134**, 316 (1999).

Goriely, A. and Tabor, M. 'Spontaneous helix hand reversal and tendril perversion in climbing plants', *Phys. Rev. Lett.* **80**, 1564 (1998).

Tobias, I., Swignon, D. and Coleman, B. D. 'Elastic stability of DNA configurations, I. General theory', *Phys. Rev.* **E 61**, 759 (2000).

Box 9.3 Chaos in different sciences

Meteorology

Kalnay, E. *Atmospheric Modeling, Data Assimilation and Predictability.* Cambridge: Cambridge University Press, 2002.

Patil, D. J., Hunt, B. R., Kalnay, E., Yorke, J. A. and Ott, E. 'Local low dimensionality of atmospheric dynamics', *Phys. Rev. Lett.* **86**, 5878 (2001).

Szunyogh, I., Toth, Z., Zimin, A. V., Majumdar, S. J. and Persson, A. 'Propagation of the effect of targeted observations: the 2000 winter storms reconaissance program', *Monthly Weather Rev.* **130**, 1144 (2002).

Geophysics

Turcotte, D. L. *Fractals and Chaos in Geology and Geophysics.* Cambridge: Cambridge University Press, 1997.

Perugini, D., Poli, G. and Gatta, G. D. 'Analysis and simulation of magma mixing processes in 3D', *Lithos* **65**, 313 (2002).

Keken, P. E. van, Hauri, E. and Ballentine, C. J. 'Mantle mixing: the generation, preservation and destruction of mantle heterogeneity', *Ann. Rev. Earth Planet. Sci.* **30**, 493 (2002).

Plasmas and lasers

Infeld, E. and Rowlands, G. *Nonlinear Waves, Solitons and Chaos.* Cambridge: Cambridge University Press, 1990.

Arecchi, F. T. and Harrison, R. G. (eds.) *Selected Papers on Optical Chaos.* Society of Photo Optical, 1993.

Electronic circuits

Ogorzalek, M. J. *Chaos and Complexity in Nonlinear Electronic Circuits.* Singapore: World Scientific, 1997.

Chen, G. and Ueta, T. (eds.) *Chaos in Circuits and Systems.* Singapore: World Scientific, 2002.

Nanotechnology

Ferry, D. K. and Goodnick, S. M. *Transport in Nanostructures.* Cambridge: Cambridge University Press, 1997.

Heinzel, T. *Mesoscopic Electronics in Solid State Nanostructures.* Weinheim: Wiley-VCH, 2003.

Ehrfeld, W., Hessel, V. and Löve, H. *Microreactors.* Weinheim: Wiley-VCH, 2004.

Acoustics

Legge, K. A. and Fletcher, N. H. 'Nonlinearity, chaos, and the sound of shallow gongs', *J. Acoust. Soc. Am.* **86**, 2439 (1989).

Acoustic Chaos. Focus issue, *Chaos* **5**, 495 (1995).

Mortessagne, F., Legrand, O. and Sornette, D. 'Transient chaos in room acoustics', *Chaos* **3**, 529 (1993).

Herzel, H. 'Possible mechanism of vocal istabilities, in Davies, P. J. and Fletcher, N. H. (eds.) *Vocal Fold Physiology: Controlling, Complexity, and Chaos.* San Diego: Singular Publications, 1996, pp. 63–74.

Chemistry

Field, R. J. and Györgyi, L. *Chaos in Chemistry and Biochemistry.* Singapore: World Scientific, 1993.

Scott, S. K. *Oscillations, Waves, and Chaos in Chemical Kinetics.* Oxford: Oxford University Press, 1994.

Epstein, I. R. and Pojman, J. A. *An Introduction to Nonlinear Chemical Dynamics: Oscillations, Waves, Patterns, and Chaos.* Oxford: Oxford University Press, 1998.

Medicine and physiology

West, B. J. *Fractal Physiology and Chaos in Medicine.* Singapore: World Scientific, 1990.

Sataloff, R. T. and Hawkshaw, M. *Chaos in Medicine: Source Readings.* San Diego: Singular Thomson Leaning, 2000.

Goldberger, A. L. 'Non-linear dynamics for clinicians: chaos theory, fractals, and complexity at the bedside', *Lancet* **347**, 1312 (1996).

'Fractal variability versus pathologic periodicity: complexity loss and stereotypy in disease', *Perspective in Biology and Medicine* **40**, 543 (1996).

Ellner, S. P. *et al.* 'Noise and nonlinearity in measles epidemics: combining mechanistic and statistical approaches to population modeling', *The American Naturalist* **151**, 425 (1998).

Biology and ecology

May, R. M. 'Simple methematical model with very complicated dynamics', *Nature* **261**, 459 (1976).

Winfree, A. T. *The Geometry of Biological Time.* New York: Spinger, 1980.

Liebovitch, L. S. *Fractals and Chaos: Simplified for the Life Sciences*. Oxford: Oxford University Press, 1998.

Costantino, R. F., Desharnais, R. A., Cushing, J. M. and Dennis, B. 'Chaotic dynamics in an insect population', *Science* **275**, 389 (1997).

Cushing, J. M., Costantino, R. F., Dennis, B., Desharnais, R. A. and S. M. Henson, *Chaos in Ecology: Experimental Nonlinear Dynamics.* New York: Academic Press, 2003.

McCann, K., Hastings, A. and Huxel, G. R. 'Weak trophic interactions and the balance of Nature', *Nature* **395**, 794 (1998).

Bednekoff, P. A. and Lima, S. L. 'Randomness, chaos and confusion in the study of antipredator vigilance', *TREE* **13**, 284 (1998).

Huisman, J. and Weissing, F. J. 'Fundamental unpredictability in multispecies competition', *The American Naturalist* **157**, 488 (2001).

Domokos, G. and Scheuring, I. 'Discrete and continuous state population models in a noisy world', *J. Theor. Biol.* **227**, 535 (2004).

Scheuring, I. 'Is chaos due to over-simplification in models of population dynamics?' *Selection* **2** 177 (2001).

Internet

Kocarev, L. and Vattay, G. (eds.) *Complex Dynamics in Communication Networks.* Berlin: Springer, 2005.

Economy

Peters, E. E. *Chaos and Order in the Capital markets: A New View of Cycles, Prices, and Market Volatility.* New York: Wiley, 1996.

Puu, T. *Attractors, Bifurcations, and Chaos: Nonlinear Phenomena in Economics.* New York: Springer, 2000.

Box 9.4 Controlling chaos

Ott, E., Grebogi, C. and Yorke, J. A. 'Controlling Chaos', *Phys. Rev. Lett.* **64**, 1196 (1990).

Ditto, W. L., Rauseo, S. N. and Spano, M. L. 'Experimental control of chaos', *Phys. Rev. Lett.* **65**, 3211 (1990).

Petrov, V., Gáspár, V., Masere, J. and Showalter, K. 'Controlling chaos in the Belousov–Zhabotinsky reaction', *Nature* **361**, 240 (1993).

Schuster, H.-G. (ed.) *Handbook of Chaos Control.* Weinheim: Wiley-VCH, 1998.

Roy, R., Murphy, T. W. Jr., Maier, T. D., Gills, Z. and Hunt, E. R. 'Dynamical control of a chaotic laser: experimental stabilization of a globally coupled system', *Phys. Rev. Lett.* **68**, 1259 (1992).

Garfinkel, A., Spano, M. L., Ditto, W. L. and Weiss, J. N. 'Controlling cardiac chaos', *Science* **257**, 1230 (1992).

Schiff, S. J. *et al.* 'Controlling chaos in the brain', *Nature* **370**, 615 (1994).

Pyragas, K. 'Continuous control of chaos by self-controlling feedback', *Phys. Lett.* **A 170**, 421 (1992).

Boccaletti, S., Grebogi, C., Lai, Y-C., Mancini, H. and Maza, D. 'The control of chaos: theory and applications', *Phys. Rep.* **329**, 103 (2000).

Ott, E. *Chaos in Dynamical Systems*, 2nd edn. Cambridge: Cambridge University Press, 2002, Chap. 10.

Shinbrot, T., Grebogi, C., Ott, E. and Yorke, J. A. 'Using the butterfly effect to direct orbits to targets in an experimental chaotic system', *Phys. Rev. Lett.* **68**, 2863 (1992).

Yamada, T. and Fujisaka, H. 'Stability theory of synchronized motion in coupled oscillator systems II', *Prog. Theor. Phys.* **70**, 1240 (1998).

Pecora, L. M. and Caroll, T. L. 'Synchronisation in chaotic systems', *Phys. Rev. Lett.* **64**, 821 (1990).

Pikovsky, A., Rosenblum, M. and Kurths, J. *Synchronization: A Universal Concept in Nonlinear Sciences.* Cambridge: Cambridge University Press, 2001.

Mosekilde, E., Maistrenko, Y. and Postnov, D. *Chaotic Synchronization.* Singapore: World Scientific, 2002.

Hayes, S., Grebogi, C. and Ott, E. 'Communicating with chaos', *Phys. Rev. Lett.* **70**, 3031 (1993).

Box 9.5 Environmental significance of chaotic advection

Jung, C., Tél, T. and Ziemniak, E. 'Application of scattering chaos to particle transport in hydrodynamical flow', *Chaos* **3**, 555 (1993).

Edouard, E., Legras, B., Lefevre, F. and Eymard, R. 'The effect of small-scale inhomogeneities on ozone depletion in the Arctic', *Nature* **384**, 444 (1996).

Toroczkai, Z., Károlyi, G., Péntek, A., Tél, T. and Grebogi, C. 'Advection of active particles in open chaotic flows', *Phys. Rev. Lett.* **80**, 500 (1998).

Neufeld, Z., López, C. and Haynes, P. 'Smooth-filamental transition of active tracer fields stirred by chaotic advection', *Phys. Rev. Lett.* **82**, 2606 (1999).

Active Chaotic Flow. Focus issue, *Chaos* **12**, 372 (2002).

Neufeld, Z., Haynes, P. H., Garcon, V. C. and Sudre, J. 'Ocean fertilization experiments may initiate a large scale phytoplankton bloom', *Geophys. Res. Lett.* **29**, 10.1029/2001GL013677 (2002).

Abraham, E. R. and Bowen, M. M. 'Chaotic stirring by a mesoscale surface-ocean flow', *Chaos* **12**, 373 (2002).

Martin, A. P. 'Phytoplankton patchiness: the role of lateral stirring and mixing', *Prog. Oceanography* **57**, 125 (2003).

Scheuring, I., Czáran, T., Szabó, P., Károlyi, G. and Toroczkai, Z. 'Spatial models of prebiotic evolution: soup before pizza?' *Orig. Life Evol. Bios.* **33**, 319 (2003).

Konopka, P. *et al.* 'Mixing and ozone loss in the 1999–2000 Arctic vortex: simulations with the three-dimensional chemical Lagrangian model of the stratosphere (CLaMS)', *J. Geophys. Res.* **109**, D02315, doi:10.1029/2003JD003792 (2004).

Grooß, J.-U., Konopka, P. and Müller, R. 'Ozone chemistry during the 2002 Antarctic vortex split', *J. Atmos. Sci.* **62**, 860 (2004).

Tél, T., de Moura, A., Grebogi, C. and Károlyi, G. 'Chemical and biological activity in open flows: a dynamical system approach', *Phys. Rep.* **413**, 91 (2005).

d'Ovidio, F., Fernandez, V., Hernandez-Garcia, E. and Lopez, C. 'Mixing structures in the Mediterranean Sea from finite-size Lyapunov exponents', *Geophys. Res. Lett.* **31**, L17203 (2004).

Box 10.1 Turbulence and spatio-temporal chaos

Frish, U. *Turbulence: The Legacy of A. N. Kolmogorov.* Cambridge: Cambridge University Press, 1995.

Bohr, T. and Jensen, M. H. *Dynamical Systems Approach to Turbulence.* Cambridge: Cambridge University Press, 1998.

Bocaletti, S., Mancini, H. L., González-Viñas, W. and Burguete, J. (eds.) *Space-time Chaos: Characterization, Control and Synchronization.* Singapore: World Scientific, 2001.

Lai, Y.-C. and Winslow, R. L. 'Riddled parameter space in spatiotemporal dynamical systems', *Phys. Rev. Lett.* **72**, 1640 (1994).

Garfinkel, A. *et al.* 'Quasiperiodicity and chaos in cardiac fibrillation', *J. Clin. Invest.* **99**, 305 (1997).

Blasius, B., Huppert, A. and Stone, L. 'Complex dynamics and phase synchronization in spatially extended ecological systems', *Nature* **399**, 354 (1999).

Earn, D. J. D., Rohani, P., Bolker, B. M. and Grenfell, B. T. 'A simple model for complex dynamical transitions in epidemics', *Science* **287**, 667 (2000).

Egolf, D. A., Melnikov, I. V., Pesch, W. and Ecke, R. E. 'Mechanism of extensive spatiotemporal chaos in Rayleigh–Bénard convection', *Nature* **404**, 733 (2000).

Winfree, A. T. 'Electrical turbulence in three-dimensional heart muscle', *Science* **266**, 1003 (1994).

Ghashghaie, S., Breymann, W., Peinke, J., Talkner, P. and Dodge, Y. 'Turbulent cascades in foreign exchange markets', *Nature* **381**, 767 (1996).

Index

I

II

III

IV

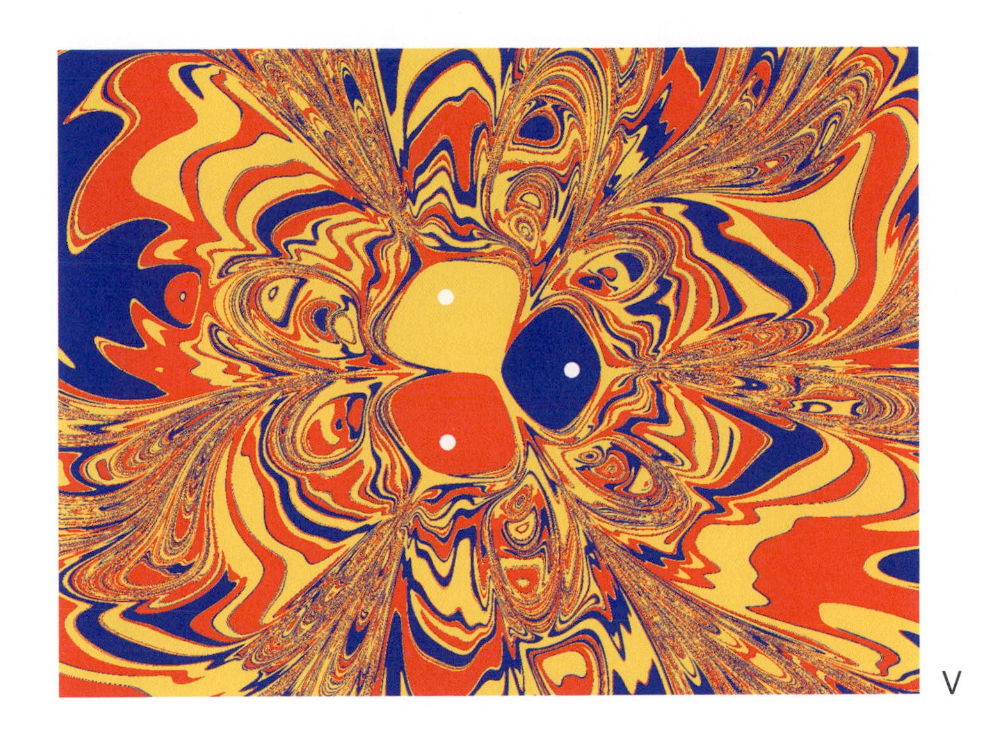

V

VI

VII

VIII

IX

X

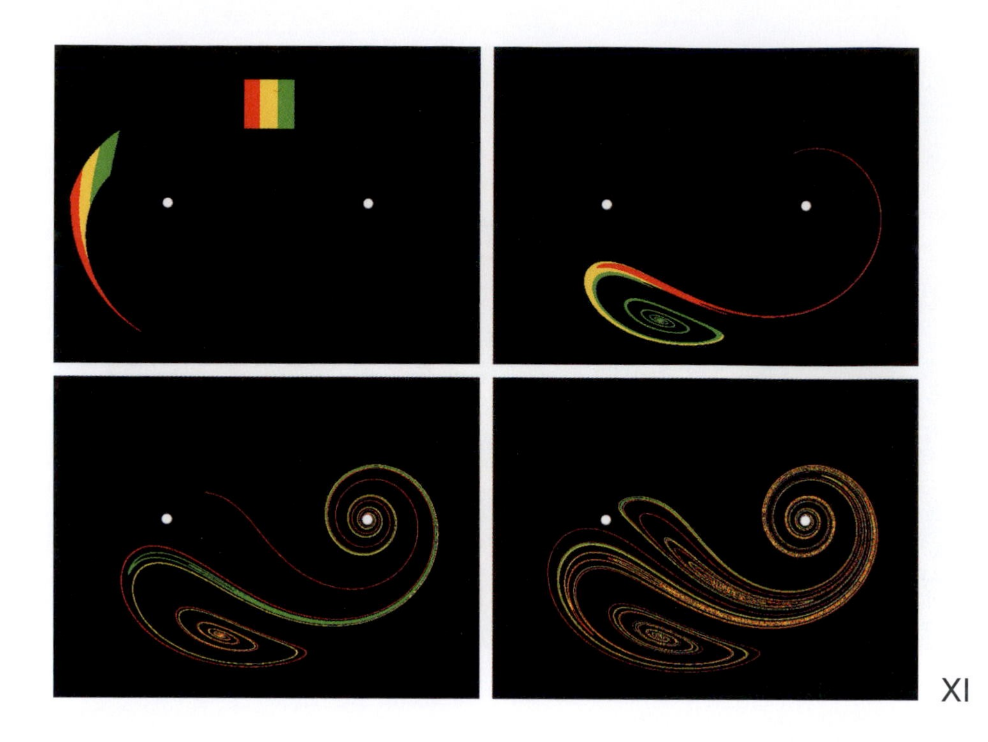

XI

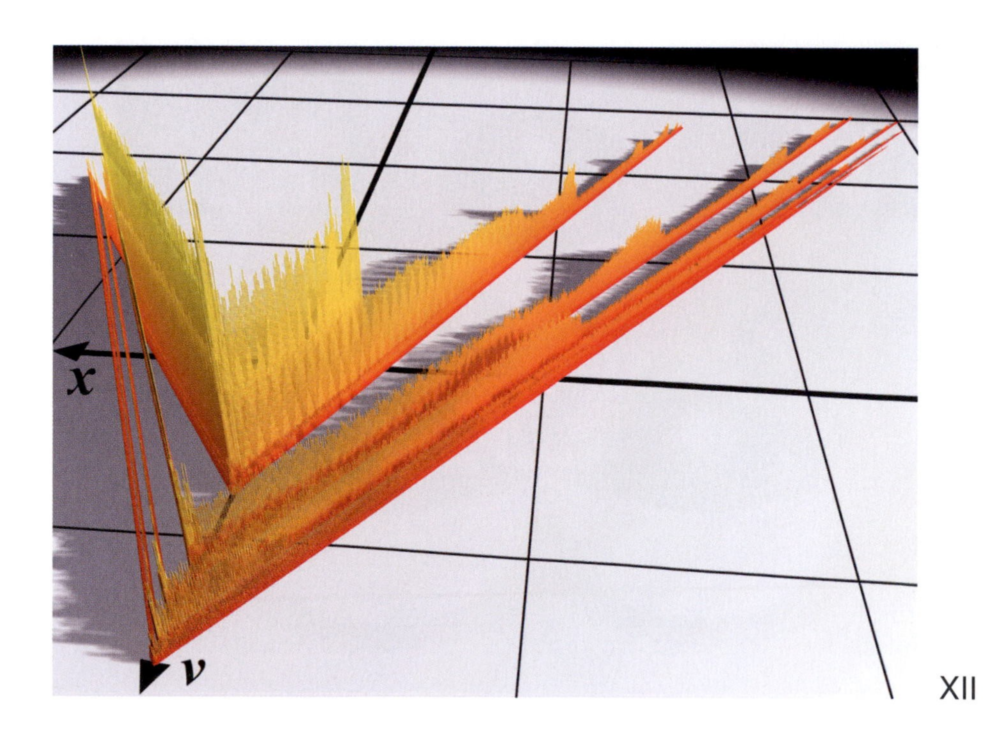

XII

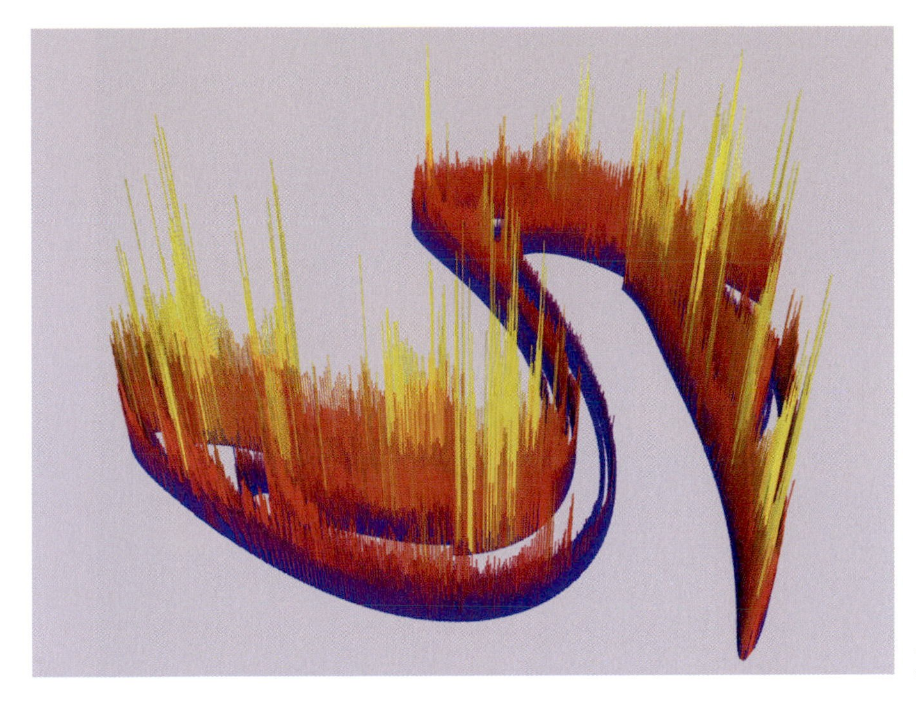
XIII

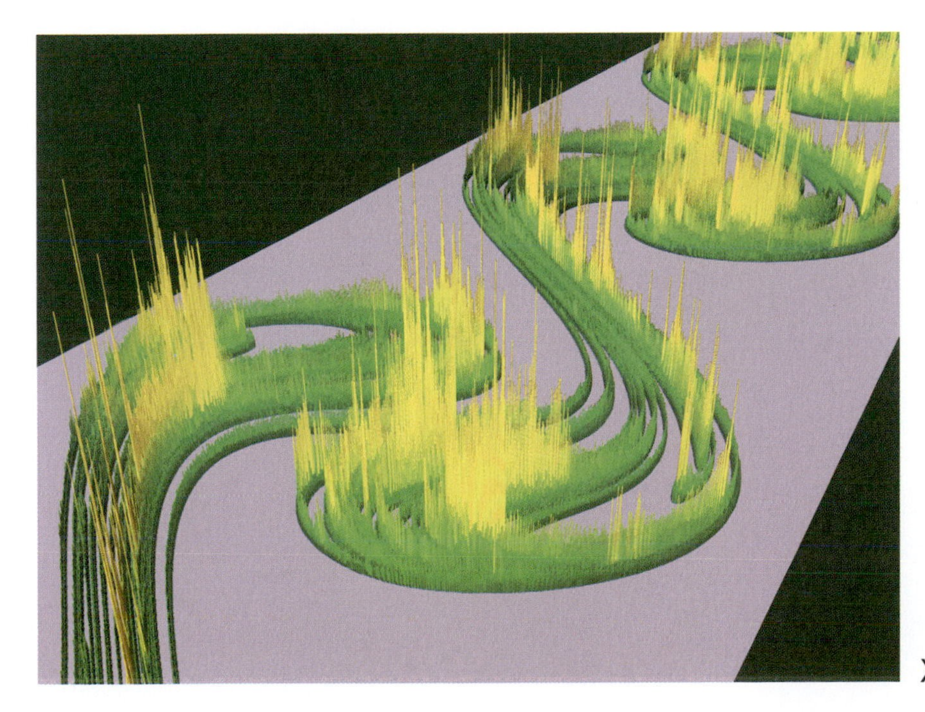
XIV

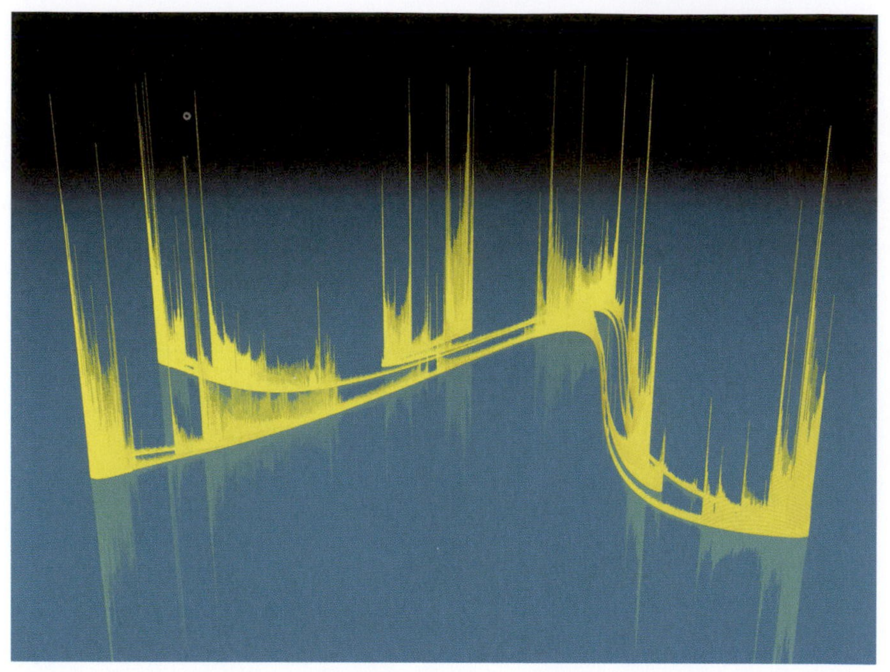

XV

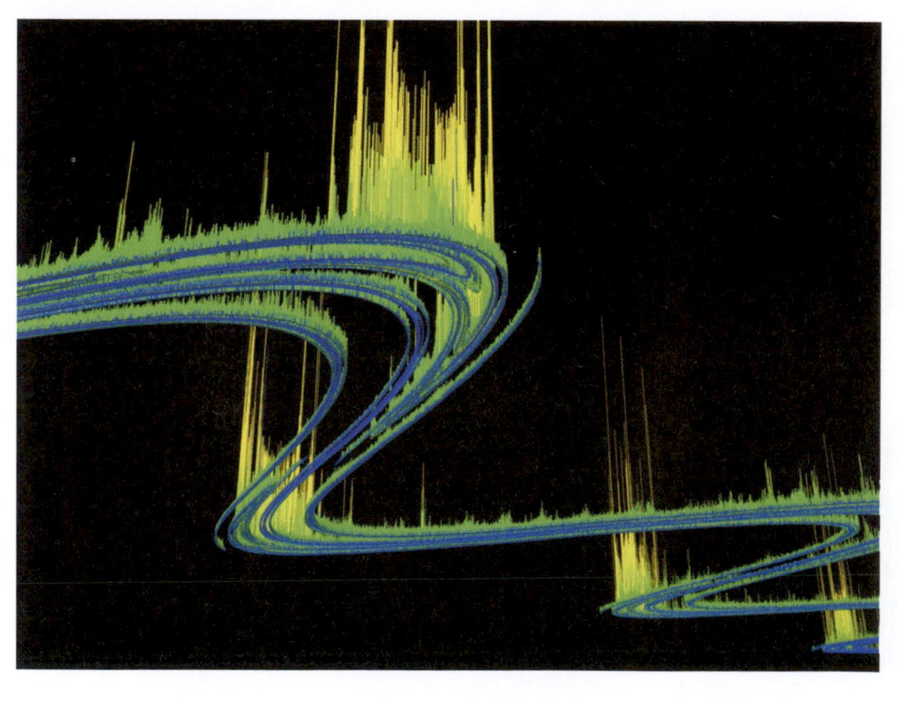

XVI

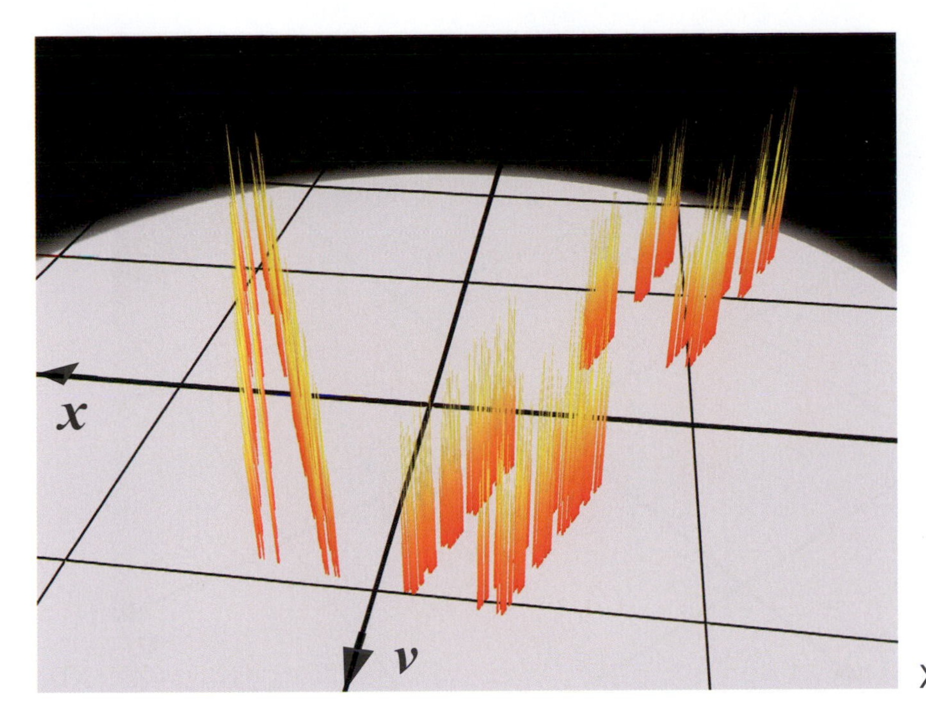

XVII

XVIII

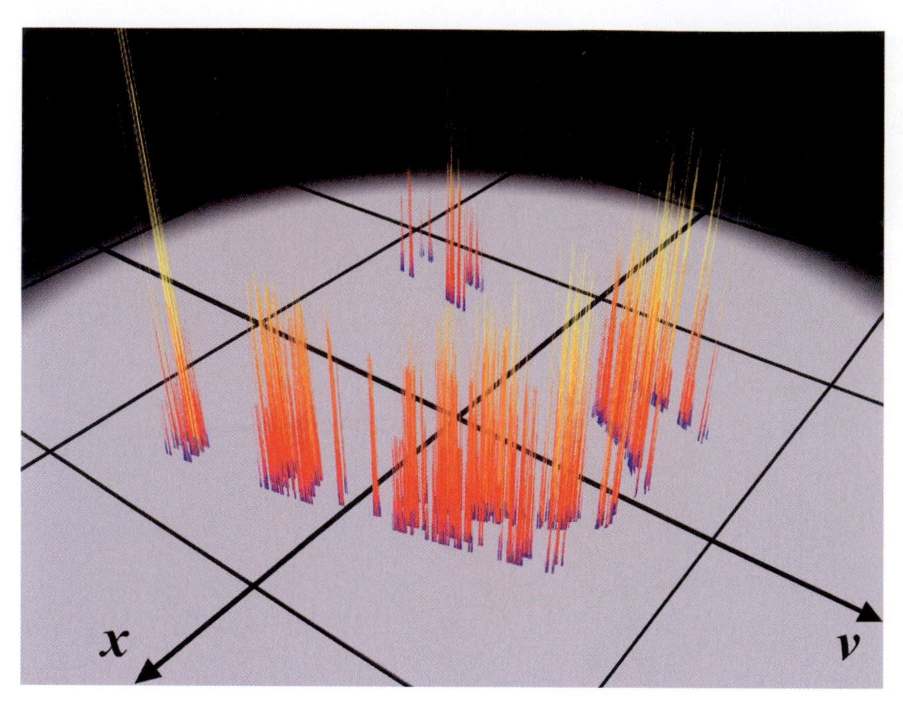

XIX

XX

XXI

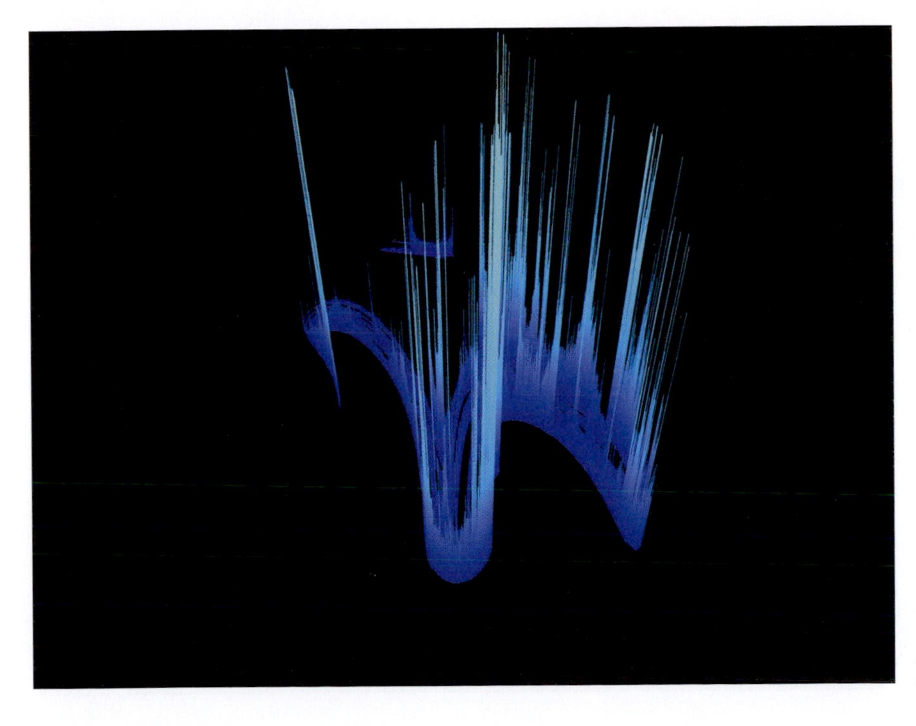

XXII

XXIII

XXIV

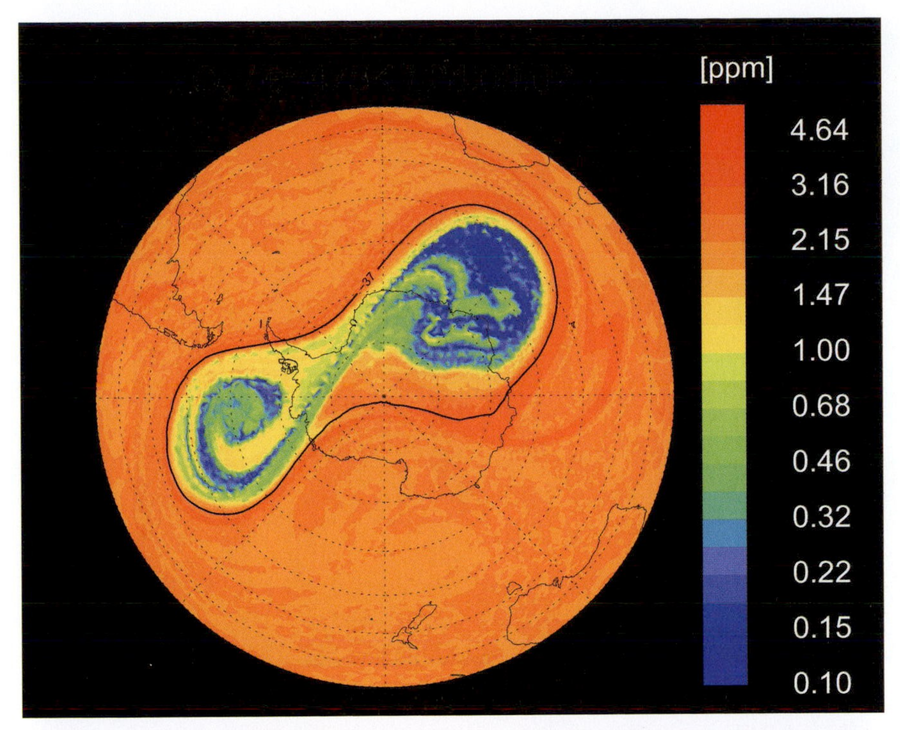
[ppm]
4.64
3.16
2.15
1.47
1.00
0.68
0.46
0.32
0.22
0.15
0.10

XXV

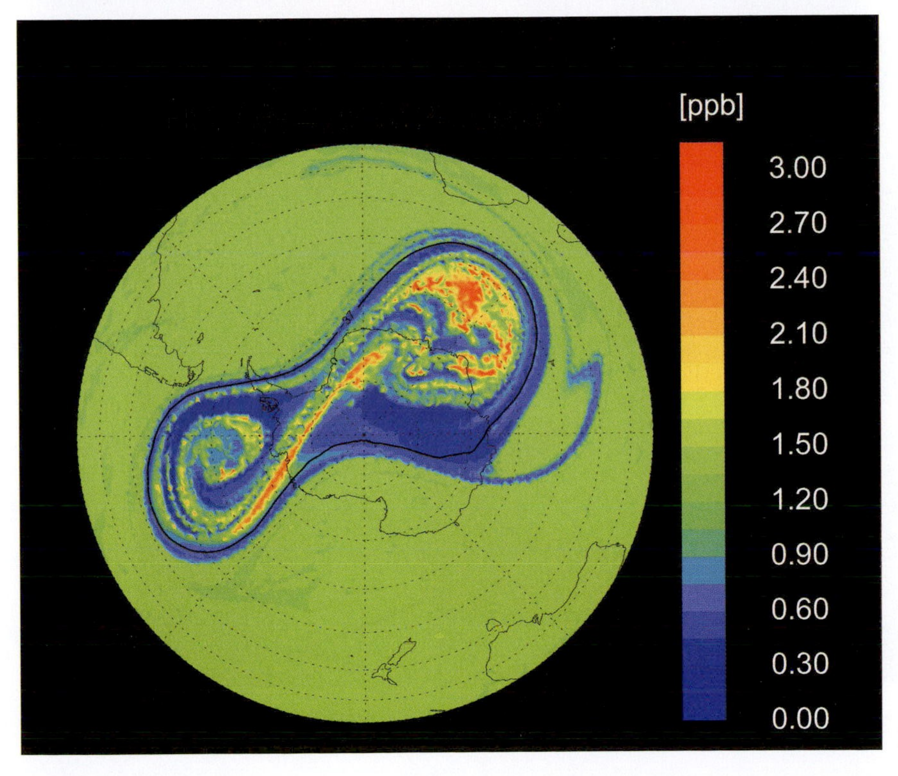
[ppb]
3.00
2.70
2.40
2.10
1.80
1.50
1.20
0.90
0.60
0.30
0.00

XXVI

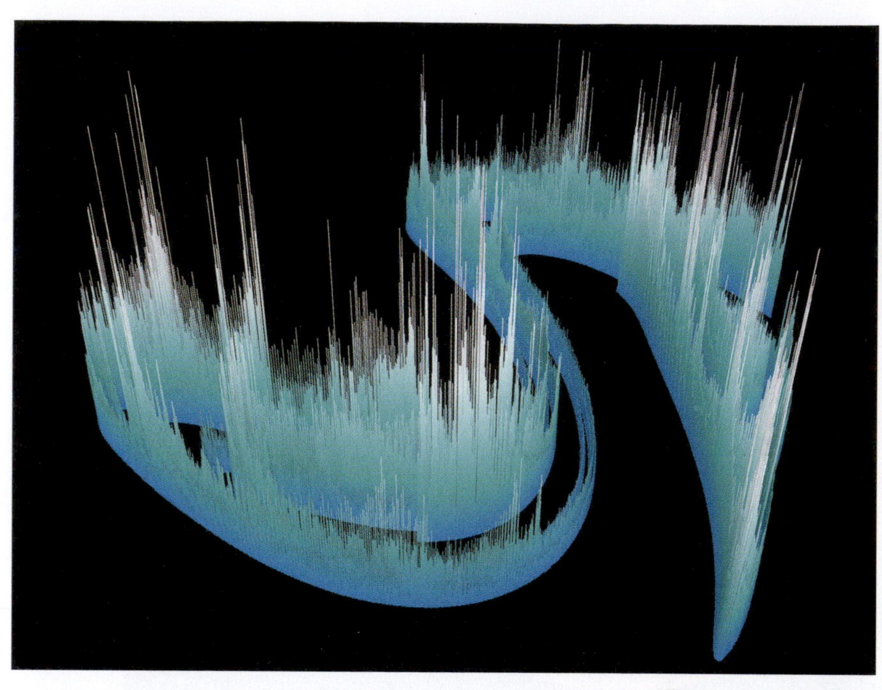

XXVII
(Front cover illustration)

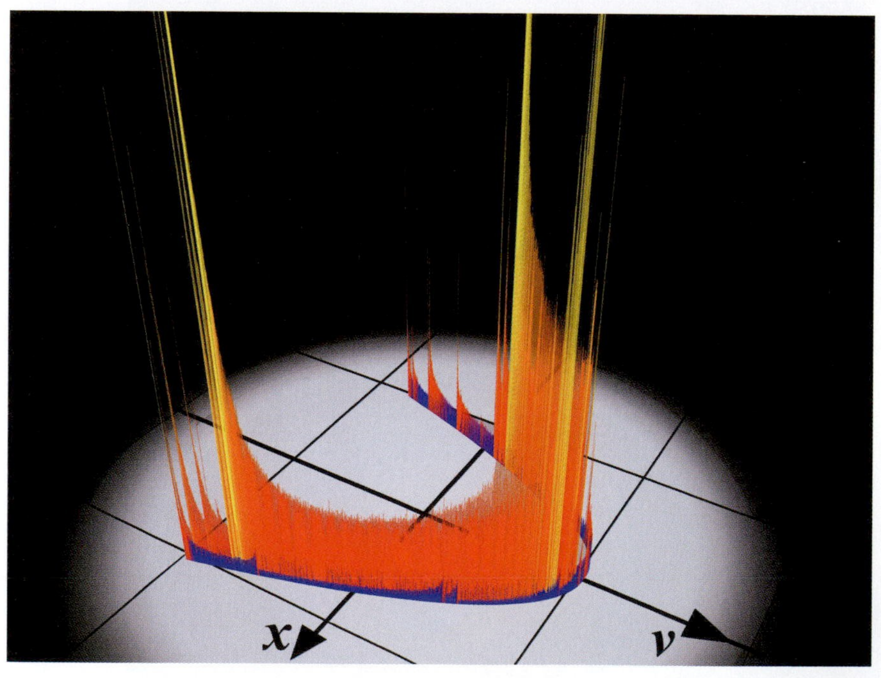

XXVIII
(Back cover illustration)